Meaning and Purpose in a Conscious Universe

A Physicist's Meditations on a Post-Materialist Paradigm Shift Integrating Science, Philosophy, and Spirituality

Marco Masi Ph.D.

Contents

"I have mapped the heavens and analysed the stars,
Described their orbits through the grooves of Space,
Measured the miles that separate the suns,
Computed their longevity in Time.
I have delved into earth's bowels and torn out
The riches guarded by her dull brown soil.
I have classed the changes of her stony crust
And of her biography discovered the dates,
Rescued the pages of all Nature's plan.
The tree of evolution I have sketched,
Each branch and twig and leaf in its own place,
In the embryo tracked the history of forms,
And the genealogy framed of all that lives.
I have detected plasm and cell and gene,
The protozoa traced, man's ancestors,
The humble originals from whom he rose;
I know how he was born and how he dies:
Only what end he serves I know not yet
Or if there is aim at all or any end
Or push of rich creative purposeful joy
In the wide works of the terrestrial power."

Sri Aurobindo – Savitri, Book VII, Canto VII

Preface

When I was eight years old, my family went camping in Sardinia, an island in Italy known for its beautiful beaches and turquoise sea. But what struck me, as a child who was born and had lived until then only in a big city like Milan, wasn't so much the sea but the starry sky. In the 1970s the light pollution in Sardinia wasn't as invasive as it is nowadays. While in big cities it was already out of the question to admire the wonders of the firmament, it was still possible to find a pitch-black sky on a dry and stony island unaffected by light pollution. It was the first time that I could see the sky as it is.

The first thing I noticed was the strange white stripe that crossed the sky. While walking with my father, I pointed at that strange "cloud." He quickly made it clear that it is not a cloud but a myriad of dense stars–that is, the Milky Way. I couldn't believe that to be true. Thousands of stars so densely packed that we could no longer discern them? I felt such a sense of awe and wonder that I could hardly sleep that night. Thanks to that sleeplessness, I made another discovery: The stars change their position in the sky! Wow! What a groundbreaking finding. Of course, the next morning my father enlightened me about how this happens due to the Earth's rotation. I quickly learned that, despite all the senses' suggestions, this was also an illusion.

Such was my introduction to the world of appearances. Though I didn't know it at the time, I can see in hindsight how my young mind went through a similar historical process that led humanity from a geocentric to a heliocentric worldview. But what most impressed that mind was the clear perception, sensation, feeling, and intuitive seeing–I would almost call it a mystic experience–that we don't live in a dead and meaningless cosmos. Because, again, appearances betray. While physically the universe is mostly a chilly and empty place, I could feel the omnipresent vibration that supports it behind a veil. Since then, my conception and perception of the universe and Nature[1] have dramatically changed. Nothing is really dead. In a sense, all is alive. Even those rocky and dry asteroids or icy comets are perceived as entities having their place in space that is not coincidental but instrumental.

Therefrom I developed, at a very young age, my interest in the meaning of life, our place in the universe, the origin of all there is, and all things about natural sciences and philosophy. While other children were playing football or hide-and-seek, I was asking myself Parminedean questions such as "Who am

[1] I will write "Nature" with a capital "N" when referring to the proper noun that represents the phenomenal world (e.g., "Hurricanes are a power of Nature") and "nature" with a lowercase "n" when referring to the common noun that designates qualities (e.g., "It is in the nature of my character to be picky").

? and "What does it mean that something exists?", or I was caught by the Heideggerian rumination "Why is there something rather than nothing?"

Later on, however, everything conjured against this philosophical style of living and grounded me back into the bubble we call "reality." I enrolled in college and university and did my Ph.D. in physics because deep down I was driven by the naïve idea that this could help me answer the existential questions that greatly interested me. As usual, life taught me another lesson. No space whatsoever was available for such contemplative topics, and I found myself fully caught up in the typical "shut up and calculate" academic environment detached from any deeper insight into reality and life.

In hindsight (the miraculous educational power of sufferance is always recognized in hindsight), I realize this was an extremely useful life lesson. I learned the dry technicalities of science, especially of physics. I learned how to work professionally in an environment that does not ask you to investigate the secrets of Nature but demands pragmatism and efficiency. Ultimately, I learned how to play the game.

But first and foremost, I learned not only what science is about but rather what science isn't. It became so adamantly clear what its strengths and limitations are when I compared my technical knowledge with the other kind of knowledge that I had nourished since my early celestial encounters but that I had to set aside for such a long time. This academic experience partially answered the how-questions but didn't answer my existential questions of what things are and, ultimately, why they are. Despite all the progress and insights of science, the mystery of existence, consciousness, evolution, and life remains essentially the same.

Could it be that the lack of progress in answering these deeper questions of a philosophical nature isn't due to a lack of information, discoveries, or technological advancement, but rather somehow inherent in science's way of seeing the world and its materialistic foundation?

Science, as a method, does not inherently require a materialistic ontology, although contemporary scientific theories often imply such an ontology. The scientific method doesn't need to assume a materialistic ontology, even though the theories we have today do. In our current cultural context, science is practiced based on the distinction between. What science as a method appears to presuppose is the distinction the subject and the object, the observer and the observed. However, this distinction need not be taken as an axiom; science could also expand and enhance the materialistic perspective without negating it, and ultimately transcend it.

The assumptions we nurture and how we look at things sometimes determine what we see and discover. What we see depends on how we see. No matter how powerful your microscope, telescope, or particle accelerator is, your belief system will determine how to interpret what you discover.

Sometimes facts are in front of our eyes, but we ignore them because they don't fit into our worldview.

At some point I began to study the spiritual traditions of East and West. For several years I was literally lost in its philosophy, metaphysical constructs, and spiritual practices. They answered several, but not all questions. I came to realize that these alone are not the key to an integral knowledge that I was, more or less consciously, seeking.

Slowly but steadily, I was able to piece things together and a more comprehensive and integral vision emerged in which both science and spirituality had to somehow be integrated into a third cognitive position, or what I like to call an "integral science," and that has yet to be born and developed.

Some could argue that we don't need a "new science;" we simply need to apply the scientific method to the claims made by spiritual traditions. However, while we shouldn't dismiss the valuable insights of the current scientific paradigm, and while empirical evidence and standard research procedures remain essential, the existing scientific method alone is insufficient. It's not only about evaluating spiritual claims supporting them with scientific evidence. It's also about developing a perspective on Nature that integrates those claims and sees them from a wider perspective. I feel that, without also adopting alternative ways of perceiving the world, we won't be able to extend our understanding beyond certain limits. Our view of the world is shaped by how we perceive and conceive it. As we will analyze, the cognitive approach influences the content of what is recognized.

Moreover, because the scientific method primarily relies—sometimes explicitly, but often implicitly—on a materialistic ontology, an intellectual perspective that challenges this assumption is essential to move beyond materialism.

It became clear how the materialistic sciences are only one way of seeing the world. Philosophy complements it with another perspective, and the mystic experience represents yet another approach. Synthesizing and transcending them into a position where they appear as the three distinctive 'windows' through which we can observe the same reality–none of which hold exclusive rights to what should be considered the "truth"–is the main objective of this volume.

I. Navigating This Book: What's It's All About and What to Expect

1. It's Time to Dispel the Myth of Naturalism

Before exploring existential questions and crafting a 'grand vision' in the upcoming chapters, I want to highlight the insights gained from my fifty years of experience: We must recognize the inherent limits of science by deprogramming our habitual materialistic thinking.

The dominant methodological approach is naturalism.[2] Naturalism is the philosophical belief that everything originates from natural properties and causes—specifically, from physical processes alone. It asserts that spiritual explanations must be excluded or discounted from the outset. Consequently, naturalism maintains that consciousness[3], mind, and life should be described solely in terms of material processes, disregarding any psychological dimensions and eliminating any consideration of sentience, feeling, or mentality. It then wonders why and how these could emerge from dead matter.

The lack of tangible progress and the shortcomings of this approach are becoming increasingly evident. There is now growing evidence challenging a strict naturalistic interpretation of the mind and consciousness. Although this new information is often disregarded for ideological reasons, the stalling of progress is persuading some scientists and philosophers that the purely naturalistic perspective is insufficient and needs revision. Additionally, several individuals have been influenced by skepticism toward neo-Darwinism as the ultimate framework for a materialistic, non-teleological[4] account of evolution, which also plays a role in this shift in thinking.

---

[2] Also known as "metaphysical naturalism," "ontological naturalism," "philosophical naturalism" or "anti-supernaturalism" that should be distinguished from "religious naturalism" or "spiritual naturalism." I will refer to the former simply as "naturalism." While the cosmology presented later could be loosely described as a form of spiritual naturalism, I will refrain from using this term, as it may suggest religious or metaphysical concepts that are not part of the integral cosmology I will discuss. Instead, I will refer to it as "spiritual emergentism."

[3] The term "consciousness" has taken on various meanings depending on the discipline and context. In this book, I will always refer to "phenomenal consciousness"–that is, subjective experience, sentience, qualitative perception–simply as "consciousness," unless stated otherwise.

[4] The term "teleology" originates from the Greek word "télos," which refers to the study of aims, purposes, and final causes in Nature. In contrast, modern science adopts a strictly non-teleological perspective.

There is a growing dissatisfaction with materialism, or more specifically, physicalism[5], as the definitive explanation, particularly in the fields of consciousness studies and evolution. This sentiment is emerging even among intellectuals who have traditionally supported a materialistic, Western analytical philosophical tradition. Although this material monism remains the dominant perspective, the notion that a strictly mechanistic, physical, and intellectual understanding of the world is the sole path to truth is increasingly unconvincing. What was once a small intellectual minority has evolved into a noticeable splinter group—still not a majority, nor a homogeneous one. However, it is likely that, over time, alternatives to physicalism will become more vocal, garnering greater attention and recognition and ultimately reaching a critical mass that will render them impossible to ignore.

A clear indication that Western philosophy is rediscovering its roots is the revival of traditional metaphysical worldviews, such as philosophical idealism, pantheism, panentheism, panpsychism, cosmopsychism, substance dualism, and substance monism, along with various developments and modifications of them. This diverse array of non-physicalist frameworks encompasses the philosophy of mind, consciousness studies, cognitive sciences, neuroscience, biology, and psychology, thus reflecting a significant trend toward both old and new ontologies that seek to explain our place in the universe.

However, this philosophical renaissance is not without its challenges. Many of these purely intellectual metaphysical speculations and conjectures rely heavily on a Western scientific, rationalistic, and third-person perspective. I argue that this reliance is a fundamental source of many inconsistencies and paradoxes that undermine their validity. This concern is one of the primary motivations for the present work. At this point, I would like to introduce this philosophical project, which, at least in theory, has the potential to transcend mere philosophy and foster a new practical approach in science, ultimately enhancing our understanding of the meaning of existence and the true and intimate nature of ourselves.

Though most scientists and philosophers would not describe themselves as supporters of a strictly materialistic worldview, nonetheless 'scientism' remains a deeply rooted perception of reality and our nature. Scientism is a philosophical attitude that adopts an absolute faith in the efficacy of science; it says that science alone can render truth about the world and reality.

Science and scientism are not the same. You can deeply value the former while rejecting the latter. While science does not necessarily claim to be the only method by which to acquire knowledge about the world, life, and mind, scientism is a single-minded exclusively left-brain understanding of reality that

---

[5] Physicalism is a broader term because it encompasses not only matter but also everything described by physics, including both massive and massless particles (such as photons), the fundamental forces, the laws of physics, and space-time.

admits only to a strictly scientific, materialistic, or physicalist worldview. Any philosophical, metaphysical, spiritual, or religious claims are branded as nonsensical and superfluous. It is an absolute view of reality in which only science, rationalism, and empiricism can open the way to truth, and it found a large consensus, especially during the first half of the 20th century in the intellectual movement of logical positivism.

Nowadays positivism has lost its appeal, and most don't subscribe to such a limited worldview, at least not officially. Or, might we say, not consciously? Indeed, I contend that, unconsciously, it remains deeply anchored in our way of thinking and seeing the world, even among people who have a spiritual or religious view of life and the world. The overwhelming and undeniable success of science and its material achievements has conditioned our mental patterns and belief systems. In a subtle and not always explicit manner, nonetheless, scientism continues to condition us in terms of how we see life, our body, our psychology, and reality. Despite the recent revival in philosophical inquiries, the rediscovery of ancient spiritual traditions, and the awakening of interest for all that is extrasensory, materialism is still the predominant ideology framing our thinking and our choices, with its decision-makings, fundings, and our policies.

The belief that external changes—such as advancements in technology, new political systems, AI, Mars colonies, or hyper-technological transhumanist visions—will save us and create better worlds without necessitating an internal spiritual transformation is a dangerous illusion. This mindset, rooted in various forms of materialism (often unconsciously), is more prevalent today than ever. It is an illusion deeply entrenched in a materialistic perception of reality that is no less ingrained in spiritually inclined minds. On the other hand, pseudoscientific claims that lack rigor and fail under critical scrutiny are also widespread.

However, my critical attitude should not be seen as an attack on science. On the contrary, it is a defense of science and an appeal to expand and complement it. Precisely because of the alarming recent rise in anti-science attitudes, this shift has become even more essential. Despite all the wonderful things science has given us it left a void: A mechanical universe without meaning and soul.

We are told that science should be kept separate from any deeper philosophical, let alone spiritual quests, and that it was never intended to tackle questions of transcendental kind, and many still believe it will, sooner or later, be able to eradicate all metaphysics. Yet, in these times dominated by a materialistic paradigm tainted by pseudo-scientific beliefs, we must rediscover and return to a deeper spiritual and post-material understanding of reality, ourselves, and Nature. This perspective does not exclude critical thinking and discernment; in fact, it emphasizes them even more than traditional analytical thinking does.

We must not abandon discernment, which relies on rigorous scientific methodology, including experiments, observations, and data collection. However, it is time to move beyond a narrow naturalistic paradigm that confines itself to a worldview of nothing but matter and the so-called 'laws of Nature.' This perspective has repeatedly failed to address several foundational questions, which will be discussed here. This form of scientific realism has evolved into an ideological belief system that significantly shapes academia, dictates the allocation of research funding, and excludes brilliant minds from academic and research positions if they propose ideas or curricula that diverge from established orthodoxy.

Whereas, we need something that encourages a culture and science that challenge the notion that everything must fit into a strictly materialistic and mechanical view of reality. It's essential to find a middle ground that incorporates some form of trans-physicalism, idealism, or spiritualism—one that transcends the often simplistic, or even pseudo-scientific approaches, of religious or New Age thinking. A good starting point could be a non-teleological idealism, as rejecting materialism does not necessarily entail embracing teleology. However, I believe that promoting a teleological perspective that acknowledges meaning, purpose, and agency in Nature will ultimately be an essential goal in the long term.

If humanity seeks a better future, science must embrace deeper and more expansive ways of thinking, moving beyond a superficial understanding of life, mind, consciousness, and the world around us. We urgently need a post-material science that recognizes the possibility that we are more than mere biological lumbering robots. It is time to question whether consciousness might be more than the result of neuronal interactions arising from a bottom-up process that once originated in a primordial soup.

One of the key priorities of our modern age is to bridge the gap between science and spirituality. We should move beyond the predominant materialistic paradigm and the exclusively rationalistic mindset to advance toward a post-material society—a trans-rational age. This new perspective should not be limited to a purely physical understanding of Nature and ourselves or confined to a religious mindset based on dogmas and scriptures. Instead, it should strive to broaden our awareness and enhance our worldview by incorporating elements such as spirit, spirituality, intuition, introspection, and a more expansive view of life.

This also implies that, if we want to avoid falling into the trap of a materialistic society that is blind to its spiritual origin, we should realize how the future of humanity will be determined much more by the individual and collective evolution of consciousness rather than its technological progress. The understanding of how the dignity of the individual will have to be harmonized with the wellbeing of the collective will play a more central role, than a conception of society that pursues the solutions to its riddles only by purely technocratic means.

The integral cosmology we are going to discover in Chapter X, describes how, while life on Earth is evolving, we are transitioning toward a new species, rising at a higher state of consciousness, though, to some, facts might suggest otherwise. We are now on an evolutionary verge and transition. Fast and abrupt changes are unavoidable. We are progressing toward a new world and are becoming aware of how, despite its superficial appearances, this is a conscious universe. Recent scientific discoveries that challenge a reductionist materialistic worldview, along with the inability to understand and control various aspects of life and human psychology, highlight a deep longing for meaning, purpose, and goals. These factors will invite us to heed the message of Nature: "I'm conscious."

We start to develop a spiritual awareness that what occurs in life is not a cruel, meaningless, or purposeless play of destiny. Instead, it serves as an external sign of a new consciousness emerging—like labor pains that herald the birth of a better humanity.

'Better' means that, despite all evidence to the contrary, we are moving toward harmony and unity, evolving into a species that is less selfish and less ego-centered, that will eventually have a collective non-dual realization, and that is moved more by an inner spiritual force rather than the outer instincts of survival, greed, and the sense of separation that still conditions us so strongly. It is an integral vision of the evolution of consciousness and an evolutionary journey leading us to the integration between Western materialism and Eastern spirituality, between Spirit and Matter, Spirit and Life, Spirit and Nature.

Many scientists feel the need to go beyond the preconceived and prescribed paradigm but don't really know what to look for and where to go. This spiritual awakening in science and philosophy still manifests itself in a confused manner, resorting to simplistic worldviews that much too often cannot go beyond one's own inherited cultural educational or religious context. The awakening is individual, personal, or expresses itself in small groups, but resists an institutional, academic, let alone political change.

To go beyond, we need to overcome our habitual thinking patterns, transcend beyond binary thinking, learn to realize the limitations of the mind, and transcend the dichotomy of science and religion in favor of the development of a more integral, new way of seeing all of Nature, the universe, science, philosophy, evolution, spiritual teachings and, especially, ourselves in a holistic complementarity. Failing to do so is one of the main reasons why this evolutionary progress, unfortunately, moves from one disaster to the next instead of becoming a more harmonious adventure of consciousness.

Here, I will highlight the necessity to overstep this binary thinking, especially in the domain of science, philosophy and spirituality. I invite people to look beyond matter and body and beyond that kind of purely analytic philosophy that doesn't allow for anything other than rigorously logical-based reasoning, obstructing further development toward an intuitive vision of Nature and our place in the universe. I reexamine many of the basic tenets of past and modern science, philosophy, and spirituality and show what kind of limitations we have, with what kind of unwarranted assumptions we work, and with what kind of fallacies modern science jumps to conclusions, especially when it comes to consciousness, mind, evolution, divine determination, and philosophical questions with a metaphysical bent.

Modern science and philosophy have rediscovered the mind-body problem, the hard problem of consciousness, the question of free will, of whether the mind is computational, and are beginning to realize that there is a plant and cellular cognition. However, this is only the tip of the iceberg. A whole conscious universe remains to be discovered but it still lingers under our surface awareness because of a short-sidedness and a too-limited perspective that we need to learn to transcend.

One of the main themes will be the introspective turn complementing the rationalistic and scientific approach. I invite readers to contemplate and look beyond the exclusively third-person perspective of science and examine what the introspective first-person approach suggests. I propose a synthesis of knowledge that integrates science, philosophy, and spirituality, rather than excluding any one perspective. This integration creates a whole that transcends each discipline, becoming greater than the sum of its parts.

However, this can be achieved only if, besides a spiritual transformation, we pursue a cognitive transformation that requires a radical change in human nature. We can't get there in one step.

We should first cultivate an awareness of not only what needs to be done but, more importantly, what should not be done. We should become aware of our biases and of the background ideology we still firmly believe and cling to and what kind of thought patterns prevent us from seeing further and proceeding toward this new state of consciousness. We should learn to connect the dots between a materialistic scientific worldview and a subtler spiritual, non-religious approach, avoiding falling into pseudo-science—that is, being careful not to throw the baby out with the bathwater.

In practice, this means that we must first become aware of our limitations; second, deprogram specific thought patterns; and third, learn to see things from different points of view at once–that is, we should become multi-perspectival.[6]

The aim is to shift away from a material, hyperrational worldview and toward a post-material, post-rational, and intuitive, comprehensive, and spiritual mindset–a trans-rational way of seeing and perceiving that makes us aware of how we live in a conscious universe. This transformation is now

[6] Not to confuse with what is nowadays called "multi-" or "inter-disciplinary."

essential, much like how pre-scientific natural philosophy required empiricism to evolve into a Galilean and Newtonian science.

2. Beyond the Naturalist's Meaningless Universe

Fundamental existential questions persist regarding the origin of consciousness and life, particularly how animated matter could emerge from the apparent chaos of inanimate substances. These inquiries delve into the meaning of life, our place in the universe, the universe's origin, and other profound mysteries that are challenging for human reason. There is even a lack of consensus on whether such questions are meaningful at all. Despite centuries of philosophical inquiry and decades of scientific exploration, these phenomena continue to elude definitive explanations.

Perhaps the solution lies not in isolating them but in embracing their interconnectedness as expressions of a deeper, universal principle. We have reached a point where we must seriously question whether the seemingly insoluble mysteries surrounding the origin and true nature of what we call "consciousness" and "life" stem from a rigid philosophical commitment to a strictly mechanistic understanding of reality. This perspective often prevents us from exploring broader possibilities and instills a fear of looking beyond the narrow horizon materialism paints.

The belief that science should rely solely on a third-person perspective—grounded in experiments, measurements, and data, along with reason—has overshadowed the value of a more intuitive and imaginative first-person perspective. This narrow focus has resulted in a multitude of highly abstract theoretical frameworks, often at the expense of deeper cognitive skills rooted in imagination, inspiration, intuition, and curiosity. Ironically, these very skills were essential to the research efforts of the founding figures of modern science.

While science is often seen as a domain governed solely by logic, observation, and empirical rigor, it also relies heavily on human faculties such as intuition, imagination, and inspiration, which stem from a deeper sense of understanding. These elements are essential to scientific discovery, as they help to bridge the gap between the known and the unknown. They enable scientists to envision concepts and frameworks that go beyond immediate evidence. Intuition, in particular, is the ability to perceive connections or anticipate outcomes without conscious reasoning. It typically arises from a deep familiarity with a subject, allowing scientists to identify patterns and relationships that might not be immediately apparent.

Many of the most transformative scientific breakthroughs originated from these intuitive insights.

Contrary to common belief, Isaac Newton might have engaged in experiments and speculations about alchemy far more than he focused on the rational and scientific inquiries in physics as we understand them today. Something compelled Newton to maintain his passion for alchemical and almost magical ideas. Modern scientific history often dismisses this as a nostalgic attachment to a bygone era. The prevailing interpretation suggests that Newton was still grappling with a pre-scientific tradition. However, his "nostalgia" might also reflect a yearning for a lost aspect of consciousness that he sensed was gradually diminishing.

Despite his lack of formal mathematical training, Michael Faraday relied on his intuitive understanding of physical phenomena to conduct experiments in electromagnetism. He discovered electromagnetic induction, which serves as the foundation for electric generators, and established the groundwork for modern electrical engineering. Most notably, he introduced the concept of the "field." By using his imagination to visualize invisible magnetic fields, he laid the groundwork for a concept that would later become central not only to electromagnetism but also to all of physics.

The German organic chemist August Kekulé famously described a dream in which he saw a snake biting its own tail. This inspired him to formulate the cyclic structure of benzene. Similarly, Dmitri Mendeleev claimed that he discovered the periodic table of chemical elements in a dream. He often remarked, *"No law of Nature, however general, has been established all at once; its recognition has always been preceded by many presentiments."*

In another example, the Scottish microbiologist Alexander Fleming observed that a mold called Penicillium notatum appeared to kill bacteria in a petri dish. Rather than dismissing this initial observation as a coincidence, his intuition compelled him to investigate further. This led to the identification of penicillin, the first antibiotic, which revolutionized medicine and saved millions of lives.

Albert Einstein relied heavily on intuitive reasoning, particularly through his thought experiments, which were deeply intuitive in nature. He used these mental images to grasp complex concepts such as time, space, and gravity. Einstein believed in the invariance of physical laws, asserting that they remain the same regardless of an observer's position or motion and that the speed of light is constant, no matter how fast one travels. However, there was no a priori reason to assume that the structure of physical laws is uniform across all contexts. The notion that the speed of light cannot vary for different observers in different states of motion is profoundly counterintuitive. Nevertheless, it was precisely a deeper intuitive insight that led Einstein to establish these principles as foundational. He accepted them without further questioning, which allowed him to explore the implications they presented.

Many other examples of this sort could be made. Many scientists were led by intuition, imagination, aesthetics, and inspiration that were harmonized with sharp analytic discrimination. Intuition often acts as a compass, pointing scientists toward fruitful areas of inquiry even before rigorous methods confirm the insights. Imagination explores scenarios beyond the constraints of current knowledge. Their synergy became the tool that allowed for creative problem-solving and conceptual innovation. The interplay between intuition and imagination often defines scientific creativity and constructs hypotheses and models needed to explore them. Together, they drive the process of discovery, whereby ideas are tested, refined, and occasionally transformed by unexplainable insights and unexpected findings.

While empirical evidence, logical analysis, and methodical validation help us to ground creative insights, a sense of reverence for the phenomena of Nature remains the bedrock of philosophical inquiry and scientific progress. On the other hand, an exclusive focus on the power of reason and an almost disregard for the exercise of our innate intuitive and creative faculties hamper the very same scientific progress. This was clearly visible in many of the fathers of philosophy and science.

Aristotle once proclaimed: *"The ultimate value of life depends upon awareness and the power of contemplation rather than upon mere survival".*

In his famous Principia, Newton declared: *"I had the intention of becoming a theologian... but now I see how God is, by my endeavors, also glorified in astronomy."*[7]

Johannes Kepler's vision of celestial mechanics was already in line with a mechanistic understanding of the world: *"I am much occupied with the investigation of the physical causes* [of motions in the Solar System]. *My aim in this is to show that the celestial machine is to be likened not to a divine organism but rather to a clockwork..."*[8] Yet, he realized that this is only one way of seeing: *"Geometry is unique and eternal, a reflection of the mind of God."*[9]

Also, French biochemist Louise Pasteur had no issues in reconciling the empirical approach with intuitive insight. *"As in the experimental sciences, truth cannot be distinguished from error as long as firm principles have not been established through the rigorous observation of facts."*[10] – *"Happy is he who bears a god within himself, an ideal of beauty, and obeys him: an ideal of art, an ideal of the virtues of the Gospel. These are the living springs of great thoughts and great actions. All are illuminated by reflections of the sublime."*[11]

---

[7] Newton, "Principia: The Mathematical Principles of Natural Philosophy."

[8] Letter to Ilerwart von Hohenburg (10 Feb 1605) Quoted in Holton, Johannes Kepler's Universe: Its Physics and Metaphysics, 342,

[9] As quoted in Epilogue, The Sleepwalkers: A History of Man's Changing Vision of the Universe (1959), 524

[10] Ésur la maladie des vers ásoie (1870), p. 39.

[11] Speech (27 Apr 1882) on his reception into the Académie Française, as translated in Maurice Benjamin Strauss, Familiar Medical Quotations (1968), 490.

However, this inner contemplation of Nature and the transcendent gradually faded with the passage of time. The rise of the positivist's mindset led to a dominant reversal in our understanding of the world and ourselves.

Arch-materialist Francis Crick, the discoverer of the DNA molecule, once declared: *"You, your joys and your sorrows, your memories and your ambitions, your sense of personal identity and free will, are in fact no more than the behavior of a vast assembly of nerve cells and their associated molecules."- "You are nothing but a pack of neurons."*[12]

Nobel physicists Steven Weinberg famously noted that *"The more the universe seems comprehensible, the more it also seems pointless."*[13]

Biologist and well know bestseller author Richard Dawkins thinks that *"The universe that we observe has precisely the properties we should expect if there is, at bottom, no design, no purpose, no evil, and no good, nothing but blind, pitiless indifference."*[14]

Historian and popular science writer Yuval Noah Harari: *"As far as we can tell, from a purely scientific viewpoint, human life has absolutely no meaning. Humans are the outcome of blind evolutionary processes that operate without a goal or purpose. Our actions are not part of some divine cosmic plan, and if planet Earth were to blow up tomorrow morning, the universe would probably keep going about its business as usual. As far as we can tell at this point, human subjectivity would not be missed. Hence any meaning that people ascribe to their lives is just a delusion."*[15]

Philosopher David Benatar: *"Earthly life is thus without significance, import, or purpose beyond our planet. It is meaningless from the cosmic perspective. Because this is true of all life, it is true of all sentient life, all human life, and each individual life. Neither our species nor individual members of it matter sub specie aeternitatis. Whatever other kinds of meaning our lives might have, the absence of this meaning is deeply disturbing to many."*[16]

How did we transition from the profound philosophical inquiries that inspired past philosophers and scientists to a science fixated on reducing everything to the cold indifference of Dawkins' universe? Was this shift a

---

12 "The Astonishing Hypothesis: The Scientific Search for the Soul," Scribner; Touchstone Edition.

13 "The First Three Minutes," Basic Books, 1996.

14 "River Out of Eden: A Darwinian View of Life," Basic Books, 1995.

15 Sapiens, pg. 331/368

16 The Human Predicament, pg. 57/289.

progression that brought us closer to the truth, or a regression? Was it a cultural evolution or a devolution?

Unfortunately, over the last century, the prevalence of a strictly third-person materialistic worldview has led to an excessive focus on analytical abstraction in science, often at the expense of deeper philosophical inquiries. While this reductionist approach has produced significant technological and empirical advancements, it has largely neglected the introspective dimensions that once connected scientific research with broader existential and metaphysical questions. Early science did not merely dissect the external world; it also engaged in a more holistic exploration of consciousness, meaning, and the subjective experience of reality. By sidelining this inward turn, modern science has become mechanistic and detached from the very human questions that originally inspired its pursuit—questions about purpose, existence, and the nature of knowledge itself. This neglect has hindered our ability to develop new and original approaches to investigating the natural world and discovering a trans-rational reality. The undervaluation of inner exploration, the subjective dimension, and intuitive forms of cognition in science have severely limited our understanding not only of the world but especially of ourselves. Worse, it has stifled our cognitive skills. We have largely become unable to go inward, to connect with the inner spirit of Nature, and atrophied our intuitive and creative skills.

I believe one reason for this shift toward an exterior focus and the elevation of the analytical mind as the ultimate authority is the unspoken assumption lurking in the background: that intuition stems from our irrational subconscious, which we should ignore. On the contrary, as we will explore, intuition is not merely a byproduct of the subconscious; rather, the subconscious, and even rationality, are pale reflections and artifacts of intuition.

For the rational mind, it can be difficult to recognize how intuition, discernment, aesthetics, and imagination are the silent architects behind many scientific revolutions. These qualities empower scientists to think beyond the confines of existing knowledge, envision the unseen, perceive integrally, and make transformative leaps in our understanding of the universe. By nurturing these qualities alongside analytical rigor, science could advance as a profoundly creative human endeavor as well. To the rational mind, it might seem paradoxical that to cultivate these faculties of consciousness, one must silence the mind. I believe that the overvaluation of rationality is partly responsible for the absence of major scientific revolutions since the early 20th century.

It is time to reassess and amend this attitude. This book has been written in that spirit.

3. Motivations, Structure and Synopsis

This work is a collection of essays complemented by a series of academic articles. I developed these essays because I sought a clear and direct way to advocate for what I believe to be the necessary and inevitable future of science: the end of materialism.

Except for the Advanced Readings each included as sections of the last chapter (Chapter XIII) in this book, I will aim to keep the reasoning less technical. The essays presented will use non-mathematical discursive language designed to be accessible to everyone. At the same time, the reader interested in a further technical account might like to resort also to my book titled "Spirit Calls Nature" (Masi, 2025a), where I offer a more academic exploration of what a post-material science might entail, along with a synthesis of insights from science, philosophy, and spirituality. Within its pages, you will find a deeper analysis of several questions we will address here, all framed within the spirit of providing "a comprehensive guide to science and spirituality, consciousness, and evolution in a synthesis of knowledge." This is why there is some overlap with the current volume, particularly in the sections V.1&2 on quantum mechanics, section IX.1 on philosophical idealism, and the integral cosmology of Sri Aurobindo in Chapter X . I chose to reiterate these parts to ensure that each volume remains self-consistent and accessible on its own.

Meanwhile, the present collection can be viewed as a call for a Renaissance—a new post-materialist perspective on the world, Nature, and ourselves. Because the contemporary worldview that seeks to address deeper existential questions is too rooted in a strictly physical and reductionist scientific approach, or resorts to philosophy alone—typically secular and often lacking the profound insights offered by mystical traditions from both Eastern and Western thought. Such approaches fail to move beyond a superficial understanding of appearances. As a result, the absence of purpose and meaning in what seems to be a mechanistic and soulless universe, as presented by conventional science, has left our society grappling with spiritual emptiness and detachment from our spiritual center and, thus, contributes to widespread mental health issues.

In this collection, I have selected distinct, self-contained topics that highlight how science must shift its perspective. While I have attempted to organize them in a logical order, you may choose to read each essay independently, without regard for the sequence provided. While the content and style of my writing can be quite technical, my goal is not to instruct or convey technical knowledge, nor to persuade the reader of a particular truth, reality, or paradigm. Instead, I aim to illuminate the unaware assumptions that often lead us to draw unwarranted conclusions. These essays share a common theme: the desire to look beyond conventional thinking.

They serve as an introduction to the more advanced readings—the technical articles published in peer-reviewed academic journals–and left as an appendix for the expert readers in the last chapter XIII. While these articles are not mandatory for understanding the essays, they are intended for readers who wish to explore the subjects in greater depth. There may be some repetition, reformulation, or similarities in wording between the essays and the articles. This occurs because my introductory essays were influenced by the published articles, and vice versa.

That being said, upon reviewing the material, you might find that although I assert a departure from a strictly rationalistic and physicalist viewpoint, I often paradoxically rely on physics and a conventional, at times even reductionist, bottom-up approach. The reader might perceive this as a contradiction that does not fulfill one of the primary objectives of the work: to argue for a trans-rational and post-material paradigm.

There are two main reasons for this seemingly contradictory approach. First, it reflects my bias as a physicist. I tend to view questions about the nature and structure of the world from a bottom-up perspective rooted in physics. I'm aware that this is not the only perspective and not necessarily the best or most insightful one. Nevertheless, given my training in certain subjects and not others, I prefer to emphasize the physical aspects I am familiar with before exploring other scientific fields, such as biology and neuroscience, which I also address thoroughly. However, this is only a secondary reason.

The main reason I insist on emphasizing the importance of examining things from a physical perspective, and at times from the viewpoint of sense-based empirical sciences, arguing for a reality that transcends it, is that science itself points to something beyond the physical and empirical. The messages that Nature conveys, even through a limited materialistic lens, hint at an inherent trans-material reality. We don't need to engage in complex metaphysical speculations to recognize that fields such as physics, biology, and neuroscience already suggest a post-material perspective. Already, scientific facts alone imply the idea that the world extends beyond its material surface and encompasses more 'subtle' and transcendental domains. If we fail to realize this, it is only due to our ideological biases. When we overlook something that is right in front of us, it is typically because we filter it out based on our philosophical preferences that reject it. However, if we strive to move beyond our presuppositions and view the same facts from a different angle, we might uncover their deeper significance, even if nothing has changed externally. Therefore, arguing for a reality beyond the current scientific paradigm, while starting from the same scientific approach, is not contradictory; rather, it is a valuable cognitive exercise that enables us to see beyond.

More generally, my approach is to ask different questions rather than focusing on clear-cut final answers. I tried to do this with a list of inquiries in which I don't provide the reader with a final answer but instead point out the kind of fallacies and misunderstandings, and on what kind of superficial awareness, the standard answer is often based. It is not necessarily the answers I suggest that you need to agree with, as the passage of time might reveal them to be wrong. Rather, it is the approach, mindset, and outlook that I propose as exercise.

This is because, first, the way to knowledge begins by asking the right question. Second, we must become aware of our possible biases and unaware premises. Last but not least, we must learn to embrace the third-person perspective of science with the first-person perspective of intuition and, eventually, go beyond both, synthesizing them into a higher-level position.

Of course, inevitably, I will also dress it with my biases and points of view, but I invite the reader to take notice much more of the "way of seeing," the intuitive feeling, and the multi-perspectival approach, without necessarily agreeing with the conclusions. Because, at this stage, what we need aren't quick and easy answers but questions that exercise the muscle of intuition and imagination. If I induced you to participate in this exercise and led you to the desire to know more, making you aware of some aspects, thought patterns, assumptions, presumptions, and fallacies that you weren't aware of previously, that would be sufficient to say that the aim of this volume has been accomplished.

How is this exercise of intuition supposed to work?

I propose that these apparent inconsistencies stem from our tendency to overlook the interconnectedness of reality. We often perceive certain aspects as mutually exclusive rather than acknowledging them as diverse yet complementary components of a broader cosmology. This perspective can help us unify them into an integral theory of consciousness, matter, life, and mind. To understand the integral structure that supports the diverse orthodox theoretical frameworks, we must recognize some of their common limitations, as follows.

Firstly, the Western philosophy of mind should be more receptive to its Eastern counterpart. The fact that the Asian tradition is primarily a first-person contemplative approach rather than a third-person scientific and analytical understanding of reality doesn't make it less valuable. To the contrary, it serves as a necessary complement to the prevailing rationalistic mindset that continues to influence Western metaphysics.

Secondly, the low-dimensional paradigms that consider reality as purely physical, purely mental, or, at best, a binary matter-mind duality must be re-evaluated. The conflation of mind and consciousness categories also requires reconsideration. This 'dimensional reduction' is a fundamental barrier that has hindered further progress.

Thirdly, theoretical frameworks that view consciousness as fundamental or extend it to a universal consciousness—such as panpsychism and various forms of idealism—often leave the nature of this universalization unexamined. Terms like "cosmic Mind," "universal Mind," or "Mind at large," as A. Huxley referred to it in 1954, are frequently used without further characterization. This Mind, often capitalized, is typically equated with a rational and analytical anthropomorphized human-like mind. In some interpretations, it is even reduced to an instinctive and blind force or a will devoid of telos and intelligence, as suggested by thinkers like Spinoza and Schopenhauer. However, if we instead consider the possibility of a trans-rational mind and propose the existence of a superconscious in addition to the subconscious, we could arrive at a much richer and more comprehensive understanding of reality.

Fourthly, the aforementioned frameworks rarely address the evolutionary perspective, if not entirely overlook it. What is evolution, and what role and purpose, if any, does it serve from a spiritual standpoint? Accepting evolution as a fact of natural history while also recognizing a non-physical substrate inherent in the material universe requires a paradigm that clarifies the relationship between the two. The emergence of a supposedly trans-physical consciousness and an immaterial mind within the context of physical evolution has been considered by only a few thinkers, such as T. de Chardin in his cosmic theology and J. Gebser in his historical view of the emergence of the 'structures of consciousness,' or in occultism, as seen in R. Steiner's Anthroposophy. Viewing biological evolution not only in terms of evolving organisms but also as evolving souls allows us to approach life from a perspective that opens new horizons.

I aim to develop a comprehensive framework that reveals how, when we remove these reductive glasses, everything aligns within a harmonious overarching vision, causing previous paradoxes to dissolve naturally. This framework is not merely a philosophical analysis; it also emphasizes inner exploration and expands these ideas into a theory of universal consciousness. It is rooted in a monistic, panentheistic, teleological, and evolutionary spiritual emergentist cosmology—a cosmology founded on a multi-dimensional and trans-rational ontology of universal "planes of consciousness".[17] This extension offers enhanced explanatory power for addressing contemporary issues in natural philosophy and science, particularly in the fields of evolutionary biology and consciousness studies, thereby providing a broader platform for reconciling science and spirituality.

My starting point is largely inspired by the vision of the Indian mystic, poet, and philosopher Sri Aurobindo. He developed an integral cosmology that presents one of the most comprehensive first-person mystical accounts of reality, incorporating the evolutionary process from a trans-rational perspective. However, Sri Aurobindo's cosmology is deeply rooted in an

[17] I sometimes refer to it as 'planes of existence,' which is similar to the spiritual 'realms' of existence mentioned in various religions. However, I typically avoid using religious terminology to prevent potential misinterpretations and confusion.

Indian spiritual tradition, which might be challenging for the Western mind—particularly those grounded in the analytic tradition—to relate to. Yet, if translated into modern scientific and philosophical language, his ontology has the potential to significantly enrich contemporary Western approaches to consciousness studies. If I place some emphasis on the spiritual cosmology of an Indian mystic, it is because, of course, his ideas resonate with my personal philosophical preferences. Nonetheless, I believe that his evolutionary spiritual emergentism and cosmology provide the most coherent framework for understanding the world. Sri Aurobindo's approach offers more credible insights into the nature of life, consciousness, and the mind than many current scientific or philosophical speculations, particularly those prevalent in cognitive sciences and the philosophy of mind.

However, I don't pretend that Sri Aurobindo's vision is an established truth or the only perspective; rather, I consider it a starting theory, teaching, worldview, or ontology—not as an absolute truth, but as a working hypothesis, a provisional conjecture, or a hypothetical framework—and I explore where this may lead us. Starting from a working hypothesis has always been the path that has brought us a step closer to truth.

This approach could provide a conceptual platform for a new synthesis, bridging the gap between East and West, as well as between science and spirituality. My project aims to outline a possible draft of this synthesis.

Below is a brief synopsis of the book's structure and how the papers I have published further develop certain arguments.

Chapter II explores the complex and enigmatic nature of consciousness, examining its origins, the so-called "hard problem of consciousness," and the relationship between brain activity and subjective experiences. Consciousness remains a deeply mysterious phenomenon that science has struggled to explain within a materialistic framework, with little progress made since the time of Descartes. Its enigma deepens when it is considered in a broader evolutionary context, prompting questions about why a supposedly purely physical system would evolve to experience subjective sensations.

The widely accepted theory that consciousness arises from brain activity is questioned, as it is unclear how unconscious particles combine to produce subjective experiences. The assumption that anesthesia renders us unconscious is challenged, suggesting that memory and subjective experience might not be entirely erased during such states. The search for the neural correlates of memory reveals that memory is not stored in specific brain areas, and various hypotheses about memory storage remain largely unproven. The conceptual conflation of mind and consciousness is challenged, emphasizing that thoughts are dynamic while consciousness is static and unchanging. At some point, I warn against the correlation-causation fallacy, emphasizing that a correlation between brain states and consciousness does not necessarily imply that neural activity is the cause of conscious experience. The transmissive hypothesis of

William James suggests that the brain acts as a filter or conduit for a broader, non-material universal consciousness, challenging the materialist view.

This chapter can be augmented by the first advanced reading in Section XIII.1 which connects to the transmissive hypothesis with a neurological background. I show how neuroscientific evidence does not point to the mind-brain identity theory as unambiguously as is commonly believed. The background theoretical assumption for the origin of consciousness is always the same: the mind-brain identity theory. It starts from the unquestioned premise that the brain is what produces consciousness. On the contrary, however, a long list of neurological correlates and their phenomenology suggests how several brain functions can best be accommodated inside a transmissive paradigm rather than a generative one.

Chapter III explores the nature of life and evolution, questioning the limitations of materialistic and reductionist views. I agree with those who argue that consciousness and life cannot be fully explained by physical processes alone and suggest that there might be a guiding force behind evolution. In particular, I challenge the rejection of vitalism, questioning whether biology has truly falsified the concept and whether life can be explained without understanding consciousness and cognition first. Here, I discuss the difficulty in defining life, noting that current characterizations focus on materialistic functionalism, which fails to capture the true nature of life. We should incorporate psychological dimensions that the materialistic sciences ignore, and that include consciousness, cognition, thoughts, emotions and sentience as fundamental primitives rather than emergent properties. This also raises the question of whether humans are the ultimate step in evolution and suggests that internal psychological forces might influence our evolutionary destiny more than outer factors. The increasing complexity of biological and physical systems suggests that our current scientific tools might be inadequate to fully understand life. Modern theories that attempt to explain life and consciousness through complex systems and dynamic processes ultimately still rely on mechanistic naturalism. We must give a chance to alternatives, such as the idea that evolution might be guided by an inner consciousness, thus challenging the idea of evolution as an unguided and accidental process. A perspective that allows itself to contemplate the idea that a "conscious design" may exist, differentiating it from creationism. The suggestion that a more inclusive view that incorporates scientific findings without rejecting the hypothesis that an intelligence in Nature is at work will be the leading idea. A vision of evolution as guided by a cosmic consciousness should be included in a philosophical and spiritual perspective that complements the scientific discourse. This leads to an integral view of Nature advocating for a trans-naturalistic vision that sees Nature as a manifestation of higher consciousness, thus emphasizing the need for a holistic approach that integrates science, intuition, and spirituality.

This chapter can be complemented by the second advanced reading of Section XIII.2 making similar points from a more academic perspective. It points out how, contrary to common wisdom, biology has not finally falsified vitalism—that is, the existence of some no better-defined immaterial or subtle element that distinguishes living matter from non-living matter. The demise of vitalism and the notion that the mind could not be a fundamental primitive but only a fortuitous emerging property of life are commonly taken as an unquestioned fact. However, the truth is that biology simply assumes cognition to be an epiphenomenon of matter and ignores the vitalist hypothesis a priori, whereas there are a good many reasons to think otherwise.

Chapter IV explores the tension between scientific determinism and the possibility of divine action in the World. It begins by outlining the classical view of a causally closed, deterministic universe, where every physical event is the inevitable result of prior causes governed by immutable laws. This mechanistic worldview, rooted in Newtonian physics and the Principle of Sufficient Reason (PSR), leaves little room for supernatural intervention or free will, suggesting that all phenomena—including human consciousness and behavior—are ultimately reducible to physical processes.

However, this strict determinism is challenged by quantum physics, chaos theory, and complexity science, which reveal inherent unpredictability and indeterminacy in Nature. I discuss "quantum theology," a perspective suggesting that the probabilistic nature of quantum events could provide a space for divine influence without violating physical laws. This approach reframes divine action not as a suspension of natural laws, but as a subtle guidance within the indeterminacies and emergent complexities of the universe, allowing for both scientific integrity and theological meaning. However, to get there, a long digression into quantum physics and the notion of free will is needed.

Chapter V discloses my bias as a physicist. It explores the philosophical implications of quantum mechanics and its challenge to materialism, emphasizing the non-deterministic nature of the quantum realm and its potential connection to consciousness and free will. The first two sections closely mirror what I've already detailed in my book “Spirit Calls Nature,” as I felt it was essential for maintaining the consistency of this work. Materialism assumes a deterministic and predictable universe—a view rooted in Laplace's demon thought experiment, which suggests that knowing all forces and positions in the universe would allow for complete prediction of the future. Quantum mechanics challenges the deterministic view by revealing the counterintuitive behaviors of particles, such as wave-particle duality, Heisenberg's uncertainty principle, quantum superposition, and entanglement challenging classical concepts of local realism. The focus is on the fact that quantum mechanics suggests that some phenomena occur without a discernible cause, leading to the idea of quantum randomness and challenging the traditional cause-and-effect paradigm. I show how this can be interpreted by taking a philosophical stance that questions the materialistic worldview and explores the potential for a non-causal paradigm including divine determination and a universal consciousness. What I call "quantum idealism" proposes that quantum mechanics reveals how reality, as we perceive it, is a mental construct based on the limitations of our sensory-based understanding of the world. I then show how there is a relationship between free will and quantum mechanics and argue that quantum indeterminacy could provide a basis for free will, thereby challenging deterministic views.

Sections V.4&5 clarify why panpsychism is frequently misunderstood, even by professional philosophers who embrace a specific version known as "micropsychism", which deviates significantly from its original meaning. While panpsychism is often interpreted as "all is mind," its true meaning is more accurately described as "all is consciousness" or "all is the World-Soul." I distinguish between micropsychism, which posits that consciousness emerges from the combination of proto-conscious particles, and cosmopsychism, which views the universe as a manifestation of universal consciousness—where the universe *is* consciousness, rather than consciousness resulting from a mix of proto-conscious elements. The latter form aligns more closely with the nondual Eastern tradition, and I will refer to it simply as "panpsychism," despite the literature often implying the former understanding. I contend that these two perspectives are complementary rather than mutually exclusive, supporting an integral vision where both micropsychism and cosmopsychism represent aspects of a singular universal consciousness. Additionally, I caution against the conceptual confusion that arises from conflating mind with consciousness.

The concept of "libertarian quantum panpsychism" is introduced, suggesting that quantum indeterminacy is the expression of a universal consciousness, integrating micropsychism and cosmopsychism into a holistic view.

In the midst of this chapter, as a physicist, I could not resist the temptation to ask whether quantum mechanics has something to do with consciousness and the longstanding question of whether we have free will or if it is only an illusion. I ended up realizing that, indeed, it does, but not in the sense in which people usually think of the subject. Here, also, we must widen our perspective and see the brain not as the source or "generator" of consciousness but as an instrument and individuation of a "Mind at large," which we will define later. Then, quantum indeterminism, free will, and self-causation begin to make sense. This view has been extended in a follow-up article on quantum indeterminacy that not only illustrates how quantum randomness can be regarded as a "backdoor" for free agency in an otherwise seemingly strictly deterministic clockwork universe but also how this can lead to a theory of universal consciousness—that is, a libertarian panpsychism.

This chapter on quantum physics and free will finds its completion in the two papers of the Advance Reading Sections XIII.3&4.

Chapter VI provides a deep dive into the mind-body interaction problem of mental causation from the standpoint of physics. The interaction problem questions how a supposedly immaterial mind could affect a physical body, thus challenging both materialistic and dualistic views. Descartes' dualism posits the mind as a non-material substance separate from the body, but modern science has largely moved away from this view. Physicalism suggests that all mental states are just brain activity, yet this view is controversial and faces the hard problem of consciousness. What is rarely recognized is that physics already deals with interactions between material and immaterial entities, like electromagnetic fields.

This, in turn, suggests that the mind-body problem might not be unique to the philosophy of mind but is inherent in material sciences as well. However, quantum field theory (QFT) blurs the line between material/physical and immaterial/unphysical, with particles seen as different modes of a single field. This hints at a unified substance, leading us to conjecture about the existence of "subtle matter," which interacts weakly with ordinary matter.

The question that arises is what the ultimate distinction between "physical" and "unphysical" entails. Anything that exists beyond space, time, and quantum fields—structures that may be emergent rather than fundamental—can be classified as beyond the physical layer. However, as we will explore, there are intermediate ontologies that do not fit neatly into either category.

The chapter can be considered a preparation to the article in the Advanced Reading Section XIII.5.

Chapter VII closes the circle started in Chapter IV, taking up again the relationship between scientific determinism and the possibility of divine action in the light of what has been discussed in the three preceding chapters. It challenges the classical view that the universe operates as a causally closed and deterministic system. It highlights how quantum mechanics and chaos theory reveal fundamental indeterminacy and unpredictability at all scales of reality, not just the microscopic. This openness in causality suggests that the future is not predetermined, and that even small quantum fluctuations can be amplified in nonlinear, chaotic systems, making long-term predictions impossible and allowing for the possibility of free will and divine influence.

Perhaps the most technical paper is the Advanced Reading Section XIII.6. It is likely beyond the grasp of readers without a physics background. However, don't worry; it is a bonus and not mandatory for understanding the rationale of the argument. The main point is that divine action is not only coherent with our modern understanding of the world, but when viewed from the appropriate perspective, we can see how it is even suggested.

Chapter VIII leaves behind the quantum perspective and integrates the Eastern and Western views on free will. Western views on free will suggest that it is an illusion created by Nature's determinism, while Eastern philosophies propose a "higher self" that transcends this determinism. The integration of both into a unitary vision offers a richer understanding that suggests that what we perceive as free will is part of a broader, more complex reality.

Chapter IX begins to explore the concept of idealism. It questions the nature of reality and whether it is an illusion or a simulation, as some modern thinkers propose. Additionally, it delves into philosophical arguments, scientific perspectives, and historical viewpoints from both Western and Eastern traditions. I recall my childhood question about whether life is just a dream and how this question persisted into adulthood, leading to my study of Indian non-dual teachings of Advaita-Vedanta. I connect this to the modern simulation hypothesis, which suggests that our reality might be an artificial simulation, and discuss it as a modern, technologically influenced version of age-old philosophical ideas like Plato's allegory of the cave. Contrary to common belief, taking the simulation hypothesis seriously does not lead to what the techno-centered mindset believes but, rather, confirms philosophical idealism, positing a self-simulating and self-causing "simulator" akin to the concept of "God."

This is the basis for discussing how the brain constructs our perception of reality and how this construction detaches us from the actual reality, thus emphasizing the illusory nature of what we perceive. This can be illustrated via a purely scientific perspective with a detailed explanation of how visual perception works, illustrating how the brain processes and interprets sensory information far removed from the actual objects it perceives. Plato's allegory of the cave is further used to explain how our perceptions can limit our understanding of reality, comparing our sensory experiences to shadows that do not represent the true nature of objects.

Spinoza's idea of substance monism, which posits that only one substance exists and that mind and matter are different attributes of this substance, is examined. This, in turn, lays the groundwork for one of the fundamental aspects of the monistic outlook inspiring the "integral cosmology"—namely, substance monism.

This connects with Eastern philosophies like Taoism, Buddhism, and Hinduism, highlighting their views on the nature of reality, consciousness, and the illusion of the phenomenal world. Sri Aurobindo's concept of the Real-Idea, which integrates idealism and evolution, thereby suggesting that a greater Intelligence and Will are at work in the universe, is discussed. The chapter concludes with the idea that understanding the illusory nature of reality and the role of a higher consciousness is crucial for moving beyond a materialistic worldview and embracing a post-material future.

Chapter X spells out the core idea of the integral cosmology.

I examine idealistic approaches with a critical eye for their failure to provide a coherent evolutionary perspective and for their coarse-grained theoretical frameworks. Idealism is a crucial step forward that allows us to move beyond a strictly physicalist worldview. However, idealism alone is insufficient; it leaves us halfway on our journey toward an integral paradigm.

This leads me to the proposal of a wider, more comprehensive and evolutionary integral cosmology that views consciousness as the fundamental primitive, with life, mind, matter, the subconscious, and the superconscious as different planes of consciousness. This is a hypothesis suggesting that the brain, rather than generating consciousness itself, instead serves as a filter and interface between these different planes of consciousness. This aligns with William James' filter theory.

The concepts of evolution and involution are introduced, with consciousness first involving itself in matter and then evolving through various forms, thus leading to a more comprehensive understanding of life's complexity. The idea of a spiritual evolution paralleling that of a purely physical evolution is introduced. This part discusses the evolution of the soul, proposing that the soul grows through life experiences and transmigrations, in turn influencing biological evolution and providing a deeper understanding of our true spiritual and physical nature. It is a conception of the world, life, and our inmost nature that reflects a spiritual emergentism, reconciling mechanistic and teleological views of evolution, suggesting that the evolution of species is guided by a top-down influence of a superconscious spirit, complementing the bottom-up unconscious emergence. This allows us to connect with the question of free will from a much broader perspective. The integral cosmology sees free will as a continuum, not a dyadic choice, with true freedom achieved by aligning with the superconscious Will rather than being constrained by the deterministic laws of the lower planes.

This chapter concludes with another essays that I published on these aspects in the Advanced Reading Section XIII.7. The paper lays the groundwork for a larger synthesis of how Eastern spiritual thought could be bridged with Western philosophical and scientific approaches: the article in the advanced reading section which illustrates the integral cosmology of Sri Aurobindo from the perspective of consciousness studies.

Chapter XI brings us back to more earthly questions that, however, are related to metaphysical worldviews. What we have learned in the previous chapters is now applied to an instance of our present age: AI. This chapter discusses the limitations of current AI technologies—particularly large language models (LLMs)—in attaining true artificial general intelligence (AGI), which cannot be achieved unless AI realizes a true semantic understanding of the world. ChatGPT and similar LLMs have no conscious experiences, no life, no sentience, and generate responses based on patterns in their training data rather than by being rooted in genuine understanding. They rely on numerical representations of words instead of human-like percepts. The claim is that achieving AGI necessitates the creation of conscious machines—a goal far beyond the current scientific understanding of consciousness in organic intelligence. Strong AI, or AGI, implies machines that truly understand meaning, whereas weak AI only simulates human intelligence without true comprehension. This discussion is meant as an exercise to highlight the enormous investment in financial and human resources we are throwing into projects that might be doomed to failure because they are based on a naturalistic and materialistic understanding of our true nature.

I argue that we need a paradigm shift and an entirely different perspective. Therefrom originated the two advanced readings at the end and that complete this chapter. The first reading is a manuscript on the relationship between consciousness, cognition, and the potential advent of the age of AGI. Here, I showed how semantics—that is, our cognitive ability to associate meaning with things and concepts—is intimately related to conscious experience. The kernel is that language is the translation into symbols of a cognitive process that does not take place in the brain but is beyond the brain. If so, the natural conclusion is that an AI system cannot achieve an agent's semantic understanding of a text or of the environment and the world if the AI system is not also conscious, no matter how complicated and powerful it might be. No consciousness, no semantics, no AGI.

The first part was taken up in the Advanced Reading Section XIII.8.

This led me to less-known but equally fascinating post-material subjects. One example is an essay left for the readers willing to dig deeper in the Advanced Reading Section XIII.9, and that summarizes Abhinavagupta's and Sri Aurobindo's perspectives on a possible spiritual explanation of the nature and origin of speech. Language is the reflection and a pale shadow of cognitive processes that take place far beyond a mere representational framing of the world. It is ultimately rooted in a spiritual domain that exclusively transcends the computational process.

Finally, chapter XII concludes discussing how we can take a step further toward a post-material paradigm. It examines the limitations of materialistic science and technology in addressing humanity's deeper issues and advocates for a post-material paradigm that integrates spiritual insights and inner wisdom. Despite significant technological progress, many scientific and technological projects have not met expectations, leading to a sense of disillusionment with their ability to solve humanity's deeper problems. Many advancements, such as the war on cancer, genetic engineering, and stem cell research, have not delivered the expected practical applications—a fact that highlights the complexity of life and the limitations of a purely materialistic approach. Scientific and technological progress was expected to ease our daily activities; however, it has not improved psychological well-being, as stress and burnout continue to rise. Material progress does not equate to inner fulfillment, which suggests that humanity's deeper issues stem from a lack of spiritual and psychological harmony.

Only a post-material science that integrates subjective, first-person perspectives, thereby complementing traditional empirical methods, can lead us further. Only something that calls for transcending the dichotomy between science and spirituality, suggesting that both must evolve toward a more integrated understanding of reality, will be able to lead us to the breakthrough that the traditional approach is unconsciously looking for. I emphasize the importance of embracing the mystical experience and propose that it offers valuable insights into reality that complement scientific understanding. Moreover, I discuss the need for an integral cosmology that sees the whole rather than just the parts, thus advocating for a multi-dimensional, evolutionary perspective of life that could lead us to a new form of psychology beyond the present naturalistic understanding of the human dimension. We need something that leads us to a standpoint calling for a Copernican shift from a matter-centered worldview to a soul-centered one, suggesting that true progress requires integrating spiritual insights with scientific understanding.

The desire and aim to put all these works together into a coherent vision led me to write this book. I hope it provides a comprehensive perspective on what I believe will become the post-material scientific and spiritual perspective on reality, life, and consciousness of the future.

II. The Mystery of Consciousness

The most iconic philosophical question that science has faced since its inception and that stands on the verge of metaphysics is the origin and nature of consciousness. Certain aspects of our conscious experience appear irreducible and resist explanation within a materialistic framework. Since Descartes' musings on the mind-body problem, science has made little progress in addressing these questions about consciousness.

Significant progress has been made in the neurosciences, largely due to groundbreaking technological advancements in brain scan technologies, not to mention all the advances in microscopy, that have enabled us to observe the structure and function of nerve cells at nearly the molecular level. As a result, the organization of the brain has been mapped with remarkable precision, revealing many of its functions in great detail. Each year, thousands of peer-reviewed scientific papers contribute to our understanding of this complex field.

However, we still know very little about the true nature and origin of consciousness. Although debates about consciousness and its emergence in the cosmos are not new, they gained renewed attention in the late 1980s and the 1990s, with the latter often referred to as the "decade of the brain," thanks to significant advancements in brain monitoring techniques. During this time, the Australian philosopher David Chalmers rose to prominence by articulating the so-called "hard problem of consciousness." He emphasized the explanatory gap between our first-person subjective experiences (qualia) and the third-person neuroscientific perspective derived from empirical observations of neural correlates. Many attempts have been made to develop new naturalistic theories of consciousness within the frameworks of modern neuroscience, cognitive science, and both computational and non-computational theories. However, these approaches have struggled to answer the deeper philosophical questions surrounding 'phenomenal consciousness'—our capacity for sentience, feelings, and subjective perceptions.

Consciousness becomes even more puzzling when viewed in its broader evolutionary context. Why would dead matter, evolve to enhance survival and reproduction, and develop the capacity to experience joy, pain, curiosity, or wonder? These aspects of our existence as living, sentient, and thinking beings remain unexplained more than ever. Despite centuries of philosophical exploration and decades of technological advancements, they are still largely resistant to conventional scientific investigation. The challenge does not seem to be a traditional obstacle that science might gradually overcome; it represents what I refer to as a "no-progress quest" —questions on which science has made little headway for centuries. The inquiries surrounding the nature, origin, and function of consciousness are among the most profound existential questions, proving exceptionally difficult to answer.

1. The Hard Problem of Consciousness in a Nutshell

One of the most fascinating and mysterious aspects of modern science and philosophy is the question of why and how a universe that science assumes to be unconscious, mechanical, and aimless nevertheless produced something that has become conscious, doesn't act predictably and mechanically, and behaves purposively like we do. The most mysterious aspect of life is that a (seemingly) unconscious Nature gave birth to conscious and sentient beings. I find this unfathomable contrast so strange, and I can't understand those who find it perfectly normal. It is like accepting without further questioning that something pops into existence out of nothing.

We are told that consciousness emerges from brain activity. It is a widely accepted theory that consciousness arises from the activity of neural networks that create, due to their electrical and chemical signals, complex patterns giving rise to a conscious experience. It is believed that some specific brain regions play a key role in the emergence of consciousness, i.e., the prefrontal cortex, the parietal cortex, and the thalamus somehow integrate the information from different sensory inputs, creating a unified and coherent experience of the world.

After all, it sounds reasonable. If your brain falls into a comatose state, you are "unconscious," right? If you take a drug that alters your brain chemistry, you will perceive the world from an altered state of consciousness. If the occipital lobes at the back of your head, which are responsible for visual perception, are impaired due to an accident, you won't be able to see anymore, even if your eyes are still healthy and operational.

So, what is the issue here? Isn't it obvious that the brain "produces" conscious experience?

The problem with this line of reasoning is that when we refer to that wet, gray, slimy matter we call "the brain," we are pointing at a material aggregate ultimately made of molecules, atoms and particles. These we believe to be devoid of any sentience, experience, sensory perceptions, feelings, emotions, thoughts. The question is: Why does a combination of insentient and unconscious particles build up a brain that, nonetheless, is endowed with sentience, experience, sensory perceptions, feelings, emotions, and thoughts?

You might object that it is due to gazillions of super-complicated and hyper-complex processes and interactions in the brain that make a subjective experience emerge. But why should more complexity make us conscious?

If you don't find this mysterious, think about it again. Why should a bunch of unconscious atoms, molecules, and neurons put together, however complicated in their arrangement and sophisticated in their interactions, give rise to a subjective qualitative experience, what philosophers call "*qualia*" of pain or pleasure, joy or grief, sensory or visual sensation, etc.?

For example, how do you know that we both see subjectively the same color when looking at, say, a red strawberry?

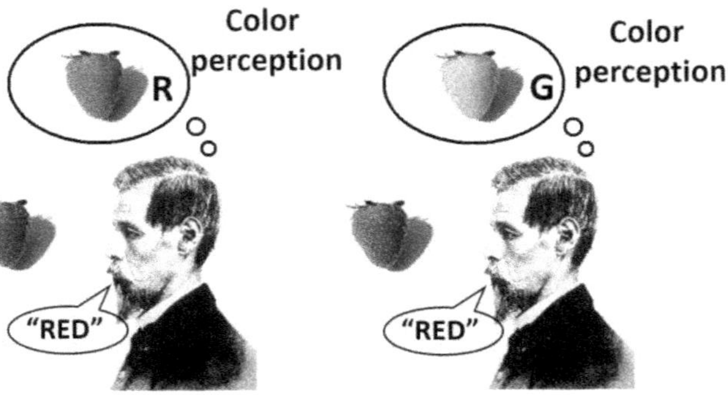

Fig. 1 The inverted visible spectra problem. [Credit: Wikimedia Commons]

How do I know that when you perceive that specific color sensation calling it "red" (Fig. 1 left), you perceive the same subjective perception, the same qualia, that I do? Perhaps, I might perceive it as green (Fig. 1 right), though we both call it "red."

Notice how this isn't something that you can understand analytically. You must become aware of the distinction between what makes a third-person description of science versus a first-person experience of a living organism–that is, "what it is like to be" seeing a color.

This, in a nutshell, is the hard problem of consciousness, sometimes also called "the explanatory gap" or the "problem of qualia," which was first formulated by philosopher David Chalmers in 1995 (though less neuroscientific formulations of it date back at least to the time of Leibniz). It refers to the philosophical problem of understanding why and how subjective experiences arise from physical processes in the brain made of subunits (neurons, molecules, atoms, etc.) we suppose be without experience. There is a fundamental gap between the objective, third-person perspective of science and the subjective, first-person experience of consciousness. Although science explains how the brain processes information, it does not address why that processing is accompanied by subjective experience. The hard problem is considered so "hard" because it resists any scientific explanation. The relationship between the physicality of the world and our undefinable, ineffable, but very concrete subjective experience remains unbridged.

The point is, if we believe only and exclusively in what science proves to exist with third-person experiments or observations and declare everything else as "non-existent" or an "illusion," then consciousness must be considered inexistent and illusory as well.

Can you prove to be conscious to someone else? You can't. Not even in principle. Why not? Strictly speaking, if we believe only in what science can demonstrate, we must brand consciousness as an irrational belief, as something that does not exist, just collective hypnosis or a superstition. In principle, science should reject the idea that you have feelings, sensations, or subjective experiences. Consciousness should be declared a pseudo-scientific concept because not the slightest empirical evidence shows that it exists. You will not find your consciousness by opening your skull and rummaging inside your brain. The first-person perspective fact that you are conscious of appears nowhere in the third-person perspective eyes of science.

You might object that we can visualize the brain activity thanks to modern brain scans or measure its electric activity with an EEG, correlating with our conscious experience. But, if you think carefully, that doesn't prove anything. It merely shows that some electrical phenomenon in the brain correlates with someone claiming to have a sensation. How do you know that it is true? It could be an imitation or the reaction of someone having no sensation at all but acting as if they have one. How could I know what you feel and if you feel anything at all? Maybe you are an insentient zombie!

This sounds silly, doesn't it? But why do even the most skeptical and materialist scientists nevertheless accept that we are conscious?

It is not difficult to see where the connection is made: Believe it or not, even scientists are humans who have subjective experiences and can access their sensations by first-person introspection. If scientists accept that subjective conscious experiences "exist" (whatever "existence" might mean in this context), it is, paradoxically, only because of that kind of introspection that science always tried to expunge, discrediting it as unreliable and misleading.

This has created cognitive dissonance to such a degree that nowadays, some philosophers, the so-called "eliminativists," deny consciousness as any type of reality and brand it an "illusion" or a "misrepresentation" of the brain to itself. But these are a minority and, in my opinion, are like the Ptolemaics who stick to an old and dying paradigm or, worse, represent yet another form of denialism among many that characterize our modern times. To claim that our feelings, perceptions, and sensations are mere illusions raises the question: who is experiencing this illusion? Responding with "you," "me," or "I" simply circles back to the very conscious individuality purported to be illusory, resulting in circular reasoning. In my view, this makes no sense. Denying consciousness is tantamount to rejecting the most self-evident and certain fact of our existence. Therefore, please allow me to happily ignore this standpoint.

We have to surrender to the fact that there remains something ineffably subjective in the first-person experience that refuses to be crammed into the narrow walls of a third-person analytic perspective. Trying to find the magic which transmutes mechanistic and insentient process into subjective sensations is reminiscent of the alchemists who believed it possible to transmute base metals into gold. The assumption was that the universe is made up of only the four elements (or qualities) of water, air, fire, and earth. In a sense, modern science believes something similar, namely, that the universe is made up only of matter, forces, space, and time. Consciousness plays the role of the elephant in the room, and it will continue to do so until we move away from this kind of mechanistic reductionist scientific materialism and a naïve realism that got stuck in the 18th and 19th centuries. I hope that how this could be done will become somewhat clearer throughout the next essays which will further clarify what this issue is all about.

2. On Anesthesia, Coma, Memory, and Consciousness

A very typical and unquestioned belief is that when we are under general anesthesia, we become "unconscious" and are completely "unaware" of anything. We simply "cease to exist."

Outside observers see how the body becomes completely unresponsive, even to what would be otherwise the most painful experience. We seemingly have all the good reasons to believe that in such a condition, one is unconscious. After all, those who receive anesthesia or are in a deep coma don't move, don't speak, have their eyes shut, and don't react to any painful stimulus.

Notice how basing our conclusions on the fact that someone doesn't move, doesn't speak, has their eyes shut, and doesn't react to external stimuli is, again, an exclusively third-person perspective assessing external facts. We look only at the external physical aspects of a patient's body, and then jump to some conclusion about his/her inner state of consciousness.

On the other hand, it is also true that, once awake, the patient reports a void of subjective experience. From the first-person perspective, when people wake up from anesthesia or a coma, they assure you that there was nothing; they didn't exist. Everyone who has been under anesthesia can testify to how one wakes up with this sensation and with an abysmal sense of temporal void, with no sense of any time flow. In a sense, many of us can relate to this. I remember when I went under general anesthetization for a surgical procedure. There seemed to be nothing between the moment I fell into a coma-like state and the awakening when I couldn't remember anything. Everything is gone, and even the sense of time disappears: Reality disappears and reappears in the blink of an eye.

So, what else should we conclude? Under anesthesia, we are unconscious.

This is the unquestioned premise with which many neuroscientists and philosophers of mind work.

However, the question is whether this rationale is sound. Are we really knocked out of existence during anesthesia?

All this seemed obvious to me. I had no doubts about it until, a couple of decades ago, an extremely vivid experience showed me the logical fallacy of this reasoning.

I once woke up from what we would call "sleep without dreams"–an awakening with the kind of feeling one has after a night of deep sleep, similar to a post-anesthesia wake-up, apparently without any dreams and no sense of time. It is a sensation of having gone "out of existence."

To my surprise, however, after a while, a plethora of (more or less fragmentary) dreams bubbled up. Suddenly, I could recall forgotten dreams: it was no longer a dreamless sleep. I focused on that remembrance and could easily compare and sense the differences between the sleep state that seemed to be without dreams and the *same* sleep with dreams. In a sense, it was a controlled experiment of popping in and out of existence!

What this experience made clear to me is how everything depends on gaps in our memory. If you remember an experience, you say, “I was conscious” during that experience. If you don’t remember any experience, you will tell yourself, “I was unconscious.” But you were not. Can you recall your experiences as a newborn baby? I’m guessing you can’t. Therefore, are babies unconscious zombies that don’t feel anything? It is unlikely that your parents would agree with such questionable extrapolation.

I don’t want to convince you that during anesthesia, we are conscious, feel the pain of the surgery, or are dreaming or having whatever experience (though some studies suggest this to be the case). From the scientific point of view, we should remain agnostic in this regard. The only point I would like to make here is that much too often, we make unwarranted assumptions that seem indisputable but which, upon closer scrutiny, are not at all obvious, and nevertheless jump to metaphysical conclusions.

On these unreasonable premises and much too simplistic reasoning, some philosophers found their careers. Their narrative is that, when waking up from anesthesia, which is a psychoactive substance modifying our brain chemistry, since we can’t report any experience, we must assume, without questioning further, that said anesthesia makes us completely unconscious and, thereby, must almost axiomatically posit that the brain produces consciousness. On this premise, people build enormous theoretical frameworks and ideological sand castles about the nature of consciousness.

This is not just a philosophical hairsplitting exercise. These kinds of misplaced assumptions and inferences have far-reaching ethical consequences that, for example, also frame the legislation over the destiny of patients in a coma or vegetative state. The possibility of being conscious during anesthesia introduces a complex ethical and legal dilemma. From an ethical perspective, even if the memory of the event is lost, the individual still experienced distress or pain in the moment — raising questions about whether harm is defined by the experience itself or the lasting awareness of that experience. The principle of non-maleficence ("do no harm") suggests that suffering, even if forgotten, is still harm. Neurologically, the body may still register the trauma, potentially leading to unexplained anxiety or other psychological effects, even if the person can't consciously recall why. Legally, this scenario raises philosophical questions about the nature of suffering—does pain that leaves no memory still matter? From a moral standpoint, many would argue it does, suggesting that safeguarding against intraoperative awareness remains crucial, even if the patient never remembers it happening.

What all this shows is that sometimes, especially when it comes to deep existential questions involving our psychological dimensions, we must be careful in walking through a potentially dangerous conceptual minefield by learning to determine what the real facts are and what our (unconscious?) assumptions are. We should always question what our premises are and how far they are justified. In some cases, it might be a choice between life and death.

3. Does Consciousness Emerge from Matter Like Wetness from H_2O?

We have seen how there must be something missing in our simplistic idea about the relation between the brain and our conscious experience, and that is summed up in what nowadays is called the "hard problem of consciousness": How can unconscious matter, made of unconscious and insentient atoms, molecules, and neurons, arrange into a brain, which has conscious experiences as a sentient being? How can a quantitative increase in the complexity of an unconscious system generate conscious qualitative sensations?

However, I hear it said over and over again, including by prominent scientists and philosophers, that there is no real hard problem of consciousness. It is a misconception, a misunderstanding, and a false argument nowadays in fashion, but it can be easily explained away. One of the typical arguments is based on an "emergentist" line of reasoning[18].

Consciousness emerges like the wetness of water, so the saying goes. When gazillions of non-wet water molecules stick together, they build up a substance that is... well, wet. Similarly, gazillions of non-conscious neurons stick together, building up a conscious brain. Consciousness emerges due to a bottom-up aggregation and activity of non-conscious neurons, no more and no less than wetness emerges due to the organization and interaction of non-wet H_2O molecules. Thus, there is nothing mysterious in consciousness. It is an "emergent property."

This neurons-consciousness relationship, compared with the molecules-wetness analogy, sounds like a reasonable explanation.

But somehow, there remains the feeling that something is missing. There is an unaware (unconscious?) conflation that we seem to have a hard time becoming aware of–or, at least, this was my case. I must confess that when I had to tackle this objection for the first time, I couldn't figure out exactly what was wrong with it. I felt it unreasonable but, at the same time, could not come

[18] Within the philosophy of science, "emergentism" analyzes how the properties and the interaction between fundamental constituents lead to new properties of the system that are not present in the constituents themselves.

up with something consistent. It was the kind of feeling in which I knew that something was deeply flawed but, nevertheless, found myself arguing with equally flawed counterarguments.

However, after thinking this through carefully, one can clearly see where the fallacy is: There is no wetness in "the world out there" in the first place, no more and no less than there are tastes, smells, colors, pains, and pleasures in "the world out there." One might even say that putting together non-wet water molecules does *not* develop wetness.

At first, this statement might sound absurd. How can someone contend that water isn't wet?

Fig. 2 Is wetness an emergent property of H_2O molecules?

But the fact is that, in physics, there is no such thing as "wetness." Rather, one speaks of viscosity (a measure of a fluid's resistance to flow). What we call 'wetness' is a subjective quality, not an objective property. It is already a subjective experience, a sensation, and sensory skin perception–that is, a qualia that emerges only *after* nerve-to-brain transmissions *in* our conscious awareness. Our language betrays this when we say that water "feels" wet, seemingly implying that it is an illusion. Wetness is a sensation, not a physical property of something, just like the sweetness of sugar is a taste, not a physical property of sugar [19]. These are experiences of something, and not the thing in itself, like shadows projected onto our sensory "screen of consciousness"–not the real thing that projects the shadow.

The problem is that we continuously forget to distinguish first-person subjective experiences (here, the wetness of water) from third-person conceptual abstractions (here, the viscosity of water).

Thus, the emergence of wetness is an interesting analogy that, however, doesn't add much to the understanding of the emergence of consciousness. On the contrary, it only deepens the problem. Nobody knows why and how the sensory information from skin receptors translates in the brain as the subjective experience of wetness. The perception of what we call "wetness" is itself already a qualia in our conscious awareness, and, therefore, its generation by a

[19] Several chemical compounds other than sugar are perceived as sweet–a fact the food industry could cleverly exploit by selling products filled with artificial sweeteners and sugar substitutes with potentially detrimental effects on our health.

bunch of insentient molecules and neurons remains as mysterious as the emergence of sentience, feelings, and all the other subjective perceptions from brain activity as well. Resorting to the analogy of the relationship between water molecules and the emergent property of wetness to explain the emergence of conscious experience from the activity of neurons in the brain doesn't clarify much. The explanatory gap between unconscious matter and conscious experience remains more abysmal than ever.

There are several other conceptual and experimental aspects to explore that challenge a materialistic paradigm. It may be worth mentioning the neural correlates of memory. It turns out that memory is not localized; memory does not reside in a specific brain area, contrasting with digital data storage. Correlations with brain size are absent: the expected correlation between memory volume and brain size or neuron complexity is not observed, as seen in cases like hemispherectomy, where memory remains intact. The ability of the nervous system to adapt may explain memory retention after significant brain alterations, but this remains a hypothesis rather than a proven fact. I have described this in more detail in the advanced reading added in section XIII.2.

In the next section, I would like to take a different approach that relies on a more contemplative and experiential dimension.

4. Are Mind and Consciousness the Same?

"*Cogito ergo sum*" ("*I think, therefore I am*") was the famous sentence that French philosopher René Descartes coined in his work "Meditations." In essence, it states that thinking is proof of our existence. Though we might behold false beliefs and deceiving thoughts, the simple fact that we think is evidence for the existence of the thinking subject.

As appealing as this might have sounded to Descartes, it also led to much confusion, especially in Western philosophy. It led to the more or less unaware semantic conflation of words indicating very different qualitative experiential phenomena– phenomena that we perceive in us instead of out of us.

Words like "mind," "consciousness," "feelings," and "emotions" are much too often thrown into the same pot and are taken almost synonymously. Suddenly, a feeling becomes a "state of mind," consciousness is labeled as a 'feeling,' or a thought is assumed to be the cause of a conscious experience. In particular, the conflation between the words "mind" and "consciousness" has become so ubiquitous that we tend to equate them with the very same thing.

But are mind and consciousness only two words synonymously indicating the same phenomenal experience? Is consciousness a "mental phenomenon"? Are feelings and emotions a "mental creation"?

To answer these questions, we must give the first-person approach a chance. If we try to quiet for a moment our exteriorizing rational mind[20] and learn to look inward, via a first-person approach with an introspective movement, an obvious fact comes to our attention: Thoughts come and go, but the subject witnessing those thoughts is always the same. The mind, in all its restlessness, continually produces a myriad of thoughts, a succession of mental images, or, to put it in Cartesian terms, an endless series of cogitations that might distract us or lead us to focus our attention on something but, nonetheless, are always changing. There is something dynamic.

On the other hand, there is something underlying and immutable in us, and that never changes (or, at least, doesn't seem to do so during the waking state.) There is a never-changing consciousness in the subject that watches, perceives, and becomes aware of the ever-changing mind with its whirl of thoughts. There is something static.

You don't need any scientific research to "know" that this dynamic versus static polarity exists. It is an introspective 'knowledge' but, nonetheless, a form of knowledge too. It is a self-evident experiential fact. One thing is the different fleeting thoughts of the mind, and another thing is the unchanging witness that apprehends those thoughts.

Unfortunately, much too often, well-trained intellectuals such as neuroscientists, philosophers, and psychologists speak of mind and consciousness as interchangeable concepts. This is an assumption that conflates all experiences and labels them as "mental events." Therefrom came the construction of sophisticated theories in psychology and neurology that, with the passing of time, encountered paradoxes or internal inconsistencies, or, worse, did not fit with the empiric assessments of modern neuroscience, leaving scientists wondering what went wrong.

This led neuroscientists Antonio and Hanna Damasio to propose a new theory of "embodied consciousness," i.e., that feelings, and not thoughts, are the source of consciousness. They claim that feelings are produced by neurological mechanisms of "interoception"–that is, the sensation allowing us to infer, from bodily feelings, our interior state. In other words, the theory conjectures that consciousness begins with feelings rather than thoughts. It argues for the existence of an interoceptive mechanism that starts from the brain's neural activity and extends to all bodily sensations.

Again, go beyond interoception and take the first-person introspective standpoint seriously. It is easy to see how this "feeling first" idea doesn't hold for the same reason that the "mind first" idea couldn't hold. Feelings can't be the source of consciousness because they require consciousness to be felt in

[20] The common misconception that silencing the mind implies "loosing" it, is another symptom of an attitude that overidentifies our true individuality with a reflection of it.

the first place. It doesn't make sense to speak of feelings without a conscious subject[21].

If, instead, we complement our analytic understanding with a subjective approach, applying it to the modern speculations of the philosophy of mind (notice how the melting pot is already evident in the word choice of Western philosophy: It is taken as synonymous with "philosophy of consciousness"), we can readily see how most of the contemporary theories about the origin of consciousness are based on a misleading third-person-only approach that refuses (again, more or less implicitly) the introspective insight.

You cannot explain consciousness starting from feelings because feelings themselves are phenomena that are meaningless without a reference to consciousness. Like with physicalism, the underlying assumption is that consciousness can be explained away by non-conscious phenomena, while we fail to see how, deep down, the non-conscious phenomena can also be described only by something requiring a conscious experience to begin with. It can't work, not even in principle, because it is a self-referential logic in which the concept that wants to be explained requires itself in the explanation. With the notion of phenomenal consciousness, this is always and inevitably the case, no matter what kind of intricate and convoluted theory you will fabricate. It is akin to the ancient symbol of the Ouroboros dragon that eats its own tail.

Fig. 3 The Ouroboros dragon eating its tail, a symbol of self-referentiality.

[21] According to some Eastern spiritual and philosophical traditions, one can attain spiritual self-realization transcending the ego, whereby even the idea of a "subject" is misplaced. Our sense of limited individuality is a thought as well, a sort of mental conception that isn't permanent either. Even the ego-individuality, the subject itself is something "seen" by consciousness: The latter must not be confused with the former. For the sake of simplicity and brevity, I will not address this aspect here. But the reader should be aware that–also adopting the first-person perspective–the way we see and understand the world depends on our state of consciousness and that things are not as straightforward as we tend to believe.

The only way out is to accept that consciousness—understood as the static element in our cognition— is a primal and fundamental entity, not an emergent property[22], and that we must make a clear distinction between consciousness, mind, and emotions or feelings.

The point I'm trying to make here isn't so much technical or about the nomenclature. You can forget about the technicalities and choose the wording you prefer. The point is that we can't speak about these philosophical and/or metaphysical questions if we don't know ourselves.

Science is a third-person approach to reality. While there are aspects of reality we can investigate only through a first-person approach. We should learn to use them both at the same time. Summing up the parts isn't enough. Synthesizing them as complements into a unified view toward a new post-material science transcending them both is the way to go.

5. Is Consciousness Fundamental? The 'Transmissive' Hypothesis

Let's face it: An undeniable connection exists between the physical state of our brains and our subjective experiences. It is widely recognized that the chemistry of our brains influences our perceptions, mental functions, and states of consciousness. For instance, alcohol can alter our awareness once it enters our bloodstream. Similarly, psychedelic drugs can produce intense subjective effects. Dopamine, a neurotransmitter, facilitates communication between neurons and plays a key role in determining the effects of drugs like cocaine. It is well-known that brain damage can lead to severe cognitive impairments; for example, disruption to Broca's area in the left hemisphere can result in loss of the ability to speak[23].

Today, sophisticated brain scan technologies provide clear evidence that each conscious experience activates specific neural correlate in the brain rather than other. For example, these tools demonstrate how the limbic system mediates emotional states. The amygdala, a small cluster of nuclei located deep within the temporal lobe, activates during negative emotions such as fear, anxiety, or stress. The hypothalamus, another small brain region, regulates functions like body temperature and circadian rhythms while also containing circuits for emotional reward, fear, and defensive behaviors.

When it comes to the emotion we call "love," neuroimaging studies reveal that dopamine is released in the brain's reward system, leading to the conclusion that love might simply be a cocktail of hormones influencing brain

[22] What is a "property"? Try to define it without making any reference to an experiential or mental quality. Can you?

[23] Interestingly, however, not the ability to comprehend language.

activity—similar to the effects of cocaine. The core-materialist sees this as evidence that love is only a neurobiological chemical mix aimed at promoting reproductive success.

These are just a few examples among many. Modern neuroscience has produced extensive literature on the correlation between specific physical, emotional, or cognitive functions and particular brain processes. [24]

This seems to suggest there is little room for any form of dualism; the mind appears to arise from matter, with no clear distinction between the two. Our personalities, identities, moods, and states of consciousness seem to depend on the biochemical state of our brains. Material monists argue that this evidence shows that mind and consciousness must emerge from a complex organization of matter, leaving no reason to believe otherwise.

However, it is important to question some fundamental assumptions.

Does the correlation between changes in brain states and shifts in mental states, consciousness, and phenomenal experiences provide logical proof that the brain causes the mind and consciousness?

Throughout the history of science, humans have repeatedly fallen into a well-known logical fallacy: the belief that correlation implies causation. We often unconsciously confuse the correlation of two observed phenomena as proof that one causes the other. The mere fact that two events frequently occur simultaneously or one after the other does not imply that the first event caused the second. If A leads to B, we cannot conclude that A is the cause of B. Such logical errors are known as post-hoc fallacies (from the Latin "post hoc ergo propter hoc" or "after this, therefore because of this") or correlation-causation fallacies.

One doesn't need to be a rocket scientist to understand that correlation does not equate to causation. A straightforward example is the correlation between ice cream sales and sunglasses sales. When the sun shines, people tend to buy sunglasses, and, especially during the summer when it is likely to be hot, eat ice cream. However, the purchase of sunglasses isn’t the cause of the increase in ice cream sales.

This example might seem overly simplistic, but when one is dealing with complex systems like the brain and its mental correlates, the nuances can easily lead us into logical fallacies, especially when they align with our expectations. This often eludes well-trained scientists because; it involves conceptual aspects that can easily go unnoticed, particularly if they have made a philosophical commitment to a specific worldview. A serious and scientifically rigorous cognitive neuroscientist should remain agnostic in this context.

[24] This is a simplified description; the reality is much more complex. Brain scans reveal that no single region is activated in isolation; rather, several areas are involved, often with varying intensities. The brain is not organized into modules dedicated to specific functions. Instead, its regions collaborate and function as a whole.

However, many scientists tend to jump to conclusions, viewing these causal relationships as obvious. They often overlook their "confirmation bias"—the tendency to seek data or empirical evidence that supports their existing beliefs and fits within their worldview.

Good science does not advance by jumping to conclusions. It is fundamentally based on the principle of excluding alternative explanations. Serious skepticism does not assume that something is true (or false) until proven otherwise, as many so-called "skeptics" tend to believe today. Instead, it involves ruling out the possibility that a specific fact or phenomenon can be explained in other ways.

What is the correlation-causation fallacy in this context? It refers to the tendency to correlate neural states with mental states, suggesting that the former causes the latter. This assumption posits that because changes in brain structure and function correlate with alterations in conscious experience, consciousness must arise from the very same chemical and material processes. However, a fundamental principle of science is that a theory is accepted only after all alternative explanations have been considered and ruled out.

So, what are the possible alternatives to the mind-body identification?

One of the most frequently cited alternative interpretations is William James's "radio metaphor" of the brain, also known as the "filter theory of consciousness." James's perspective challenges the materialist view that the brain generates consciousness. Instead, he proposes that the brain functions like a receiver that "tunes into" or "filters" a universal field of consciousness, similar to how a radio picks up electromagnetic waves to produce sound.

In this analogy, just as a radio does not create music but instead receives and decodes radio waves from the environment, the brain serves a similar role as a "conduit" or "filter" for a broader, non-material universal consciousness. This implies that consciousness may be an ever-present, non-local phenomenon, with the brain merely facilitating individual experiences by "channeling" and limiting this expansive awareness.

This viewpoint does not fall into the correlation-causation fallacy. It would be incorrect to conclude that the existence of the ability to modify the radio circuitry to change or disable the receiving signal means that the radio itself generates the signal; the same reasoning applies in this case.

James viewed the brain and consciousness as a prism that separates white light into colored beams. If a broken prism fails to reveal these colored beams (the correlation), we should not fall into the logical fallacy of believing that the prism itself produces colored light (the causation). The prism is simply an object with a transmissive function; it doesn't generate anything.

His thesis could be summarized with his own words:

"My thesis is now this: that, when we think of the law that thought is a function of the brain, we are not required to think of productive function only; ***we are entitled also to consider permissive or transmissive function.*** *And the*

ordinary psycho-physiologist leaves this out of his account"(James, 1898) (Emphasis in the original text.)

The English writer and philosopher Aldous Huxley shared a similar perspective after undergoing mystical experiences with psychedelic drugs. He suggested that the brain acts as a "reducing valve" for what he termed the 'Mind at large,' a universal or cosmic Mind that encompasses all of reality, including all ideas and thoughts. Huxley argued that our minds filter reality under typical conditions; without this filtration, the vast knowledge of the universal Mind would overwhelm us. However, psychedelic drugs can lift this filter, thereby allowing us to connect with the Mind at large and experience the states that many mystics have described.

In his words: *"To make survival possible biologically, Mind at large has to be funneled through the reducing valve of the brain and nervous system."* (Huxley, 1954)

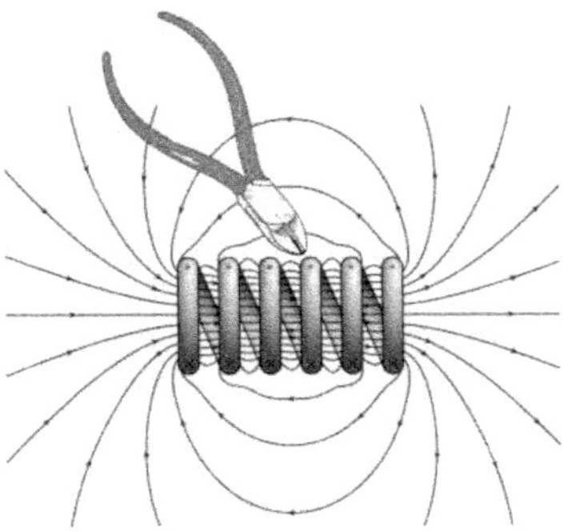

Fig. 4 A current flowing through a solenoid coil produces a magnetic field. Cutting its wire will shut down the field. Does the coil generate magnetic fields?

According to James and Huxley, the brain functions as an instrument of consciousness, while the mind transcends mere matter. The brain does not "generate," "produce," or "create" consciousness; rather, it enables consciousness to express itself in and through the physical realm.

James's and Huxley's perspectives are perfectly legitimate, especially when we consider the correlation-causation fallacy. Despite common belief, their ideas align with contemporary neuroscientific findings. These perspectives are particularly compelling because they allow for alternative interpretations that avoid a rigid commitment to naturalism. While their concepts might initially seem speculative, they provide a thought-provoking counterpoint to the equally speculative reductionist view that consciousness is simply an epiphenomenon—a mere byproduct of neural activity.

The previous sections introduced the mind-body problem. While their primary goal was to explore unconventional questions through deeper intuitive reasoning, they also aimed to lay the groundwork for my article in the

Advanced Readings section XIII.1 titled "**An Evidence-Based Critical Review of the Mind-Brain Identity Theory**" (Masi, 2023a). This theory posits that the mind and brain are identical, with no distinction between them. I examine whether current neuroscientific evidence supports this identity, as many believe, or if much of the evidence is viewed through a lens of confirmation bias.

III. Life and Evolution: Beyond Naturalism's Straitjacket

1. What Is Life?

Something I repeatedly hear scientists and philosophers saying is that, sooner or later, we will be able to explain the emergence of consciousness in the same way we were able to explain life. It was once believed that some immaterial "vital force," an "élan vital," as French philosopher Henri Bergson called it, stands behind the emergence and processes of life. But nowadays, so the theory goes, we know this is all wrong: Vitalism has finally been falsified and is only a relic of naïve and outdated speculations. The same will be true of consciousness: It is only a matter of time before we will be able to explain the emergence of consciousness starting from the elementary processes in our brain, without any need for "woo woo theories" of immaterial or "subtle substances."

Besides the wetness analogy I discussed in a previous essay, this historical analogy between vitalism and consciousness is another of the most preferred lines of reasoning that philosophers of science, biologists, neuroscientists, and physicists present as a killer argument to defend a naturalist standpoint and conception of life.

At first, it sounds reasonable, doesn't it? After all, we don't need any non-material force to explain how our heart pumps blood, our muscles contract, our cells process chemicals, and how hereditary information is encoded in the DNA molecules. Our body is generally a complex biochemical machine, and no immaterial energy or force is needed to explain how it works. One can reduce everything to metabolism, homeostasis[25], growth, reproduction, etc. No need to invoke "vital spirits." Problem solved. Case closed, right?

My objection isn't so much that the conclusion is fallacious but, rather, that the premise hides an unaware social conditioning that has led us, again, to embrace unquestioned dogmas.

Did biology really falsify vitalism? If we can't answer the question "What is life?", can we nevertheless explain it? And can we claim to be able to explain life without explaining consciousness in the first place?

Let us look at the same question from another perspective.

While scientists still can't agree on a universally accepted definition, there is a general consensus that the notion of "life" refers somehow to any complex self-organizing, and self-sustaining homeostatic system able to grow, reproduce, adapt to the environment, metabolize, respond to stimuli, and, most

[25] In biology homeostasis refers to the tendency of a living organism to maintain a stable internal state.

importantly, undergo Darwinian evolution. Some will go a bit further and tell you that life is an information-processing, problem-solving and self-referential system of self-sustaining cells in equilibrium with the environment. Theories of "autopoiesis" and "enactivism" also add up to this overall view.

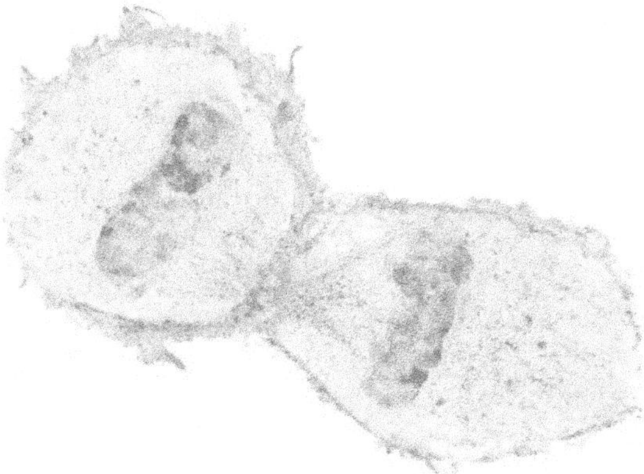

Fig. 5 The living cell during the process of mitosis is an example of an autopoietic system.

This is, in broad lines, the characterization by which biologists describe life.

While these approaches might illuminate some functional aspects of mind and life, it is evident to me that they do not address the fundamental questions surrounding the true nature of life. The explanatory gap will remain wide open. Any definition or explanation of life grounded solely in a materialistic functionalism invariably fails to encompass the essence of what it seeks to define or explain. Why is it so difficult to define and explain something whose existence is undeniable and whose distinctiveness is apparent? Perhaps it is for the same reason that consciousness eludes definitive understanding. Could it be that life, like consciousness or the mind, is unphysical? Is it because we attempt to exclude, at any cost, our inner experiences while focusing exclusively on the external world?

If someone asked you, "How is your life going?" or "What does it feel like to be a living being?", would you answer, "I am a complex self-organizing and self-sustaining homeostatic system able to grow, reproduce, adapt to the environment, metabolize, respond to stimuli, and undergo Darwinian evolution"? Or would you describe yourself as an "information-processing, and self-referential being in homeostatic equipoise"?

I hope not. But why not?

The point is, an abysmal hiatus exists between what we subjectively perceive from a first-person perspective of what we, in contrast to what we observe from the third-person perspective of science.

Does biology explain why life is imbued with desire, passion, love, hate, fear, pain, and pleasure? Does it tell us why, how, and what emotions, feelings, and sentience are?

Not only has life evolved morphological traits, but it also expresses volition, cognition, aim, creativity, purpose, will, and a whole set of psychological or, in less complex living organisms, "proto-psychological" traits from the first protozoa. How does that infamous "instinct of survival" in every living organism come into being in the first place? Why and how does an insentient, instinct-less, aimless, feelingless, emotionless, volition-less bunch of molecules organize into life forms with sentience, instincts, aims, feelings, emotions, creativity, a will to live, etc.? Can science, biology, or philosophy pretend to be able to explain life if they don't have a clue what consciousness, mind, sensations and all those unfathomable subjective experiences are, let alone wherefrom, why, and how they emerged? There is a vast internal, psychological, and cognitive dimension in life that we call "consciousness", "mind", "emotions", "sentience", and "subjective experience" that even the most complex and sophisticated theories of chemistry and biology not only can't capture but don't even begin to address.

Thus, what makes the difference here, is that one cognitively switches from an external third-person perspective to an internal subjective first-person view of life. One no longer sees the organism only in its purely material morphological embodiment, or its molecular genetics, but recognizes also by an "inward psychological turn" that what is so special in life isn't only its material aggregation and biological functions but is, especially, its psychological dimension. The very essence of life, or at least of complex organisms, which is made of feelings, desires, sensations, intentionality, agency, mind, and consciousness, for some reason remains as mysterious as ever.

One might legitimately doubt that a plant, let alone a single cell, such as a bacterium, can feel or perceive anything. Their behavior may only mimic or misleadingly resemble purposeful, goal-directed, sentient agency. There may be no feelings, let alone thoughts, and all its activities could ultimately be complex mechanisms of adaptation, regulation, modulation, reproduction, defense mechanisms, and instinctive responses aimed at maintenance and self-preservation, just homeostatic feedback loops or merely complex biochemical reactions emerging from an evolutionary process. We should avoid anthropomorphizing these rigid automatisms and jumping to conclusions that attribute human-like psychological traits to other living organisms. So the story goes.

Leaving aside the question of whether denying psychological properties to other living organisms and reserving them exclusively for humans might also be a form of anthropocentrism, the question then is where the dividing line lies between these purely mechanical responses and the emergence of some form of elementary sensation, subjective awareness, or experience. Is it in a life form that has a brain? Say, an insect? If so, this does not explain the emergence of consciousness; rather, it renders it even more mysterious. In what sense does the machinery of a brain differ qualitatively from that of a mechanical feedback loop? All the complex neuronal firings in a brain are also bioelectric and chemical reactions. Why does that dividing line emerge at all?

In my view it would make more sense to assume that consciousness, with all its phenomenal and experiential psychological dimensions, is the starting point, rather than a more or less fortuitous emergent by-product. Even a plant and its cells might nonetheless express an indwelling, elementary, proto-conscious phenomenon. Of course, plants don't think, perceive, or feel like humans do. However, if we allow them to possess some form of extremely basic but non-zero awareness and primitive cognizance, which some scientists call "basal cognition" or "minimal cognition," the issues with the dividing line between the sentient and insentient world would disappear. We could think of it as a continuum rather than a binary conscious vs. unconscious dichotomy. Yet, we do not allow ourselves to take this step. Consequently, inevitable inconsistencies and conceptual dissonances arise.

That's the reason why, despite centuries of research, the attempt to answer the question "What is life?" with a rigorous definition based on purely objective and life-less properties of something that has a high degree of subjectivity must inevitably fail or remain an incomplete and ambiguous characterization. In other words, contrary to what most biologists might contend, biology is *not* the science of life but, rather, only the science of those physical processes that sustain, support, and maintain the bodily functions manifesting life. It is a subtle but decisive difference and a completely different view of life itself.

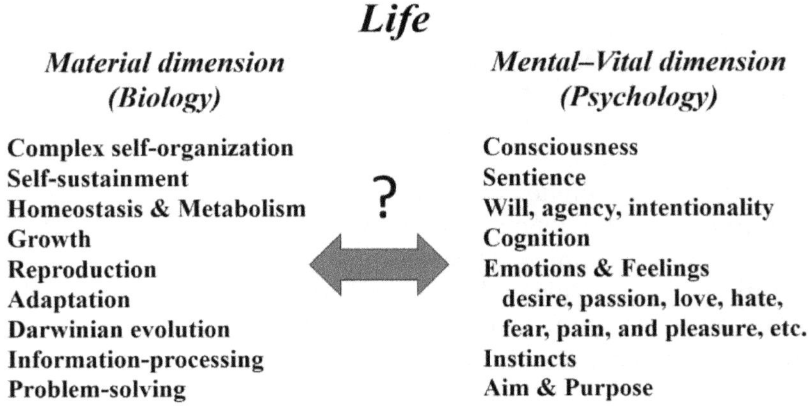

Fig. 6 The explanatory gap between life and matter.

However, the typical objection that comes up at this point is that science is supposed to explain only how and not why and what things are as they are. Its job is to explain the machinery of Nature, not to elucidate deeper philosophical or metaphysical questions.

I say: All right, I have no issues with that. But then one can't make the initial claim. Science is far from having removed metaphysical speculations involving vital forces, vital energies, élan vitals, or whatever immaterial entities standing behind the manifestation of life.

So, no. Science hasn't explained life, and it has not expunged immaterial vital forces either. While I can safely say that a steam engine does not need vital forces to spin the flywheel, this is not so for life. This is because the steam engine does not behave with agency, purpose, will, desires, instincts, etc. Life does. Contrary to common belief, Bergson's hypothesis is far from having been falsified. Vitalism isn't dead, it is just asleep and subsumed in a scientific culture unaware of its underlying scientism. And, at any rate, the analogy that so many scientists make by predicting that neuroscience will explain away consciousness like biology eliminated vital forces only betrays how they don't know themselves and how we unreflectively mistake the limited third-person perspective as furnishing us with the view on things in themselves, as a sort of God's eye view from nowhere. This is what I call the "illusion of knowledge."

2. Biological Evolution from Within?

We often overlook or give little consideration to the question of whether humans are the ultimate step of evolution or, eventually, whether our species might transcend itself or be surpassed by another.

I rarely hear evolutionary biologists pondering this question. While it has become a widespread fashion to predict a transhumanist future in which humans have hybridized into a mix of flesh and machine, a scientistic mindset that envisions the endosymbiotic fusion between nanobots and our brain cells, only a few seem to be interested in what Nature might hold in reserve for us. I always perceived this as a cognitive dissonance. Didn't science tell us, for about one and a half centuries, that natural Darwinian evolution is true? Of course it is. Right? So, why do we perceive it as so difficult to think in terms of the future natural evolution of mankind?

What follows are the kind of objections and counterarguments I got from smart and scientifically-minded people.

Several of them would tell you that predicting the future of human evolution is highly speculative and uncertain. It's difficult to predict the specific direction and outcomes of such changes.

That's correct. But this misses the point. We know that evolution is influenced by various unpredictable factors, including genetic, environmental, and cultural factors, but the question is not what exact future is waiting for us but, rather, if evolution has stopped with the human being or if we will, like every other species, be able to go beyond ourselves.

Then, once this has been clarified, the following arguments begin to surface: Advancements in technology, such as gene engineering, medicine, AI, brain implants, etc., could lead to intentional modifications in the human genome, our body, and our cognitive abilities. Thus, we will take up our evolutionary fate by ourselves, Nature will soon have no say, and it makes no sense to ask questions about our natural evolutionary destiny.

Might be. But wherefrom comes this overconfident belief that we will be able to do so? Despite all the hopes and hype surrounding biotechnologies, medicine, AI, stem cells, neural transplants, etc., not only are we light years away from being able to realize any of those transhumanist dreams, but things seem to become increasingly complicated and out of reach the more science progresses.

For example, when the human genome was mapped, we thought this would automatically pave the way to the development of personalized medicine that could treat genetic diseases. Genetic engineering has long been hailed as a revolutionary technology with immense potential. However, despite the initial excitement and high expectations, while genetic engineering has made some significant progress, it has been disappointing, with several failed promises. Reality has proven to be far more complex than expected. Most of the predictions scientists made a few decades ago have not lived up to their initial promises. The complexity of biological systems, and the limitations of current technology, have revealed a huge gap between what we initially envisioned and what can be realistically achieved.

Anyway, the ability to alter the genome of living organisms to create new species with higher cognitive and physical skills remains something we will continue to see only in sci-fi films for a very long time. Whatever the pervasive sensational announcements from the media might tell us, we are nowhere near being able to alter in a controlled manner the direction of our natural organic-genetic evolution.

Another deeper reason why scientists and philosophers look at the future of humanity only in terms of technological progress, and only rarely in terms of mental, spiritual, and natural biological evolution, is that the evolutionary changes occur gradually over very long periods. We are talking not of thousands but possibly hundreds of thousands of years. Therefore, according to this line of reasoning, it makes no sense to ponder what will happen to humanity in the year 100,000 AD. Let us leave that to posterity.

This is, indeed, in my view, a more serious objection. It is a fact that evolution is a slow process compared to a human lifespan. According to modern evolutionary biology, our species, the Homo sapiens, emerged about 300,000 years ago in some African savanna. Since then, we have made some cultural and technological progress, but one might legitimately doubt that we evolved much. There has been prevalently technological and material advancement, but, from the physiological and psychological perspective, we are to a large degree still an "animal-like" species scarcely able to detach from its instinctive nature. Greed, violence, wars, injustice, selfishness, etc., are the hallmarks that characterize our history. From the scientific biological perspective, there hasn't been much of a genetic and phenotypic change either. Accordingly, if we transcend ourselves, that might happen in about 100,000 years, or even more.

While this reasoning seems plausible and is in line with what modern science suggests, it relies, again, on premises we rarely question.

We have already seen how biology is not really the "science of life" but only the science of the material manifestation of life. Modern biology still conceives of evolution as something guided by purely material processes (natural selection, genetic variation, inheritance, etc.) and the impact of the environment on the adapting organism (environmental conditions influencing survival and reproduction chances, etc.)–that is, it is a process acting from the outside to the inside of the organism.

What modern biology doesn't see (or, more precisely, prefers to ignore) because it believes that it has no or only negligible effects, is how the organism might affect its own evolution from the inside out. We are not just automatons reacting to the stimuli of the environment. Psychological forces–that is, desires, passions, ideas, visions, and dreams–drive our behavior, choices, and actions. The underlying assumption has always been that these don't affect the biological, cognitive, and spiritual evolution of a species. Whatever we think, feel, desire, and ultimately do has no connection to what we will become, at least not on the morphological and biological scale. It's up to genes, the environment, and more or less random physical events.

But what if these inner forces of psychological nature also determined our evolutionary destiny? What if they not only defined an evolutionary path but also molded our biology?

This is not an entirely new idea. For example, French naturalist Jean-Baptiste Lamarck conjectured that the skills acquired due to the use of an organ could be passed by inheritance to the offspring. In this sense, behavior in one lifetime could influence the direction of evolution. Modern biology didn't find any evidence for this and ignored Lamarck's ideas for a long time. While, more recently, it partially reevaluated this view with the discovery of epigenetic changes[26], so far, the orthodox view prevails: Inheritance is a matter of genetics, the environment, and chance.

However, the hypothesis that inner psychological forces also drive evolution isn't something we must necessarily find confirmed only in a macromolecule, like DNA. For example, the fact that ancient civilizations were driven solely by material necessities is questionable. It is true that if certain environmental conditions aren't met, these will determine new behaviors. The first humans might have been forced to migrate due to environmental changes from the African continent to the rest of the world, but the migration of Homo sapiens might have also been conditioned by curiosity

[26] Whether psychological factors also determine such changes remains a matter of debate.

and an innate desire to know what is behind the visible horizon. There is a polar interplay between what we can do and what we like to do, between material obstacles and our desire to overcome them, and between the limitations of the body and the will of the mind. This dichotomy didn't change much until our days.

You might object and say that this is too focused on a single lifetime. On periods of hundreds of generations, population genetics has the say.

I doubt that. Consider further how, if we take a first-person perspective, one could make sense of aspects of evolution that otherwise remain unexplained. For example, from the fossil records, we know that evolution is marked by that strange property of making sudden "jumps" (or, at least, by a short period of gradual but rapid changes), which in biology is called "saltations"–that is, sudden macro-mutational changes in a few generations–followed by long periods of "equilibrium" in which, apparently, nothing happens in evolutionary terms. These abrupt phenomena have been observed repeatedly in the natural history of our planet and were recognized by the famous evolutionary biologist Stephen Jay Gould with his hypothesis of "punctuated equilibria." While scientists are convinced that, sooner or later, this will be explained away by genetic, molecular, or other physiological mechanisms, to date saltation remains controversial.

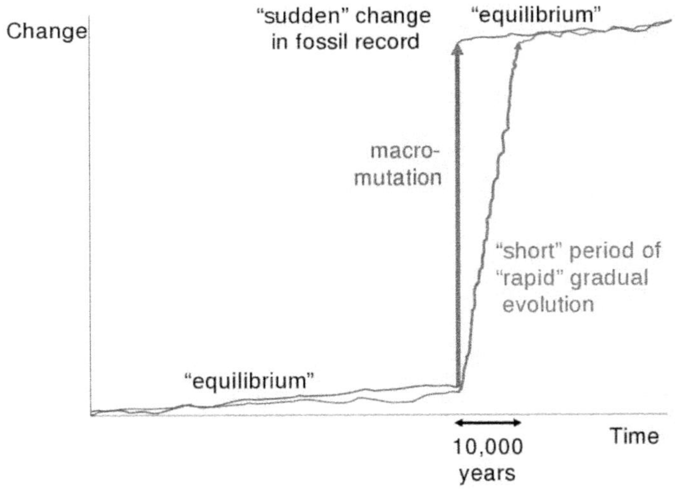

Fig. 7 Stephen Jay Gould's theory of punctuated equilibria.
[Credit: Wikimedia Commons]

To see how this might relate to our inner psychological dimensions, take an introspective stance and ask yourself whether your life history was a linear and gradual experience in which changes occurred incrementally and altered your everyday life only stepwise? Or, rather, was your life occasionally affected by sudden changes followed by less dramatic periods of equilibrium? I guess that unless one has a monotone, grey, and seemingly unchangeable lifestyle, or, to the contrary, one resembling Indiana Jones, most of us would opt for the second case. Abrupt shifts and transformations are quite common in both our individual as in the collective life of nations and societies.

We could also step out of such binary thinking and take a point of view that comprehends both the first and third perspectives, seeing both options as true: The outer and inner changes do not always (and perhaps never) go hand in hand. Of course, physical limitations, such as illnesses, pandemics, financial issues, and present material and technological contingencies, define the ups and downs of our life. Nonetheless, there are (conscious or unconscious) inner changes in our psychological dimension that do not necessarily erupt in appreciable changes in our material life. It is our inner will, mental state, mindset, aims, purposes, goals, desires, inner drives, etc. that determine how we behave and what choices we make, without necessarily being completely determined by outer external and material conditions.

The fact is, we know from our everyday experience how our life isn't ruled by exclusively outer conditions but is an interplay of mutually dependent and bidirectionally interacting forces acting from the outside environmental conditions on our internal psychological status, and the other way around, starting from internal mental and emotional drives acting on the external world. Sometimes, there might be long periods when seemingly few changes occur either outwardly or inwardly. Yet, even these periods might be only an apparent stagnation in which a lot is going on subconsciously and subliminally without manifesting at the level of our surface awareness. It may be preparing for the future unexpected visible transformation.

This is how we know our life unfolds. Thus, there is no reason to believe the natural biological evolution of life did and does work differently limited to a unidirectional process that makes us slaves to the environment or selfish genes. Evolution might be pretty much dependent on our inner mental, emotional, and spiritual state of development as well. This internal evolution might determine a breakthrough on a material level–that is, specify a change in phenotypic traits–in an extremely short time even though the outer environmental conditions may not even favor such an abrupt revolution.

We don't need peer-reviewed scientific papers or particularly complicated experiments to know how our being is, roughly speaking, dependent on physical, emotional, mental, and subconscious factors. Eventually, for those who believe in it, our essence is represented by an immortal soul, even though this is something we might be less certain about.

Yet, the idea that what we think and feel might influence the very physical aspects of evolution remains anathema. According to modern biology, no evidence supports this conjecture. Perhaps this is because there is no effort and interest in looking for such evidence. The assumption that only physical and stochastic factors determine our evolutionary path, while emotional, mental, and, perhaps, "soul-factors" play no role, is a widespread belief that has become a sort of self-fulfilling prophecy. Because scientists don't believe this to be possible, they don't search for any possible evidence and, when asked, declare that there is none.

Whereas, if we assume evolution is driven by these internal factors too, the sudden changes, the "saltations" in human and natural history begin to make much more sense. The idea of a gradual change can be harmonized with that of a sudden change: Our inside nature develops gradually and the outer "vessel"–that is, our body, the genetics, its biology, and morphology–may follow later, but then, eventually, express and crystallize that inner shift by an outer physical transformation that, in evolutionary and geological terms, appears to be almost instantaneous.

Now, turning back to the question of the future evolution of humankind, in my opinion, there are good reasons to believe this evolutionary shift is already taking place. However, I don't envisage it in a technological "singularity." Many describe having attained higher states of consciousness, a more intuitive vision of reality that goes beyond a mere intellectual understanding, and openly speak of a "spiritual transformation." This is also the evolution of consciousness, and it might well contribute, behind the veil of appearances, to the acceleration toward a physical transformation. We might not need to passively wait another 100,000 years to see this evolution taking place.

The most important power we call the "analytic mind" or "reason" isn't so much to predict, create, organize, calculate, and build spaceships. On the contrary, its greatest power is that it can learn to know what it doesn't know.[27] If it is honest with itself and doesn't retire to forms of ideological denialism (such as scientism), it can become aware of its own ignorance. This can give it the power to ascend the ladder of evolution by finally transcending itself. Contrary to other non-human animals, we are the first species that can participate willingly and with an inner effort in the evolutionary process. This is the first time in terrestrial natural history that a species can become aware of the existence of this process and assent to accelerate its own evolution not only with technology but also by sanctioning the call that comes from within.

I guess the next evolutionary step expressed by the new species that will go beyond the human, first by an inner trans-mental, trans-emotional transformation, and an inner "soul-change," is already taking place. The physical aspect might still be blind and deaf to this inner transformation but may well follow with a sudden unexpected physical transformation, perhaps even earlier than the transhumanists' dream could become a reality. Paradoxically, while our technological marvels seem to us to be changing our lives so fast, with the pace of Nature too slow, reality might surprise us by

[27] This is the line of reasoning I'm inviting you to adopt throughout all these essays: Look closer to see what is unknown to reveal that which was mistaken for known.

putting things upside down. In evolutionary terms, this inside-out transmutation might be just around the corner, say, a few centuries or millennia.

Whatever the case, try to see for a moment evolution not as a mere organismal change but as the expression of the evolution of consciousness *in* the organism. Would our present idea of guiding evolution by genetic engineering, or by other transhumanist dreams, still make sense? In my view, they appear as naïve attempts of a species that doesn't know itself and can't conceive of reality beyond superficial appearances, and especially itself, as other than a clump of neurons with bipedal locomotion. This physicalist worldview has led us to the unaware assumption that the rational mind and the power of reason are the ultimate tools of knowledge. Somehow, science–or it might be more correct to speak of scientism–posits implicitly that if something can't be explained by reason, it can't be true. This is yet another anthropocentric way of understanding evolution, the natural history of this planet, and reality as a whole. Mind, reason, and science are transitory means of a transitional being that will be transcended by an inner evolution, the adventure of consciousness in the manifestation, leading us to trans-rational forms of life and cognition.

If we view things from this perspective, we can see how a much brighter–and, frankly, much more interesting–future awaits us than that predicted by the current physically obsessed hype which is unable to look beyond its own limited scope and where all problems must be fixed by some external world-saving hyper-technology. Nature will not allow us to cage ourselves into this sort of Star Trek Borg civilization.

3. Is Evolution Unguided?

We are told that evolution has no direction, no goal, no aim, no purpose. And how do evolutionary biologists know that? Because, so the theory goes, evolution is about natural selection, variation, and inheritance, plus some other material and stochastic processes that determine life's destiny. Thus, there is no need to invoke final causes or whatever "divine actions or determinations."

In particular, the word "random" is the magic wand that allows us to jump to conclusions. It is about random mutations, random selection processes, random events in the environment to which the organism adapts, etc. Thus, if something is random, it is unguided, has no aim, no purpose. Right?

As we are going to see in more detail in a coming chapter, in science, the word "random" is only a label that stands for the uncertainty or (un-) predictability of future events and is measured with probability functions. It highlights our ignorance and lack of knowledge. It tells us nothing about final causes in Nature.

So, how far is the claim that evolution is "unguided" or "undirected" an objective truth because of whatever stochastic processes? Is this a scientific fact or an unwarranted assumption?

To clarify my position, I want to emphasize that my scientific mindset is far removed from any naïve biblical creationist view of the world, life, and the cosmos. I have no issues in accepting evolution as a genuine process (i.e., acknowledging that humans descend from primates), the Big Bang theory, and modern scientific findings in biology and cosmology. However, this does not necessarily compel one to adopt a materialistic and atheistic worldview that embraces the concept of unguided evolution. Neo-Darwinism, with its perspective, essentially reduces everything to natural selection, random mutations, and a few other processes that are said to account for all phenomena, thereby rendering hypotheses of "divine guidance" unnecessary. It begins with well-established truths, such as evidence of evolutionary processes in molecular biology, fossil records, and homologous features (which are scientific facts) but then assumes that the same processes must also explain everything else in biology, such as the emergence of complex phenomena like consciousness and the mind (a mere assumption, not a proven fact). This attitude stems from a common tendency of the human mind: Once it uncovers a limited truth, it instinctively generalizes and extrapolates that truth beyond its observed boundaries.

Despite centuries of progress in the sciences, the unanswered question remains more than ever: If the universe is purely deterministic, with no consciousness, no agency, no freedom, no volition, no aim, and no purpose, why did it produce beings such as us, having the sense (or at least the illusion) of being conscious, with agency, freedom, volition, and aim, and acting with purpose?

To speak of "evolutionary advantages" or the "instinct of survival" only restates the fact without explaining it. Why does this instinct exist in the first place? Where does it come from? Pointing at selfish genes only kicks the can down the road.

If the universe is nothing "but blind, pitiless indifference" (Richard Dawkins) or is "pointless" (Steven Weinberg), and if we are "nothing but a pack of neurons" (Francis Crick), well, so be it. But as long as these are only metaphysical speculations, we have the right to speculate the opposite. We are allowed to adopt a teleological standpoint.

What if these psychological properties of life might not be emergent but inherent? Did science falsify this possibility? No, it didn't. It simply ignores it. Thus, we have the right to pursue such a line of reasoning and see where it leads us.

The American analytic philosopher Alvin Carl Plantinga, summarizes this point as follows.

"...As far as I can see, God certainly could have used Darwinian processes to create the living world and direct it as he wanted to go; hence evolution as such does not imply that there is no direction in the history of life. What does have that implication is not evolutionary theory itself, but unguided evolution, the idea that neither God nor any other person has taken a hand in guiding, directing or orchestrating the course of evolution. But the scientific theory of evolution, sensibly enough, says nothing one way or the other about divine guidance. It doesn't say that evolution is divinely guided; it also doesn't say that it isn't. Like almost any theist, I reject unguided evolution; but the contemporary scientific theory of evolution just as such—apart from philosophical or theological add-ons—doesn't say that evolution is unguided. Like science in general, it makes no pronouncements on the existence or activity of God." – Alvin Plantinga (2010)

What about going beyond our present way of seeing the world and expanding our vision, stopping this unaware (but, at times, very aware) self-censorship that might not favor progress but, rather, hamper it?

4. Toward a Post-Material Interpretation of Evolution

While I do not support a traditional creationist view of evolution, I tend toward a perspective on Nature and evolution that recognizes purpose and direction. To clarify this perspective, I will first discuss the advantages and disadvantages of the so-called theory of "Intelligent Design" (ID).

ID posits that natural causes alone cannot fully explain the origin of life and suggests that a higher intelligence is responsible for the formation of the universe and the process of evolution. Its supporters argue that ID rests on scientific evidence-based reasoning supposedly demonstrating *"that certain features of the universe and of living things are best explained by an intelligent cause, not an undirected process such as natural selection."*[28]

Many dismiss ID as creationism in disguise, viewing it as another pseudo-scientific belief system rooted in an untenable literal interpretation of the Bible. However, I believe this perspective oversimplifies a complex cultural phenomenon.

Traditional creationism is a religious ideology that, for instance, outright rejects evolution, posits that the Earth was created around 10,000 years ago (the so-called "young-earth creationists"), and asserts that humans were created directly by God, as described in the Book of Genesis.

I wouldn't categorize ID as "creationism" in this traditional sense. Most ID advocates acknowledge the existence of evolutionary processes, do not necessarily reference sacred scriptures, and have moved beyond the notion that

[28] Behe, M.J.; Meyer, S.C., Discovery Institute, https://www.discovery.org/v/what-is-intelligent-design/

Eve was created from a rib taken from Adam. The ID movement accepts most findings of modern science and attempts—whether convincingly or not—to build arguments based on those findings. Those who equate the ID movement with conventional creationism clearly have not taken the time to educate themselves on the subject. In this regard, I view the ID intellectual movement as a somewhat more credible defense of a metaphysical evolutionary framework that has evolved beyond its original, untenable formulation.

However, despite this fact, I soon discovered that it did not have the boldness to fully embrace its potential. Instead, it became mired in a predominantly Christian and anthropocentric worldview, portraying humans as the ultimate goal of the evolutionary process—a stance known as 'human exceptionalism.' Moreover, only Western philosophical and metaphysical frameworks were acknowledged.[29]

So, yes, unfortunately, ID is still a movement upon which one cannot convey the status of a scientific organization. It is unfortunate that such an intriguing argument in natural philosophy has been monopolized by them. In the end, this situation benefits a mechanistic, orthodox neo-Darwinian culture that presents unguided evolution as a scientific fact and readily dismisses ID as merely another version of conservative creationism.

A spiritual perspective on evolution that transcends both the physicalist, reductionist, atheist paradigm and the constraints of religious or New Age interpretations urgently needs an upgrade. Our society requires a framework that goes beyond the still-dominant mechanistic, reductionist, and soulless vision of Nature and life, as well as ecclesiastical worldviews and simplistic pseudo-scientific theories. The goal is not to impose yet another dogmatic "ultimate truth"—we already have had enough of those—but to establish a cultural counterbalance that demonstrates another way of thinking and perceiving the world, one that aligns seamlessly with modern scientific discoveries and, nonetheless, calls for a trans-naturalistic and trans-religious paradigm of evolution, open to broader perspectives.

I see the potential contribution of religion and its mainly Christian scripture-based theological ruminations as being quite limited with regard to teleological questions regarding meaning and purpose in the universe or debates between theistic and non-theistic evolution. Even though, if there is anything we should preserve rather than discard, it is the profound insights of mystical insights across all civilizations—not just those of the Christian tradition.

[29] Even more concerning, their websites feature articles with clear far-right Christian-evangelical political biases. Officially, they assert their non-partisan stance and independence from religious or political influences, but the evidence contradicts this claim.

I hope that ID will evolve into a genuine inquiry that integrates science, philosophy, and spirituality within a more trans-cultural and less ideological framework.

In the meantime, I prefer to use the term "conscious design" (CD) to convey the idea that evolution is a guided process, life is the creation of a higher power, and the universe is not merely a blind clockwork devoid of purpose and meaning but, rather, is a manifestation of an inherent divinity. I choose "consciousness" over "intelligence" because the latter implies a form of human reasoning, an anthropomorphic perspective that shapes the universe. However, terminology is not the main issue here. I suggest we set aside our religious, cultural, and ideological biases to acknowledge evolution as a scientific fact while also recognizing the presence of a conscious design in nature—a purposeful, goal-driven force that organizes matter, life, and the universe.

Would such a theory of CD, as described above, be considered a 'science'? My answer is still no; it does not qualify. At least, it does not represent a conventional form of Newtonian science, which is based solely on a third-person worldview that admits only empirical inter-subjective facts and regards analytical reasoning as the only form of cognitive understanding. However, the question of whether a creative power in Nature underlies evolution falls within the realm of the philosophy of Nature. While this inquiry might be a speculative approach to fundamental existential questions, it cannot be dismissed as pseudo-science, provided it does not claim to be scientific.

Should ID or CD be taught in biology classes? No. Because, as they stand in their current formulations, they cannot be recognized as having a scientific status. It should not because it's not science. However, why can't metaphysical questions—such as the mind-body problem, free will, fine-tuning, life after death, and whether evolution is guided or unguided—be discussed in a philosophy class?

Proposing the question of our origins within a philosophical conversation that doesn't deny scientific facts is not educationally problematic; in fact, it can enhance young minds' critical thinking skills. This approach promotes a first-person understanding of the world and broadens the often overly materialistic perspective to include a critical thinking framework that is more philosophically sound. This framework sees reason and intuition as complementary aspects of our cognitive abilities rather than irreconcilable opposites. However, I would advise against introducing this discussion in a religion class, as these topics extend far beyond our religious traditions and local cultural contexts. It is fundamentally a question of natural philosophy, not religion.

Should evolution be taught in schools as an 'alternative' theory? My answer is no: Evolution is a scientific fact and should not be presented as mere opinion. However, biology teachers should not depict evolution as an unguided process. This perspective is also a metaphysical statement and should not be treated as scientific fact in the classroom. The question of whether evolution is guided or unguided remains a topic of philosophical speculation.

In the U.S., the attempt to introduce ID into biology classes faced significant backlash from the scientific community, and rightly so. However, there seems to be less indignation when evolution is taught as an unguided process. The question of whether evolution has a telos—that is, an aim, purpose, final cause, or ultimate goal—is fundamentally a matter of personal belief shaped by one's religious or (non-religious) philosophical views rather than scientific evidence.

Some might argue that the burden of proof lies with those asserting that evolution is a guided process. However, in science, the burden of proof rests on anyone making a claim. If you assert that evolution is an unguided process and present this as a scientific truth, you must also provide evidence to support your statement, just as those making the opposite claim must do. It is true that biology, like other branches of Newtonian science, cannot include divine determination in its theoretical frameworks, which assumes only natural causes. However, it is misleading to present something as a scientific truth when it is, in fact, a metaphysical assumption. This nuance is often overlooked in debates about the role of ID in education. Science should maintain an agnostic stance on these issues.

The idea of an evolution of consciousness within matter, reflecting an indwelling, concealed, yet self-revealing Spirit in Nature, cannot be framed as a scientific fact. However, it remains a viable hypothesis that does not contradict established scientific knowledge. We are allowed to propose that matter itself may be a form of "veiled consciousness"—that, despite its apparent inertia, conceals a universal Consciousness-Force operating within it. Even if matter appears purely mechanical and unconscious, it could be working toward the realization of a vast, hidden Plan.

I might also go so far as to state that if present science ignores or struggles with these subjects, a more advanced form of 'integral science' might one day address these questions, moving beyond mere philosophical speculation. In the meantime, I suggest we embrace these speculations and step outside a purely naturalistic mindset to explore where this might lead us, including within the scientific context. Our perspective on the world and our understanding of reality shape our actions, our scientific approaches, and, ultimately, what we discover. This represents a shift from viewing the world as a lifeless machine to recognizing it as a vibrant web of interconnected relationships, where everything is alive and in a constant state of flux and renewal. This shift in perspective is what will truly make a difference. As the Catholic priest and cultural historian Thomas Berry used to say: *"The universe is a communion of subjects, not a collection of objects."*

5. Can Theories of Self-organizing Complexity Explain Mind, Consciousness, and Life?

Scientists and philosophers, including Descartes and the French mathematician and philosopher Laplace, once believed that a day would come when all phenomena could be explained by elementary particles—tiny pieces of matter interacting like billiard balls in a vast game. While this perspective might not have enabled the prediction of every event, it offered a simplified view of the world.

However, this naïve understanding eventually gave way to the modern standard model of particle physics, which represents one of the most complex and successful intellectual and mathematical frameworks ever developed by science. This isn't a trend peculiar to particle physics. Molecular biology and genetics have become even more intricate than particle physics undergoing a similarly remarkable evolution. Over the past century, biologists have discovered an exponential increase in complexity in the organization of living organisms. Today, we understand life is much more organized, functionally complex, and information-rich than Darwin could have ever imagined. There is no indication that this trend will slow down; we can expect each decade to reveal increasingly astonishing levels of organization and dynamic functionality in living things.

The traditional view of a living cell as a simple cytoplasmic bubble containing a nucleus with a DNA molecule (believed to hold all the information about what we are), along with mitochondria and a few other organelles performing basic energetic and functional tasks, has evolved into a much more complex understanding. While a reductionist, gene-centric perspective on life persists in the collective narrative, it is no longer a tenable scientific framework.

The enormous complex multicellular machine with fine-response multi-step environmental signaling cascades, structural variations, functional canalization, emergent properties, redundancies, and who knows what, makes it clear how the single gene-single trait idea is outdated.

We have barely scratched the surface of a highly complex and integrated system of interacting genes, protein complexes, metabolites, secondary products, etc. It may be no coincidence that the Human Genome Project, personalized medicine, stem cells, the war on cancer and, more recently, CRISPR gene editing, struggle to meet expectations. Perhaps it's because life doesn't work as we assume and isn't what we would like it to be?

As Denis Noble writes (Noble, 2024): *"It's time to admit that genes are not the blueprint for life. The view of biology often presented to the public is oversimplified and out of date."*

We are far from mastering the terrifying complexity of life and, to a large extent, even understanding it. The belief is that by delving deeper into the microscopic structures of living things, we will gain greater understanding and control. However, in many cases, the opposite has proven true. Our discoveries often reveal a complexity in Nature that outpaces our ability to comprehend it. The structure and organization of complex natural phenomena are far more intricate than we can imagine or simulate. As a result, our sciences are becoming increasingly complicated, leading to a fragmentation that passionate calls for multidisciplinarity cannot overcome. We are rapidly losing control of our own models of Nature's vast organization and processes.

Yet, the prevailing belief persists that it is merely a matter of time—requiring more mappings, deeper microscopic analysis, greater computing power, or advancements in AI and, eventually, AGI—to understand better the functions and processes of life.

I believe this is wishful thinking, often dismissed by ongoing evidence and discoveries. Every scientist understands that life is complex, yet we still have no clear understanding of just how complicated it truly is. A general trend becomes visible: We consistently underestimate the complexity of Nature and life.

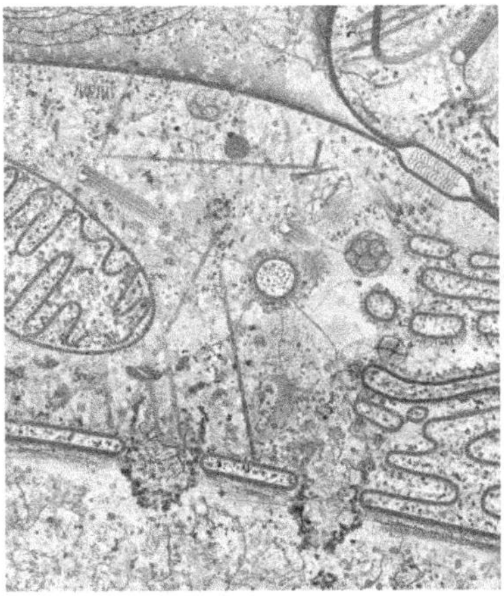

Fig. 8 Computer-generated artistic rendering of the cellular landscape cross-section through a eukaryotic cell modeled according to X-ray, nuclear magnetic resonance (NMR), and cryo-electron microscopy datasets. By Evan Ingersoll & Gael McGill - Cell Signaling Technology, Inc. https://www.digizyme.com/cst_landscapes.html

I argue that our primary investigative tool—the mind—struggles to grasp the full structure and interconnectedness of Nature's processes. This challenge extends beyond gaps in knowledge or progress; it lies in the very core of our rationalistic and naturalistic perspective, which cannot fully comprehend the intricate complexities of life's origins, from molecular and biochemical levels to the structural dynamics of a cell, not to mention the workings of the brain. Such complexity exceeds the natural cognitive capabilities of the human mind.

This limitation may also apply to the increasing complexity of our society and technology-driven world. The rapid pace of modern life, driven by technological advancements and globalization, the constant influx of information, and the interconnected nature of global economic dynamics—alongside migration and potential crises like climate change, cyber threats, and societal unrest—add new layers of complexity. Navigating this ever-expanding landscape demands continual adaptation and, particularly, non-linear complex thinking skills.

The conventional scientific mindset is inadequate for navigating this vast, interconnected, and entangled reality. It is only when we adopt a perspective in which reductionism and holism, multiplicity and unity, and the One and the many coexist harmoniously, we gain a better understanding of the complexity inherent in Nature and life. We have not yet fully arrived at this understanding. However, there are signs that we are striving to move beyond the overly simplistic linear thinking that prevails in our culture. Fortunately, we are beginning to recognize the need for a more complex, non-linear, and systemic view of reality. This awareness is contributing to a revival of complexity sciences, which, originating in the 1940s, adopt a more holistic perspective on life, mind, and consciousness.

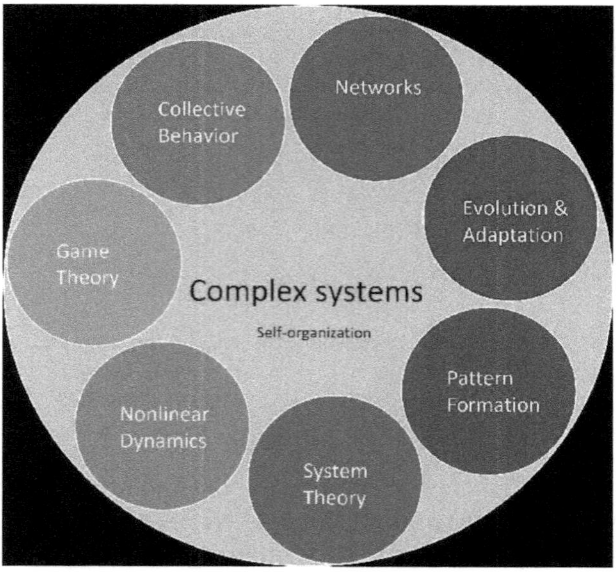

Fig. 9 Different types of complex system theories.

Currently in vogue are ideas that claim to transcend a simplistic, reductionist view of life, mind, and consciousness. These theories imagine processes occurring not only in our brains and cells but also because of our interactions with the environment. By shifting from a bottom-up reductionist view to a systems dynamics approach, we expand our perspective to encompass the idea that “the whole is greater than the sum of its parts.“

This broader viewpoint considers not just the microscopic processes at the cellular or network level but also the interactions and functions within a cell, in the brain, and in relation to the environment and the rest of the body.

However, they ultimately do not escape the mechanistic naturalism they purport to surpass. Here is a collection I have compiled over the years of processes that people believe possess the explanatory power to revive consciousness, agency, cognition, and life from…

‘Representations,’ ‘meta representations,’ ‘higher/lower order mental states,’ ‘self-representation,’ ‘attentional amplification,’ ‘recurrent and reentrant processes,’ ‘neural Darwinism,’ ‘multi-level predictive inferences,’ ‘self-generated actions,’ ‘homeostatic routines generating affects,’ ‘multilevel reciprocal communication,’ ‘informational interactome,’ ‘neurally encoded models of selective attention,’ ‘encoded body maps,’ ‘signaling integration,’ ‘electromagnetic fields encoding,’ ‘quantum collapse in microtubules,’ ‘global workspaces,’ ‘non-linear self-organizing complexity,’ ‘unpredictable autonomous self-regulating autopoietic systems,’ ‘informational autocatalytic hypercycles,’ ‘self-referential feedback loops,’ ‘re-entrant or feed-forward architecture’, ‘resonances in dynamical systems,’ ‘radical emergence,’ ‘far from equilibrium non-linear thermodynamics,’ ‘phase transitions,’ ‘entropy theories,’ ‘free energy principles,’ ‘garnering information game that maximizes the evidence for my own existence,’ ‘adjacent possibles,’ ‘Markov blancket,’ ‘bayesian evidence maximization,’ ‘enactive,’ ‘extended-, embodied-, or distributed cognition,’ ‘interoceptive and exteroceptive body representations,’ ‘cellular automata,’ ‘Turing non-computable,’ ‘Gödel incomplete,’ ‘undecidable,’ ‘irreducible,’ ‘impredicative,’ ‘unformalizable,’ ‘non-algorithmic,’ ‘non-prestatable,’ ‘an adaptive and emergent evolutionary dialectic,’ ‘a metabolic and ecological co-constructive dynamics,’ ‘hierarchical and self-referential far-from-equilibrium thermodynamic autopoiesis, anticipation, and adaptation,’... and ‘I am a strange loop!’

The list of fascinating concepts and processes could go on, but I believe you understand the main idea. This certainly is a step forward compared to the conventional Cartesian way of seeing Nature. It helps us to relax that reductionist view that has conditioned so much our understanding of the world and is so characteristic of our species. A viewpoint suggestive of a new way of seeing life and cognition. It might also turn out to be useful to embrace a more ecological and sustainable approach to the environment: It helps us conceive of Nature beyond a naïve understanding that thought of it as a predictable entity made of independent parts we can exploit without limits, but, to the contrary, sees Nature as a complex networked system from which we are not separated but of which we are an integral part, where any imbalance reverberates throughout the whole involving, sooner or later, us as well.

However, this being said, when it comes to more fundamental questions, I don't think this approach will lead us much further than the past ones, such as the "cybernetics" of the self-controlled systems of Norbert Wiener in the 1950s, or the theory of "autopoiesis" of Humberto Maturana and Francisco Varela, which they introduced about half a century ago in their bestselling book, "Autopoiesis and Cognition." A fascinating approach[30] to life and cognition that, however, didn't really produce the expected breakthrough. Nowadays, these theories have been refined further in the light of new findings and are going through a new renaissance.

For example, the so-called "enactive theory" or "enactivism"—essentially, a combination of complexity theory 2.0 and autopoiesis 2.0—is currently one of the most popular theories in the philosophy of biology. It aims to naturalize agency, cognition, life, and consciousness. However, I believe it remains an experience-blind naturalism that is unlikely to succeed in its goals, just like previous attempts to explain life and consciousness through dead matter, abstract functionalism, and intricate dynamical processes.

Adding that cognition is the result of a homeostatic process, or that it is not the brain but the whole organism that is conscious, or that cognition must be seen as the dynamical relationship between the organism and its environment drawing boundaries between self and other, and from which all higher levels of cognition ultimately derive, are insights that don't have much explanatory power from a philosophical standpoint. At best, they explain the how but not the what and why of things. The good old problem of consciousness, the inherent agency of biological organisms, their purposeful behavior, and its cognition, sentience, and will remain impenetrable existential domains.

And it couldn't be otherwise. Because deep down, these theories maintain a mechanistic perspective that still hopes to explain mind and life in terms of network dynamics and more or less abstract functions and are only the modern version of the gears, cogwheels, and clockworks of the mechanistic Newtonian paradigm. It has just become "complex" instead of "reductionist," "functionality" has replaced equations, or "system-dynamics" has replaced "particle dynamics", and words like "teleology" are banned while "teleonomy"[31] replaced the heretical nomenclature. Extending it to networks, functional organizations, or more abstract levels of conceptualizations, dressed

---

[30] Roughly speaking, autopoiesis conceives anything as living any system capable of producing and maintaining itself by creating its own parts.

[31] While teleology claims that Nature is purposeful, teleonomy maintains that this is only an appearance: Nature seems to pursue an aim, but doesn't. It is only the anthropomorphizing human mind that likes to believe to see an aim and a purpose in Nature, where in reality, there is none.

up with labels such as “non-linearity,” “enactive,” “recursive,” “self-organizing,” “self-X,” “self-Y,” or “self-whatever,” with a bit more holistic thinking, enlarges the descriptive level but not so much the explanatory one. Pointing out that Nature is “non-computable,” underlining that it is conceptually irreducible, does not change the background ontology, which remains (more or less unconsciously) a strict materialist, mechanistic, and reductionist conception of Nature, life, and cognition–not turtles all the way down but still functions, processes, or dead “selves” all the way down. There is no theory that clarifies how sentient, mental, agential, and conscious beings could arise from non-sentient, non-mental, non-agential, and non-conscious matter.

Thus, while theories of self-organizing complexity could shed light on life’s processes, they don’t tell us much about fundamental and existential questions. I think we aren’t aware of the immeasurable complexity of Nature. There is still a large movement of intellectuals who truly believe that a science of self-organizing complex system, supplemented by mathematics, chaos theory, network theory, computer simulations, and God knows what, will lead us toward a paradigm shift. But these approaches are another attempt of the human mind to cram everything into a naturalistic explanation, trying to avoid metaphysical speculations at any cost. They are still entrenched in the belief that, finally, we will be able to naturalize life with all its psychological dimensions as something emerging from dead matter constituents and their interactions. These conceptual frameworks might eventually tell us something more about *how* life and the mind work, but, so far, I haven’t seen anything that could convince me that they have any chance to shed more light on *what* life, mind, and consciousness are, let alone *why* they are.

Perhaps, also in the sciences with all its analytic thinking, we should learn to see the world going beyond mere mental constructions and learn to perceive the Spirit that is at work behind the appearances. Then, all these questions might appear in a completely new light, and all the world’s processes with those apparently so mysterious phenomena like mind, life and consciousness, would reveal themselves not as “emergent” properties of matter, but rather as its “manifestations” in and through matter. The question is not so much what we should do, but how we should see, feel, and perceive Nature and the World.

Perhaps, if we allow ourselves to assume the unspeakable, we might make a more decisive step forward, than clinging on to paradigms that reframe the old way of seeing and perceiving. It is about allowing ourselves to perceive, apprehend and comprehend something that goes beyond our ordinary sensory and analytic awareness, and then see where it leads us.

Nature is much more complicated than we believe and can imagine. It is not willing to be apprehended and comprehended in terms of the simplistic human one-dimensional thinking patterns and reductionist materialistic processes. I contend that a different approach is necessary in order to be able to go beyond.

This chapter introduced a different approach to understanding life, Nature, and evolution. In the Advanced Reading section XIII.2, my article titled "**Vitalism and Cognition in a Conscious Universe**" (Masi, 2021) revisits the question "What is life?" from a more academic perspective. It proposes a way of seeing that challenges an exclusively naturalistic viewpoint while staying within the framework of analytical and empirical sciences, such as biology. By questioning the premises and unwarranted assumptions of established science, we can reform, improve, and ultimately transcend a limited mechanistic view of life.

6. The Twofold Aspect of Nature: Toward a Trans-naturalistic Vision

> *"So long as I myself am identical with Nature, I understand what a living Nature is as well as I understand my own life; I apprehend how this universal life of Nature reveals itself in manifold forms, in progressive developments, in gradual approximations of freedom."* – Schelling, "Ideas for a Philosophy of Nature" (1797).

While scientists, philosophers, and inquisitive minds ponder the questions, "What is Life?" and "What is a law of Nature?", we rarely reflect on the question, "What is Nature?" Depending on our backgrounds and perspectives, there are various interpretations.

One instinctual and survival-oriented view asserts that Nature is simply the resource on which we depend for survival—everything from the air we breathe to the food we eat and the water we drink. This perspective places humans at the center of creation, granting us the right to exploit its resources as we see fit. Unfortunately, this morphed into consumerist and capitalist mindset viewing Nature primarily as a resource to be exploited, with humans positioned at the top of the food chain.

The scientific perspective offers a broader and less primitive view. It sees Nature as a vast and interconnected tapestry of processes, encompassing everything from the smallest particle and microorganism to the largest landscape. It includes the intricate ecosystems that support diverse plant and animal life, as well as the atmospheric, geological, and hydrological processes that shape our planet. Beyond, Nature can be understood as the cosmos—the physical universe made up of planets, stars, and galaxies, analyzed through the empirical and quantitative methods of science. This scientific approach, however, still represents a limited view, one that predominates in educational institutions. This narrow perspective has also influenced many environmentalist movements, which often believe that all environmental issues can be resolved solely through technological means. This technocratic approach to Nature, while not wrong in and of itself, is insufficient for transcending a naturalistic worldview that is still deeply rooted in a Cartesian mechanical clockwork universe with living beings regarded as automatons—except, of course, for the "ensouled" human being.

An attempt to delve beyond superficial appearances involves the poet, the artist, and the aesthetician, all seeking a deeper connection with Nature. In this light, Nature becomes the embodiment of a sunset's beauty, the raw power of a thunderstorm, or the gentle rustling of leaves in a forest. These emotions manifest as poems, paintings, or musical compositions. This inner connection engages with something tangible that science often dismisses as "just poetry." We tend to view it as a product of human imagination—merely a figment of the mind or irrational emotions—rather than as a real component of the visible universe. It is curious, then, how many scientists have also been artists. For example, Einstein, who played the violin, once said, *"If I were not a physicist, I would be a musician. I often think in music. I live my daydreams in music. I see my life in terms of music... I get most joy in life out of music."*

The perception of the natural philosopher is closely related to this understanding. Such a perspective offers a different yet complementary approach to Nature, blending elements of science and the arts, and is rooted in a specific tradition of continental philosophy. Historically, it represents one of the earliest approaches to studying Nature. From Aristotle, who laid out a general theoretical framework for the study of Nature in his work 'Physics,' to Spinoza's pantheism, which equates Nature with God (or God with Nature), this tradition has evolved. One can view Spinoza's cosmology as a response to Descartes's desacralization of Nature and Francis Bacon's notorious insistence on "torturing Nature to extract knowledge." This unsettling analogy was not coincidental; Bacon, also a statesman, advocated for the use of torture on humans during legal proceedings for the purpose of discovery rather than for obtaining evidence. Such a cultural backdrop fostered a detachment from the deeper inner realms that inspired writers, artists, and intuitive-minded natural philosophers.

Among those whom I find particularly interesting is Wolfgang von Goethe, the renowned German poet and writer. He developed a phenomenological and non-reductionist perspective that can be described as a form of "spiritually minded" natural philosophy. This approach recognizes the multiplicity within the unity of Nature and could serve as a foundation, or at least a source of inspiration, for a discipline that bridges science and intuition.

Goethe viewed life through a lens of unity, interpreting reality as an interconnected whole. He first "sensed" this entirety, recognizing its elemental components through a top-down approach. This perspective focuses not on summing parts but on understanding how they manifest from the whole. While

Newton's scientific method seeks the fundamental laws of Nature that govern existence from the microscopic level to the macroscopic level, Goethe pursued the "primal phenomenon" (the "Urphänomen") from which all other phenomena emerge.

Goethe's scientific approach is rooted in this "twofoldedness," allowing us to perceive reality as "reading the phenomena in the book of Nature," as Galileo famously stated. However, whereas Galileo interpreted this book in mathematical terms, Goethe challenged conventional reductive quantitative science, with a holistic qualitative way of seeing. He advocated for viewing Nature as an open book—not by analyzing individual letters but by quieting the mind and grasping the significance of the phenomena it presents. He focused on the qualitative aspects of the world, employing a more intuitive and holistic understanding. Thus, Newtonian science and Goethe's conception—or, rather, perception—of Nature's wholeness are not opposing approaches. Instead, they represent two complementary forms of cognition that enrich our understanding. Goethe, the poet, seized the opportunity to express his unique vision of nature through his poetry.

"Nature! We are surrounded and embraced by her: powerless to separate ourselves from her, and powerless to penetrate beyond her.

Without asking or warning, she snatches us up into her circling dance, and whirls us on until we are tired and drop from her arms.

She is ever shaping new forms: what is, has never yet been; what has been, comes not again. Everything is new, and yet nought but the old.

We live in her midst and know her not. She is incessantly speaking to us, but betrays not her secret. We constantly act upon her, and yet have no power over her.

The one thing she seems to aim at is individuality; yet she cares nothing for individuals. She is always building up and destroying; but her workshop is inaccessible.

Her life is in her children; but where is the mother? She is the only artist, working up the most uniform material into utter opposites, arriving without a trace of effort at perfection, at the most exact precision, though always veiled under a certain softness.

Each of her works has an essence of its own; each of her phenomena a special characterisation; and yet their diversity is in unity.

She performs a play; we know not whether she sees it herself, and yet she acts for us, the lookers-on."

Johann Wolfgang von Goethe, "On Nature" - (November 1869)

Thus, our journey to connect with Nature must also be contemplative rather than merely intellectual. We need to rediscover how to truly "see"—not just with our minds but also with our hearts and intuition, which grasp the wholeness of Nature.

Joseph Schelling approached this idea through a phenomenological lens, adopting a spiritual perspective echoing the animistic beliefs of indigenous cultures. He made a radical choice to unify Spirit, Mind, and Nature, viewing them as different aspects of a single identity. For Schelling, Nature represents the visible aspect of Spirit, while Spirit remains the invisible aspect of Nature. He elevated Kant's transcendental idealism to a monistic ontology where Nature and Reality share an "original unity" ("ursprüngliche Einheit"), the Absolute. Our disconnection from Nature leaves us spiritually empty because everything is a living being imbued with consciousness. In his 1798 treatise, "On the World-Soul" ("Von der Weltseele"), Schelling describes the human spirit as a somnambulist yet as fundamentally part of the World-Soul, which manifests as an object of contemplation within living organisms. This World-Soul is the universal principle that integrates every soul. It is an Absolute that, by manifesting itself, becomes part of Nature and, through successive stages of development, rises in consciousness to reach human status. Any problems, tensions, or disconnects we experience with Nature stem from our sense of separation and lack of connectedness.

Schelling recognized that it is our separation from Nature that makes it appear to us so differently, not how Nature presents us to our senses. Nature and Spirit are two complementary components of a Whole, with Spirit striving for consciousness within Nature. The various properties of things—such as form, substance, quantity, and quality—represent different aspects of an internally differentiated "primordial totality" ("ursprüngliche Ganzheit") in its opposites. What we perceive through our senses and interpret in our minds as objects, forms, and symbols constitutes only a partial and incomplete expression of a dynamic Whole. Schelling envisions the "World-Soul" as the foundation of all reality within which we are, so to speak, embedded. This situation compels us to differentiate between a subjective mental world, influenced by our feelings and perceptions, and what we refer to as an "objective world" outside ourselves.

This distinction arises from the internal differentiation of an "undivided Unity" ("ungeschiedene Einheit") of an absolute ideal or absolute subjectivity within an "eternal act of cognition." For Schelling, although the essential reality of the world is beyond the reach of our ordinary reasoning and mental cognition, we can still conceive of it as a "state of being" independent of cognition. This Unity may be cognitively inaccessible to us, but it undeniably exists as "being-in-itself"—that is, as the only pure state of existence in which perceptions and mind are non-essential additions. The Universe is viewed as an organism, a whole that embodies the organic principles of mind and spirit. The life-principle is inherent in the entire Cosmos. This "all-organism" represents the manifestation of the Divine, with all things contained within this divinity.

Goethe was a poet and natural philosopher, while Schelling was a philosopher; however, neither considered themselves mystics. Nonetheless, their approaches can help us move beyond a mere intellectual understanding of Nature. The phenomenological approach of the Romantic philosophy of German idealists like Goethe and Schelling represents a significant step toward overcoming our crisis in relation to Nature and the Earth.

However, while their perspectives are important, they remain insufficient. Philosophy can connect with the spirit but ultimately remains an intellectual activity. Poetry can immerse us in the atmosphere and vibration of inner spiritual knowledge, but the mind and ego of the poet often overshadow it. What is necessary is a deeper shift that allows us to turn inward, connecting the outer and inner realms, the physical with the immanent, and the immanent with the transcendent. This deeper understanding can be articulated only through philosophical or mystical language, which, in itself, does not guarantee an inner turn. Words are intrinsically limited, yet they can serve as pointers that guide us beyond those limitations.

Goethe asked:

"Her [Nature's] life is in her children; but where is the mother?" - "She performs a play; we know not whether she sees it herself, and yet she acts for us, the lookers-on."

We understand that Nature performs a play, but we question whether anything or anyone is consciously experiencing that play aside from the human observers. What this natural philosopher feels and perceives is a vague uncertainty, marveling at whether he is grasping the edge of a deeper truth or if a trick of the mind has deceived him.

An inner movement or psycho-spiritual process enables us to shift our perception from viewing something as dead to recognizing it as alive—or perhaps we should say "someone." Nature possesses a duality that we still find challenging to fully grasp. This realization doesn't come from new information or knowledge; rather, it arises from an intuitive, heart-based cognitive shift that transitions between modes of understanding. Something reminiscent of the vase-face Gestalt figure that remains constant yet presents itself in a complementary duality.

That's why simple hearts, unburdened by complex intellectual preparation, can see clearly beyond the veil of sensory appearances, while an educated scientist might perceive Nature as only a lifeless, universal clockwork.

Fig. 10 Nature: A dead vase, or two alive faces?

Schelling goes further.

"All merely embryonic life is in itself full of yearning and desires to be elevated out of the mute and inactive unity and into the expressed and acting unity. In this way we see Nature yearning. In this way the Earth so ardently sucks the force of heaven to itself. In this way the seed strives toward light and air in order to gather a spirit for itself. In this way flowers sway in the rays of the sun in order to shine the rays back as fiery spirit and as color." - Schelling (Ages of the World, p. 72).

There exists a deep yearning and desire that subconsciously suggests a need for elevation—something that seeks to evolve, persist, and continue existing. On the one hand, there is a sense of muteness and inactivity; on the other, a vibrant expressiveness and vitality. This creates a divide between the earthly realm and a higher, seemingly unattainable heaven. We can observe this strange polarity in the relationship between a seed and a flower; the seed is rooted in dark, lifeless soil yet strives toward the light as if in search of a lost spirit. This phenomenon reveals a clear sense of two-foldedness, a dichotomy that eludes complete rational explanation. It is this gap that we must seek to bridge and this duality that we must learn to engage with.

The first step is to detach ourselves from a materialistic understanding of the world—a perspective that cannot see beyond the mechanism and is reluctant to explore anything beyond the physical sense-mind viewpoint. We could recognize, feel, and deeply perceive that consciousness is the primary essence of all that exists. Why not allow ourselves, at least provisionally, to embrace the idea that the foundation of all existence is Spirit, the Conscious Soul, the Divine, or God, whatever we choose to call it? This understanding does not stem from theological discourse or logical reasoning; rather, it becomes self-evident when we move beyond a strictly analytical perception of reality and transcend the limitations of a separative physical consciousness.

The second step is to reconsider the nature of Nature. In contemporary culture, the word "Nature" has taken on many meanings, but its deeper significance in an integral view is quite different. Here, we are not referring to a simplistic understanding of Nature from a purely biological or material perspective. Instead, we are discussing something that transcends the idea of life and the physical cosmos as merely a display of "nature." This interpretation, with a lowercase "n," often leads us to mistakenly view ourselves as separate from it. Such a conceptual separation between humankind and Nature is at the core of the environmental issues we face today.

Nature, with a capital "N," encompasses all manifestations, from the most material to the most subtle planes of existence or consciousness. It includes everything from stones to the individualized mind as part of the universal consciousness in its dynamic interplay. In this broader sense, humanity is an integral part of Nature.

By adopting a phenomenological approach to Nature (establishing not only an intuitive perception but also a heartfelt connection), we begin to see and understand that beyond the surface of a seemingly mechanistic and physical expression of blind material processes, an active and conscious force is at work. When we engage in this soul-perceptive awareness, we can clearly "see" the Knowledge and Will that underpin Nature's creative impulse. A higher Idea operates behind the scenes in an apparently idea-less world.

We can (re)start from our relationship with the Spirit in and of Nature. From an integral perspective, we recognize that the mechanical aspects of Nature, which we perceive through our physical senses, are merely one pole of existence. The other pole is accessible only through more "subtle senses." This duality is not separate from the Spirit; instead, it is the Spirit manifesting itself in different forms. Nature is the Absolute that presents itself to itself. On the lower hemisphere of a semi-conscious existence, it might seem distinct from itself. Adopting this viewpoint allows us to see, feel, and understand that a fundamental Unity lies within Nature's diversity.

On the one hand, there is an aspect of Nature characterized by its temporal, spatial, and material manifestations within an evolutionary process marked by diversity and multiplicity. This aspect seems to have been lost in an unconscious and mechanical universe, reduced to the superficial existence that science studies and technology exploits. On the other hand, we find the Nature perceived by poets and artists, who sense a different vibration and connection with the same reality. However, this perception often eludes perfect expression, remaining limited to a sentimental impulse that aspires to grasp the entirety of the phenomenon and its powers yet ultimately falls short.

Instead, from an integral perspective, both viewpoints are not contradictions but, rather, complementary aspects of the same reality. The two approaches serve as distinct, albeit narrow, windows through which we observe this shared reality. They appear irreconcilable and mutually exclusive due to the mind's tendency to separate and exclude, admitting only matter, or only spirit, or only itself. Integration is impossible while the mind remains our primary tool for understanding, leading us to choose between the perceptions of scientists and engineers and those of intuitive poets and artists. However, through an integral lens, one can engage with these perspectives simultaneously. By embracing both intuitional and spiritual cognition, we can recognize them as foundational insights of a higher form of cognition.

We can recognize the presence and action of a "Nature-Soul" that manifests as a Will within the cosmic play. Nature serves as an instrument for expressing a divine consciousness intimately connected to rather than separate from it. It embodies the Power of the Spirit, the executive Consciousness-Force, and cosmic Consciousness-Energy, which are the origin and cause of all dynamic activity in temporal manifestation. Here, one can sense a boundless feeling of commonality between the inanimate and the animate, which highlights a deep interconnectedness with everyone and everything. This is about a profound, intuitive, heart-based connection with all life in a state of embodied oneness. Everything becomes an emanation of Nature while simultaneously radiating a light and force that transcend ordinary perception. Nature transforms from a lifeless, mechanical clockwork into a vibrant pulsation of life and something beyond it. Everything is infused with an energy, consciousness, and intelligence that we often overlook, even when it is right before us.

Something that even some physicists, like David Bohm, felt to be true and expressed through the idea of the existence of an "implicate order" and "explicate order." Bohm might have sensed something profound and sought to express it through the language of particle physics.

"Ultimately, the entire universe (with all its "particles," including those constituting human beings, their laboratories, observing instruments, etc.) has to be understood as a single undivided whole, in which analysis into separately and independently existent parts has no fundamental status."[32]

However, we are not discussing a universe composed solely of particles interacting holistically. The dual aspect of Nature extends far beyond a model of discrete chunks of matter interacting with one another, regardless of whether they are holistic, non-local, or entangled. Matter and particles represent only the final rung of a ladder that reaches much deeper than the realm of isolated entities in space and time. From this perspective, the "lower hemisphere" of mechanical Nature serves as the alter ego of its "higher hemisphere," which embodies a conscious, intelligent, and goal-directed aspect of Nature.

Our inner perception has become deaf and limited because we learn about Nature primarily from biology textbooks, which focus only on the "exterior Nature" characterized by mechanical forces, molecules, chemical processes or morphological aspects. This approach emphasizes division and limitations; we are not encouraged to engage with or connect to the "interior Nature." This interior Nature operates through unification and transcends limitations,

---

[32] *David Bohm, "Wholeness and the Implicate Order."* Routledge and Kegan Paul; London, UK: 1980. p.221.

manifesting a unity in diversity. One aspect of Nature is the super-conscious, which supports the other aspects: its own mental, subconscious, and inconscient aspects. The interior Nature reconciles and unites these primal elements—life, mind, and matter—to elevate them into their inherent light, force, and joy. One Nature possesses a broader self-guiding consciousness and a greater force, while the other Nature has become more mechanical and deterministic, seemingly governed by blind forces due to its involvement in a more basic form of itself. This duality in and of Nature is also reflected in a twofold evolutionary process—one that unfolds from below, driven by a spirit in ignorance that ascends the vertical ladder of consciousness within a still unconscious or semi-conscious realm, and another that dynamically acts from above, led by a Spirit of knowledge descending the ladder from a superconscious higher plane.

This inner connection with the two-folded Nature provides us with a comprehensive view of both polarities. On the one hand, we find a seemingly unconscious universal clockwork operating within a realm of duality, division, separation, and limitation. On the other hand, there exists something we can sense and connect with that operates through this polarity but that is not rooted in fragmentation. This is the silent perception and reception of experiences that touch us—whether it be admiring a landscape, observing the vastness of the celestial sphere, or simply listening to a bird's song.

This is not merely an abstract philosophy; it can become a lived experience and a tangible reality. After all, this very concept subconsciously sustains environmental movements despite numerous setbacks. It also captivates many who seek to understand how indigenous people relate to Nature. We recognize that they maintain a connection to a deeper aspect of Nature. This inner sensitivity seems to be something that modern "homo technologicus" has lost, while indigenous cultures still connect with it, albeit faintly. We might have lost this connection as long as 100,000 years ago when the Earth entered its last ice age. At some point in our evolutionary journey, our prehistoric ancestors might have been more "spiritual" and "intuitive," existing in a heightened state of consciousness compared to our current state. However, the development of the mind, intellect, and rational analytical faculties was still in progress, which, I believe, was the root cause that necessitated an evolutionary relapse.

Yet, we can re-learn what we have always known, integrating it with our minds and transcending the mere intellectual understanding of reality. This allows us to inwardly grasp how Spirit and Nature are two equally real poles of the same Reality. Although we might not be fully conscious of its purpose and are still mostly guided by an instinctual Nature-dynamics, we can perceive that these movements are influenced by an active Will within, stemming from a Nature of Knowledge-Force. It involves a concealed Consciousness and Wisdom present in all things, driven by Will-power and a Knowledge-sense.

We can recognize the presence of a conscious Intelligence in a sunrise or sunset, in the cyclical nature of the seasons, in the play of ocean waves, in mountains and landscapes, or in the benign and violent interplay of forces.

We can offend, torture, and destroy Nature, but we will never be able to extinguish it. Nature will always remain a hidden expression of the Spirit, operating through unconscious forces and a subconscious life that nonetheless embodies a conscious power. This power executes a conscious grand design aimed at linking the finite with the infinite. The Spirit works through a dual nature that formulates a plan and intention within the cosmos. What might seem like a brute machine on the surface is, in fact, the movement of an unseen God within creation—a powerful cosmic force operating behind the veil. Creation itself is the manifestation of a creative force, driven and guided by the intelligence of the creative Spirit. It is the Eternal that has cast itself into time, the One that organizes itself into the many.

Thus, we should reconsider our conception—or, more precisely, our perception—of Nature. When we refer to it as "Mother Nature," we touch upon something indefinable and intangible. This title signifies not only the act of giving birth but also an innate understanding of the deeper Mother Nature that underlies the child of Nature. It is a cosmic play, revealing a dual aspect of the Divine. We can learn to nurture various "modes of knowing," "modes of feeling," "modes of perceiving," and "modes of connecting."

Fig. 11 Trees in a forest: The superficial appearance of separation, the deeper truth of unity.

Waking up to this inner union is essential for our survival and for avoiding environmental catastrophe. It is time for science to move beyond naturalism's mode of cognition, which observes and analyzes the physical cosmic plane only through the outer mind and sensory perception—the "physical-mind." This externalized state of consciousness regards matter as fundamental because it lacks contact with the "beyond-planes" of Nature. Naturalists often do not recognize the principle of increasing division from Unity to multiplicity or the origin of atomicity in matter. They mistake the latter for the fundamental starting point—essentially confusing the endpoint with the starting point.

While the materialist perspective captures the bottom-up evolutionary aspect of a mechanistic Nature, it remains limited by a physical sense-mind form of cognition. This limitation prevents awareness of the Spirit's dynamic action in manifestation and the deeper nature of matter as a result of an involution. Consequently, the materialist misses the top-down evolutionary aspect of a vaster movement of Nature and fails to acknowledge the consciousness behind mechanical material processes, which often appear unpredictable—merely coincidences or random events. Such a lack of awareness, rooted in subconscious metaphysical assumptions, leads most biologists to assert that evolution lacks aim, goal, or purpose. This kind of spiritual blindness prompts materialists to dismiss the opposition between organicism and mechanicism as merely a social construct. They label the Gaia theory as pseudo-scientific nonsense and view the concept of the unity of Nature as an emotional self-delusion.

Only by becoming aware of reality in its entirety can we perceive the universe as an indivisible and undifferentiated whole characterized by infinite variety and multiplicity. The One and the infinitely many, the objective and the subjective, the personal and the impersonal, Nature and Super-Nature—all represent different aspects and modes of the same Being. All are real, and none is an illusion.

Only then can humans reconnect with Nature and rise above their current fantasies of AI, AGI, transhumanism, and colonies on Mars. The pressures of urbanization, technology, and modern living have distanced us from the natural world, resulting in a growing sense of disconnection and environmental neglect. Rekindling our relationship with Nature requires not just knowledge but deeper feelings and intuition. While ecological education in curricula—fostering greater understanding of ecosystems, wildlife, and conservation efforts—is beneficial, it alone will not transform us into environmentally respectful beings. This transformation must be accompanied by a spiritual connection that promotes a different way of seeing and feeling, as only this can truly reshape our thinking. We can and should actively work to preserve natural spaces, reduce our ecological footprint, and support sustainable practices. This includes advocating for policies that protect the environment, reducing waste, and embracing renewable energy sources. However, true change will come only from an inner alignment with the recognition that Nature is two-fold: superconscious and subconscious, spiritual and material, indeterminate and determinate, free and mechanical.

However, if this is the case, how can the superconscious shape and influence the subconscious aspects of life? How can something spiritual interact with its own self-forgetful material appearance? How can the indeterminate become determinate while still determining its own determinations? How can freedom be reconciled with a world seemingly imprisoned by unchangeable mechanical laws? It is challenging enough to answer these questions using philosophical arguments, but framing a coherent vision within the current scientific context adds an additional layer of complexity. Nonetheless, it is possible to adopt a perspective that coherently encompasses all these particular positions, if you're willing to join me on the lengthy journey through the next five chapters.

IV. Divine Action in a Conscious Universe

Science and philosophy will likely never conclusively answer questions about God's existence or divine action. These theological issues transcend the capacity of analytical reasoning. Every sound religious and spiritual tradition teaches that we cannot find God outside ourselves. Rather, we must seek Him within.

Yet, a widespread misconception posits that science and spirituality are incompatible. This is especially true regarding the belief that God could not act in the mechanistic and mostly lifeless world that the physical sciences describe. Many hold the conviction that science has proven this to be an impossible fantasy or mere wishful thinking. Either science or religion must be correct; we cannot have it both ways.

The purpose of this chapter and the next is not to prove the existence of divine action in the world, but to demonstrate that nothing in science—especially in physics—prevents us from holding such a belief. In fact, in my opinion, science even suggests it, though not conclusively. While a definitive answer isn't necessary or even desirable, as that would distract us from seeking the divinity in us, we should at least break the spell that confines scientific and theological thinking to two irreconcilable perspectives on the world.

To get there, however, we must honestly confront the dilemma between a seemingly unconscious and deterministic universe governed by strict laws that discourage belief in anything transcendent and the undeniable reality that we are conscious, purposeful, and sentient beings. A good starting point is addressing the powerful objections with which physical reductionism challenges any notion of divine action.

1. No Divine Action in a Clockwork Universe

Anyone who believes in God believes that, directly or indirectly, somehow God acts in the world. Through guidance and inspiration, God is thought to influence human decisions, be the cause of the origin of life, and, eventually, intervene through miracles or less obvious but physically concrete phenomena.

However, for the philosophical and scientific minded, a fundamental question arises: If so, how is this supposed to happen? The universe, as far as we can tell, operates according to consistent principles that don't need divine actions. From atoms to galaxies, the idea that every event unfolds according to precise natural laws allowed scientists to build a coherent, unified picture of the universe. Classical physics, in particular, provided a framework in which the same natural and mathematical principles could explain the motion of planets, the fall of an apple, the nature of light, or the workings of a steam engine. This vision reached its peak in the 19th century, when Nature was seen as a grand clockwork, predictable and ordered. It fostered technological progress, promoted confidence in the power of reason, and shaped modern science.

A strictly mechanistic view of Nature in a mechanistic universe contains no place for divine action, because every event is seen as the necessary outcome of prior causes governed by immutable laws. That is, a strict determinism reigns. If the universe unfolds like a clockwork mechanism, then miracles, interventions, or any form of supernatural influence would either violate those laws or be entirely redundant. Philosophers say the world is "causally closed" from external non-natural causes. In such a framework, scientific explanation replaces appeals to divine causality, leaving little or no room for God to act in the world without disrupting the very order that determinism so strongly affirms. This led many to believe that either divine action doesn't exist, or it cannot simply override or ignore these laws without plunging reality into chaos.

To see whether this idea is sensible or sidesteps fundamental aspects of physical reality, let us first briefly examine what "causality," "determinism," and "causal closure" actually mean.

Causality

We will explore the notion of causality in more detail in the upcoming sections. To introduce the topic, it can be said that throughout the history of philosophy, causality has been one of the central and most contested concepts. In antiquity, Aristotle's doctrine of the four causes provided a comprehensive explanatory framework. These four causes are: the material cause (the matter out of which something is made), the formal cause (the structure, pattern, or essence that makes it what it is), the efficient cause (the source of motion or change), and the final cause (the purpose or end for the sake of which something exists).

Modern science primarily recognizes Aristotle's efficient cause—the agent, process, or mechanism that brings about a change—as the central notion of causality. For example, a chemical reaction is explained by the interactions of molecules (efficient causes), and the motion of planets is explained by gravitational forces. To a lesser extent, science also makes use of the material cause, as the composition of a thing (for example, the atomic structure of a metal) is crucial for explaining its properties and behavior.

By contrast, the natural sciences have largely abandoned the formal cause and the final cause. Formal causes—essences or defining forms—were replaced by structural descriptions in physics, chemistry, and biology, without appeals to metaphysical "forms." Final causes—purposes or goals—were explicitly rejected in modern mechanistic science after the scientific revolution, as Nature was no longer thought to act "for the sake of" an end. However, some vestiges remain. In biology, teleological language persists in the description of functions (e.g., "the heart is for pumping blood"), though most biologists reinterpret this in terms of evolutionary processes rather than intrinsic purposes.

Meanwhile, medieval thinkers, such as Aquinas, integrated causality with theological concerns, especially in arguments for the existence of God as a "first cause." In the early modern era, philosophers like Descartes and Spinoza sought to ground causality in metaphysical necessity, whereas Hume famously challenged the very idea of causal connection, arguing that we perceive only constant conjunctions, not necessary links. Kant responded by re-establishing causality as a category of understanding, a condition for the possibility of experience rather than a property of things-in-themselves.

Across these shifts, causality did not find an ultimate and coherent definition. However, in modern science, it is the Newtonian and Cartesian mechanistic view that imposed itself more than any other notion of causality. Causality was reframed within scientific laws, from Newtonian mechanics to probabilistic accounts in quantum physics.

While all my essays reveal a teleological inclination, I will begin from a perspective of physical causal realism, as described next. Most of the analysis will demonstrate that the eminently teleological character of Nature is not only compatible with contemporary scientific knowledge but is even suggested by it. Chapter IX will also address "essences" and "Platonic Forms"—concepts that can be related, albeit somewhat vaguely, to formal causes in the Aristotelian tradition.

Determinism

The search for the meaning of determinism has a long history, with many proposed definitions, conceptions, and nuances.

According to the Stanford Encyclopedia of Philosophy, causal determinism is *"the idea that every event is necessitated by antecedent events and conditions together with the laws of Nature."* The Encyclopedia Britannica describes it as *"the thesis that all events in the universe, including human decisions and actions, are causally inevitable."* Meanwhile, the Oxford English Dictionary defines determinism as *"the doctrine that all events, including human action, are ultimately determined by causes regarded as external to the will."*

Notice that the latter definition characterizes determinism as incompatible with any form of will from the outset: All events must be external to the will. Otherwise, by definition, the concept does not qualify as determinism.

However, in the contemporary philosophy of science, what philosophers and scientists—especially physicists—have in mind is "physical causal determinism," whose roots lie in the advent of Galilean and Newtonian science. In physics, physical causal determinism (which, from now on, I will simply call "determinism") is a fundamental concept that represents the laws of Nature—such as Newton's law of universal gravitation and the laws of motion—as mathematical differential equations dependent on time. Once the initial and boundary conditions are established (that is, the actuality of the present moment is known), the motion of a particle or body (that is, its trajectory through space and time) can be calculated uniquely and with, in principle, infinite precision. This means, if the current "state"—that is, the current actuality—of a particle or body is known, its future dynamical behavior (that is, the future actuality) can be predicted indefinitely. In other words, these differential equations can reveal everything that has occurred and everything that will occur. Within the framework of classical mechanics, the universe has only one actual past and one distinct actual possible future. These mathematical equations can completely describe (pre-)determination.

Therefore, we can define something as "determined" when the current (or past) actualities of a system uniquely determine(d) its future (or present) actuality.

An actuality refers to what is realized or exists in fact, rather than what is merely possible, potential, or theoretical. It denotes what actually exists as the realization of an event that actually occurs, in contrast to a potentiality or disposition that *could* be or *might* happen but is not yet realized, existing only as a state of possibility, uncertainty, or virtuality among many.

In physics one refers to the physical "state" of a system rather that "actuality," "event," or a "fact," but in the present context, it is crucial to make this distinction. In classical mechanics, these are synonymous—the physical state uniquely determines with certainty the measurement outcome we designate as an actuality or event. However, as we will see later, in quantum mechanics, the notion of a single "quantum state" can describe the probability or potentiality of possible measurement outcomes—that is, different actualities or events–and the realization of which remains uncertain. In the former case, a single cause determines a single effect, while in the latter case, a single cause can lead to potentially different events or effects. Thus, in this treatise, we will consistently differentiate between the "state" of a system and its "actuality," or the "events" it potentially entails.

This deterministic perspective reflects a mechanical and materialist view of reality in which the universe operates like a machine, our bodies function as machines, and our minds resemble computers. There can be no souls, spirits, or immaterial minds, nor can there be a transcendent God. This is a deterministic view of the world that cannot be dismissed as mere ideology, as numerous scenarios have validated it.

The simplest scenario involves planetary motion. Once we know a planet's initial position and velocity at a specific time, along with its boundary conditions, such as the sun's position and, eventually, the dynamic information about other surrounding planets and celestial bodies in the solar system, we can uniquely determine the planet's future path. With sufficiently powerful tools for calculation, we can compute this trajectory with high precision.

In fact, Newton's law of universal gravitation is encapsulated in a clear formula that leads to Kepler's laws of planetary motion. These laws neatly describe how planets orbit the Sun and form the foundation of our understanding of celestial mechanics. They reveal that planetary orbits are ellipses governed by elegant rules defined by precise mathematical relationships.

In the 17th and 18th centuries, because of the incredible success of Newtonian mechanics, especially celestial mechanics, with its ability to mathematically formalize the motion of all bodies, and the powerful ability to predict their future dynamical evolution in time, as observed through telescopes, the conviction that all the universe is a huge deterministic clockwork was soon extended to all phenomena in physics and, to a good degree, in chemistry and biology, as well. Ultimately, the scientific and industrial revolutions stemmed from the idea of expressing all that happens in the universe in mathematical equations that capture the present actuality of a system and uniquely predict its future actuality.

Pierre-Simon Laplace famously summarized this strictly mechanistic view of reality, declaring that if a being—the so-called "Laplace's demon"—knew all the particles, their properties, and their initial positions and velocities at one moment, it could predict the entire future of the universe: a claim often taken as the manifesto of determinism. In such a clockwork model, divine action after creation is impossible and even unnecessary. From a philosophical and theological standpoint, the only way to save the belief in God was "Deism," the idea that God created the universe with its laws but no longer intervenes.

Scientists often use the term "laws of Nature" without clearly defining its meaning. This is one of those concepts that people discuss without fully understanding what it entails. In fact, there has been a longstanding philosophical debate that has not reached a definitive conclusion. The notion of a "law of Nature" remains vague and somewhat subjective. However, in physics, we can arrive at a relatively clear understanding of what a law of physics actually means. A law of physics is a concise, general rule—usually expressed mathematically—that describes a fundamental and consistent relationship between the initial and final states of a system. These states are connected by a set of differential equations, the solutions of which tell us how a system behaves over time. For example, the motion of a planet in space during a time interval is described by mathematical rules. These rules express quantitative relationships that allow us to predict the state of a system at a future time. In short, the laws of physics are a mathematical abstraction about how Nature behaves, discovered through observation and experimentation, that can predict outcomes under defined conditions. This may not satisfy everyone's intuition or understanding of what a "law of nature" should mean; indeed, there are many alternative definitions—but it suffices for the purposes of the present context.

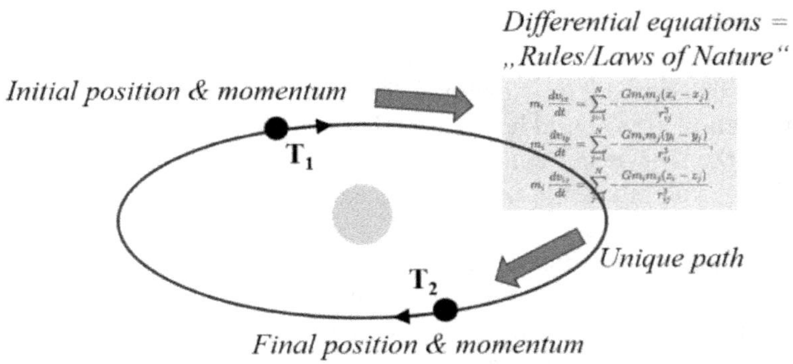

Fig. 12 An example of what defines a "law of Nature" in physics. The motion of a planet around the Sun: Once the initial position and velocity are known at time T1, the solution of the differential equations will uniquely determine their position and velocity at time T2.

However, the extrapolation of this strictly deterministic view of the universe from particles, or inanimate bodies, to complex systems such as biological cells or even a human brain was much less successful and straightforward. The universe is not the rigid clockwork that Enlightenment thinkers imagined. In the 20th century, the rise of quantum mechanics, chaos theory, and complexity science suggested that the cosmos might not be a rigid machine after all. It is complex and unpredictable—features that some interpret as providing space for divine creativity.

Do these aspects of physical reality provide a backdoor for divine action? How could a spiritual being interact with physical matter? How can an immaterial God interact with a material world? How can God influence the world while keeping its lawful structure intact?

Causal closure

From a scientific standpoint, the key difficulty is causal closure: the assumption that every physical effect has a physical cause. It is the philosophical principle according to which all physical events have sufficient physical causes within the physical domain. In other words, the physical universe is "closed" under causation: If something happens in the physical world, its cause can be fully explained by physical laws and entities, without intervention from non-physical factors. This principle underpins the success of modern physics, as it assumes that every effect in the physical realm can, in principle, be accounted for by prior physical conditions and the laws of physics.

Causal closure is often invoked in debates about the relationship between physics and other domains, such as the philosophy of mind or theology, because it implies that if physical causation is complete, then an immaterial mind or supernatural cause cannot independently alter physical outcomes without violating physical laws.

No exceptions to this rule have been observed, at least not at the most mechanistic level of reality, such as the orbits of planets, which follow trajectories determined by Newton's laws, chemical reactions, the laws of statics that allow us to build houses, the laws of thermodynamics that allowed us to build the first steam engines, etc. God seems to have no authority over these things or no interest in them. If, instead, that is the case, then the burden of proof rests with the believers.

If energy and matter interact according to strict laws, how can an immaterial agent, such as a transcendent God, or an immaterial mind or soul, determine anything in our worldly life without violating those laws?

Therefore, in a deterministic and causally closed universe, all the physical actualities of the future are predetermined by the physical actualities of the past—that is, there is no under-determined "open future." For every single cause in the past, only one history leads to a single effect. Every physical effect in the future had its prior physical cause in the past, and the transition from cause (the past actuality) to effect (the present actuality) follows a unique history.

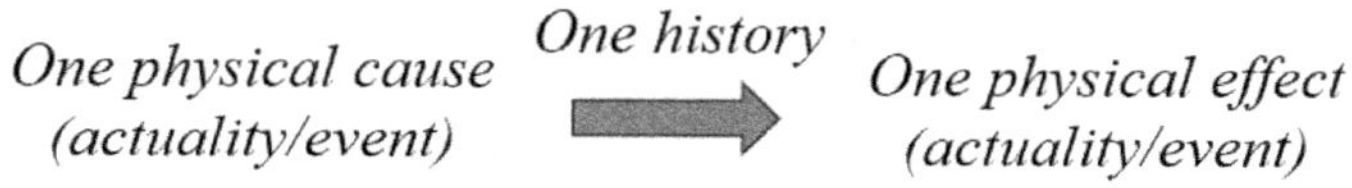

Fig. 13 Principle of causality in a deterministic and causally closed universe.

A frequent objection to determinism considers the following scenario.

If I let an apple fall to the ground, it will always fall in the same manner, provided it starts from the same initial conditions. No "causal openness" to metaphysical forces will be acting upon it—that is, there is no open future. This is not a metaphysical assumption; it is an empirical fact.

However, one might object that this is an intentional restriction of the parameters of the experiment. Perhaps a hungry person will run into this scenario by snatching the apple before it hits the ground and eating it. Physics and chemistry intentionally restrict their observations and explanations to limited parts of the universe. However, causal closure is the thesis that the entire universe, also with its hungry people eating apples, is causally closed, not just the delimited phenomena that chemistry and physics study. Including all the physical and biochemical reactions in the brains of hungry individuals, Laplace's demon could, at least in principle, predict their behavior and the apple's fate.

Ultimately, this idea also rests on the assumption that we are allowed to extrapolate everything, from simple to complex systems, up to a clockwork universe from the outset.

On the one hand, in the domain of macroscopic and inanimate objects like planets and asteroids, we are confident that the law of universal gravitation works flawlessly. On the other hand, we still cannot predict which asteroids will travel toward Earth in the distant future and whether they will hit New York, Rome, or Beijing. We know, however, that once an asteroid is near enough to allow us to observe its dynamics with sufficient precision, the calculations would allow us to trace its unique and predictable orbit by means of ordinary physics. It can be convincingly shown that this failure is due to our limited knowledge—that is, an epistemic limitation. We could also call it "*epistemic indeterminism*" or "*pseudo-indeterminism*"[33] because we are dealing with extremely complex phenomena which future evolution we can't predict, and, thus, we must expect multiple and unpredictable outcomes, though ones that remain inherently and ontologically deterministic. If we have sufficient computing power and accurately know the initial and boundary conditions of all the asteroid's surroundings, we can calculate its dynamical evolution over a certain period. With even greater computing power, we can extend this prediction further, and so on.

However, in practice, the numerical prediction of the behavior of complex systems with computers, breaks down due to the exponential propagation of internal numerical uncertainties regarding the system's state. This is not related to "doors opening to spiritual powers." Planets, asteroids, or comets move along precise trajectories described by differential equations and seem to not be subjected to anything other than purely mechanical and deterministic laws.

Weather forecasts are difficult to make for the same reason. The laws of thermodynamics that regulate the atmosphere are precise and well understood. However, predicting the weather with precision—say, a month in advance—remains impossible due to the extreme complexity of the Earth's atmospheric motions, turbulence, temperature gradients, and numerous uncertainties. These factors prevent us from making exact predictions. Such unpredictability arises from our inability to fully understand and account for these uncertainties and to calculate the evolution of a complex system. The unpredictability of weather in complex systems is an epistemic uncertainty; there is no under-determinate open future, but we do not know which future will unfold, and there is no reason to believe this uncertainty implies supernatural causes. It is another example of epistemic or pseudo-indeterminism, not true, inherent, and "*ontological indeterminism.*"[34]

Thus, at least in those restricted domains, we observe strictly deterministic, mechanistic, and mathematically formalizable behavior. By expanding the domain a bit, we still see such determinism hold until we have sufficient observational and computing power. Beyond that, we cannot predict the future

---

[33] Indeterminism is frequently also called "indeterminacy." We will use it interchangeably.

[34] Sometimes also called "*ontic indeterminism*" and related to an inherent uncertainty or randomness, both also called "*ontological*" *or* "*ontic uncertainty and randomness.*" We will use them interchangeably.

with sufficient precision, as we are no longer able to follow the entire chain of causes and effects, and all the dynamics become non-computable. However, nothing indicates that the unpredictability is related to causal openness.

Notice how this brings up another problem for the theist. If we believe God has no causal powers over inanimate objects like planets and rocks, then this implies an extremely limited kind of divinity because, as far as we can observe, 99.99999...% of the universe is composed of inanimate matter. Moreover, how could God have made life emerge in the first place without having power over inanimate matter? If we admit divine influence only in living beings, but deny God's power over lifeless things and big objects like planets, their atmospheres, molecules, and all the physical reactions that brought life into existence, the only remaining way out I can see is a return to the mechanistic Darwinian "blind watchmaker" account of the origin of life.

Many believers respond to this issue by appealing to mystery. They say, "God's ways are beyond our understanding," or "God acts by Grace." This might satisfy the religious-minded, but it cannot be an acceptable argument for philosophers, let alone scientists. I believe this answer avoids the question. It is vague thinking that offers psychological relief but no firm answers in terms that a clockwork account would accept. The clockwork account has every right to demand answers. If the Divine interacts with the physical world (answering prayers, causing an illness to heal, or guiding history), then something must happen at the physical level. Energy, matter, or information must be affected. Something about our notions of causality must be inherently flawed. Some purely deterministic mechanism based on our conventional notions of cause and effect cannot be the whole story.

One option is to break the laws—a classic interpretation of miracles. However, this presents immediate problems. Why would God create an orderly cosmos only to disrupt it?

That being said, we don't know whether the universe is causally closed. Despite abundant evidence that causal closure seems to be in effect at the level of planets and all things inanimate, down to the level of molecules and beyond, an extrapolation to the entire universe fails at some point. For example, can we reduce the activity of our brains to mere chemical clockwork? Are the immensely complex biological processes occurring in our bodies reducible to the interactions of molecules that, in principle, could be calculated and predicted like the motion of planets?

The answer is: We simply don't know.

At the psychological level, assessing causal closure—that is, knowing if all that we think, feel, and do emerges only from a chain of biophysical causes and effects in our brain—is difficult, if not impossible. Meanwhile, at the chemical and physical levels, causal closure is more evident. There is no indication that a chemical or physical phenomenon is caused by anything other than chemical and physical factors. At the biological level, the situation is less clear. We don't know whether this chain of causation is disrupted in the higher layers, such as the working of a cell and a brain. If so, we must point out where and why this occurs.

The scientific rationalist might resort to Occam's razor and contend that there is no reason to believe otherwise. Making supernatural forces intervene and forcing God into the gaps of our theories isn't a parsimonious hypothesis.

However, the real divide emerges because of the dichotomy between inanimate and animate matter, between non-living things and living creatures, between unconscious and conscious entities, between non-mind and mind. If we were unconscious machines, lacking sentience, feelings, thoughts, and subjective perceptions, we would not ask these questions. This hiatus between the apparent inanimate, unconscious, and mechanistic realm and our everyday experience as conscious, animated beings requires clarification.

In conclusion, we must admit that, if the causal structure of the universe is purely deterministic, operating like a clockwork mechanism governed by fixed natural laws, then the materialist is correct: It must also be causally closed and, thus, no divine causation is possible. This is because the very notion of determinism is, by definition, rooted in a rigid, mechanistic conception of cause and effect that does not allow for the interference of any will, whether human or divine. There is no room for free will, and any notion of will is an illusion. We would simply be machines living the illusion of consciousness and freedom of choice, and there would be no reason to speculate about the existence of a superior Will determining the fate of its creation.

To be fair, one should mention that a longstanding philosophical tradition, known as "compatibilism," argues that determinism and free will are compatible. Proponents of compatibilism often rely on convoluted arguments grounded in sociological, psychological, and ethical reasoning, while neglecting physical realities. At best, they may invoke complexity sciences and ideas based on top-down causation processes (moving from the macro to the micro), rather than bottom-up causation that begins at the micro level. However, in my opinion, these arguments amount to little more than word games and obfuscating jargon, obscuring the fact that determinism—specifically, physical causal determinism as understood in the physical sciences—implies, *by definition*, the absence of any interfering will. This can't be undone by making things more complicated, intricate, and convoluted. In my view, compatibilism lacks logical coherence and appears reasonable to some only due to a lack of conceptual clarity. I will revisit the compatibilist standpoint later in sections V.6&7.

Once we have clarified what determinism and causal closure mean and imply, the question is: What is their opposite? What does it mean that something is indeterminate? What would a universe with an open causality look like? Do other meaningful principles of causality exist?

2. Ontological Indeterminism in a World With an Open Future

Determinism and the principle of causal closure are (more often than not implicitly) rooted in Leibniz's "Principle of Sufficient Reason" (PSR), which states that every fact, everything that exists, and every event must have a sufficient reason or sufficient cause—that is, an explanation—for why it is the way it is, rather than otherwise. Events and facts can't pop into existence out of nothing; they must have always had a prior reason or cause to exist or take place. Philosophers tell us that the PSR should not be confused with the principle of causality that in every effect has one antecedent cause. The former relates to the intelligibility of facts in general (that is, why anything is the way it is), while the latter is about events in time (that is, why one thing produces another). However, while the PSR applies to broader domains than physical sciences, causality is one special case of it when applied to the domain of physics.

After all, even without knowledge of these principles and philosophical subtleties, we regard the PSR as self-explanatory and intuitively true. If an object can no longer be found in a specific place, there must be an explanation for what determined its disappearance, such as someone picking it up and moving it. No one would believe it simply dematerialized into thin air without cause. If one's bank account unexpectedly gets emptied, one would hardly be convinced that this could happen without a reason. One would speculate about theft or unexpected bills. Stones don't fall to the ground for no reason, and nobody would pretend that we should stop asking silly questions about the world. Aristotle speculated about the teleological causes of gravity, suggesting that all bodies have a telos—that is, an aim or purpose—to fall toward the center of the Earth. It was the wrong reason, but still a reason. Even those who argue in terms of miracles or divine interventions use as reasons and explanations regarding causes and effects of something we deem otherwise inexplicable and without reason or cause.

Ultimately, our mind is inherently hard-wired and incapable of conceptualizing processes over time without a causal relationship expressed through some form of deterministic principle involving causes and effects.

The question then arises: What is the opposite of a deterministic process? What does it mean when we say that some phenomena are inherently stochastic, random, and indeterministic?

An instinctive response might be to reference the throw of dice. Its outcome cannot be predicted with certainty and can only be described probabilistically. The system's dynamics allow for multiple possible results rather than a single fixed outcome. That's why we tend to think of it as "indeterminate." However, upon closer examination, we can see that this unpredictability does not indicate a lack of physical causal determinism; there is still an underlying chain of mechanistic causes and effects that make dice bounce and roll in one or the other direction and ultimately determine the result. From our human and ignorant perspective, we can't tell which of the six possible facts will become an actuality, but once the initial conditions of a die—that is, its present actuality—are set, its dynamical history, which will lead to one future actuality, is uniquely determined. From the perspective of physical causal determinism, even the throw of dice is a strictly deterministic process. It represents yet another form of pseudo-indeterminism.

Therefore, it is a common misconception to conflate unpredictability with under-determination. As previously outlined, the unpredictability of complex systems, such as Earth's atmospheric dynamics and the resulting weather patterns, does not imply an indeterminate process. Instead, it reflects our human epistemic limitation due to the extreme complexity of the motions of bazillions of particles and molecules in the air, which prevents us from making exact predictions. Thus, we perceive, as random, multiple possible outcomes. The PSR still applies; there is a reason—albeit unknown—why today is a rainy day rather than a sunny day.

These processes are referred to as "stochastic" because tracking all the particles and calculating their trajectories is impossible, necessitating the use of statistical analysis in terms of probability functions.

What, then, is a "truly" or ontologically indeterministic event?

Some might argue that human behavior is "truly" and ontologically indeterminate. We experience a sense of free will and the freedom to choose, regardless of what others might say.

The answer is that, as we have seen with hungry people snatching apples, we don't know. We will address the complex issue of free will in more detail later, but it may be said only that we must admit we have no certainties about the extent to which our behavior is rooted in deterministic processes in our brains or whether there is something more. Even human behavior might ultimately stem from purely deterministic processes that are so complex they create the illusion of free will. For example, a physicalist and reductionist neuroscientist like Robert Sapolsky argues, with down-to-earth scientific evidence, that free will does not exist. An overwhelming amount of data demonstrates that our personality is, if not completely determined, significantly influenced by genetics, biological factors, the environment, culture, and more (Sapolsky, 2023). However, no one can definitively say whether our behaviors, desires, thoughts, and actions are entirely determined from the bottom up by genes and molecules. Ultimately, we must humbly acknowledge that we simply don't know.

The question is, then: Do we know at least one single phenomenon in our daily life that we could label as being "truly" indeterministic? If you carefully think about it, you must admit that we don't even know what this question is supposed to mean. We can define determinism in a clear-cut manner, at least the physical causal determinism of physical sciences, as we have seen in the previous section. However, the notion of something that is "truly" indeterministic is vague.

Nonetheless, there is a way to frame a form of ontological indeterminacy through the PSR, as follows.

Physical causal determinism posits that the future actuality of a system is uniquely determined by its past actuality. In other words, every physical effect has a single physical cause. This concept can be framed using the PSR, which asserts that every event must have a sufficient reason. Thus, every actuality that comes into existence must have one specific cause. "Sufficient" implies that the explanation or reason is both adequate and necessary to fully account for why something is the case rather than otherwise.

We can reformulate this using the following diagram of Fig. 14. The present single physical actuality evolves deterministically along a unique path in history, leading to another future single physical actuality. However, in a complex system whose future evolution we cannot predict from time T_1 to T_2—such as the zigzagging of a molecule in air turbulence—we might have to rely on estimates and probabilities. That is, we estimate "*virtual potentialities*," such as the many possible future positions of the particle, even though only one actuality will be realized at T_2. These potentialities are considered "virtual" because the actuality at T_1 has already determined the actuality at T_2. However, due to our ignorance, we can only make educated guesses. These potentialities lack inherent reality and arise solely from epistemic randomness. Essentially, they result from pseudo-indeterminism. We are limited beings, and the future might seem open and undetermined to us, but it is not.

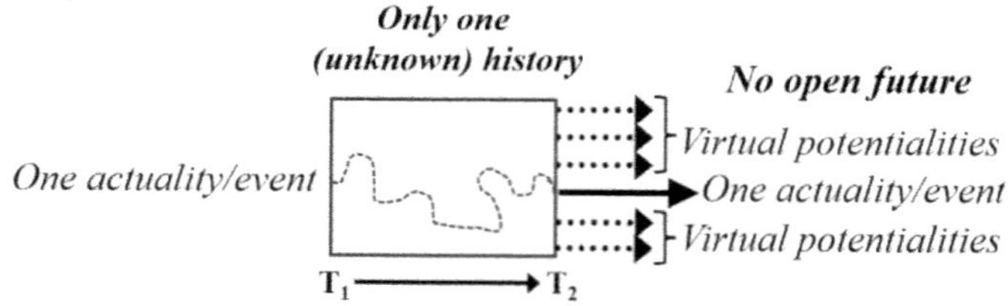

Fig. 14 Diagram for determinism or pseudo-indeterminism.

To address "true" and "real" uncertainty (specifically, ontological indeterminism), we must abandon the PSR, or at least relax the notion that every physical event must have a sufficient reason for existing. This requires us to accept what might seem strange or almost impossible to our everyday human understanding: that there are phenomena in which no reason exists for why something is the case rather than otherwise. It simply actualizes without reason, cause, or explanation.

This ontological uncertainty decisively differs from epistemic uncertainty. In the case of epistemic uncertainty, when something appears to happen arbitrarily, there is still a reason why it occurred in that particular way, even if we are unaware of it. The uncertainty resides in our minds, not in the natural processes themselves.

In contrast, ontological uncertainty is not due to our ignorance; it exists because there is no inherent reason for one possibility to actualize over another. Such uncertainty would be both inherent and real in Nature.

This can be visualized with a diagram depicting the dynamical evolution of a system from the past to the present (or from the present to the future), where, at time T_1, a single physical state can evolve along all possible histories simultaneously until it actualizes one among the "*real potentialities*" at time T_2. These potentialities are "real" because they also had the opportunity to be physically actualized, but did not. At time T_2, there is no necessary or sufficient reason for one actuality to be realized over another. The realization of the fact at T_2 is *not* predetermined by the physical state of the system at T_1, and it makes no sense to question why one actuality came into existence instead of another. That is simply the way it is, with no reasons, neither sufficient nor insufficient. The effect self-actualizes or "*self-causes*" among multiple permitted potentialities. Those potentialities might eventually be constrained by a physical law; for example, a certain event will have to occur within a domain. Nonetheless, the exact location of its actualization within that domain remains inherently uncertain.

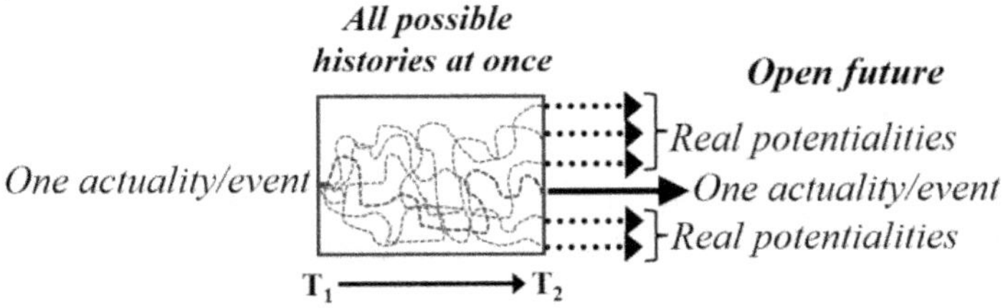

Fig. 15 Diagram for ontological indeterminacy. All potentialities could actualize, but only one does so.

Moreover, one could extend the same ontological indeterminacy, leaving uncertain not only which physical event will occur and where, but also when. For instance, the time it takes for a system to transition from actuality A to actuality B may be indeterminate. The realization of B could potentially occur after a time interval ΔT_1, or after longer or shorter intervals ΔT_2, ΔT_3,..., and so on. These points in time are also potentialities; however, only one specific instant will see state B physically come into existence. From the perspective of ontic uncertainty contrary to the PSR, there is no necessary reason or cause for its actualization at one instant in time rather than another.

In other words, from the standpoint of this ontic uncertainty, the future is open in the sense that it is factually and temporally "truly" undetermined.

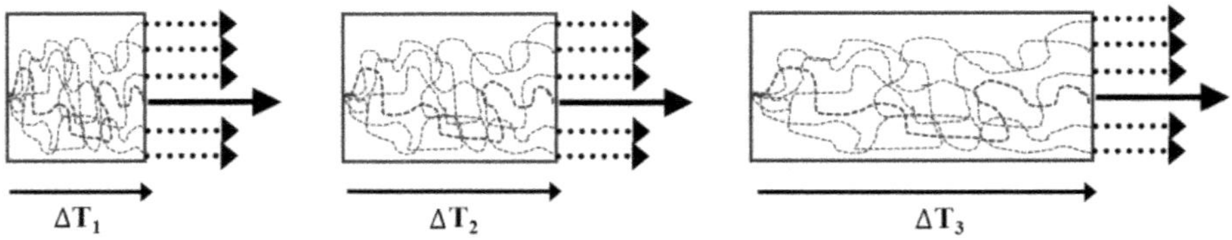

Fig. 16 Extension of the previous diagram: ontological uncertainty over time regarding event actualization.

Some might consider this a ridiculous principle of "non-causality." However, readers who have some familiarity with quantum physics might recognize that this type of ontic indeterminacy underlies Heisenberg's uncertainty principle. The following sections of the next chapter will explore the quantum physical effects relevant to the present context in greater detail.

Meanwhile, let us ask: Does this type of ontological indeterminacy challenge causal closure? Would it suggest a causally open universe?

Not necessarily. Notice the previous emphasis on considering physical events. If we relax the notion that the PSR does not apply to physical events, we are still discussing a causally closed universe. We must abandon our instinctive desire to find a reason why one physical event leads to a specific physical outcome among many possibilities and why it occurs at a particular instant. Physicality does not require a deeper reason or cause for certain physical facts to be true. Actually, this is the argument of most materialists: If certain facts come into existence rather than others, at specific times rather than others, with no reason whatsoever, then this is simply how it is. Physical facts do not always require humanly intelligible reasons for their existence. Things can be ontologically random without a source for their randomness and still be considered "physical." After all, one might question why our intuitive everyday principle of deterministic causality should be the only one. Determinism might simply be a human construct with no inherent validity compared to any other principle of causality. So, why not?

However, if this kind of modern magic doesn't sound right, the only alternative is to consider a causally open universe. We should recognize that the PSR is a very general principle that does not necessarily align with physicalist worldviews. We can postulate that the "reason" need not be physical but could instead be metaphysical. This means there might be forms of causality that do not rely on physical events causing physical effects in a linear succession in time, as our minds would like them to be, but rather metaphysical causes that elicit physical effects.

There could exist an unpredictable Will that, while not physical, interacts with the physical world to actualize events in ways that are not reducible to any form of determinism or pseudo-indeterminism. To our physical senses and linear, time-bound minds, this would manifest as ontological indeterminism.

Such a perspective might recall Arthur Schopenhauer's concept of the "*World as Will*." In fact, Schopenhauer contended that the PSR applies solely to causal explanations within the phenomenal world of Nature (the world as representation), while the role of the Will transcends the PSR. The PSR pertains to how the world appears to us but cannot reveal the true nature of reality. The ultimate reality—the Will—exists beyond reason and, consequently, beyond the PSR. The PSR governs all appearances of the world as representation across four domains: becoming (causality), knowing (logic), being (space/time), and acting (motivation). In contrast, the Will transcends rationality and reason. And, most importantly, we should recognize that our intuitive principles of causality—rooted in the understanding of causes and effects unfolding in a linear succession of events—may represent a limited perspective and could ultimately be entirely incorrect.

I will demonstrate how this aligns with modern quantum theory, although I will diverge from Schopenhauer's view that the Will is blind, aimless striving, and irrational. The universal Will is supra-rational, not sub-rational.

If so, this might immediately raise several conceptual difficulties that have been thoroughly discussed in the philosophy of mind. For instance, how can an unphysical entity interact with physical matter? Would this not violate principles of energy conservation? Would it not contradict the physical laws that we know to be true? The following sections and chapters will address these questions in detail, illustrating how quantum indeterminism elegantly resolves them.

Thus, if reality is governed by an ontological indeterminism that cannot be reduced to an underlying deterministic process and can be interpreted as the activity of a universal Mind, this might be a causally open universe, and several mysteries in the naturalist account of reality could begin to make sense. Perhaps we should consider that, despite all the evidence that seems to suggest otherwise, even for inanimate macroscopic matter, there might be some causal openness with divine action somehow "percolating through." If one could prove that the history of every event in the universe, even the most mechanical and apparently deterministic one, is open, and ultimately originates from and is driven by an "a-causal" or "self-causal" spontaneous activity of a universal field that embeds everything, from atoms to galaxies, I would argue that this could be a compelling argument showing how the hypothesis of divine action is perfectly compatible with modern science.

3. Toward a Quantum Theology?

About two decades ago, the theories of theoretical physicist, theologian, and Anglican priest John Polkinghorne gained popularity, suggesting that Heisenberg's uncertainty principle opens the door to divine action. This uncertainty in quantum physics implies a universe governed by probabilistic behavior and inherent randomness—an idea that did not resonate with

Einstein, who famously stated, "God does not play dice." According to this perspective, today labeled "*quantum theology*," the inherent unpredictability of quantum phenomena is not merely a reflection of ignorance but rather an indication of intrinsic agency. It conceals other causal principles that operate in the universe. If the universe is fundamentally random, divine influence could potentially operate within quantum probabilities, guiding events in ways that align with natural laws. This influence would be subtle yet genuine.

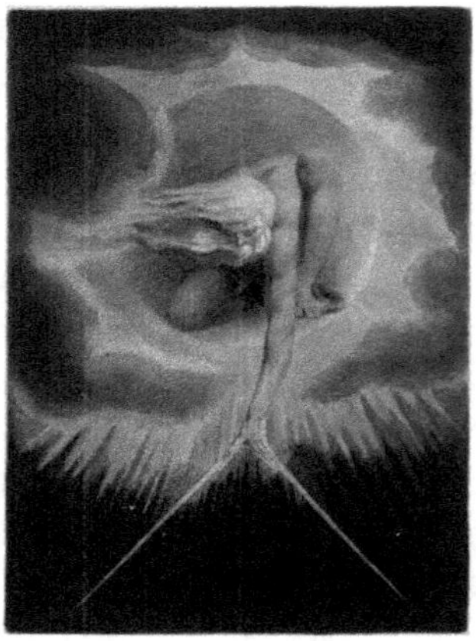

Such an approach to divine action emphasizes a different understanding of God's role in the universe. We should not envision God as a supernatural "intervener" who occasionally suspends natural laws. Instead, we should perceive the universe as fundamentally causally open and layered, with new properties emerging at higher levels of complexity—such as life, consciousness, and culture—that cannot be reduced to mere physics. In this perspective, divine action unfolds through these emergent processes, functioning within the indeterminacies and flexibilities that nature inherently offers, rather than overriding physical laws. Quantum theology states that quantum uncertainty, chaotic dynamics, and the openness of complex systems allow for divine influence without disrupting causal chains. God acts in every event, yet never in a manner that contradicts scientific principles. This grounds God as being both immanent and transcendent. Thus, God can be intimately present in creation, influencing it from within rather than manipulating it from the outside.

In my view, this is, so far, the theoretical model that provides the most coherent affirmation of divine action in a scientific age, portraying a God who operates through the dynamic creativity of the universe itself.

Others, such as Philip Clayton, Ian Barbour, Arthur Peacocke, and Robert John Russell (and what about good-old A.N. Whitehead or T. de Chardin?), made similar claims. They claimed that God does not need to "break the rules" of physics but can instead operate through the flexible structures inherent in the cosmos: operations that are intrinsic indeterminacies to us.

The catch is that this is easier said than done. To my knowledge, none have provided specific details that could be considered a scientific theory; their ideas remain vague. Intrinsic limitations and logical reasons stand in the way of these speculations, and they cannot be ignored. If divine intervention were to occur, with God determining which of many possible quantum outcomes takes place in an "open" universe that is not causally closed, it would alter the probabilistic laws of quantum physics, resulting in observable anomalies.

It is like trying to cheat at a casino. In a fair lottery, every number has an equal chance of being drawn, and over time, their frequencies should roughly balance out. Cheating—that is, intervening in the random process—disrupts these patterns and becomes evident because it violates fundamental statistical principles. Certain numbers may appear more frequently than expected, while others may show up less frequently, resulting in imbalances too significant to be random. These anomalies would quickly become apparent as the number of draws increased, making systematic manipulation easy to detect through statistical analysis. However, nothing like this has ever been observed. Those who use quantum randomness to justify divine agency never address the question of how God intervenes in the cosmic lottery machine without getting caught red-handed.

Another persistent weakness among most of these philosophers, theologians, and scientists is their adherence to an exclusivist Christian theological perspective. To be taken seriously when making metaphysical claims related to science, one must transcend one's own religious affiliation and early educational conditioning, adopting a worldview that surpasses individual religions or local cultural belief systems. For reasons beyond my understanding, we remain utterly unable to achieve this in the globalized 21st century.

Today, these ideas are further complicated by a plethora of pseudo-scientific theories that invoke quantum physics to explain concepts like quantum laws of attraction, telepathy, and quantum healing, among other forms of "quantum woo."

Conversely, naturalists also fall short when they uncritically accept the peculiar randomness of the quantum world. The simple assertion that it is "just random" does not lend them greater philosophical credibility. Physicists will tell you that quantum mechanics is a "theory without hidden variables," which means nothing, and ultimately translates to "shut up and calculate."

Thus, to my knowledge, no satisfactory (classical or quantum) theory has been proposed that can explain how divine causation, if any, operates within the universe as described by science.

Why is this important? Because the credibility of faith in a scientific era hinges on intellectual honesty. If believers assert that God intervenes in history, heals the sick, or answers prayers, they must confront how these beliefs coexist with a universe governed by physical laws. If the universe operates according to established laws and God acts within it, believers have a responsibility to provide an explanation—at least in principle—of how such divine actions could occur.

Conversely, when philosophical naturalism asserts that the universe is causally closed, fundamentally meaningless, aimless, without agency and purpose, and with no divine action, it is also making a statement based on faith. It ignores the startling fact that it produced something opposite to its nature—

namely, conscious beings that essentially act with agency, purpose, and aims. Ultimately, it also lacks substantial philosophical depth.

What is needed is a dialogue that respects both theological integrity and scientific understanding. Up to this point, there has been little more than wishful thinking and wordplay.

I (provocatively) contend that the naturalist has significantly better arguments than the metaphysician regarding causal closure in cells, brains, and life. As described in the previous section on determinism and causal closure, this can be interpreted as strong evidence that, as complexity increases, the principle of causality becomes more challenging to adhere to. However, this difficulty arises not because of evidence of an alternative principle of causality that begins to intervene, but rather because the extrapolation from simple deterministic laws on the microscopic scale to human-sized bodies with the complexity of living beings fails due to our ignorance. Things are much too complicated to be described by the enormous number of extremely complicated but deterministic and causally closed processes of a living cell, let alone an entire body. The belief in divine action simply assumes this interruption due to the unpredictability of life and human actions. However, this is not in and of itself proof of the interruption of a deterministic causal chain of causes and effects. While theories of (weak or strong) emergence, which seek to explain our supposedly emergent behaviors with bottom-up or top-down causation, complex systems, chaos theory, autopoiesis, etc., are all interesting and fascinating, they ultimately remain rooted in the classical mechanics of perfectly deterministic physics.

Thus, causal closure appears to be powerfully operative in physics and chemistry and cannot be waved away. I could not find, from the opposing side, a similar counterargument demonstrating the possibility of "interjection," "insertions," or "subtle deviations." Where is this better-suited model of divine action?

I submit that quantum theology is the way forward. However, the argument must be framed within a coherent framework that aligns with our scientific understanding and is supported by at least theoretically valid proof. To achieve this, we must first delve into the strange realm of quantum physics and grasp the basics that will enable us to construct such a hypothesis.

V. Quantum Physics, Panpsychism and (Free?) Will

1. Quantum Mechanics: The Basics

Materialism is based on several key assumptions, one of which was and remains, to a large extent, the belief in a strictly deterministic and, at least in theory, entirely predictable universe.[35] This worldview hinges on the principle of physical causal realism that can be traced back to the thought experiment known as "Laplace's demon," introduced in 1814 by the French mathematician and philosopher Pierre-Simon Laplace. It serves as a (frequently unaware) philosophical foundation of modern materialism and much of contemporary science, embodying a form of strict determinism and reductionism in its conceptual framework.

Laplace's famous line of reasoning was the following:

"We may regard the present state of the universe as the effect of its past and the cause of its future. An intellect [the demon] *that at a certain moment would know all forces that set Nature in motion, and all positions of all items of which Nature is composed, if this intellect were also vast enough to submit these data to analysis, it would embrace in a single formula the movements of the greatest bodies of the universe and those of the tiniest atom; for such an intellect nothing would be uncertain and the future just like the past would be present before its eyes."*

Fig. 17 Pierre Simon Laplace (1749-1827)

In the 20th century, quantum mechanics emerged as a theoretical framework in physics that reveals the indeterministic behavior of particles at subatomic and molecular scales, thereby fundamentally challenging previous conceptions. Despite this, the deterministic mindset continues to influence reductionist scientific thinking.

To provide a clearer understanding of quantum physics, I will first offer a brief and straightforward introduction to the topic, thus allowing you to grasp its essence intuitively.[36]

---

[35] Not all individuals who identify as materialists, naturalists, or physicalists subscribe to mechanistic determinism. This is likely due to their awareness that such a position might ultimately necessitate the admission of being soulless biological robots—a conclusion that even the most radical materialists find challenging to accept.

[36] For much more in-depth literature, the reader might resort to my two-volume book on quantum physics (Masi, 2019a) and (Masi, 2019b).

Quantum physics is rife with counterintuitive aspects. For instance, the well-known "wave-particle duality" explains the behavior of particles such as electrons and photons, which exhibit characteristics of both waves and particles.

In the famous double-slit experiment, particles exhibit wave-like behavior, seemingly passing through both slits simultaneously. However, when one measures their position on a detecting screen, they manifest as discrete spots creating an interference pattern, even when fired one at a time. Put plainly, the experiment suggests that until the particles' position is measured (that is, "observed"), they behave like waves; what traverses the slits isn't represented as a particle but rather as a wave. Yet, once you attempt to determine a particle's location, say on a detection screen, it "collapses" into a single point in space–that is, interacts with and gets absorbed in one of the pixels of the screen.

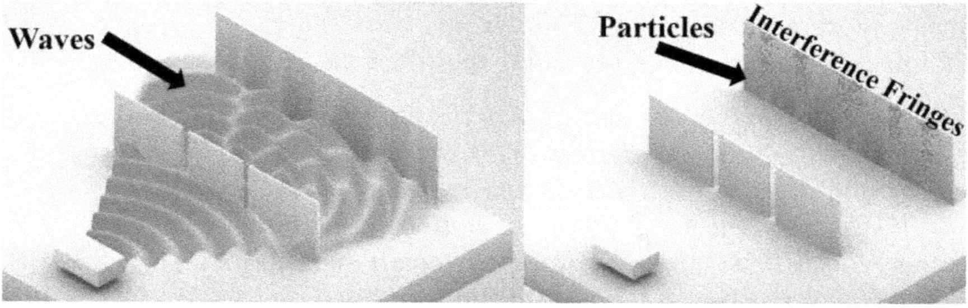

Fig. 18 The wave-particle duality is shown with the double-slit experiment. Left: Before the measurement. Right: After detection on a screen.

The question arises: Is it a wave or a particle?

There's also the equally well-known "quantum superposition principle,"[37] which suggests that a single particle acts as though it could exist in multiple physical states—for example, in several places or even in a combination of different energy states simultaneously (see Fig. 19). Translating this to our everyday macroscopic world implies that, for instance, one could not only occupy various locations at the same time but also move at two or more different speeds simultaneously. Yet, once one measures the position or energy, only one of several possible results is obtained. The quantum state describes a probability—specifically, the possibilities or potential outcomes of measurements—that we subsequently label as "events," "actualities," or simply "facts." Notice how, unlike classical physics and our everyday experiences, where a physical state uniquely defines an actuality or event, a single quantum state embodies multiple potential outcomes, of which only one ultimately manifests as a reality.

---

[37] Here, I won't discuss the famous 'Schrödinger's cat experiment,' which relates to the notion of quantum superposition.

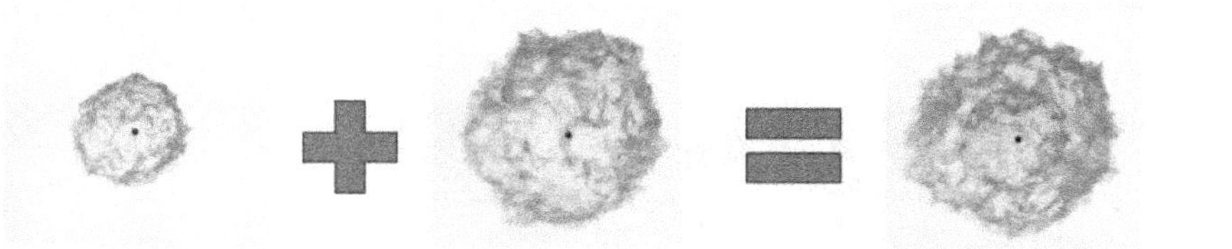

Fig. 19 An atom in energy superposition: Not E1 or E2, but E1 AND E2!

And who has never heard about "quantum entanglement"? When two particles become entangled, they behave as if they are no longer separate entities but, rather, a singular, undifferentiated whole. However, upon measurement, this "entity" appears to "collapse" into two distinct particles with correlated properties. This phenomenon suggests that the state of one particle in one location instantly correlates with the state of the other particle, even if vast distances separate them—an idea that Einstein discredited as *"spooky action at a distance."* This concept is known as "quantum non-locality": Two or more particles can act as a singular and interconnected entity, no longer defined by their individual properties and correlated over considerable distances.

Here, the non-reductionist facet of quantum physics emerges: A system of entangled particles cannot be simply reduced to the sum of its parts. Under specific conditions, these particles behave as if they exist in a physical state lacking individuality—that is, a precise, independent location in space and time. This challenges the conventional understanding based on local realism, which relies on our reductionist viewpoint of particles as distinct entities with specific positions and properties.

Most depictions of quantum entangled particles portray two particles as two "marbles" connected by an odd string or energy beam (Fig. 20 (a)). However, this representation doesn't accurately capture the situation. According to quantum physics, there are no two individual particles; instead, there is a singular "wave function" in one unified quantum state. The wave function is an abstract mathematical description of the state of a quantum system. The most common symbol for a wave function is the Greek letter Ψ. Through the application of some mathematical procedures, its amplitude furnishes the probability with which a certain quantum event will take place.

For a better illustration of this notion, let's consider the decay of an atomic nucleus emitting two entangled particles, as visualized in Fig. 20 (b).

Before measurement, the wave function represents what physicists refer to as a "probability wave". The probability wave is represented by a spherical wave extending throughout space. There are no interconnected particles or separate entities yet; rather, there is a unified and undifferentiated whole that gradually expands in space isotropically.[38]

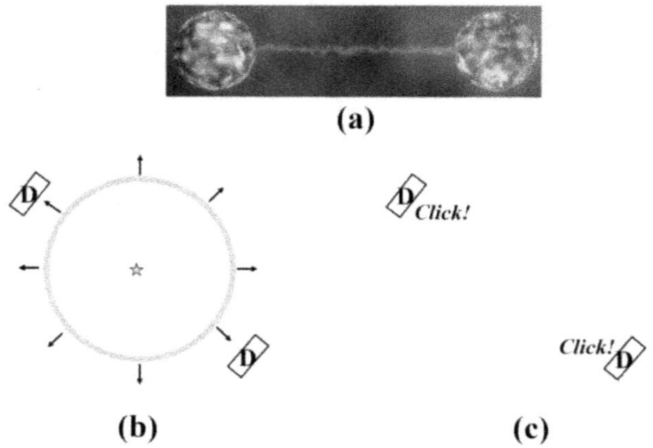

Fig. 20 (a) Wrong understanding of quantum entanglement as two particles interacting with each other. (b) Correct quantum entanglement depiction of a probability wave expanding throughout space: there is no "particle" to begin with. (c) Collapse of the wavefunction. The detectors "click."

Beware! It's important to note that this is *not* the probability of where the particles are located; rather, it is the probability of measuring the particles at particular locations. Consider this distinction carefully, as it will make a significant difference in your understanding of quantum mechanics!

Once a measurement takes place, the detectors indicate two locations: What we conceive of as particles "showing up right there," on some points on the spherical surface (Fig. 20 (c)). As a result, we no longer need to discuss the probability of finding the particles elsewhere. The spherical probability wave has "collapsed" to two single points—the location that the measurement devices indicated. This phenomenon is referred to as the "collapse of the wave function."

Again, beware! This does *not* indicate where the two particles could potentially be, with us not knowing where. Instead, the depiction of a spherical probability wave expanding from the initial position of the decaying atom represents an abstract mathematical concept that allows us to calculate where particle detectors *could potentially register* a "click." Until a measurement is

---

[38] More precisely, in quantum mechanics, probabilities and their evolution in time are mathematically determined by the Schrödinger equation. Once the initial and boundary conditions are established, this equation constrains the structure of the wave function, while the Born rule quantifies it in terms of probabilities. Readers who are unfamiliar with these technical details need not be concerned, as they are not essential for understanding the philosophical rationale presented later.

made, we cannot even assume that particles exist as two separate entities. When detectors click within the spherical volume, it signifies that a measurement has occurred, interacting with something we conceptualize as a particle. At that moment, the probability wave is said to "collapse" into two distinct particles located on the expanding spherical surface. There is no longer a sphere, only two points. These two particles are always correlated, meaning they exhibit properties, such as spin, that instantaneously become opposite to each other, even if light-years separate them.

What about Heisenberg's uncertainty principle? It states that the position and momentum (the quantity of motion or simply imagine the speed) of a particle cannot be known with absolute precision at the same time, not even in principle—hence its designation as a "principle." The more accurately you try to determine a particle's position, the less accurately you can know its momentum and vice versa.

It's crucial to understand that this principle is not a consequence of measurement errors from a device, nor is it influenced by a conscious observer, as we will explore later. Instead, it reflects an inherent quantum indeterminism imposed by Nature, regardless of how meticulously you attempt to define a particle's position and momentum. This uncertainty arises not from our ignorance about the particle's location but because the particle itself does not possess a definite position and momentum to begin with. In the quantum realm, properties do not have inherent definiteness. There is no conventional understanding of a point particle having definite and precise properties. Instead, everything seems somewhat nebulous and uncertain, conflicting with our intuitive notion of particles moving along specific deterministic paths.

In essence, every measurement reveals an inescapable element of "quantum randomness" and unpredictability. Each time a measurement is made, the particle appears in a slightly different position, with certain probabilities depicted by the "probability cloud" in Fig. 21 (right). Contrary to popular belief, this variability does not stem from the interaction with a measurement device that, due to the interaction with a particle, supposedly "kicks" the particle away; rather, it is an intrinsic property of the particle itself. In philosophical terms, the uncertainty principle reflects not an epistemic lack of knowledge but, rather, an ontological or "ontic uncertainty."

The term "epistemic," derived from the Greek word ἐπιστήμη (epistēmē), refers to matters related to cognition or knowledge. For instance, an object might possess certain properties or exist in a specific physical state, yet we might lack the means to determine those properties or the state in which it resides. Unfortunately, the uncertainty principle is often misinterpreted as a mere lack of knowledge. This misconception leads some—including, at times, physicists—to believe that the uncertainty principle is about our inability to accurately measure a particle's position or momentum. This epistemic interpretation assumes that the particle must have a definite position and momentum, but we simply cannot observe it accurately. In this view, the particle possesses an unknown and unknowable position and speed. However, this interpretation represents a fundamental misunderstanding of the uncertainty principle.

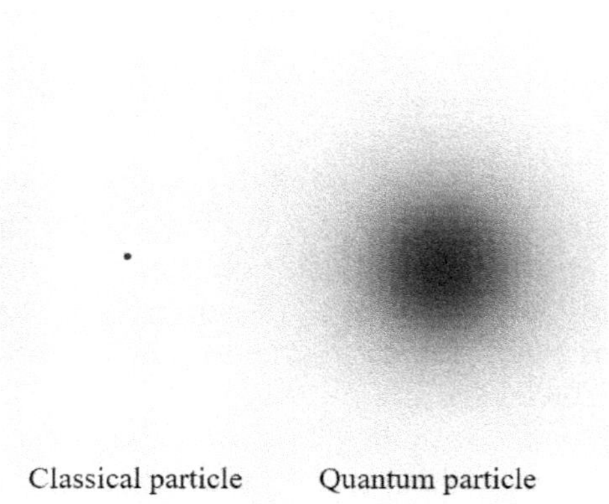

Fig. 21 Classical (point-like) particle vs. quantum (fuzzy) "particle."

Quantum uncertainty represents an ontological lack of knowledge. The term "ontological," derived from the ancient Greek words ὤν (ṓn) and ὄντος (óntos), refers to "being," "existing," or "essence." It pertains to the fundamental nature of things as they truly are—independent of our knowledge, perceptions, or conceptual frameworks, which can be misleading.

Quantum uncertainty extends beyond just the position of a particle; it is a fundamental aspect of quantum mechanics. Energy levels of an atom, the decay time of a nucleus, or the spin of a particle, are properties which measurement outcomes aren't unique but show up as random fluctuations inherent to the fabric of reality. Consequently, in quantum mechanics, phenomena are expressed in terms of probabilities rather than certainties. The wave function determines the likelihood of various measurement outcomes that, however, shouldn't be reified as definite properties of a point-like particle which state represents a definite actuality.

For instance, the measurement outcomes of the spin of two entangled particles is inherently random, with a 50% likelihood of being recorded as either "+" or "-" spin. In physics, a particle's spin represents its intrinsic angular momentum. To visualize this concept intuitively, imagine the particle spinning counterclockwise (spin |+>) or clockwise (spin |->) at two distinct angular speeds.[39]

[39] This is only an intuitive understanding and should not be taken literally, but we don't need to go into the conceptual details here.

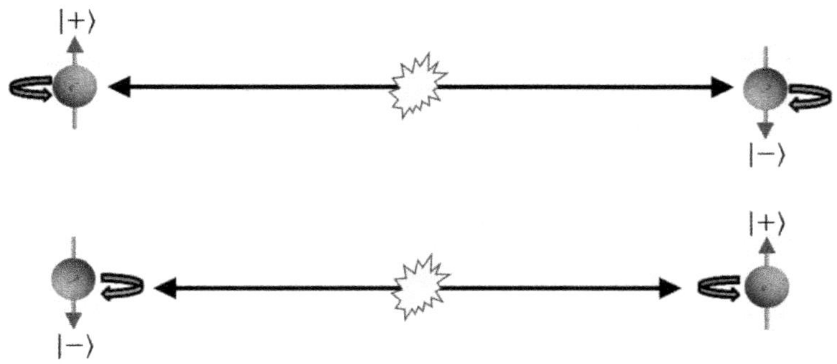

Fig. 22 The two possible random spin states measured upon two entangled particles.

When measuring the spin of two entangled particles, as illustrated in Fig. 22, we will only observe two possible states: spin-up or spin-down. There are no intermediate states. The measurement of only one particle will instantly—that is, faster than light—"collapse" the undifferentiated entangled quantum state to a differentiated pair of localized probability waves that, upon interaction with the respective detectors, will always show up with opposite spin values. It is as if a "spooky action" at a distance has taken place without interaction; whereby physicists prefer to talk in terms of "correlations" instead of "actions" or "interactions." However, which particle exhibits the spin-up and which exhibits the spin-down is entirely random, with each outcome having a 50% probability. This aspect is entirely beyond our control and is solely a choice made by Nature.[40]

Thus, the essential point to keep in mind during this philosophical exploration is that, in the quantum realm, ontological uncertainty indicates the uncertainty of a measurement; it doesn't tell us what is being measured. The notion of a particle is merely a fictitious model, a human-made conceptual construct that lacks a corresponding entity in the microscopic physical reality. Discussing the position or the spin of a particle is a sort of fiction we create to make sense of the world, but the concept of a point-particle is a human-created mathematical abstraction. This idea traces back to Descartes, who introduced his Cartesian grid system with infinite coordinate points. However, the true nature and essence of what we refer to as a quantum "particle" has nothing to do with points.

---

[40] People sometimes entertain the notion that this "spooky action" could be used to instantaneously transmit information faster than the speed of light across the universe. Unfortunately, the reality is more complex. Primarily, no direct "action" or "interaction" takes place; rather, it's a correlation (or anti-correlation) of particle properties that manifest only at the moment of measurement, not before. Additionally, since the determination of which particle will assume a particular state is random and beyond our control, quantum entanglement cannot be harnessed for transmitting information.

Similarly, we might question whether the collapse of the wave function and the various strange phenomena and properties described by quantum theory represent merely an epistemic description of reality. The mathematical framework of quantum mechanics provides all the necessary information for making measurements, but it doesn't tell us much about the ontology of the reality it describes in terms of measurements. For instance, is the wave function a real entity "out there" in the world, or is it merely a mathematical abstraction?

The wave function is a tool used to calculate the probability of measurement outcomes, but it does not reveal the true nature of the measured object itself. From this viewpoint, the wave function is primarily epistemic. However, it also describes interference phenomena associated with something "wavy" that appears to exist in the world—a fact that suggests this wave might possess an ontic existence, like in Fig. 18 left.

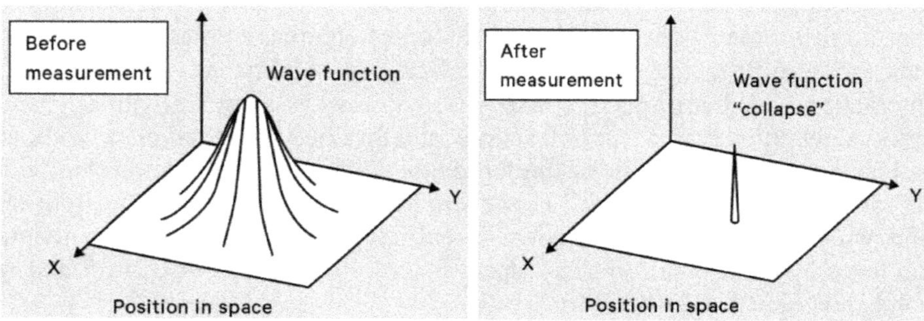

Fig. 23 Is the "collapse" of the wave function an epistemic or ontological description of reality?

Despite this inherent uncertainty, quantum mechanics still enables predictions by employing statistical calculations. Although we can't accurately predict specific outcomes (such as precisely determining when the next atomic decay will occur), we can make statistical predictions using a powerful and successful mathematical formalism. For instance, we will never know the exact time of decay of a single carbon-15 atom isotope or which of the N atoms in a sample will decay next, but we can predict that in approximately 40 minutes, about half of all these N atoms will have decayed.

This raises significant questions about our understanding of cause and effect. What prompts atoms to decay at one moment instead of another, or what causes particles[41] to seemingly "quantum fluctuate" erratically throughout space without any discernible external cause? From where does this quantum uncertainty, randomness, or "fluctuation" stem if no apparent triggering cause exists?

[41] In the following, I will use the word "particle" only for linguistic convenience. Keep in mind that the word doesn't reflect the ontology for which the word stands.

As per the traditional formulation of quantum mechanics, it's considered a "theory without hidden variables." This implies that no intrinsic yet unknown reason or concealed underlying cause, such as some unknown microscopical details–that is, so-called "hidden variables"–determine determine a particle's unpredictable behavior. The effect merely comes into existence without a discernible reason and cause.

Moreover, there are other perplexing quantum phenomena, such as the "quantum tunneling effect."

This effect is about particles that traverse barriers even without possessing the energy that classical mechanics would traditionally deem necessary to overcome these obstacles. Instead, they "tunnel" through potential barriers (in Fig. 24 the "hill") that, in theory, should be impenetrable.

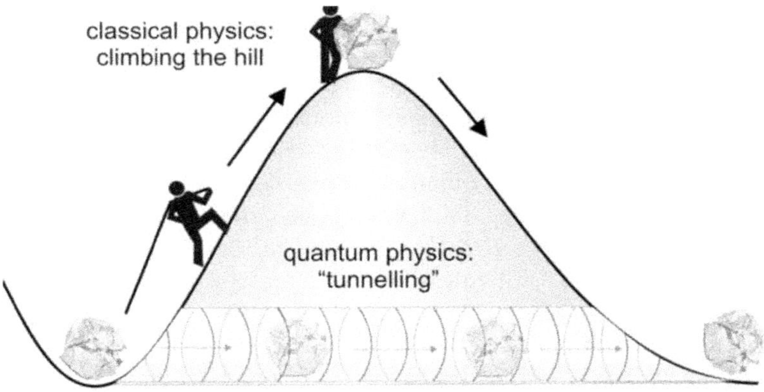

Fig. 24 Illustration of the quantum tunneling effect with a classical analogy.

The particle tunnels through a thin barrier, but sometimes it doesn't and is reflected back. Essentially, this means if a particle does not have sufficient kinetic energy (speed) to overcome the "potential wall" (here, the hill), we'll observe it being reflected (rolling back) in most instances, as is the case in our everyday experience in a macroscopic classical world. However, there's also a smaller probability that the particle will traverse or "tunnel" through the obstacle even though, according to classical mechanics, it lacks the necessary energy to do so.

This is completely counterintuitive and at odds with our ordinary experience of the world. The important aspect to keep in mind is that one can't predict when the single event of the reflection vs. the tunnelling process will take place. This is a choice left for Nature—that is, for us, it is a purely random or stochastic phenomenon that nobody can predict, not even in principle, other than in terms of probabilities and statistical considerations based on a large number of measurements.

So, what is left of Laplace's idea *of a universe* where we can know *"all positions of all items of which Nature is composed,"* and that we could *"embrace in a single formula"* with nothing of past, present, and future remaining uncertain? And is the quantum universe *"the effect of its past and the cause of its future"*?

These unusual quantum effects challenge our traditional human-centered macroscopic worldview. Our understanding of the world as composed of concrete objects with definite properties, occupying specific locations, and capable of jumping over obstacles only with sufficient energy—following precise paths like pinpointed particles in a mathematical diagram and governed by strict principles of causality—has been upended. This Cartesian, Galilean, Newtonian, and Laplacian mechanistic realism, based on a reductionist, deterministic, and local perspective of unconscious matter adhering to classical laws of physical causality, has been remarkably successful for nearly three centuries. However, this vision of reality as a completely predictable, unconscious, and mechanistic clockwork universe is no longer tenable.

The quantum revolution that came about a century ago, led by scientists such as Max Planck, Werner Heisenberg, Erwin Schrödinger, Max Born, Albert Einstein, and several others, suddenly tells us a very different story about reality. The inherent stochasticity of quantum mechanics—that is, quantum randomness—suggests that particles can behave as if they are making independent "choices," regardless of our attempts to precisely measure or control them. The traditional principles of cause and effect—in which a specific initial condition leads to a single, predictable outcome—are called into question. For instance, what determines why one particle passing through a double slit strikes a specific point on a detection screen when another particle, starting under exactly the same initial conditions, hits a different point? If quantum mechanics is a theory without hidden variables, the answer seems to be simply: nothing. It's just the way it is.

2. Quantum Mechanics: Lesson Learned?

Now that we have covered the few technical aspects of quantum mechanics necessary for the present work, let's turn to its sociological implications. Nature has a message for us, one that we should recognize and understand. Are we paying attention? Have we grasped the lesson it offers?

When speaking with physicists, you will rarely encounter anyone who still subscribes to the simplistic mechanistic views of Newtonian and Laplacian realism, which envision elementary particles as marbles bouncing around like billiard balls. There is a widespread acknowledgment that our everyday macroscopic perception of reality is no longer valid.

What is the primary emotional response when a scientific fact challenges your belief system and worldview? You might enter a state of denial, choosing to ignore the facts (a trend that is quite common today), and you might attempt to reinterpret the new findings within the framework of your existing belief system rather than accepting a new paradigm. History provided an example of this struggle during the transition from the geocentric to the heliocentric model of the universe. For a long time, people tried to explain the unusual looping motion of celestial bodies by developing complex theories that included deferents and epicycles, all in an effort to keep the Earth at the center of the cosmos. They struggled to accept the notion that we are not the focal point of the universe and tried to "save appearances" by rescuing the geocentric system.

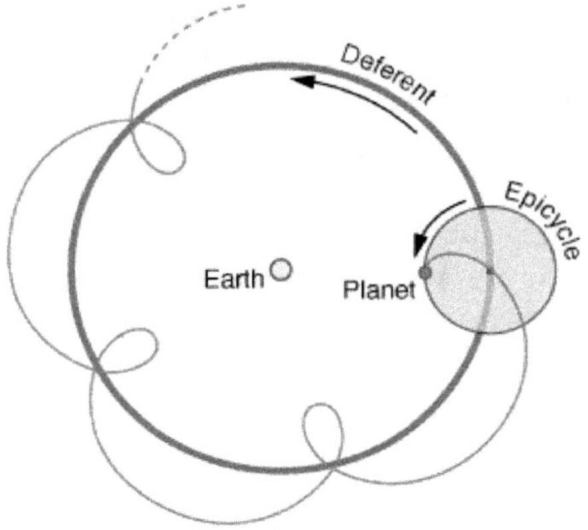

Fig. 25 The Ptolemaic geocentric model with its deferent and epicycle.

Indeed, it works! By carefully selecting the appropriate size and angular speed of the circles moving on other circles, you can accurately represent the planet's motion.

The same situation is recurring with indeterministic and non-local realism, which has always been foundational to our materialistic understanding, not only in physics but also in biology, medicine, and neuroscience. Many scientists and philosophers acknowledge how quantum physics challenges our everyday perceptions, yet several of them strive to uncover deeper layers—namely, the ontology—beneath quantum mechanics that might restore a sense of classical realism aligned with our human-centered experiences. For decades, exceptionally intelligent physicists have developed numerous interpretations of quantum mechanics in an effort to partially revert our understanding to a 19th-century form of realism.

Indeed, it works again! Alternative explanations in the form of a plethora of "interpretations of quantum mechanics" have been developed, and they pretend to save appearances. These interpretations construct models that could, at least in principle or in part, restore our everyday understanding of reality composed of solid objects that behave predictably–that is, without existing in superposition states, collapsing into infinitesimal points, tunneling through barriers like ghosts, or moving randomly in zigzag patterns. That is, they "save appearances," rescuing a semi-classical deterministic and/or local realism.

For example, Louis de Broglie and, later, David Bohm developed a theory that maintains a non-local framework while restoring determinism by proposing that tiny point particles have definite positions. This is achieved by envisioning a universal “pilot wave” that enables particles to exhibit wave-like behavior. In this model, particles can still be treated as mathematical points while behaving as waves.

Another example is the “Many Worlds Interpretation,” which imagines a deterministic and locally realistic universe but at the cost of multiplying it in innumerable “worlds” every time someone makes a measurement. This means no superposition of states occurs; rather, the present universe branches into two or more twin “worlds,” with one state in one world and another in a different world—and with yourself being multiplied and copied in all these worlds simultaneously as well! I don’t know about you, but I don’t like the idea of someone or something being able to multiply me. And, ultimately, this worldview is much more absurd than the quantum weirdness it wanted to eradicate. Determinism or locality are recovered but with a very questionable ontology.

What about “superdeterminism”? This interpretation arises from a loophole in Bell's theorem, which uses an inequality to assess whether quantum mechanics can coexist with a local hidden-variable theory. Numerous experiments have demonstrated that quantum mechanics violates these inequalities, suggesting that it must be a non-local and/or indeterministic theory. However, Bell’s theorem relies on several fundamental assumptions, one of which is the statistical independence between the particles and the detectors that measure them. If we assume that the future settings of the measurement devices are correlated with the particles that determine the outcome of the measurement, we could explain the observed violations without invoking non-locality or hidden variables. This perspective might allow us to reconcile quantum mechanics with our local and deterministic understanding of reality. However, if this is true, it remains uncertain whether experimentalists truly have the freedom to choose detector settings, as this assumption implies that there are no independent events in a pre-determined universe. Similar to deist theology, in superdeterministic cosmology, everything was predetermined at the time of the original creation—what we understand as the Big Bang. After that, everything unfolded in a completely deterministic manner, like in a clockwork universe.

Another interpretation such as “quantum Bayesianism” (in short, “QBism”) starts from an assumption with which I agree—namely, that what we refer to as the “quantum state” (for instance, the quantum state of an atom) cannot be linked to any objective property of the external world that we perceive through our senses and interpret with our minds. Quantum mechanics is a theory that reflects our understanding of interactions with the quantum world in statistical terms. For instance, consider the superposition of quantum states. When we make a measurement and obtain a single result (such as detecting an electron in a specific spin state), we are essentially "updating our beliefs" based on this new information. Previously, we didn't know where it was, but now we do.

In a sense, I appreciate this approach because QBism highlights the subjective nature of the observer in the measurement process. It also challenges the notion of an objective reality that exists independently of our observations—a concept I will discuss in Chapter IX on philosophical idealism.

However, QBism stops short of providing insights into the ontology of the world beyond what we already know, as it teeters on the edge of a subtle form of solipsism. It emphasizes that we should think solely in terms of "degrees of belief" regarding the outcomes of measurements, but it fails to explain what is actually being measured or why those outcomes fluctuate like mad. Ultimately, QBism is yet another "shut up and calculate" approach in disguise. It refrains from posing deeper philosophical questions and focuses solely on the calculations.

Anyway, the primary shortcoming of these interpretations of quantum mechanics is that they do not make predictions. Instead, they offer various models of reality that fail to provide testable predictions. This makes it impossible to verify or differentiate between them. While they reframe our existing knowledge within different "theaters of reality," they ultimately replicate the same quantum theory without introducing anything significantly new.

Despite all these speculations, little progress has been made since the first interpretation of de Broglie in the 1930s. Most advancements have come from applied physics, and the conceptual foundations have remained largely unchanged. As a result, most physicists adopt a pragmatic stance, often expressing disdain for philosophical inquiry: "Shut up and calculate! Who cares about the ontology of quantum physics?" [42]

Why, then, have generations of physicists invested so much effort into developing interpretations that seek to restore a deterministic and/or reductionist worldview? The answer lies in a deep-seated discomfort with the strangeness of quantum mechanics, which, even if not openly acknowledged, troubles the naturalist mindset.

This quantum "weirdness" contradicts their (often unspoken) hope and belief that one day everything can be explained in mechanistic terms—free

[42] "Shut up and calculate" is a phrase commonly used in the context of quantum mechanics. It conveys the idea that instead of getting caught up in the philosophical or conceptual implications of a theory, one should concentrate on using mathematical tools and formulas to make predictions. This expression reflects a pragmatic approach, prioritizing empiricism over efforts to grasp the deeper meaning or reality behind the equations. It is particularly linked to the Copenhagen interpretation, which emphasizes using the theory to predict outcomes, regardless of whether the underlying reality is completely understood.

from uncertainty, indeterminacy, and "spooky" and unpredictable behaviors, while re-enthroning the inert, unconscious, lifeless, and ultimately meaningless universe of classical mechanics. This desire has a psychological and subconscious impetus: It offers a way to dispel speculations about supernatural causes. At its core, this urge reflects a wish to eliminate concepts of final causation, purpose, and goal-directedness in Nature, confining everything to the traditional framework of an Aristotelian material and efficient causation.

Unfortunately for them, the long-awaited triumph of physicalist science over religion and spirituality, which seemed so close at the end of the 19th century, has not become reality. This is why most physicists discuss theories without hidden variables or describe certain phenomena in quantum physics as simply "random," carefully avoiding deeper implications.

This sociological perspective explains why most materialists attempt to reinterpret quantum physics in simplistic terms of human realism. In contrast, those with a more spiritual or idealist worldview often show less resistance to embracing concepts like quantum superposition, "spooky" correlations, and randomness.[43] The obsession that some philosophers and physicists have with preserving determinism and locality in quantum mechanics stems from sociological factors. It reflects a desperate effort to maintain our naive, human-centered realism and models of reality, and it parallels the attempts of those who sought to uphold geocentrism due to religious beliefs, clinging to their human-centered view of the universe. Ultimately, this is more of a psychological reaction than a scientific one.

However, Nature's message is loud and clear: *"No matter how hard you try to imagine yet another ontology, no matter how long you look for yet another loophole, no matter how many times you come up with yet another interpretation: I am indeterministic; I am non-local;, I am irreducible. Get over it!"*

I find it more reasonable to accept this state of affairs and seek to understand the true message that quantum physics conveys. The essence of the matter is that our perception and conception of reality, even on the macroscopic scale of Newtonian physics, are reconstructions—mere figments or "symbols" shaped by mental representations within our conscious experience.

In quantum mechanics, the distinction between the epistemic and the ontological is reminiscent of Plato's cave allegory or Kant's distinction between noumenon and phenomenon. This is in line with what we will see in more detail in Section IX.1: there exists an inherent essence of things that elicits in our senses and mind a perception and conceptualization of the world which, however, are mere "shadows."

---

[43] I say "most" because a notable exception was David Bohm.

I refer to this idea as "quantum idealism." Idealism is a philosophical worldview that suggests the reality we perceive is a construct in our minds because it is filtered through our limited human senses. It might not actually exist as it appears to us. It is like a shadow that "exists" only in relation to the object casting it. While the object itself exists, any attempt to understand its true properties from the perspective of a being confined to a world of shadows is futile.

Quantum physics reminds us that the "shadows" we experience in our everyday life (chunks of hard matter, particles with precise positions and momenta, etc.) do not exist "down there" in the microscopic quantum realm because they don't exist, as we perceive them, also not in a human's macroscopic everyday reality in the first place. We try to inject our human-centric mental constructs into all of Nature in terms of space, time, and causality when these mental constructs exist only in our minds. That kind of mental reality doesn't exist even when one observes it. It was and will remain a "shadow" in our minds.

Thus, it isn't surprising that the microworld doesn't reflect our macro-world conceptions, as there is no reason to believe our physical-mind reality has any "ontological superiority" that supposedly must stand above the quantum reality.

However, why does this mental construct seem to jitter and wobble uncontrollably—appearing to be governed by blind chance—every time we take a measurement? Could it be that what we interpret as "random" is actually Nature making "choices"? "Choices" we interpret as being "random" because, due to our limited cognitive ability, we fail to grasp all phenomena in their entirety?

3. The Myth of "Consciousness Causing Collapse"

Other interpretations of quantum mechanics do not imply any form of determinism or locality. However, they introduce other ideas that are more the product of fantasy than of real evidence and concrete facts. One of these is the extremely popular notion that quantum phenomena supposedly suggest the existence of the observer's role. An observer's mind or consciousness collapses the wave function, so the story goes. According to this interpretation, the outcome of an experiment involving quantum objects depends on the observer's mental state.

This idea dates back to the "Von Neumann–Wigner interpretation" of quantum mechanics and is described as "consciousness causing collapse." It is based on one of the many peculiar aspects of quantum physics that has sparked numerous speculations, conjectures, and controversies.

First, without delving too much into the technical details, let us see the facts and what this "observer effect" is really about.

In section V.2, we explored the strange and unintuitive indeterministic nature of the microscopic world of quantum physics. According to Heisenberg's uncertainty principle, it is impossible to measure, with absolute precision, both the position of a particle and its momentum. Attempts to more accurately determine a particle's position result in increased uncertainty regarding its speed and vice versa. In practice, this means that when multiple measurements are made, the outcome of each measurement—such as the position of a quantum particle like an electron—will always vary slightly from the previous one. This variability suggests that the particle's position is somewhat indeterminate and uncertain, appearing fuzzy, although the overall results will cluster around an average expected value. Consequently, we can replace the concept of a particle with that of a probability cloud, as illustrated in Fig. 21.

We have observed that this uncertainty is not a result of measurement errors; it is ontic rather than epistemic. This uncertainty is not related to the imprecision of our measurement devices. Instead, it is an inherent uncertainty that will persist regardless of the precision of our measurements.

Another common but misleading belief is that the interaction between the measurement device and the particle causes quantum indeterminism. When we observe microscopic objects, such as through a microscope using light particles, we inevitably perturb and alter the system. This tiny yet unavoidable disturbance—known as the "observer effect"—changes reality only because we observe it. In this sense, reality can be seen as subjective. This interpretation has gained traction for a simple historical reason: It was Werner Heisenberg's initial interpretation of his own uncertainty principle. However, we now know that it ultimately represents an overly simplistic understanding of what the uncertainty principle truly entails.

First, it's important to clarify that in physics, the term "observer" refers to a measurement device, such as a photodiode, a CCD camera, or a photographic plate. The "observer" does not require consciousness or a mind to "observe"—that is, to perform a measurement. In physics, an "observer" is simply a mechanical apparatus that registers data without any need for human awareness of the result. Furthermore, if we consider the interaction between a measuring device and a quantum object to be an "observation," should we then view every physical interaction between particles in the universe as a form of "observation"? For instance, should we regard a cosmic particle colliding with a rock on Mars as an "observation" as well?

Second, today, modern electronic and quantum optical devices enable what are known as "interaction-free measurements. This innovative approach allows researchers to test quantum effects, such as the uncertainty principle, without directly interacting with the system.[44] However, uncertainty remains ineliminable. No hidden causes or "hidden variables" can account for it—a fact demonstrating that uncertainty is a fundamental and intrinsic (that is, ontic) aspect of quantum nature. It is not merely a consequence of a noisy observer effect.

In other words, every measurement is subject to unpredictable random fluctuations, which are beyond our control. Einstein couldn't come to terms with the randomness and unpredictability inherent in quantum phenomena, which notoriously prompted him to declare that "*the old One doesn't play dice.*"[45]

Third, quantum theory is a "contextual theory," meaning the results of our measurements depend on how we set up an experiment to gather information about the quantum system, such as the position of a particle, which slit it passed through, its momentum, or its energy.

For example, reconsider the wave-particle duality of the double-slit experiment of Fig. 18.

If you don't attempt to determine which slit an electron passes through in a two-slit setup, the electron will appear on a distant screen according to a statistical distribution that reveals interference fringes. In this case, it behaves like a wave.

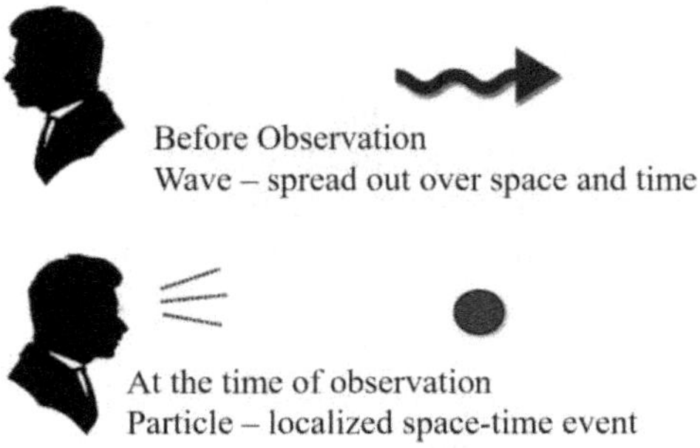

Fig. 26 The "observer effect"? Replace the human with a measurement device.

If, however, you position a detector at one of the two slits to control whether a particle passes through it, then the interference fringes disappear, and the particle appears to behaves like a point moving in space.

---

[44] You might wonder how it's possible to take a measurement without interacting with the object being measured. Explaining this fully in just a few sentences is impossible. For a more detailed analysis, please refer to the first volume of my book on quantum physics (Masi, 2020).

[45] In the popular literature, Einstein is cited as saying, *"God doesn't play dice."* Niels Bohr promptly answered, *"Don't tell God what to do."*

This illustrates the concept of "quantum contextuality."[46] In specific experimental contexts, we observe particle-like behavior, while in others, the same object—such as a photon or an electron—exhibits wave-like behavior even when the initial conditions are identical.

The wave-like or point-like behavior of a system depends on the information available to us. This distinction highlights the fact that quantum theory is a "contextual theory" in contrast to Newtonian classical mechanics, which is "non-contextual." To put it in terms of the famous physicist Eugene Wigner, in the quantum realm, Nature answers according to how we frame a question. In the macroscopic world of classical mechanics, the properties of objects do not vary based on how we observe them. For instance, an object's size, color, or shape remains consistent regardless of the angle from which we view it.

Physicists have demonstrated quantum contextuality through countless laboratory experiments. How should we interpret the inherent uncertainty and contextuality? Is it an effect of a conscious observer?

The key point is that these quantum effects are always present, even when no one is observing the experiment, the particles, the slits, or any part of the quantum system or measuring device. Furthermore, these effects were already in play at the moment of the universe's creation. Quantum effects were the first phenomena that shaped the fate of the universe, long before any physically conscious observer existed.

However, Wigner and the mathematician John von Neumann, a pioneer of quantum formalism, sought to preserve the concept of the "observer effect" by linking it to our mind or consciousness. They argued that the outcome of an experiment holds no meaning unless someone interprets it. Ultimately, a conscious individual must observe every measurement result that a device produces. A measurement lacks significance unless someone is present to interpret it—such as by reading a number on a pointer. Furthermore, since measurements yield fluctuating and context-dependent results that are neither due solely to the measurement interaction nor explained by unknown hidden variables influencing the dynamics of the measured object, we must conclude that it is the observer's mind that determines the particle's behavior. This leads to the idea that the observer's consciousness plays a crucial role in causing the particle to move in one direction at one moment and another direction the next, resulting in the particle acquiring a definite property, such as position, momentum, or spin. This view regards the collapse of the wave function—such as the transition from wave-like behavior to point-like interaction observed in the double-slit experiment—as being triggered at the time of our conscious observation of the measurement device and the experimental setup.

---

[46] And an aspect of 'complementarity:' Here, the wave and the particle aspects complement each other.

This raises an important question: Why do we not observe these peculiar effects in our macroscopic, everyday lives? Where is the boundary between the microscopic domain of particle physics—governed by these quantum phenomena—and our macroscopic experiences involving rocks, living organisms, mountains, airplanes, and stars? Why do measurements reveal superpositions, entanglements, and uncertainties in the microcosm but not in the macrocosm? Why, and at what scale does this transition—that is, wave collapse—occur, and what mechanisms drive this shift?

This issue is known as the "measurement problem." It is a conceptual loophole in quantum theory, which is otherwise a well-established framework. Furthermore, we lack a clear understanding of what constitutes a measurement. As odd as it might seem, there is no definitive explanation of what a measurement physically entails. This problem has not yet achieved a complete resolution and is central to the measurement problem.

Advocates of the idea that "consciousness collapses the wave function" often cite this gap as evidence for the notion that reality depends on our observations and the contexts in which they occur. They suggest that there is a connection between our minds, or consciousness, and the mechanism that "collapses" reality, a phenomenon described as the "observer effect." According to this perspective, nothing truly exists until a conscious entity observes it.

This reflects, in simple terms, Wigner's perspective. He demonstrated this concept through a thought experiment known as "Wigner's friend," [47] which has since become a prominent interpretation of quantum mechanics in popular literature. While the Von Neumann–Wigner interpretation of quantum mechanics finds almost no support among physicists, it became so popular that people tend to believe that the observer effect is an established fact of science. Unfortunately, some physicists perpetuate this myth when addressing popular audiences, even though they would never mention it in front of an academic audience. Please refrain from presenting this concept as a scientific fact; it is wild speculation and always has been.

Nevertheless, the popularity of this idea among supporters of an anti-materialist worldview has good reasons to exist. It stems from its suggestion that consciousness and the mind play a fundamental role, existing as aspects of reality that precede the material realm. This perspective also opens the door to speculation about paranormal phenomena, often referred to as PSI phenomena. If our conscious will can influence the outcomes of non-local events, it could enable instantaneous transmission of information across the universe, thus potentially providing a scientific basis for telepathy and other extrasensory phenomena.

[47] I won't describe this in further detail here. For more information, refer to Masi (2019a).

The implications seem to overreach. For instance, should we really argue that the entire universe came into existence only when humans—or at least some conscious beings in any galaxy—began to observe it? Do I (or you) come into existence only when you (or I) look at me (or you)?

Frankly, in my view, this appears to be a desperate effort to uphold an interpretation of reality for the purpose of maintaining a paradigm that seeks to link quantum physics with human mental processes or consciousness, regardless of the cost.

Recent reports suggest that participants in a double-slit experiment can influence the statistics of the measured outcomes. This finding is claimed to illustrate how the mind affects quantum phenomena—specifically, by collapsing the wave function—and is often cited as evidence for the quantum observer effect. However, even if the experiment's results are valid, they indicate only that the mind might possess telekinetic abilities, not that it directly causes quantum effects. It is also possible that similar phenomena could be observed with non-quantum objects, such as moving a grain of sand or a feather. If such psychic abilities do exist, there is no compelling reason to link them unequivocally to quantum physics, as they could just as plausibly operate within our everyday macroscopic, non-quantum Newtonian realm.

We should always remember that our belief systems, ideological backgrounds, and personal preferences significantly influence our interpretation of scientific facts. For instance, if you identify as a materialist or reductionist who believes only blind and purposeless phenomena—essentially random events—govern the universe, you might favor the interpretation that there is no observer effect at all. In quantum physics, phenomena can arise from sheer chance, described as "random" events. There is no need to introduce concepts such as "observers," "minds," or "consciousness" into a robust quantum theory that does not mention them and that operates within a sound mathematical framework. Any further philosophical extrapolation is simply unnecessary metaphysical speculation.

Conversely, if you find the reduction of mind, life, and consciousness (essentially, ourselves) to mere mechanistic processes to be unconvincing, you might be drawn to more metaphysical interpretations. The von Neumann-Wigner interpretation appeals to those who adopt a form of philosophical idealism, suggesting that observers shape reality and that the universe exists due to our observation.

While I identify as an idealist, I view this perspective as an extreme form of philosophical idealism—almost a solipsistic interpretation of reality that lacks substantial backing in quantum physics. There is no equation, principle, or theorem in quantum physics that references "mind," "consciousness," or even "observer." Instead, quantum mechanics consists of mathematical equations describing the statistical outcomes of unconscious measurement devices that operate independently of observers, minds, or consciousness collapsing anything. This is one of the few instances in which I align with the orthodox naturalistic perspective. The idea that our minds are responsible for collapsing wave functions and thereby determining reality is suspiciously anthropomorphic. Again, we have fallen into the temptation of placing ourselves at the center of the universe.

Much more interesting is the question of whether reality, as we perceive it, is inherently existent or merely an artifact of our cognitive system. We will explore this question in Chapter IX.

Meanwhile, we can explore the origin of the "causeless," seemingly self-determining, and intrinsic quantum randomness. Simply labeling quantum theory as a "theory without hidden variables" does little more than describe the situation. In reality, compelling arguments suggest that quantum physics might be related to consciousness—but not by giving prominence to human observers. To understand this connection, we must adopt a universal and panpsychist perspective. The next section will begin to unpack this idea.

4. Will, Force, and Agency: Introduction to a Panpsychist Perspective

The inanimate material world is generally not considered to be dominated by will, agency, desire, and volition. Physics describes the universe from the elementary particle to galactic clusters, in terms of a play of matter and forces, or fields, in space and time. There is no consciousness or mind that thinks and acts with a purpose, a direction, or an aim. All is a blind mechanistic sequence of causes and effects. There is only space, time (or spacetime), and causality. It might be unpredictable, non-computable, or intrinsically uncertain, but this doesn't imply the existence of any soul or mind, let alone a "God" (whatever this word means), that determines and orders the events leading to the emergence of life. Shortly after the Big Bang, it must have initially appeared as a quite do-nothing universe: Nothing and nobody was willing to do anything. Yet, at least on our planet Earth[48], something pretty much active and driven by will, purpose, and desire emerged: Life.

Most biologists (or, at least, those biologists interested in the more philosophical aspects of life) try to explain the emergence of will and purposeful action in Nature–that is, in organisms that are conscious agents, like us humans–starting from a bottom-up approach. Willful action is explained as the "adaptation of the organism to the environment," the "feedback of a non-linear complex system," the "necessity to maintain homeostasis," or the drive for gene transfer, etc. All these mechanistic processes that, at the fundamental level, are nothing other than biochemical reactions or organic functions and

[48] At the time of this writing more than 6000 planets around other stars have been discovered, many of which are Earth-like. My statement is presumably extremely conservative.

cycles, lead to a behavior that we call the "will to live'–that is, the famous "instinct of survival."

I always wondered where this "instinct of survival" comes from. Why does the agglomeration of instinctless molecules bring into existence an instinctual cell? I don't know about you, but for me, there is no rational or logical connection. The seemingly all-powerful instinct of survival is a fact, but it does not shed any light on why the combination of elementary biochemical processes and functions leads to that subjective feeling of desire, an urge to act, the will to do or avoid something in a living organism.

When confronted with these questions, many will tell you that the emergence of will in Nature is like the wetness of water, which emerges from the interaction of H_2O molecules. We have already discussed, in essay II.3, how questionable this point of view is. Wetness is not an emergent physical property but rather a subjective perception of a conscious agent. The analogy not only doesn't hold but should give us only more reason to doubt the mechanistic narrative.

Moreover, the notion of "will" and "agency" in biology is somewhat different from what we imply by those terms in our everyday parlance. We associate will, desire, agency, or intention with a feeling, a sensation, or an inner psychological urge that is felt, again, by a conscious and sentient agent. Reducing will to a complex or unpredictable self-regulatory, self-referential, self-organizing, "self-whatever" system explains, eventually, how the bodily aspects of a living organism work but not why they should give rise to anything resembling a willful subjective feeling. The nature and emergence of will and agency in Nature remain an utterly unexplained phenomenon no more and no less than consciousness, the universe, or life itself.

If you (instinctively) feel like reaching for the "evolutionary advantage" magic wand, that doesn't answer the question either. With speculations based on evolutionary advantages, one could explain everything and the contrary of it, always. If pigs could fly one could come up with some seemingly plausible explanation in terms of evolutionary advantages as well, and make the impossible possible.

Nonetheless, this psychological aspect of ourselves we call "will" has one thing in common with the most fundamental aspects of physics–namely, the ability or potentiality to act, to perform a work by expressing some force and causing some change. Will is always about the ability to determine, cause, and change something.

Poetically, or at least metaphorically, we could say that physical forces–that is, electromagnetic, gravitational, and atomic forces–act like the expression of a "blind will." By "blind," we mean that seemingly nothing and nobody is knowing and planning anything by expressing that "will." The action and resulting dynamics from these forces can be captured by abstract differential equations that don't presuppose anything living, volitional, or desiring in the sense that we as conscious organisms do.

But what is a force in physics? In physics, a force is defined as the ability to cause a change of motion in time. For example, whatever can impart a change in momentum–or, equivalently, a change in velocity–to a material object is a force. It could be of electric, magnetic, gravitational, or nuclear type, but it is always a force. A static force—such as the weight of a stationary object—does not set anything in motion. However, it can potentially do so once the state of equilibrium is broken. This is true in classical Newtonian mechanics, relativity, and quantum mechanics. One might add the interesting fact that in quantum field theories, the weak interactions responsible for all nuclear decays, always implies a change of one type of particle into another. So, the notion of force in physics is always something about (actually or potentially) setting matter in motion or about a causal power of transformation in time.

And, what is matter? As is well-known from Einstein's mass-energy equivalence, matter is a form of energy. (That's what his famous formula $E=mc^2$ states.) What, then, is energy? It is the ability to perform work, such as lifting a heavy body, etc. And what is work? It is what you have when a force acts upon an object moving it throughout space.

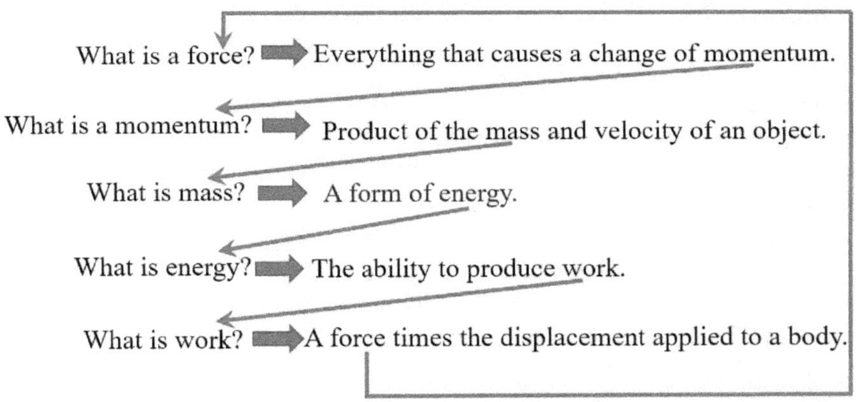

Fig. 27 Self-referentiality of the concept of force in physics.

Thus, matter is a form of energy, and energy, through work, expresses itself as a force in action in space and time.

So, by trying to understand what force is, we can't avoid referring to force itself. The logic of physics is, in this sense, a wheel of self-referencing entities inside a circular logic.

Despite this, it works if you don't bother about the existential questions we are dealing with here. However, it shows how, when it comes to deeper questions, the so-called "exact sciences" don't tell us much about what the world ultimately is and why it does something. Ultimately, they boil everything down to forces, space, and time. Physics won't be able to tell you more than that.

At this point, instead of always asking how all this subjectivity could arise from mechanistic unconscious forces in a spatial and temporal universe, could we, perhaps, for one time, allow ourselves to think the other way around and conjecture that physical forces might not be entirely unconscious? Could the universal causality that all these forces express be, perhaps, not just metaphorically or poetically but literally the expression of a blind will? How would the same problem appear if, instead of positing an unconscious force as the fundamental primitive, we posit a consciousness or subconscious force, will, and agency as the basis of all there is?

After all, there is in every particle, atomic nucleus, atom, and molecule something prone to act or react, bind or dissociate, produce or resist a change and a work, that determines or effectuates something, no more and no less than someone driven by will does. Only a fortuitous analogy and play of words?

The most fascinating definition of will that I found is that of the 20th-century Indian mystic and poet Sri Aurobindo, who defined will as "*consciousness applying itself to a work and result.*" He also called it the "*force of being in conscious action*," pointing out, "*Will does not act, but causes and determines actions.*"

In what sense does this differ from the forces that physics describes?

The only difference is that science expunges consciousness, will, and "beingness" a priori from the world and then wonders how it could emerge in the very same world. It doesn't allow itself for a moment to assume that, perhaps, also cells, let alone particles, might be driven by more or less primitive motives, desires, or cravings, and which could make choices–that is, "proto-consciously" will and wish something. We are then puzzled about how those psychological properties could come into existence by piling up unconscious and unwilling entities, apparently out of nothing.

Modern thinkers resort to neural networks, self-organizing complexity, holistic theoretical frameworks, self-regulating autopoietic systems, etc. but are always left with the same question: Why should all this give rise to a subjective experience that wills, desires, and purposefully acts?

If we don't create that wall of separation in our conceptual constructs between the intentional action and causation by a mental agent and a physical force, things might make more sense. If we postulate consciousness, mind, and life as the fundamental primitives of the universe, then the emergence of intentionality, purposefulness, volition, choice, decision-making, and self-determination at complex levels of organization no longer appears mysterious but, to the contrary, becomes a natural and self-evident outcome.

5. What is Panpsychism? When Words Matter

The advanced reading XIII.4 is a technical paper; therefore, it might be beneficial to summarize the rationale of the entire argument in more accessible language. To introduce the subject gently and clarify the context, I will begin where it ended—specifically, with panpsychism.

Unfortunately, even among professionals, panpsychism is often understood to mean "all is mind." Not only is this an inaccurate translation, but it also fosters a misconception by conflating "mind," "consciousness," "spirit," and "soul" as synonymous terms, thus leading to enormous semantic and categorical confusion. On this shaky foundation, people construct complicated metaphysical theories and then wonder why paradoxes arise and where they come from. Therefore, let's first obtain some etymological and historical clarification.

The original meaning of the word "psychic" comes from the Greek word "psychē" (ψυχή), which translates to "soul" or "spirit." It can also signify "breath" or "breath of life," although the more accurate Greek term for this concept is "pneuma" (πνεύμα), also related to "air." The etymology of "psychē" traces back to the verb "psychō" (ψύχω), meaning "to blow." (In modern Greek, it refers primarily to cooling something.) Some suggest it is onomatopoeic—a word that imitates a natural sound. [49] The connection between the soul or spirit and the act of breathing is evident in many other Indo-European languages. For instance, in Sanskrit, "atman" signifies the "self" or "soul," rooted in "atma," meaning "breath," "principle of life," or "essence," according to the context. This concept is also reflected in modern German, which retains the root "Atem," meaning "breath." Similarly, in Latin, "anima," a cognate of the Greek "anemos" (άνεμος), meaning "wind," also translates to "soul"—a meaning that modern Italian preserves as "anima." The Christian tradition also draws parallels between "breath" and "soul": *"And the Lord God formed man of the dust of the ground and breathed into his nostrils the breath of life; and man became a living soul"* (Genesis 2:7).

Caution is necessary when one is interpreting some contemporary languages that are not Greek. For English speakers, "psyche" can be used in various contexts, often deviating from its original Greek meaning. Since the mid-19th century, the term "psychic" has been used to refer to a person believed to possess occult powers or to act as a medium.[50]

---

[49] This aligns with Sri Aurobindo's understanding of the nature and origin of languages, which suggests that language originated from 'seed-sounds.' For more details, refer to the advanced reading on "The Nature and Origin of Language in Abhinavagupta and Sri Aurobindo."

[50] In preparation for the sections on Sri Aurobindo's cosmology, we will refer to the 'psychic being' as the 'soul' or 'divine spark.' It is unrelated to mediumship or occult powers.

The primary source of confusion in this context is that people use the term "psyche" to refer to the "mind," "intellect," or "intelligence." The Greek word for "mind" is "nous" (νοῦς), which is unrelated to "psychē," meaning the soul or spirit in both ancient and modern Greek.

Unfortunately, the equation of "mind" (or "intelligence," "reason," "logical thought"—essentially, the rational and mental faculty) with "soul" has become pervasive, particularly in Western culture. This stems from the modern conflation of mind and consciousness. In the cosmology that I will outline in detail later, mind and consciousness are related and intertwined entities, but their natures are very different, as I have described in detail in section II.4. It's like equating vapor, liquid water, and ice without any distinction. Even professional academic philosophers almost always translate the term "panpsychism" as "all is mind-like" (with "pan" meaning "all" and "psyche" being mistranslated to "mind"), rather than correctly translating it as "all is soul-endowed" or "all is consciousness."

The Age of Enlightenment, despite triggering the scientific revolution and enabling us to build better material conditions by technological means, has diminished our inherent ability for spiritual discrimination. Thus, it has led to a blending of concepts. As a result, we tend to mix or make only vague distinctions between "soul," "spirit," "consciousness," "emotions," "feelings," "mind," "reason," and other aspects of our inner psychological world.

This isn't a minor linguistic detail; it represents a significant conceptual distortion introduced by Western analytical philosophers due to their conflation of mind and consciousness. Historically, Italian Renaissance philosopher Francesco Patrizi (1529–1597) introduced the term "panpsychism" ("pampsychia") into Western thought in the 16th century. He used the term to emphasize that all things possess a soul or spirit as individual components of a "World-Soul."

Thus, I will use the term "panpsychism" in its original sense. Panpsychism refers to the idea that all is the "World-Soul" or "World-Spirit," animating and permeating the objects and beings of the world. This perspective can be described as semi-animistic, with everything regarded as a form of consciousness but not reducible to mind only.

In the integral cosmology that I will describe later, there exists both a universal mind and a universal life, as well as other planes of existence. These realities are, as in a Matryoshka doll, nested within the ultimate nature of all that exists—a universal spirit or universal consciousness. This ultimate universal nature is distinct from both universal mind and universal life. However, it is not necessary to agree with this viewpoint in order to appreciate the rationale of this chapter.

Panpsychism has various forms,[51] but when people refer to it, they typically mean "micropsychism." This viewpoint asserts that all matter and its constituent particles possess a form of elementary "proto-consciousness." For instance, every elementary particle, such as an electron, is thought to have some degree of basic conscious awareness, primitive perception, or fundamental sensation. The full manifestation of conscious awareness in organisms, including humans, emerges from the combination of these proto-conscious entities into atoms and molecules, ultimately forming complex structures like brains.

Notice how micropsychism implicitly assumes that consciousness is not necessarily fundamental. The distinction between matter and consciousness persists, with consciousness being merely a property of matter. Every microscopic particle possesses a tiny elemental form of minimal consciousness, while the conscious awareness of complex organisms like us emerges from a combination of these proto-conscious particles.

On the other extreme is the concept of the universe as a manifestation of universal consciousness, with consciousness being a fundamental principle from which everything else, including all matter, is derived. This is known in contemporary philosophy of mind as "cosmopsychism."[52] In this view, the entire cosmos represents 'psyche'—a form of undifferentiated and holistic universal consciousness analogous to what the Indian Vedic tradition identifies as the non-dual Brahman or the Absolute.

Therefore, while micropsychism presents a reductionist perspective on the origin and nature of consciousness, positing that individual consciousness arises from the combination of proto-conscious particles through a bottom-up process, cosmopsychism adopts a universal and holistic top-down approach, suggesting that each individual is an individuation or "de-combination" of the universal soul. Despite their differences, both viewpoints agree that consciousness is not merely a neuronal process of the brain. This is why both perspectives have found favor among idealists and non-physicalists.

However, the distinction between micropsychism and cosmopsychism is frequently viewed as an irreconcilable dichotomy. This compels individuals to choose one perspective over the other. It highlights the inherent challenge of the human intellect, which often struggles to move beyond a binary view that demands definitive "yes" or "no" answers.

---

[51] Such as panexperientialism, pancognitivism, constitutive vs. non-constitutive pan/micro-psychism, panprotopsychism, etc. For a good summary, see the entry for "panpsychism" in the Stanford Encyclopedia of Philosophy: https://plato.stanford.edu/entries/panpsychism.

[52] As we will see, cosmopsychism can be extended from a pantheistic conception (in which the universe, or Nature, is the one substance, or God) to a panentheistic idea (in which Nature is contained within this universal consciousness or God, but God encompasses more than Nature).

Instead of viewing cosmopsychism and micropsychism as mutually exclusive, I consider them complementary. If we envision the One becoming the many and acting in and through them, the apparent dichotomy disappears. The consciousness of the singular World-Soul expresses itself in the plurality of the individuals, and the individual's cells, molecules, and particles. The electron isn't an isolated little proto-consciousness randomly zigzagging around, as the conventional micropsychist views it. Rather, it is a self-concentration of the universal consciousness in itself and, thereby, reflects or "funnels through" a somnambulant version of it in an extremely limited materiality. Not only is the physical universe that science describes a manifestation of a universal consciousness, but every event, every phenomenon, and all microscopic processes are expressions of that cosmic consciousness. It is a Consciousness that is both global and local, working as a whole and yet intervening in each part of itself, determining the macrocosm, and, simultaneously, instantly determining and working out all the processes in the microcosm.

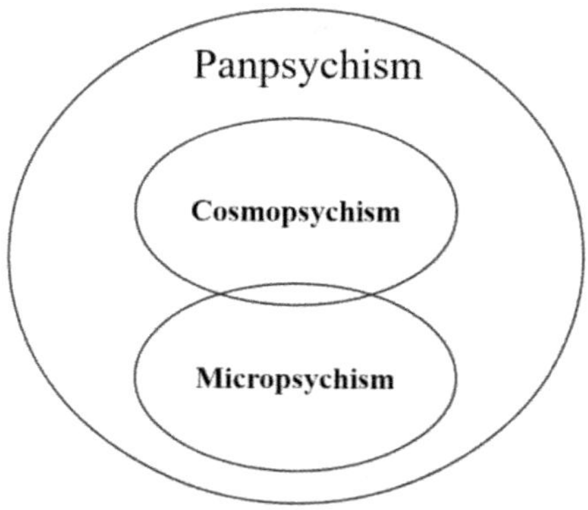

Fig. 28 Panpsychism: "All is Soul." An integral vision from microscopic particles to the cosmos.

An aspect that might be worth anticipating is that this consciousness is not a Mind in itself, but it acts, works, and organizes the cosmos through a universal Mind or a Mind at large in and through every bit of its own creation as well. Here, we get the best view of the relationship between mind and consciousness. Universal consciousness is a transcendent and immutable entity that exists beyond matter, forces, spacetime, and mind. In contrast, the universal Mind is a "delegate" within the spatio-temporal realm, functioning through dynamic self-concentrations. This is mirrored in our individual experience of mental activity, characterized by fleeting thoughts that are sustained by a constant conscious awareness of a witness-identity, as emphasized in II.4.

In summary, the kind of panpsychism adopted here is an integral vision in which micropsychism and cosmopsychism are not mutually exclusive but, rather, are two complementary aspects of the same universal consciousness. Only when we see them as irreconcilable views of reality do paradoxes arise.

This was only a superficial review, as I didn't go into the many other panpsychist theoretical frameworks in the literature. However, for our purposes, the distinction between micropsychism and cosmopsychism, together with that of mind and consciousness, are sufficient.

6. The Problem With Free Will: An Overview

One question that nowadays seems to be in fashion again, is whether we have free will. This is not a new philosophical issue but, with the advent of increasingly intelligent machines and the advancement of neuroscience, it gained renewed attention. Because it isn't clear if our actions are free and self-determined or, to the contrary, we only live in the illusion of possessing free agency but, in reality, are conditioned by subconscious and unconscious physiological processes. Maybe we are just biological robots that believe ourselves to be free but aren't at all. On the other hand, if this is the case, wherefrom then does that undeniable sensation of having the freedom of choice come from?

I was once utterly disinterested in these kinds of questions. Who cares whether we have free will? I was happy with the feeling of being able to make free choices. Whether this feeling was truthful or only an illusion, I didn't need to know. I've always wondered why philosophers are so obsessed with this question. The issue didn't seem particularly interesting to me until I embarked on this project of exploring the connections between science, philosophy, and spirituality in pursuit of a post-material future. It was then that I realized how deeply this question is intertwined with the nature of consciousness. Ironically, the question of free will has become central to my inquiry—not so much because of the topic itself, which may seem like abstract philosophizing to many, but because it reveals how we think and, more importantly, why we believe what we believe about reality and ourselves. After all, our lack of awareness of how we think and why we've adopted certain ideas and worldviews is one of the root causes of human suffering, especially in the era of social media.

The question of whether we are just sophisticated machines guided by purely "physical causal deterministic"[53] brain processes, and what this could imply for moral responsibility, has been discussed among philosophers and scientists for a long time. There are different philosophical currents on the subject (see Fig. 29.)

[53] The word "determinism" can have different significances in different contexts. Here we assume the "physical causal determinism" of classical physics which essentially captures the causally closed Laplacian worldview described in Section IV.1.

The so-called, "compatibilists," believe in a sort of "weak determinism" whereby, despite reality being ruled by deterministic laws, we nevertheless have freedom of action and choice. How physical causal determinism could be reconcilable with the idea of free will is beyond my cognitive abilities. It makes no sense to me; you can't have both. If you start with the assumption that everything is ruled by a strictly deterministic causality, it makes no sense to re-label it "weak" or "soft" determinism. You are positing from the outset and per definition that there can't be any freedom of choice. In my opinion, all the attempts I've seen trying to show the contrary end up in an obscure wordplay that desperately tries to defend what is logically indefensible. Anyway, that might well be my own cognitive bias or weakness, because the fact is that compatibilism has a large basis of supporters also among very smart people.

While the "incompatibilists" will tell you that if our will depends only on brain processes and if "hard determinism" is true, namely, that we live in a purely deterministic and causally closed universe where for every effect there is only one specific cause, and everything is determined mechanically like clockwork, then we can't have free will. Consider the universe as being a gigantic material clockwork, like in an immense billiard table, where the trajectory of the balls is uniquely determined once the initial and boundary conditions are set. Every event is caused by a physical causal determinism where every event of the present was caused by another event in the past, and every event in the future is strictly dependent on the physical state of everything that exists in the present moment. And, if you are a physicalist who believes that there is no distinction between your mind and your brain, then there is no way out: We also, as physically embodied beings, are the result of what we were in the past. Every action we take now is only the result of a past physical state of our brain. While, our present brain state will determine our next action, thought, emotion, etc. Ultimately, we are also a clockwork, nothing more. Therefore, if this physical causal determinism is true, then free will is an illusion. We are just machines (pre-)determined by the laws of physics, chemistry, and the biological processes in our brains.[54] This is the standpoint of the hard determinist.

On the opposite camp stand the "libertarians," like me, who don't believe that determinism is compatible with the idea of free will, not even in principle, and invoke indeterminism as a possible source for our volition. Libertarians

---

[54] Some thinkers like to point out that the mind isn't reducible to brain activity alone but must be considered a synergy between the brain, the rest of the body and their interaction with the environment. There is not only a bottom-up micro-causality that starts from particles, atoms, and molecules up toward the whole organism, but also a top-down causality from the environment onto the organism, brain, cells, etc. Quite so. But this doesn't change anything with regards to hard determinism. It only extends the determinations from the brain to the rest of the universe. Any local deterministic process determined by global deterministic processes remains deterministic. No freedom of choice can be smuggled in because of this.

try to save free will, speculating that the universe might not be so deterministic as the classical theories of physics seem to suggest. In fact, we have seen how in quantum physics, this naive form of classical physical causal determinism has been shattered into pieces. In the quantum world, events can be stochastic, with no apparent cause determining a specific outcome. As we are going to see, this standpoint isn't free from conceptual issues.

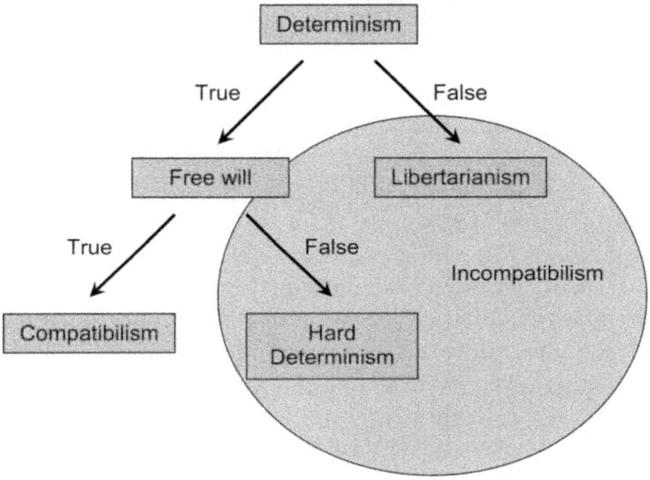

Fig. 29 Different schools of thought on free will.

I'm not going into the subtleties of the different schools of thought here. Instead, I would like to make some general but important remarks that introduce the two final (non-mandatory) advanced readings XIII.3&4. [55]

Some say that free will is, per definition, "the freedom to choose otherwise." Sounds reasonably clear, doesn't it?

But think about it carefully. How do we know if we could have chosen otherwise?

You can't, not even in principle, and you will never know because you can't make two opposite choices at the same instant. Once you make a choice you opt for one unique timeline, it makes no sense to ask whether you could have chosen otherwise–that is, opted for another timeline–unless you have a time machine that beams you back to the time before you made that choice, and then try to make a different one. Thus, I find this definition a sort of oxymoron in disguise.

[55] For more philosophical musings you may also like to read my book "Spirit Calls Nature" (Masi, 2025a).

Others define free will as the freedom to follow one's desires. But what distinguishes "desire" from "will"? Some desires may be so overwhelming that we simply can't resist them (food, sex, smoking, drugs, etc.). What will you do if you win 100 million dollars in the lottery? Say, "No, thank you?" And is there a distinction between desire, instinct, motivation, intentionality, and will? Do desires lead to a will to act or, rather, is there a will to desire first? And what determines that desire? If there is something determining that desire, aren't we slaves to that determining factor leading to that desire? Or, in other words, are we free to desire something rather than something else?[56]

By the way, can there be action without desire? Just asking, because there are people who claim to have attained higher states of consciousness through contemplative practices in which the action becomes more effective, precise, and conscious than in the ordinary intellectual states of awareness, precisely because one acts without desire, motivations, preferences, or plans. If so, the causal relationship that we assume to be so self-evident between will and desires or motivations is a very questionable one. We are using psychological labels, such as "will," "desire," "intention," "agency," "motivation," etc., but without knowing what they really mean and are, other than through some superficial subjective experiences bubbling up from our subconscious meanders, impelling us to do something.

Ultimately, all the concepts of the philosophy of mind are based on these fuzzy waves of our surface personality. If we inquire more carefully, including also a first-person perspective, and especially without jumping so instinctively to conclusions (that we desire?), we realize how the issue of free will is utterly unclear from the outset. It isn't entirely clear what free will is supposed to mean, and to what it relates to. Perhaps we can't go further than that, but at least we should become aware of the limitations of our present state of consciousness and how it obfuscates the mental categories and, thereby, our discussions. That might not necessarily lead us to a final answer, but it would clarify at least what we *don't* know.

One aspect that I strongly feel needs much more attention is whether the answer to the question of free will could be only binary. Whatever the truth is, must the answer be either "yes" or "no"? Might not reality be a bit more complicated and admit another possible alternative in between a purely affirmative or negative reply?

Even among professional philosophers, one finds these questions only rarely addressed. In most cases, the debate is dominated by the pro- and contra-free will factions, and what it entails for moral responsibility. The hard determinists say we have no free will, while compatibilists and libertarians, even though following very different lines of reasoning, claim we have the

[56] I guess this was what Schopenhauer meant when he stated: *"Man can do what he wants, but man can't want what he wants."*

freedom of choice. Why can't there be a more nuanced answer to this perennial question?

You might ask what could it possibly mean that we have only a half, or a quarter, or "three-quarters of free will"? Let me unpack this.

One thing is for sure: No, we can't do whatever we will. I cannot materialize out of nowhere a ton of gold in my cellar. I cannot become immortal simply because my instinct for survival wills me to do so. However strong my will is, I can't fly by flapping my arms. In other words, the environment, my bodily structure and, in general, the physical conditions constrained by the laws of Nature, impose limitations that strongly constrain my freedom of agency.

You might object that this is not what people mean when they frame the question of free will. We are talking about the freedom to choose, say, between buying a chocolate or vanilla ice cream, dressing in one or the other way, marrying mate X instead of mate Y, etc.

But, upon closer inspection, one can see how this objection originates from our too outwardly focused consciousness that looks upon things with an exclusively third-person perspective and an intellectual inability to look inward by a first-person approach. It rests on the more or less unaware assumption that our choices dictated by our mental preferences, while limited by the environment, aren't constrained by bodily and emotional limitations as well.

To clarify, let me start with the easy part.

Suppose you have strong physical pain–say you suffer from a terrible toothache. Would your mind be free to float harmoniously from thought to thought on whatever it wishes without any interference from the toothache? Or would it rather be compelled to focus on how to relieve that pain? The obvious answer makes it clear that our thoughts are not independent of our bodily sensations, feelings, instincts, and appetites.

But let us suppose you are perfectly healthy, and no organic necessities must be addressed. Would your mind be free to think and choose whatever it wills?

The answer is no, again. One has only to realize introspectively how our emotions, passions, and desires condition our thinking. If we honestly look inward, we can clearly see how a constant whirl of more or less strong or subtle emotions, desires, affections, passions, and even social conditioning continuously determines our thought patterns. The mind is not completely free from emotional forces either, and that compels it to return to certain thoughts.

This is to say, it would be naïve to believe that our choices make an exception to this state of affairs. They are not free from mental and emotional influences. Like it or not, our choices are strongly affected by our emotions, our feelings, our desires, most likely by genetic, hormonal and other biological factors and, certainly, plenty of conditioning as well. Inside us, psychological forces determine how and what we think and, consequently, how we act–that is, what we choose.

We shouldn't forget how our mind is highly dependent on more or less subtle forms of cognitive biases that determine what and how we perceive and how we interpret things. We see the world through cognitive filters and intellectual predilections that create interpretations and mental formations influenced by emotional impulses over which we have little control. How much contact do we have with the truth of things in itself, without any deformations, independently of our mental constructions, ideologies, desires, and attachments and whatever kind of preferences filtering and coloring reality? Modern psychology has amassed numerous experiments showing that we translate reality into a fiction. And isn't our mind strongly influenced and conditioned by our past education, memories, experiences, and cultural context? Does our religious affiliation, non-religious or anti-religious identification, and political and ideological background not determine, or at least influence, our choices? Do you believe that commercial advertisements or political propaganda leave your thoughts, emotions, and worldviews unaffected? Do others' opinions have no influence on your behavior? Are your personality and the way you think free from the influence of society? Are you sure that your sound conclusions weren't guided by an unaware bias?

Answering all these questions would probably require a doctoral dissertation, but I hope you get the point. There are good reasons to conclude that these internal psychological conditionings don't leave much space for genuine free will.

If we see things from this perspective, another rarely addressed but decisively missing point becomes clear: When we ask if we have free will, who is supposed to be free, and free from what? Many philosophers will answer that "you" alone make choices, not anyone else. But who is that "you," or my "me," "I," or "ego" that supposedly makes free choices? The mind? My emotions? My body? My brain? Or are we talking about a "soul" or "spirit"? And I don't even begin with opening Pandora's box of the subconscious, unconscious, subliminal, or superconscious (whatever these words might mean). How far are we conditioned by these unaware layers of our personality? This is a whole chapter apart that would only add to what I'm trying to point out here. There exists an enormous, submerged part of ourselves that defines "me," that "I" and "my personality" which is not one single block but an amalgam of very different personalities which ultimately make one choice instead of another.

If it isn't clear who is meant to be free and from what, and what the word "will" itself is supposed to mean, how meaningful is it to ask whether we have free will to begin with? In other words, we don't know ourselves.

Thus, if we train ourselves to take the first-person perspective, we don't need a Ph.D. in neuroscience or psychology to realize the self-evident fact that we have no idea what we are really talking about. Moreover, it is also clear that if we have any free will, it must be limited.

Nonetheless, we might still assume we have a certain amount of self-determination–that is, we might be able to choose freely if we make an inner effort not to be conditioned by our emotional states and mental preconceptions. That's, after all, what the Age of Enlightenment, also called the "Age of Reason"–that is, the period in the 18th century that ignited the scientific revolution–tried to achieve: Let rationality, analytic reason, and crude facts determine dispassionately and "objectively" (whatever that might mean) with the empirical third-person approach of the scientific method what the truth is, without emotional and ideological interferences. This was, and still is, the great ideal. The truth, however, is that this also is something that scientists, who are humans like everyone else, cannot always do either.

Maybe we can become free if we learn to detach ourselves from the physicality of our bodies and transform our passions from enslaving impulses to something that enriches and delights us but without subjugating us. This would also make more sense from an evolutionary perspective. The stone is subjected to purely mechanistic and blind material laws. The animal is only scarcely beyond it and must obey its instincts of survival. Humans are, in principle, able to take some control over the impulses dictated by Nature but can do so only limitedly. Maybe a future, more evolved species will be less determined by outer and inner physical and psychological forces that govern their thoughts, choices, and behaviors.

Whatever the case, the bottom line is that the question of free will can't be answered in binary terms. There are different "degrees of freedom" between the completely constrained and automatic behavior of the machine and a completely self-determining entity. We are neither the former nor the latter.

Failing to realize this has led people to construct philosophical castles on very shaky grounds. Let us go beyond this black-and-white thinking by looking inside ourselves, going beyond an anthropocentric worldview, and honestly admitting what we are: semi-animal forms of life, still, and quite strongly driven by the outer and inner forces of Nature.

7. Free Will and Indeterminism: An Impossible Alliance?

Once we have clarified the subtle but necessary distinctions to be made regarding the questions surrounding free will—specifically, what it cannot be simplified to—and acknowledge that there may not be straightforward answers, let's see how our concepts of causality and (in-)determinism shape our understanding of self-determination and freedom. I will approach this from a more physical perspective rather than that of ethical and moral responsibility.

The question is whether all our thoughts, desires, and actions are predetermined by our brain states, together with the outside factors that we can't control? Or do we need some sort of indeterminism to save the concept of free will?

As we have seen, the point is that, if the universe is ruled by purely deterministic laws, it implies, per definition, that if two identical physical systems are set into exactly the same initial conditions and are subjected to exactly the same environmental conditions, they must behave in the same manner. If this is the case, there is no reason to believe that things would work out differently for brain processes.

On the other hand, compatibilists and libertarians maintain otherwise. While the compatibilists believe a deterministic universe is compatible with the concept of free will, the libertarians reject the possibility of this idea even in principle and maintain that free will is a logically coherent concept only inside a non-deterministic universe. If some form of indeterminism rules the universe with its laws of Nature, our brain may "harness" this indeterministic aspect to become free from an otherwise inescapable determinism. Because the lack of determinism implies that even if two identical physical systems are set into exactly the same initial conditions and are subjected to exactly the same environmental conditions, they can, nevertheless, behave differently.

However, both points of view–that of the compatibilist and that of the libertarian–are plagued by a dilemma summarized in the so-called "standard objections against free will", which goes roughly as follows.

a) The **"determinism objection'**: If all of reality is ruled only by deterministic laws, then all the <u>choices</u> we make must be causally determined by our brain activity and environmental factors that we can't control. Therefore, we can't have free will.

b) The **"randomness objection'**: If, however, the universe is ruled by some indeterministic laws any random or chance occurrences that would break the causal chain of events, would make us behave randomly in an uncontrolled way that couldn't be the result of rational deliberations and intentional actions. All our life would end in uncontrollable chaos, it would be ruled by mere chance and luck. Ergo, self-determination is not possible.

Thus, neither determinism nor indeterminism is compatible with the idea of free will. It must be an illusion.

The standard objections against free will are a subject of ongoing debate in philosophy. You will find plenty of diverging opinions.

Let me briefly unpack my standpoint.

I agree with the determinism objection. As outlined in the previous section, the lack of free will is inescapable if everything is purely deterministic. There is no room, not even in principle, to break a strictly causal chain, allowing for any sort of freedom to creep in.

While the randomness objection seems equally inescapable, one can show that it is not. To see why, we need to shift our viewpoint from a local to universal perspective and that needs to be specified more precisely, as I will do in the coming section.

Meanwhile, let me outline the reasons why the claim that indeterminism could be the backdoor for allowing the brain, or living cells to "harness" (classical or quantum) indeterminism in Nature allowing for free agency, is fundamentally flawed.

The libertarian claim is that it is this underdetermination that leaves freedom for self-determination. It is a point of view that also some biologists and neuroscientists have recently adopted.[57] However, while this is a first step in the right direction, it falls short of completing the circle. This is why.

It still fails, or at least remains vague, in answering the randomness objection. The assumption of randomness being the origin, or "source," or a sort of "fountainhead" for freedom of choice, doesn't explain where that freedom of choice is supposed to come from. Tipping into a random sequence of events and exploiting whatever kind of indeterminacy makes you become a slave to these stochastic events as well. It remains something external to you. Ultimately it is still the external source with its unpredictability that determines your choice. It is not the subject—that is, "you" (whatever that "you" could be)—that makes a choice. There is no agent having agency, but a passive entity that can only wait for the random event to occur and then have no other choice than being subjected to it. "Harnessing" indeterminism does not confer any freedom of choice, it only delegates to the physical stochastic external source the "choice." It is like tossing a die. Try to "harness" its outcomes as a source for your freedom of choice. You will quickly understand that you are determined by something other than you. Not by your own free will but by the capricious "will" of a die.

The idea of harnessing stochasticity as a source of freedom is a standpoint which is still rooted in the notion of the so-called "event causation." It envisages a duality between the agent and the events determining its actions. It is the event that causes you to act, not you yourself. Once we make this separation between the stochastic event and the agent, we posit, again, almost by definition something other than a self-determining entity and fall back into the impossibility of any form of free agency.

So, is there no way out? Must we finally surrender to the fact that we live an illusion to be the agents responsible for their actions? I will take up the randomness objection in a following section showing how it resolves naturally if we adopt a panpsychist perspective.

---

[57] For example, physiologist and biologist D. & R. Noble (Noble, 2018) suggests that organisms may choose by 'harnessing stochasticity.'

In Section IV.1 we made the distinction between three forms of causality: determinism, pseudo-indeterminism, and ontic indeterminism.

Notice that unpredictability does not necessarily mean something is indeterministic. Some argue for an ontic indeterminism based on the second law of thermodynamics, which states that a system's entropy must always increase. This law renders every natural process irreversible, identifying it with a so-called "memoryless Markov chain"—meaning the next state depends only on the current state and not on any of the past states. Thus, it is impossible, even in principle, to compute the system's exact state in a distant past. For similar reasons, one cannot predict the exact system's state in a distant future other than by averaging statistical methods.

However, the notion that we cannot predict a physical event or the exact state of a system at time T_2 based solely on its state at time T_1 does not imply that every process between T_1 and T_2 was indeterministic. A "memoryless" Markov process is a statistical concept indicating that the system's past history has been stochastically "washed out." It does *not* imply that the particles making up the system can randomly shift to any state, regardless of their current physical state.

Others reference Gödel's incompleteness theorem to illustrate the idea that the complexity of living systems is "unprestatable." To predict the state of a system interacting with its environment, one would need to account for its interactions with the environment and, eventually, the entire universe, including oneself and the energy and machinery necessary for making those calculations. This leads to a self-referential logical loop. Consequently, some claim this results in a form of "weak" indeterminism that is loosely associated with ontic indeterminism.

However, the inability to predict an outcome due to natural, mathematical, or logical principles does not imply that the physical processes themselves are ontically indeterministic. Even if the interactions of a system's particles render its future state unpredictable, unprestatable, or Gödel incomplete, they are not less deterministic in the Laplacian sense. What remains is a lack of knowledge—or, more precisely, an impossibility of knowledge—resulting in epistemic indeterminism.

Thus, according to the presently known laws of physics, the only true indeterminism we believe to exist in Nature is in quantum mechanics.

However, we never observe it in our everyday experience (other than, perhaps, in human behavior?). It is an indeterminism proper to the world of elementary particles.

If quantum mechanics is a theory without hidden variables, quantum stochasticity is not considered pseudo-randomness but, rather, an ontic type of indeterminism. How should we interpret this other than saying that it is all "random"?

There are only two ways out.

The first is to admit a sort of "magic" that rules Nature in which things just happen for no reason and physical events manifest out of nothing. For example, QFT admits that even particles can quickly pop in and out of existence, or that empty space isn't really empty but a "quantum foam" of energy fluctuations from nothing–the so-called "zero-point energy'. Yes, that's weird, but the orthodox particle physicists that isn't interested in philosophical questions prefers to stop questioning further this state of affairs. We should limit ourselves to the math: "Shut up and calculate," so the saying goes.[58]

However, there is another possible interpretation that is rarely considered nowadays but was seriously debated by some great minds of the past. The other possibility is to admit how a principle of "self-determination" or "self-causation" is at work in Nature. This is something some philosophers of the past (such as Spinoza, Descartes, Hegel, and others) already considered even without any knowledge about quantum phenomena and called "causa sui.'

The term "random" is a buzzword that lacks meaning when one is addressing more profound philosophical questions. The act of clinging to the notions of "randomness" or "stochasticity" or to a theory without "hidden variables" reveals a subconscious desire to avoid confronting reality: Quantum mechanics suggests a "non-causal paradigm." Some quantum effects can occur without a cause. When asked about this, many physicists reject the idea, often quite vehemently, insisting that "quantum mechanics has no hidden variables, it is random, but not "non-causal'!"

The truth is that if quantum mechanics lacks hidden variables, the implication is that certain aspects of the world are "non-causal"—or perhaps we should call them "self-causal"? Ultimately, this could even become the backdoor for divine determination. Scientists, more or less unconsciously, smell the metaphysical implications and, as a result, often feel compelled to change the conversation.

If you believe that to be a too-illogical or too-weird concept to be acceptable, you are left only with a principle of creation from nothing–that is, something that didn't exist and suddenly appears ex-nihilo. If you prefer that, go for it. But, in my opinion, the idea that in Nature there are things that could self-determine themselves has more explanatory power. This standpoint does not ask for a creation out of nothing. Rather, it allows for a modification of something by itself and in itself, without an external natural necessity imposed by a law or external natural phenomenon leading to the change from the outside. Something that determines itself independently from outside causes might not be the kind of causality we are familiar with in Newtonian

[58] Criticizing this purely pragmatic and 'philosophically blind' attitude may seem easy, but consider that it allowed modern physics to frame one of the most successful theories of all times: The standard model of particle physics.

macroscopic everyday life, but it makes perfect sense as a definition for something having free will.

After all, in our everyday semantics, we associate the notion of freedom pretty much with that of self-determination. If we have a problem with that in physics, it is because we are accustomed to thinking in these self-determining terms only in sociological and political contexts, not in the physical world that we still conceive in a naturalistic manner as a world of marbles kicking around each other according to a push and pull mechanics. However, quantum physics tells us a very different story.

From the perspective of self-determination, or self-causation, or causa sui, the randomness objection acquires a different meaning. Here, in the self-causative and self-determining paradigm, the indeterminacy, or what to us appears ontic randomness, is the expression and manifestation of a will and an agency.

In this view, an individual's will and agency are considered contingent on neither deterministic or pseudo-random processes and even not on indeterministic quantum phenomena. The logical fallacy one makes with the randomness objection is that we imagine our mind, consciousness, or brains somehow tapping into a "reservoir" of indeterministic processes. Then the conclusion that we would be driven by erratic and uncontrollable behaviors is unavoidable. But if we admit the uncertainty of quantum phenomena themselves being the expression of a self-determining will or mind, or eventually a universal mind in all atoms and particles, the paradox resolves.

Self-causation reveals itself to be the opposite of what in our normal parlance randomness suggests. My actions are neither caused nor uncaused but self-caused. I don't do things accidentally, I'm not ruled by blind chance and I'm not a slave to probabilities. Will, desire, planned action, or my preferences are not subjected to a whimsical destiny. They originate in free choices or, we should call them more appropriately, "self-deliberations.'

At any rate, it is time that science transcends its self-imposed classical anthropomorphic understanding of the law of causation and explores other venues. Then, finally, free will and self-determination may find their way back to themselves.

Does quantum randomness circumvent the randomness objection?

It does. But to see why and how it elegantly dissolves it, we must shift our perspective from the individuality of a subject and single-minded agent, like we are, to a universal perspective.

8. The Libertarian Quantum Panpsychism

Let's take a less anthropocentric perspective and shift from the human-centric observer to a universal-centric one. If we take a cosmic perspective and connect this to what we saw in the preceding sections about quantum mechanics, randomness, panpsychism, will, force, agency, and purposeful action we can elegantly reconcile the mechanical and apparently unconscious exterior Nature with a goal-oriented and conscious interior Nature. To understand that strange "non-causal" quantum uncertainty without hidden variables, which determines one event to become actual instead of another, seemingly just by chance and no reason, we need only to question our ingrained principles of mechanistic causality and shift our reference frame from the little human perspective to a universal and teleological point of view. We don't need to pursue weird conjectures about observers collapsing wave functions with their consciousness making the universe. Neither our human mind nor consciousness determines the random quantum outcomes, but a Will, or a working universal Consciousness in things, that, by an "exclusive concentration" in space and time, causes one quantum event to become actual rather than another one.

Without pretending to have a God's eye view of reality, it is plausible to conjecture that all those mysterious quantum fluctuations are "potentialities," "latent possibilities", a "power of choice," that come into being by the works of a "self-determining Mind at large.'– a "super-mind" that works in and throughout the whole universe, in and through the tiniest elementary particle in every point in space and time. It is not a human will but a universal will that applies itself to a local work and result. It is a universal conscious action that causes and determines those actions that in modern scientific parlance is called "physical forces.'

This is something that would be well in line with modern quantum field theories working with the notion of a universal quantum field. Every particle is nothing else than a vibration of a universal particle's field. We, as cognitively limited beings, can interpret all its workings, in no other way than as being "random" and ruled by what we call "chance."

The forces physics describes as localized actions in space and time are the lower rung of the ladder of a universal will and agency. Even every physical interaction between particles in the universe is a sort of conscious action as well. But it isn't a human observer that determines collapses of wave functions, or some observer that meddles with the double-slit experiment. It is a universal consciousness that acts by an exclusive local concentration in itself.

If we find it hard to relate to this standpoint it is because we are like little ants that, from the top of their anthill can't see in the universe other than a random patch of "coincidental" and meaningless events and phenomena. If, instead, we allow ourselves to take an integral view many apparent paradoxes resolve naturally.

One might ask why we should come up with such a metaphysical theory and not leave things as they are in the naturalistic frame: Quantum uncertainty is just what it is, phenomena happen without a reason and a cause, events pop in and out of existence, just out of the blue from nothing.

Personally, I don't find this worldview less weird and intellectually more satisfying. On the contrary, it is a form of "magic" in disguise that pretends to be "scientific" only because it refuses, for ideological reasons, to even consider the hypothesis of divine determination.

At any rate, nothing in science disallows us to entertain such a hypothesis.

To see further how and why this idea is perfectly in line with the known laws of physics, let us see how this could help us in relating a world of apparently blind and mechanical phenomena to a realm where will, cognition, agency and consciousness play a role.

An interesting objection to panpsychism comes from theoretical physicist and prominent YouTube influencer Sabine Hossenfelder. She argues that, according to quantum physics, elementary particles—such as all the electrons in the universe—must be indistinguishable. One of the foundational principles of quantum mechanics is the concept of indistinguishability, which states that it is impossible to differentiate between two particles of the same type, even in principle, because they share identical properties. For instance, two electrons are entirely indistinguishable; they can be thought of as two reflections of the same electron.[59] This indistinguishability is essential for explaining many quantum phenomena.

If that's the case, quantum particles cannot exhibit any form of proto-consciousness, such as intrinsic volition, agency, instinct, or goal-directed behavior. If they are truly identical, no intrinsic properties could lead to differences in behavior. Physical indistinguishability must result in behavioral indistinguishability. Even if particles possess some form of micro-sentience, they would be unable to express it. Therefore, panpsychist theories regarding the conscious or mental properties of particles are inconsequential. An electron, for instance, cannot be conscious.

If 'panpsychism' is defined as the exclusive micropsychist view, which posits that particles possess 'psychic' properties independently of one another, then Hossenfelder's objection holds merit. However, if we consider the 'micropsychic' properties of localized particles as being the result of a localized inflow of a universal spirit, the problem dissipates. From this perspective, particles themselves are not conscious entities; rather, they serve as an "instrument," a "conduit," or a local "channel" for the influx of a larger, more expansive universal consciousness. There is no need for particles to be distinguishable from one another. It is a cosmic volition that governs the behavior of each particle rather than an inherent agency or distinct intrinsic structure. The electron itself is not conscious, but something is conscious

[59] Quantum indistinguishability differs from the classical concept of indistinguishability in that it involves ontic uncertainty rather than epistemic uncertainty. This means that two objects are not indistinguishable due to our lack of knowledge about their minor differences; instead, they are fundamentally undifferentiable from one another.

through it. By harmonizing the local aspects with the global ones in a unified vision, we eliminate any paradox.

We are now in a position to answer the second of the standard objections against free will—that is, the randomness objection we outlined in the previous section.

The only logical alternative to accepting a deterministic and mechanistic worldview—which ultimately reduces volition and freedom to illusions—is to recognize the inherent indeterminacy within Nature as the basis for conscious and free action. Your choice is not contingent upon a random sequence of events; rather, the indeterminacy of Nature itself constitutes a conscious choice through you. This concept can be understood as a universal consciousness, a cosmic mind, or, as Aldous Huxley referred to it, a Mind at large. Through an act of self-concentration, this consciousness self-determines the sequence of processes.

For instance, if an atom decays at time t_1 rather than at time t_2, the conventional interpretation suggests that this is merely a random occurrence. Or, if the spin of a particle appears in either the spin+ or spin- state with equal 50% probability every time it is measured (essentially appearing "randomly"), physicists assert that no underlying reason, hidden cause, such as some unknown microscopical details–that is, so-called "hidden variables'–determine why a particle assumes one state over another upon measurement. This phenomenon is considered "acausal." In quantum physics, events occur 'randomly,' out of the blue and for no reason at all.

Whereas, we could also adopt the panpsychist perspective and conceive of the observed indeterminacy as arising from a volitional act of Nature manifested through the particle. The decay of an atom or the spin of a particle can be seen as the choice of a universal will. No human observer is collapsing wave functions (as proposed by the Von Neumann-Wigner interpretation of quantum mechanics, as we have seen in Chapter V.3.), nor is it any consciousness within the particle (as micropsychists claim). Rather, it is the action of a universal consciousness acting upon itself.

As strange as it might sound, nothing in known physics is preventing us from taking this perspective seriously.

This serves as the first response to the randomness objection: Our brains, minds, consciousness, and wills do not need to rely on indeterminacy or harness whatever (quantum or classical) randomness in order to be free. The intrinsic quantum indeterminacy that we observe throughout the cosmos is itself the result of a free, self-determining volitional act. Quantum indeterminism represents will, choice, and the expression of the universal through the particular—manifested in atoms, molecules, or particles and fields. After all, it is sheer logic: If something encompasses all and is the totality of the quantum processes it contains, it must determine those same quantum processes, as there can be no external cause influencing those processes.

This is my quantum interpretation of the concept of "agent causation'—the idea that agency originates from a being rather than a process. Many philosophers and scientists remain ambiguous about the identity of this "agent" or "being." I propose what I call "libertarian panpsychism," which flips this notion on its head. The "process of becoming" that Alfred North Whitehead described in his "process philosophy' is actually the manifestation of a universal agent that facilitates it. In this framework of "quantum agent causation," the cosmic agent makes a choice, and then the process unfolds, rather than the reverse. This macroscopic being actualizes microscopic processes within itself. It represents a space of pure self-sufficient creative potential. It is its own cause—what theologians refer to as "self-causation" or "causa sui," as discussed in a previous section.

Once we adopt this perspective, we can see why these microscopic processes seem 'random' to us. A deeper intelligence underlies reality, and it is universally active at every point in space and at every moment in time. The hidden variable *is* universal agency.

Notice how the paradigm of a conscious universe, in which each individual is a projection or self-concentration of this universal consciousness, has two important consequences.

Firstly, the primary illusion is the sense of singular selfhood. Our individuality, with its mind and consciousness, is merely a "partition," a "speck," or a "projection" of the universal Consciousness and Mind. There is no "subject" in the conventional sense, only a mental demarcation of the cosmic mind within itself.

Secondly, because we are projections of the One, this universal context must frame all discussions about free will. If free will exists, it does not belong to the illusionary subjects with their distinct personalities and characters; rather, it is the will of the One. In a sense, we might conclude that we are not truly free because there is no one to be freed to begin with. Fundamentally, our will is not our own. Yet, this universal Will cannot be anything other than free, as it encompasses all existence and cannot be subject to anything from which it could be freed.

This might initially seem contrived and overly metaphysical. However, once you become familiar with this universal perspective, it leads to a coherent worldview. Many aspects that once appeared mysterious or resulted in paradoxes and logical inconsistencies become almost natural and self-evident.

By the way, the hypothesis of a conscious universe acting within itself isn't entirely unfamiliar in modern physics. For instance, the American theoretical physicist John Archibald Wheeler introduced the idea of "it from bit" in the late 1980s. This notion posits that every physical phenomenon, or "it," derives its ultimate significance from bits—binary yes-or-no answers. According to Wheeler, everything in the physical world, including particles, fields of force, and spacetime itself, is fundamentally rooted in information theory. He argued that the act of measurement creates reality and that the observer plays a crucial role in shaping the outcome. This perspective challenges the traditional view of physics as a means of discovering an objective reality. Instead, it suggests that the act of observation shapes reality. Ultimately, it implies that the universe comes to know itself through the perceptions of conscious entities, such as ourselves, despite our cognitive limitations.

The "it from bit" metaphor represents a positive step forward, but it still fails to fully encapsulate the universal paradigm. We must also consider the complementary perspective, which is the "bit from IT." Every bit can be viewed as a self-concentration of IT.

One can refine this classical view of quantum physics by incorporating the laws of Einstein's special relativity, thus resulting in the QFT of the standard model of particle physics. This theory begins with a model of reality that posits a universally "fluctuating" quantum field of particles in spacetime at microscopic scales. Every particle is represented by a little "bump" in this field (see also Fig. 23).

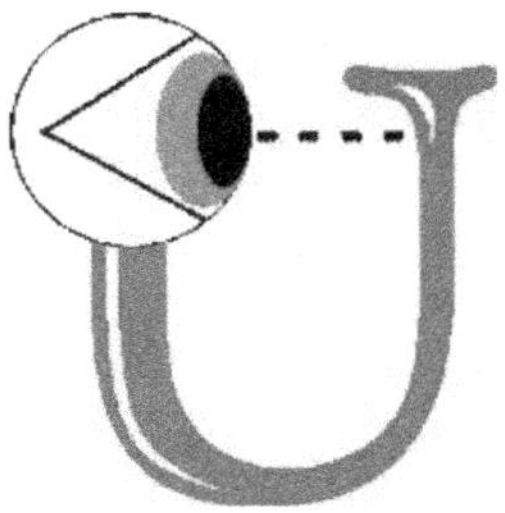

Fig. 30 Graphical representation of Wheeler's "it from bit': The universe becoming aware of itself.

It also visualizes the creation and annihilation of "virtual particles," which are particle-antiparticle pairs that briefly emerge and vanish due to the fluctuations of this quantum field. At the smallest scale, known as the "Planck scale," the vacuum energy density of empty space can become extremely high, leading to significant fluctuations in the curvature of spacetime. Fig. 31 provides an artistic representation of this randomly fluctuating field. While it serves more as a metaphor for what is occurring, it offers an intuitive understanding for the present context.

The libertarian quantum panpsychist perspective considers quantum fluctuations, Heisenberg's indeterminacy, the random manifestations of particle spins, and all the stochastic quantum processes to be expressions of a universal consciousness exerting its will and agency at every point in space and time.

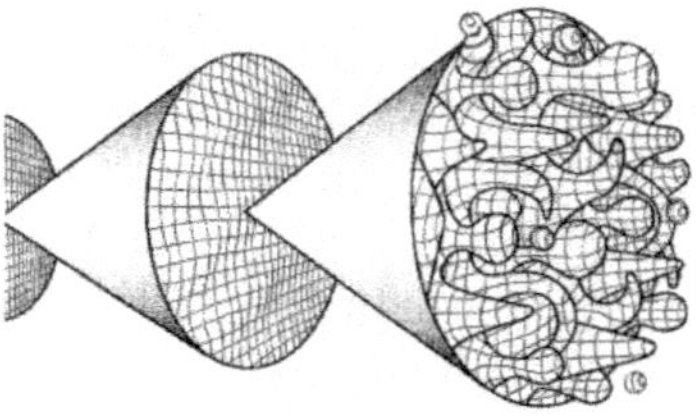

Fig. 31 Artistic renderings of the vacuum quantum fluctuations, or the "quantum foam," permeating all the universe.

9. The Randomness Fallacy

The words "randomness" and "chance" are more buzzwords than clearly terminologically defined terms. In science, we use them to indicate something statistically unpredictable, and we shouldn't assign them more significance than that. There isn't a definition of randomness that relates it to a lack of agency or (free or unfree) will, a lack of motivations, desires, or planned action, or even a lack of goal-directedness. We are now going to see how a phenomenon can appear random and yet be the manifestation of a willed and purposeful agent. There is no logical and semantic connection between randomness and purposefulness or control over something. It is these sorts of unaware conflations that we must first identify and deprogram from our thought patterns that instinctively imply or link one notion with the other when there is none. Once we are able to do so, things appear much more clearly.

In life, there are unpredictable situations we describe as happening "by chance" or "randomly", meaning having no intention or a planning laying behind them. A few examples of events happening by "pure chance" could be winning the lottery, the outcome of a dice toss, coincidentally finding a lost item, meeting someone in an unexpected place at an unplanned time, etc. In a more natural context, one could also consider natural disasters like earthquakes, hurricanes, and tornadoes as "happening by chance" because, at least for our current scientific means, especially over long periods, they remain largely unpredictable events. We don't consider an earthquake morally responsible for killing people because we don't believe there was any conscious and willful agency behind it with a goal-directed intention or premeditation to cause harm. These events just happen and are beyond our control. That's why we speak of "luck" or "bad luck", good or bad "fortune."

Thus, unguided, purposeless, aimless processes can express themselves in the form of a series of events that, for us, are unpredictable and that we label as being "random" or governed by "pure chance.'

The question, however, is: Is the converse true? If a phenomenon is unpredictable and appears even to the best of our statistical analysis to be purely random, or just "white noise,'[60] can we infer that it must be an unguided, aim- and purposeless process?

Fig. 32 White noise image.

Usually, we tend to believe so. However, it is easy to show that this isn't necessarily the case.

For example, think of the ups and downs of the financial market. Nobody can predict the value of company stocks or the exchange rates of the dollar in the coming years. Fig. 33 shows the Dow Jones index between 2017 and 2020.

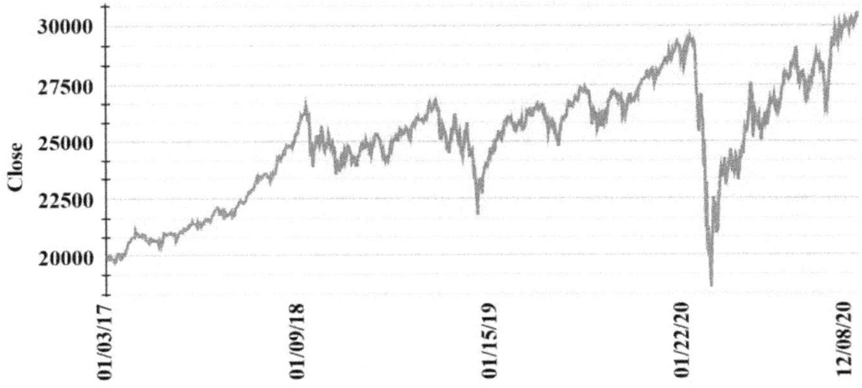

Fig. 33 The Dow Jones Industrial Average between 2017-2020.

One might not perceive it as completely random, as it is not pure white noise, but it looks pretty much unpredictable, as was its crash in the first quarter of 2020. Nevertheless, the stock market isn't governed by blind, mindless, and aimless forces of Nature. On the contrary, it is the sum of a large number of very purposeful and goal-directed agents–that is, lots of profit-oriented stockbrokers.

One might object that this is an example of the clash of many independent intentional sources. It doesn't represent the action of one single volitional agent. Instead, it is like the (apparent) "chaos" of a beehive.

Quite so. But the example shows that what appears to an external observer as "random chaos" doesn't necessarily indicate a lifeless, meaningless, and undirected blind process. It might well be the result of the sum of many goal-directed agents.

Anyway, it isn't difficult to find other examples of random behavior that hides conscious and willed action. Think, for instance, of how the letters you

[60] The older generation might recall the typical white noise image and sound produced by a CRT TV when no channel was available.

are reading are encoded in digital format. At the machine language level, everything is expressed with the two binary symbols "1" and "0." For example, the sentence "Randomness is a myth" is encoded as:

"0101001001100001011011100110010001101111011011010110111001 10010101110011011100110010000001101001011100110010000001100001 0010000001101101011110010111010001101000"

In this string, there are 76 ones and 84 zeros–that is, about 47.5% vs. 52.5% chance each. One can do even better by encrypting the same sentence with the secret encryption key "Randomness is a myth" and obtains: "uMaLt3sS8/qlPgoy8Bl3ENQVVW9CjwgOHY4w3tZYYuI=" which, when converted into binary digits, gives:

"011101010100110101011000010100110001110100001100110111001101 010011001110000010111101110001011011000101000000110011101101111 01111001001110000100001001101100001100110100010101001110010100 010101011001010110010101110011100101000011011010100111011 10110 0111010011110100100001011001001101000111011100110011011 1010001 01101001011001010110010111010101001001001111101".

In this string, there are 177 ones and 175 zeros–that is, about 50.28% vs. 49.72% chance each. The ones and zeros appear with almost the same frequency–that is, the sequence is almost perfectly random.

This demonstrates that even a completely random and unpredictable sequence can represent a meaningful sentence created by a conscious agent. Without knowledge of the code, language, or decryption password, one could easily mistake it for a meaningless sequence produced by a random generator. The randomness or appearance of white noise, as shown in Fig. 32, tells us nothing whether there is any meaning, purpose, or intention behind it.

We will consider another example that will be useful later.

Let's say one rolls a die six times and obtains the following sequence:

"6 3 5 1 2 5"

This is a sequence of outcomes that doesn't look particularly meaningful unless one agrees to posit that it has a particular significance. People would say that it is just a random outcome of numbers. In fact, the probability of obtaining this sequence with the first six throws is (1/6)*(1/6)*(1/6)*(1/6)*(1/6)*(1/6)=1/6^6=1/46,656. That is, this ordered sequence has the probability to show up, on average, only one time each 46,656*6 =279,936 throws.

Then, let us assume that, after a huge number of throws, we obtain:

...... 3 1 4 1 5 9

where the dots represent the preceding throws of the dice. This might or might not be a meaningful sequence of numbers, depending on your knowledge. Those with a little math background will recognize the first six digits of π. For those who do not, it still looks "random." However, what is the probability that this sequence shows up? There is no reason to believe it could be different: Again, only one over 46,656 times throwing the die six times.

Finally, let us assume that, after another huge number of throws, we get:

...... 1 2 3 4 5 6

Of course, nobody would consider this to be a random sequence. Yet, the probability of getting precisely this sequence is no different than the probability of getting the previous ones. If you roll the die a sufficient number of times, you will, sooner or later, end up with that sequence as well.

Therefore, one thing we must become aware of is that the notion of "randomness" is in the eyes of the beholder. If a sequence doesn't make sense to you, you will call it random. If it makes sense or represents something particular, like a known number such as π, or a sequence of letters that form a known word, you will no longer call it "random." Meanwhile, someone else who doesn't know the value of π or who speaks another language will see only a random occurrence of meaningless events.

Many more examples could be provided that highlight how any inference from statistical notions, such as "chance" or "randomness" to the supposed existence or inexistence of a volitional agent, or a meaning and purpose are unwarranted extrapolations. What words like "random", "chance", or "coincidence" really signify is only our ignorance. They point out that the causal principles governing it are much too complicated for us to follow in their chain of causes and effects and the best we can do is that of resorting to statistical analysis to make probabilistic predictions, but they tell us nothing about supposed unguided phenomena and cannot serve as evidence for the absence of a final cause, intention, or purpose in Nature.

One might, again, object that, after all, this is self-evident. No news here. And yet, we always tend to associate the concept of random chance and unpredictability with something indicating a lack of agency or purposefulness. We instinctively conflate the lack of predictability with the lack of a predetermining will, agency, and intention. This way of seeing and perceiving the world is deeply rooted in our mindset.

Also in science, when it tackles more philosophical questions, such as the hypothesis that evolution could be a guided process that eventually conceives of divine determination and intervention in the unfolding of evolutionary history, one of the most frequent objections is that this would contradict our scientific knowledge of evolutionary biology, which tells us how evolution is largely determined by random mutations, coincidental environmental factors, and other unguided phenomena from the ground up. As shown above, this is not at all a coherent argument. Arguments based on unpredictability that jump to conclusions regarding final causes are logically incoherent. Random arguments about randomness will not eliminate divine determination in Nature. Let's call this the "randomness fallacy.'

This was just an appetizer. Let's revisit this topic in more detail and explore another aspect of the randomness fallacy that goes even deeper.

10. Free Will, the Laws of Physics and Combinatorial Causality

One aspect that requires deeper understanding is the notion of randomness in relation to conscious agency. Here, another fundamental question should be addressed. The key point is that the laws of quantum mechanics impose strict and precise constraints on quantum phenomena. Rigorous mathematical laws constrain the probabilities governing quantum events. The wave function is not arbitrary; rather, it is shaped by the system's initial and boundary conditions. A system of differential equations and additional calculations[61] strictly determines the probability law. While individual events are unpredictable, one will, after a large number of events, observe that, overall, they conform to a defined probability distribution. For example, the interference fringes observed in the double-slit experiment (see Fig. 18 right or Fig. 36) do not come about randomly but are fixed and determined and can be predicted with precise mathematical tools. Nonetheless, when and where each dot will appear on those interference fringes is completely indetermined.

Nobody expected Nature to be so subtle. Quantum mechanics is a nuanced theory that intricately weaves together elements of indeterminism and determinism.

Therefore, here comes a powerful objection: If quantum indeterminacy affects the entire cosmos at all times, and if the so-called "zero-point field," which, in QFT, is the "vacuum filed" or "quantum foam" that permeates all of space governing all interactions in the universe, then this cosmic mind in action must also adhere to its own self-imposed laws of quantum mechanics. If a universal mind were to manipulate particles, such as their spin, we would expect a change in the inherent 50-50 chance rule. It would be like creating a biased coin that, when flipped multiple times, statistically favors a specific outcome (e.g., obtaining more heads than tails), thus enabling you to win repeatedly. In theory, you could easily beat a casino. However, due to the clear violation of the 50-50 probability law for heads versus tails (recall Section V.1 and Fig. 22), it wouldn't be long before the casino owners detected the fraud.

This is the quantum statistical perspective on the randomness objection. We cannot invoke quantum indeterminacy to support conjectures of free will, as that universal Will is itself constrained by the probabilistic laws of quantum mechanics. If it were entirely free, it would fundamentally alter the laws of Nature. In other words, either our will—more precisely, the will of the individuations of the universal Will—is bound by self-imposed natural laws,

[61] The Schrödinger equation and the Born rule.

which means it cannot be considered free because it must adhere to these probabilistic laws, or our will is indeed free, in which case we would expect to observe violations of those natural laws.

A century of measurements and observations demonstrating the success and effectiveness of quantum theory has produced no evidence to suggest the latter case. It appears that while God might play dice, those dice can't be of any use and serve no purpose.

This led the 20th-century Austrian physicist Erwin Schrödinger to ask,[62] *"Would physical indeterminacy give free will a chance?"* His answer was: No. According to Schrödinger, it is impossible because *"the direct stepping in of free will to fill the gap of quantum indeterminacy does amount to an interference with the laws of Nature."* (Schrödinger, 1951)

The aim is to demonstrate that this conclusion is incorrect. It is based on a common fallacy: We often overlook the fact that probabilities are derived from time series. The temporal aspect of a stochastic event is frequently treated as an irrelevant detail, leading us to erroneous conclusions. To reveal what this means, I will offer a few examples.

In sections IV.2 and V.9, we learned about the importance of distinguishing between epistemic and ontic indeterminism. A common example of epistemic indeterminism is the unpredictability of dice toss outcomes, which stems solely from our ignorance of the various forces acting on the dice. If we knew the initial physical conditions, the material structure of the die, and the precise direction and magnitude of all the forces involved, we could, at least in principle, predict the outcomes—much like Laplace's demon.

Conversely, in sections V.1-3, we observed that Heisenberg's uncertainty principle is not merely epistemic but ontic. Quantum randomness is not a result of our ignorance, imprecise knowledge of the conditions affecting a quantum system, inaccurate measurements, or intrusive observations. Instead, the unpredictability associated with Heisenberg's uncertainty principle reflects a form of inherent, cause-less randomness—what we called ontic randomness. This type of randomness would still exist even if we had complete knowledge of the system we are observing. Thus, we can meaningfully refer to this as 'real' or "true" randomness.

While the distinction between epistemic and ontic randomness might not be crucial for calculations, it is highly significant for the philosophical implications surrounding the nature of agency and final causes. The clockwork universe deterministic machinery of the former leaves no space for free agency, whereas the latter allows for a cosmology where forces are viewed as expressions of true free agency in action. The shift from a Laplacian deterministic view of classical mechanics to a universe governed by inherent

[62] As far as I know, this objection was originally expressed for the first time by Erwin Schrödinger.

indeterminacy has been, and for many still is, a difficult reality to accept. This struggle led Einstein to famously declare, "God doesn't play dice." To illustrate how Einstein himself fell into a more nuanced aspect of the randomness fallacy, consider the following situation.

Let God roll a die. Suppose that God's rolls adhere to the probability law requiring each number to show face-up with an average frequency of one in every six tosses. This means each outcome has a probability of $p = 1/6$, or approximately 16.66%. It seems like a fair die, doesn't it? However, as we know, God is quite clever and could easily deceive us, as follows. Nothing prevents Him from selecting the outcomes in any order He likes.

For example, consider the following series of outcomes:

"5 1 3 6 2 4."

Fig. 34 Does God play dice?

Each of the six numbers appears with the same frequency (once in six throws), indicating randomness.[63]

A different permutation[64], such as

"6 4 3 2 1 5,"

does not alter the frequency of each number. The probability of each outcome holds true regardless of the order. The act of shuffling the outcomes in a time series doesn't affect their relative frequencies. Thus, the probability law stays constant, with $p = 1/6$ for each outcome. The same principle applies to any other probability law. This means that though one might be constrained by statistical laws, one can still retain the freedom to determine the temporal order of the outcomes.

---

[63] Of course, many more throws would be needed to ensure they appear with a similar frequency, but let's keep it simple.

[64] Let me repeat this, since this is a common objection. You might object that it is very unlikely that throwing a die six times it would spit out the first six numbers "123456." You might suspect that it isn't random at all. However, this coincidence is no more and no less probable than immediately obtaining the sequence "5 1 3 6 2 4" or "6 4 3 2 1 5," etc. Wait and see. If the sequence "1 2 3 4 5 6" doesn't repeat more than $1/6^6=1/46656$ times on average, you can safely say that it was a real coincidence.

What practical purpose could this serve? How do we transition from stochasticity to purposeful agency?

Consider a scenario in which, if the outcome is the number “5” at time T, a switch activates a light; if not, the light remains off. One could freely turn on the light by rolling a “5” precisely at time T, should one choose to do so. If the frequency of outcomes remains consistent over an extended period, no one would notice any difference, and the time series of events would still appear entirely random or in line with the prescribed probability law.

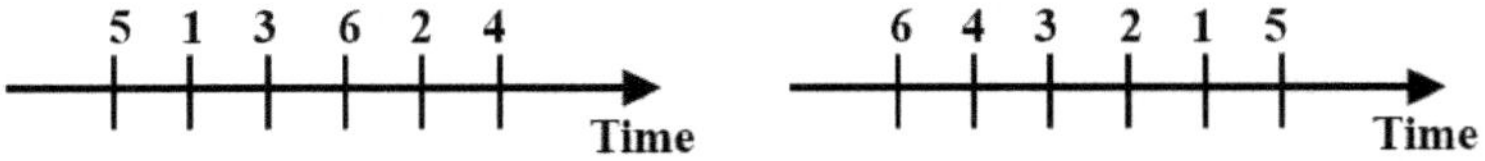

Fig. 35 Two time series of a die toss outcomes, with two different orderings in time, but obeying the same probability law (each outcome has probability 1/6=16.6% to occur.).

Although each toss is governed by a probability law (here, the likelihood of any specific outcome must be 1/6), there is still a degree of freedom of choice–in terms of *how to order* the events—that is, of *when* a specific event is manifested. This ability to determine when a particular outcome manifests also confers an agential power.

The mechanisms behind God's proclamation of "Let there be light" are undoubtedly more complicated than that. However, this metaphor illustrates that randomness does not equate to being 'enslaved by probabilities.' This highlights another facet of the randomness fallacy. We might be bound by natural statistical laws, but we retain the freedom to choose the permutation–that is, the sequence of events.

It is essential to remember that, in statistics, the temporal order of a time series—or, more generally, the arrangement, configuration, or combination of events—is irrelevant when assigning a probability law. A conscious choice that recombines the order of events does not change their frequency of occurrence and, therefore, does not alter the probability function. However, choosing one combination of events over another may confer an agency that alters the history of actualized facts. This is what I refer to as “combinatorial agency,” that relies on a “combinatorial causality.”

Notice how the number of permutations increases dramatically with the number of dice. The number of permutations of n distinct outcomes is represented by n! (n! is pronounced “n factorial”), which is calculated as $1 \cdot 2 \cdot 3 \cdot ... \cdot (n-1) \cdot n$. For example, with six throws, there are 6! = 720 possible permutations—that is, different time series. With ten throws, there are 10! = 3,628,800 permutations. For twenty throws, the number of permutations is 2.43×10^{16}. And so on.

I argue that this type of freedom is the fundamental basis of free will. Schrödinger's objection has been refuted. Physical indeterminacy not only gives free will a chance without interfering with the laws of Nature, but it also creates a virtually infinite array of possibilities.

This also solves the randomness objection. If a cosmic consciousness is not constrained by random events but, rather, serves as the source of the stochastic phenomena that Heisenberg's uncertainty principle describes, then despite being governed by probabilistic laws, it possesses significant freedom in determining how these events are actualized over time—that is, it possess a combinatorial agency. To external observers, however, its workings will always appear entirely random and devoid of meaning. Regardless of how advanced your statistical analysis is, it will never reveal such volitional acts. This distinction is crucial: Agency is maintained without leading to random behavior, and with no violation of natural or statistical laws.

Therefore, I agree with Einstein, but for very different reasons than those he had in mind. In fact, God doesn't play dice. God *is* the dice and even tosses Himself.

11. Quantum Free Will: A thought Experiment

Let's connect the rationale from the previous sections to a concrete example that demonstrates how microscopic quantum events that are considered ontically random can scale up to macroscopic agency. To illustrate this within the realm of quantum physics, I will reference the classic macroscopic quantum device: Young's double-slit experiment. As outlined in section IV.1, this experiment involves shining light through two closely spaced slits onto a screen, resulting in an interference pattern of bright and dark fringes. This pattern indicates that light, specifically the particles known as 'photons,' behaves like a wave, with the waves from each slit overlapping and interfering on the screen. When particles such as electrons are substituted for photons, they too produce an interference pattern, even when fired one at a time. This phenomenon reveals the dual wave-particle nature of matter, suggesting that what effectively passes through both slits is a 'probability wave.'

Fig. 36 illustrates the interference fringes alongside the corresponding probability function of the distribution profile. (I will clarify the meaning of the arrows shortly.)

The fringes indicate the areas where a particle is most likely to be detected on the screen, which is typically a CCD detector or a traditional photographic plate. The highest likelihood of detecting the photon or electron occurs at the peaks of the probability distribution, represented by the maxima of the 'wobbling curve.' The center fringe, marked with the two up/down arrows, has a significantly higher probability of particle detection, while the two secondary fringes on the left and right, indicated by the left/right arrows, have a lesser probability. In contrast, there is almost zero likelihood of detecting a particle in the dark areas between these fringes.

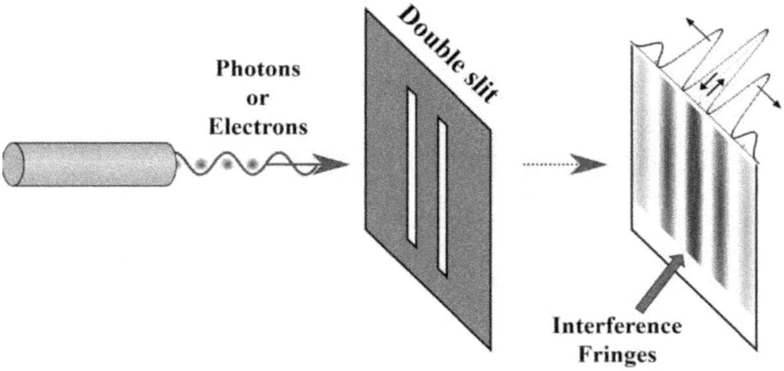

Fig. 36 The double-slit experiment.

The key point is that the double-slit experiment highlights the probabilistic nature of quantum mechanics. The probability law—determining the number of fringes, their intensity (the amplitude f the fringes), and their positions on the screen—is uniquely defined by the physical properties of the apparatus and the particles, in accordance with the laws of quantum mechanics. However, the order in which particles arrive on the screen is unpredictable. Which particle will hit which fringe and at what time is a completely random process. Each particle randomly strikes one fringe or another, representing a "caprice of Nature" that cannot be altered.

For instance, consider that the time series of particles "fired" through the double-slit reaches the screen as follows: The first particle hits the right center fringe, the second hits the left center fringe, the third hits the right second fringe, and the fourth hits the left second fringe. We can represent this with the corresponding sequence of arrows, as shown in Fig. 37.

The detection of particles at specific locations on the screen occurs in a *particular temporal order*, moving from left to right. We will refer to these as the "arrow-up," "arrow-down," "arrow-right," and "arrow-left" vectors, which correspond to particles displaced across the central and two lateral interference fringes on the detector screen.

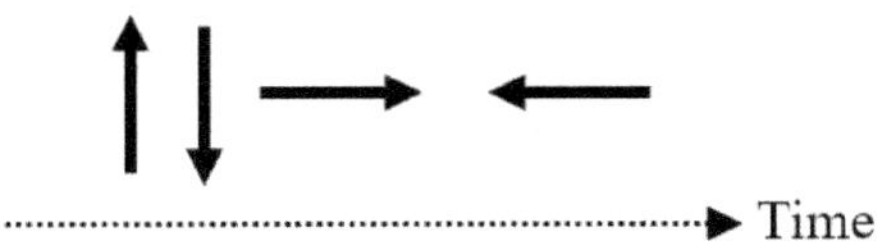

Fig. 37 Sequence A: Four events form the double-slit according to Fig. 36.

However, similar to the time series of dice outcomes, if Nature had "chosen" a different temporal arrangement, i.e., permuting the order of particle arrivals on the screen, it would not alter the profile of the fringes. In other words, the quantum probability law remains consistently the same.

It could just as easily have resulted in:

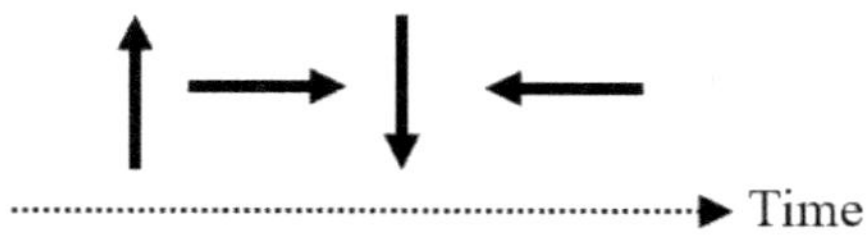

Fig. 38 First permutation of sequence A of Fig. 37

or also

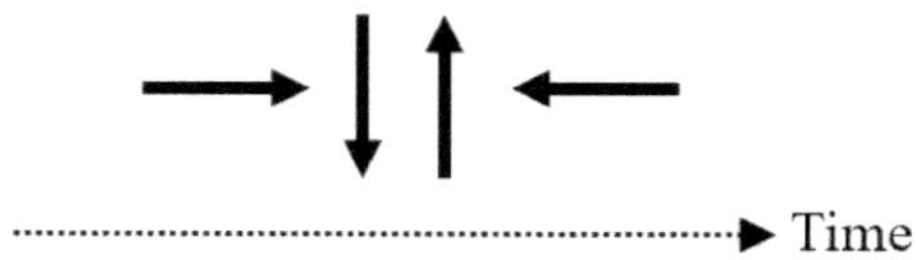

Fig. 39 Second possible permutation of sequence A of Fig. 37.

and so on.

In other words, on the one hand, the probability laws deterministically imposed by Nature must be respected: Particles must distribute themselves according to the likelihood represented by the shape of the mathematical probability function (the accumulated particles resulting in the fringe-amplitudes in Fig. 36). On the other hand, Nature retains complete freedom in determining the timing and order of their arrival of each of the particles, meaning it can choose the order in which they appear. Nature is still allowed to "shuffle the whole deck of cards" whenever and however it likes, without modifying the probability function. And nobody can notice, not even in principle! This is what I call "combinatorial agency" or "combinatorial causality."

The question arises: How can we use the freedom to alter the order of events by means of such combinatorial causality in a meaningful and creative way?

The answer lies not with human agents or individual conscious beings but, rather, with the principle of quantum uncertainty. This principle is inherent to the laws of Nature, and it is beyond our control. However, if we adopt an agent-causation perspective, with Nature itself as the agent, we find that nothing in the known laws of physics prevents it from exercising freedom of choice in determining the order of events. This very freedom opens up a vast array of options. To illustrate this concept, let's consider a thought experiment using a 'toy model.'

Imagine connecting the double-slit device to a PC. Each time the detector measures an 'arrow up' particle, it will move a black dot on the computer monitor one step upward. Conversely, whenever an arrow-down, arrow-right, or arrow-left particle is detected, the system will move the dot downward, rightward, or leftward, respectively.

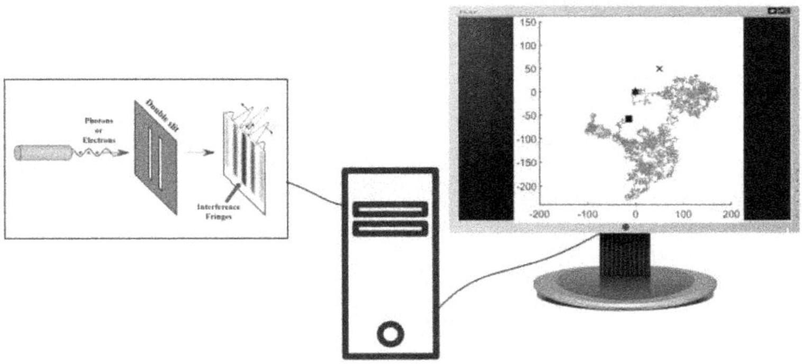

Fig. 40 Thought experiment connecting the double-slit outcomes to a path generating computer.

We utilize a stochastic quantum phenomenon—the stream of particles from the double-slit—as a source for generating paths.

This process can be simulated on a computer. Writing a program to generate a sequence of probabilistic events based on the probability distribution of the double-slit experiment using a random number generator is relatively straightforward.

The result is as follows: the (virtual) dynamics of the sequence of events is zoomed-in in Fig. 41 depicting what is happening.

The star symbol marks the starting position at coordinates (0,0), while the cross symbol represents the target: the aim or goal, such as a "food source" for our tiny, virtually artificial-life entity. Once the data stream is extracted from the double-slit device, the dot moves according to the measured directions indicated by the arrows. If we allow this simulation to run for a while (specifically, 20,000 steps), the result is a random walk within an area of approximately 400x400 pixels, culminating at the black square.

As you can see, there appears to be no clear direction; the "walk" is chaotic—and that's the essence of quantum randomness. Achieving the cross symbol is challenging if you are at the mercy of random events governed by strict probability laws. You must be exceptionally fortunate to obtain the correct sequence from the double-slit experiment, which might randomly guide you toward the target by chance. Alternatively, you might need to wait a long time and expend significant energy to produce enough photons for Nature to coincidentally provide the right sequence to lead you there. This is why "harnessing stochasticity" does not support the concept of free will.

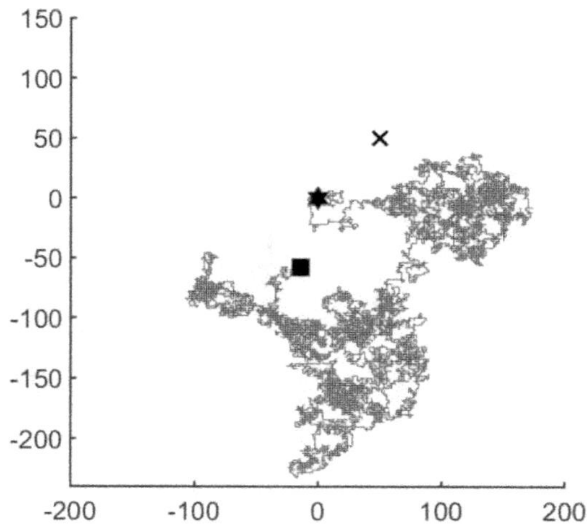

Fig. 41 Random walk starting from the star and ending in the squared position. Target x missed!

On the other hand, if you possessed a causal superpower capable of creating a time series of quantum events that enabled you to travel in a more direct manner, ultimately tracing a straight line from the star to the cross symbol, you would generate a sequence of events that disregards the probability distribution of interference fringes. Consequently, this would violate the laws of quantum physics, which was Schrödinger's key argument.

However, if we allow Nature to act, viewing the series of quantum events as intentional and volitional actions (an agent-causation perspective), the situation changes. Naturally, Nature must adhere to its own laws and cannot produce whatever kind of phenomena at will. It is required to generate quantum events that align with the probability distribution and, after a significant number of particles strike the screen, to create the interference pattern.

The useful causal power—the combinatorial agency—is the ability to change the order of events which doesn't change the probability distribution. The random walk described above represents just one possible path corresponding to a specific temporal arrangement of the series of 'arrows.' By considering the entire time series (the arrows)—which, in the context of a computer simulation, refers to the time series that the random number generator provides—and carefully permuting their order, one can still achieve the desired outcome. There is no need to change the elements themselves; one simply "reshuffles the deck" of the very same "cards" (the arrows are the same, but their order is changed.) The underlying probability law remains unchanged, meaning the interference fringes stay the same. However, this reshuffling alters the path of the random walk.

For instance, Fig. 42 illustrates one such path among countless possibilities.[65]

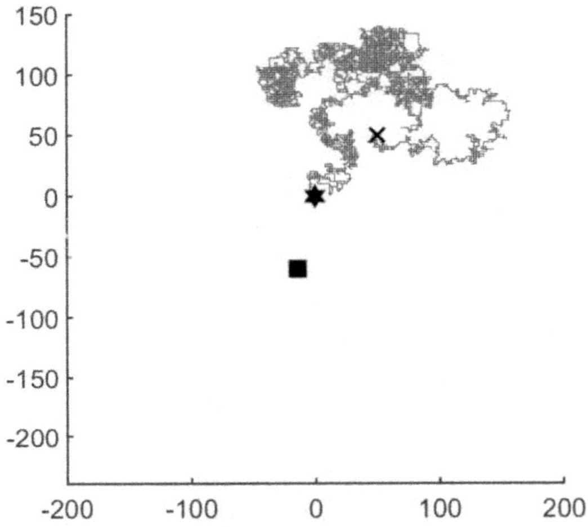

Fig. 42 Random walk with the same events of Fig. 41. Target hit!

This specific sequence of events achieves the desired outcome. It hits the targets after a relatively short sequence of steps. The random path remains statistically 'random,' as it still represents a series of events that correspond to the same probability distribution of interference fringes on the double-slit detection screen. If one observes only the final distribution of photons (or electrons) on the screen, without considering their arrival times, there would be no noticeable difference compared to the earlier random walk. Without knowledge of the arrival times, the final dataset of events appears indistinguishable from the previous one. Yet, the intentional act of rearranging the sequence enables an agent to follow a goal-oriented path.

In summary, the objection regarding randomness does not hold up under examination. In fact, suggesting that quantum randomness is the foundation of free will is a valid hypothesis that aligns with scientific understanding. The libertarian perspective remains consistent, particularly when integrated into a panpsychist framework.

Another frequent objection is that quantum indeterminacy and various microscopic quantum effects are too small to have a significant impact on mesoscopic scales, such as that of the human brain, let alone on the macroscopic scales of planets, stars, and galaxies. Furthermore, quantum noise is much smaller than many classical random effects, such as the noise induced

[65] The sequence in Fig. 42 stops once the target has been hit, which is why it shows fewer steps than Fig. 41.

by the random Brownian motion of particles due to thermal effects. Even if an agent could utilize this kind of quantum combinatorial agency, it would quickly be “drowned” or “wiped out” by the much more intense noise of classical deterministic stochastic processes. These are legitimate and seemingly powerful objections. However, Chapter VII will demonstrate that they rely on oversimplifications and provide compelling reasons to conclude that, on the contrary, mesoscopic and macroscopic objects are also governed by quantum causality with an open future.

12. Quantum Free Will: A Matter of Credibility?

Finally, let us take up the central theme: What is the relationship between quantum physics and concepts like consciousness, free will, or even the idea of a universal consciousness or divine determination? Most physicists would assert that there is no connection, dismissing these ideas as pseudo-scientific nonsense. According to the orthodox naturalistic interpretation, the peculiar quantum “fluctuations” and strange quantum indeterminacy—often imagined as bubbling spacetime foam—lack metaphysical implications. From the perspective of theoretical physics, particularly the “shut up and calculate” approach, quantum phenomena are simply what they are, and we shouldn't delve deeper than our ability to express them mathematically, as outlined in the standard Copenhagen interpretation favored by Niels Bohr. Naturalists view these metaphysical inquiries as absurd and magical.

However, modern science pays a steep price for its reluctance to explore the nature and ontology of things. In avoiding deeper inquiries, it replaces one perceived absurdity with another equally absurd belief system. Concepts such as particles popping in and out of existence, the notion that quantum mechanics is a “theory without hidden variables” implying that everything is fundamentally random without an underlying cause suggest a worldview that is just as metaphysical—and arguably even more magical—than those it seeks to dismiss. They label any non-naturalistic speculation as miraculous thinking or accuse you of invoking "God of the gaps." Yet, they propose many-worlds interpretations that ask us to believe we exist in countless parallel worlds, suggesting that a strange copy-and-paste operation occurs every time a measurement is made. To what extent do these speculations rely less on miraculous thinking? Alternatively, they contend that everything was already encoded in the billion-degree-hot plasma of the Big Bang billions of years ago and subsequently unfolded in a super-deterministic manner. How much less "godly" is that? Are these incredible naturalistic interpretations more credible?

Despite all the pushback, several serious physicists, including some Nobel laureates, are intriguingly drawn to the idea of a connection between the ontology of quantum physics and the nature of the mind. In the last century, foundational figures in quantum physics, such as Arthur Eddington, Arthur Compton, and Pascual Jordan, speculated that microscopic quantum stochastic phenomena could be amplified to a macroscopic scale, influencing brain processes. Roger Penrose developed a slightly different model, yet he, too, believes that consciousness and brain activity could be linked to quantum mechanics.

In a broader context, Heisenberg suggested that inherent quantum indeterminism—the characteristic stochastic aspect of quantum mechanics that influences all phenomena in the universe—can be viewed as the dynamic possibilities of Nature. This concept resembles Aristotle's idea of "potentia," which describes the process of transforming potentiality into actuality.

Why, when discussing the nature and origin of consciousness and mental processes, do some feel compelled to reference quantum physics?

Although they did not explicitly advocate for non-material interpretations and maintained their intellectual discourse within a naturalistic framework, I believe they intuitively recognized that the inherent indeterminacy of the quantum world provides an answer similar to the one I have discussed here. What is a "potential" that transforms potentiality into actuality if not a form of "idea"? Due to its universality, this "Idea" can only be that of the Universal Mind. Indeterminism does not "generate" free will; rather, it represents the expression of a free and creative universal spirit within all things, and that I posit acts by a combinatorial agency. The hidden variable in quantum mechanics is the agency of the Spirit.

However, it is important to note that in the model I presented, this does not suggest that the quantum level of reality is the fundamental ontological primitive. While quantum indeterminacy is ontic, the ultimate ontological primitive is a transcendent consciousness that exists beyond spatio-temporal relations. This idea aligns more closely with Whitehead's "process philosophy" and "process theology." From this perspective, everything is process-oriented, characterized by a continuous act of creative transformation. It presents a teleological view in which agency, activity, action, and motion define the nature of the universe.

Nature might be trying to convey an important message. It encourages us to view reality from multiple perspectives—macroscopic and microscopic, universal and particular, global and local—simultaneously. Often, we don't need new data or groundbreaking discoveries; simply shifting our perspective can transform our worldview. No scientific, logical, or philosophical barriers are preventing us from adopting this viewpoint.

Embracing a dual vision that recognizes both the macro and the micro, the whole and the part, as two aspects of the same dynamic and eternally creative entity allows us to align our perspective with modern physics. This holistic libertarian panpsychist view reveals that the micro- and cosmopsychist natures of reality are two Yin-Yang aspects of universal consciousness. This perspective naturally resolves the hard problem of consciousness and the question of free will and conveys a deeper significance to quantum uncertainty.

When we consider the actions of an agent (whether a human, another living being, or even particles, molecules, or atoms), we must ask: Are these actions a reflection of free choice? The answer is that free will exists, but it is not our own. Our will is intertwined with the will of universal consciousness, which acts through us. However, we are limited beings, constrained by our physical bodies, emotions, and subconscious minds.

This marks the initial step toward a deeper understanding of reality, consciousness, and free will. Later, we will expand on this into a comprehensive cosmology.

Two articles in the Advanced Readings section XIII.3&4, titled "**Quantum Indeterminism, Free Will, and Self-Causation**" and "**Quantum indeterminacy and Libertarian Panpsychism**" (Masi, 2023b; 2024) reiterate the arguments presented in this chapter, particularly those from the last two sections on free will, from a more technically rigorous perspective. The first article illustrates the metaphor describing the relationship between seemingly random agent causation and letter frequency in a text. The second scholarly article also employs mathematical formalism and encourages greater engagement from readers willing to explore the subject. It applies the same principle through a thought experiment based on the double-slit experiment further extends the physical scenario from a local human perspective to that of a universal agent–that is, to a panpsychist view.

VI. The Interaction Problem and Mental Causation: A Physicist's Perspective

1. The Interaction Problem

One of the core assumptions of a Cartesian worldview is that the mind and its mental states cannot be reduced to mere epiphenomena of the brain. René Descartes famously proposed that the mind is a non-material, unphysical substance distinct from the body—a view known in modern philosophy of mind as "substance dualism." Cartesian dualism posits the existence of two types of substances: an extended, non-thinking substance ("res extensa"), which refers to matter, and a thinking, non-extended substance ("res cogitans"), which refers to the mind. Descartes identified the pineal gland as the seat of the immaterial soul.[66]

However, this simplistic form of dualism has lost credibility today, as we now understand that the pineal gland is an endocrine gland responsible for producing the hormone melatonin, and there is no evidence to suggest it has any connection to non-material trans-cerebral activities.

Nevertheless, we have seen how the opposing physicalist doctrine, which claims that all our thoughts, feelings, and perceptions can be reduced to mere biochemical activity in the brain, remains a highly controversial topic among philosophers, scientists, and theologians. While more sophisticated models than Cartesian dualism have been proposed, the idea that the emergence of consciousness and the mind in evolutionary history cannot be attributed solely to material processes in the brain continues to be debated.

However, dualists who argue that mind and matter are fundamentally distinct "substances," with the mind being composed of something "non-physical," implicitly propose a form of "interactionist dualism" that faces the so-called "interaction problem." This problem raises the question of how an immaterial and non-physical mind can causally interact with a physical entity, such as a brain made up of material cells, molecules, and atoms. If the mind consists of an unknown, distinct, and independent non-physical substance, it is hard to understand how it could interact with material objects like the physical neurons in our brains. How could the mental state of an immaterial mind exert any causal influence on a material body, including its brain, nerves, and cells? In this view, matter and mind would be unable to affect one another.

The issue dates back centuries. Already in 1643, Princess Elisabeth of Bohemia presented a pointed challenge to Descartes regarding his

[66] In this chapter, I will use the term 'mind' in a general sense. However, the same issue applies to any entity considered unphysical, such as 'soul,' 'spirit,' or 'vital energy.'

"Meditations on First Philosophy": *"Given that the soul of a human being is only a thinking substance, how can it affect the bodily spirits, in order to bring about voluntary actions?"*

In its modern formulation, according to the Stanford Encyclopedia of Philosophy, "interactionism" or "interactionist dualism" is *"the view that mind and body–or mental events and physical events–causally influence each other."* The objection to interactionists *"is that, in so far mental properties, states or substances are of radically different kinds from each other, they lack that communality necessary for interaction."*

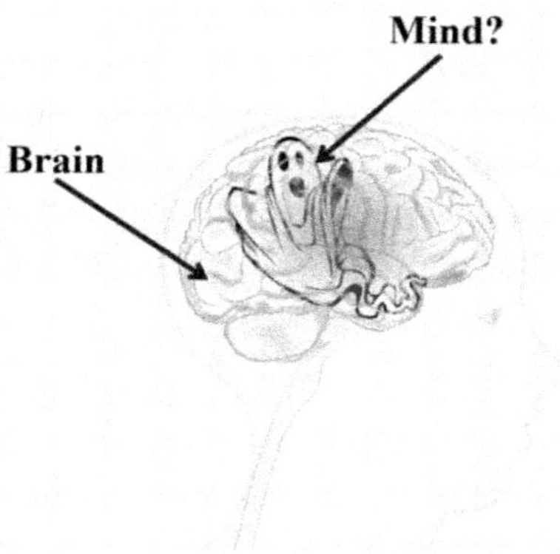

Fig. 43 How can a (ghostly) mind interact with a physical brain?

Furthermore, if an unphysical entity were able to interact with physical particles, it would contradict the established principles of energy conservation. Because, if an unphysical mind were to set a physical particle in motion, that particle would gain momentum and a small amount of kinetic energy. This energy would not originate from anything within the physical universe, thereby violating the fundamental principle of physics that energy cannot be created or destroyed, only transformed.

Notice also how one could come up with the same objection to speculations about divine action: If God, or a divine cosmic Mind assumed to be the essence of something transcendent and immaterial, were to interfere with the particles in the universe, altering the trajectory of a deterministic evolution, the result could be physical phenomena that violate principles of energy conservation. This would imply that the physical world is not "causally closed," meaning not all physical events can be explained by other physical events.

Thus, if the mind or some universal mind, as a non-physical entity, interacts with the body and Nature, it would challenge the causal closure of the physical realm by introducing non-physical causes into physical processes.

Because nothing like this has ever been observed in any experiment, and because the principle of energy conservation is one of the most robust and empirically verified concepts in all of science, the physicalist concludes that the notion of an immaterial mind, divine action, or any kind of "unphysical" field can't interact with the brain and body, if it exists at all. Anything alike interacting with the brain, or any other form of matter in the universe, is not only logically incoherent but also contrary to the laws of physics.

This idea sounds plausible and aligns with our current scientific understanding, leading some to view it as a conclusive argument against substance dualism.

I will demonstrate that this isn't a real issue by presenting a solution based on the model discussed in previous sections. However, before we delve into that, some conceptual clarifications are necessary. I believe that taking a moment to explain these concepts can enhance our understanding of the topic by establishing clear boundaries and contexts that are often overlooked.

2. What Does "Immaterial" Mean?

First, we must clarify the terminology. What does it mean for something to be "immaterial"? If we define as "material" the hard stuff against which we can bounce our heads (that is, anything having mass, such as particles, atoms, and molecules constituting the physical world), then physics also addresses "immaterial" things. For instance, electric fields, magnetic fields, or, more broadly, electromagnetic fields like light, are considered physical entities that possess no mass. Light is described as a massless wave or particle traveling at about 300,000 km/sec through empty space, and it can interact with material bodies. This phenomenon is common in our daily lives. For example, when light strikes a material object, it can cause that object to warm up. On a microscopic level, "immaterial" photons can scatter or be absorbed by material particles such as electrons or atoms (transferring their momentum and energy) and set these particles in motion.

And what about gravity? How does the Earth through an action at a distance interact with the Moon through empty space? The gravitational field between the Earth and the Moon is a force field that extends throughout empty space without any medium, unlike sound or water waves, which require a material medium such as air or water. The kinetic energy exchanged between two objects, like the Earth and the Moon, occurs through the gravitational field without necessitating direct contact and anything in between. A gravitational field is massless; it is an immaterial yet physical entity that affects material bodies. In general relativity, this concept gains deeper significance. The gravitational field is represented by a curved spacetime manifold, and the curvature of this manifold describes the strength of the gravitational field.

Why do material objects cause an immaterial spacetime to bend? There isn't an answer other than to pragmatically state the fact that an immaterial curved spacetime interacts with a material object. What we know is that matter curves spacetime and then moves along the shortest path in that curved spacetime—the "geodesic"—as defined by Einstein's field equations of general relativity. As Wheeler summarized, *"Matter tells spacetime how to curve, and curved spacetime tells matter how to move."* While this might seem like self-referential logic that doesn't clarify the relationship between matter and spacetime (or what they fundamentally are), it does not deter physicists, who are often uninterested in these philosophical nuances, to accept this state of affairs.

Fig. 44 "Matter tells spacetime how to curve, and curved spacetime tells matter how to move."

This prompted Isaac Newton to famously declare, *"Hypotheses non fingo," ("I don't frame hypotheses")* when he was questioned by someone overly philosophical about how gravity interacts through empty space.

While, in his private letters, he admitted:

"It is inconceivable that inanimate brute matter should, without the mediation of something else which is not material, operate upon and affect other matter without mutual contact, as it must be, if gravitation in the sense of Epicurus, be essential and inherent in it. And this is one reason why I desired you would not ascribe innate gravity to me. That gravity should be innate, inherent, and essential to matter, so that one body may act upon another at a distance through a vacuum, without the mediation of anything else, by and through which their action and force may be conveyed from one to another, is to me so great an absurdity that I believe no man who has in philosophical matters a competent faculty of thinking can ever fall into it. Gravity must be caused by an agent acting constantly according to certain laws; but whether this agent be material or immaterial, I have left open to the consideration of my readers." – (Cohen, 1978)

Yet, for centuries, nearly all scientists with a "competent faculty of thinking" have taken pride in accepting this "great absurdity" without further questioning it. They often dismiss the issue of action at a distance as a minor philosophical detail that rarely comes up in conversation.

The same phenomenon presents itself at the microscopic scale in the realm of nuclear forces. The stability of matter relies on the strong nuclear force, which binds quarks through so-called "asymptotic confinement"[67] to form protons and neutrons in atomic nuclei. This interaction is mediated by massless gluons, the gauge bosons responsible for the strong nuclear force.

Inside a proton, what we imagine as a simple structure of three quarks (two up, one down) is actually a "boiling" and dynamic environment often described as a "quantum foam." This term refers to the constantly fluctuating sea of virtual particles—quark-antiquark pairs and gluons—that spontaneously appear and vanish due to quantum uncertainty. These fluctuations are governed by quantum chromodynamics (QCD), the theory of the strong force. The gluons, not only couple with the three quarks but also interact with each other, creating a dense, non-linear web of energy. Virtual quark-antiquark pairs (called the "quark sea") pop in and out of existence, momentarily influencing the proton's mass, spin, and charge distribution. This complex quantum foam contributes the majority of the proton's mass—not from the quarks themselves, but from the dynamic energy of the gluon field and virtual particles, which are themselves massless (a manifestation of energy-mass equivalence via Einstein's famous equation $E=mc^2$.) In other words, your brain and body is mostly made of immaterial stuff, it's weight results prevalently from these interactions mediated by massless particles.

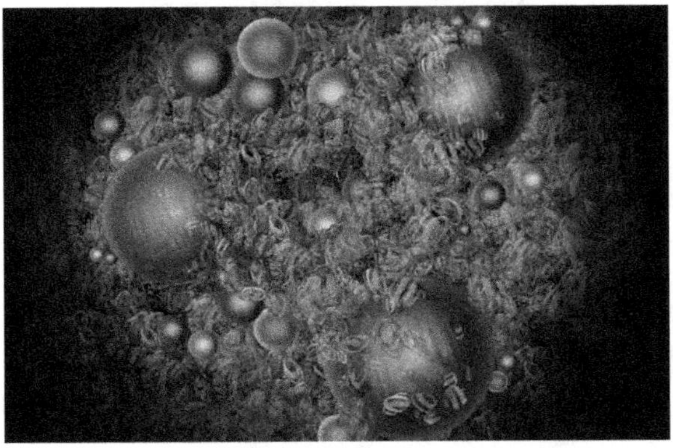

Fig. 45 An artist's impression of a proton – the large red spheres are up quarks and the large blue sphere a down quark - CERN

Therefore, the interaction between two distinct "substances"—one composed of material particles and the other of immaterial ones—is a completely ordinary state of affairs that is rarely questioned. No modern

[67] Asymptotic confinement in QFT is the peculiar phenomenon whereby the interaction strength of the strong nuclear force grows—instead of decreasing—without bound at long distances. Thereby, quarks cannot exist as asymptotic free states.

physicist considers this to be a problem that could undermine well-established theories such as electromagnetism, gravitation, and particle physics.

Consequently, we must acknowledge that the question of how the "immaterial" can interact with the "material" remains unresolved within the realm of physics itself. Physics does not provide answers to these inquiries; instead, it accepts them as facts, formalizes them in mathematical language, and verifies whether they align with experimental observations.

Those questioning how two so different "substances" like a mind can interact with a material brain might not realize that this issue is present in all physical sciences. It is not a dilemma that pertains solely to dualism. Yet, we instinctively raise this question only when it presents itself in the mind-body problem.

The interaction problem stems from the common assumption that the distinctiveness and independence of substances render causal influence impossible. For a physical interaction to occur between two substances, there must be some shared properties or states. Without this commonality, reciprocal causality would be unfeasible, leading to the conclusion that one of the substances is "unphysical." However, physics demonstrates that this is not true. Electrons and photons, though very different, can interact with one another. The real question is not whether these two distinct physical entities can interact but how they manage to do so.

Why do we feel so compelled to raise this problem for the mind-body interaction and even present it as the final nail in the coffin of dualism? I suspect it stems from a preconceived notion about speculations involving entities beyond the reach of naturalism. When enigmas or paradoxes arise within a materialistic and mechanistic framework, we tend to overlook or downplay their significance, instead adopting a pragmatic and instrumentalist perspective. In contrast, when inconsistencies emerge in contexts that challenge naturalism, we not only accept these contradictions as valid but often regard them as decisive arguments against alternative worldviews, rarely probing deeper.

Anyway, I won't use this as an excuse to avoid the issue. Let us take a step further.

3. What Does "Unphysical" Mean?

One might argue that we are discussing something that transcends the physical realm. Interaction dualism not only claims the existence of an immaterial mind but asserts that the mind is an entirely "unphysical" substance.

The question that challenges substance dualism is: How can a supposedly unphysical entity like the mind interact with a physical entity like the brain?

Therefore, the first important question to address is: What constitutes the boundary between "physical" and "unphysical"? What does it mean for something to be "unphysical" and to exist beyond the laws of physics?

Conventional physics textbooks lack a clear distinction between the physical and the non-physical, other than by reminding us that the term "physics"–and, thereby, "physical"–originates from Aristotle's treatise on Nature, titled "The Natural Things." It comes from the Latin "physica" ("study of Nature"), which itself is borrowed from the Greek phusikḗ (φυσική) the "Natural science", a term derived from phúsis (φύσις), meaning the "origin," "Nature," or "property".

This, however, doesn't help us much in making the distinction with its opposite significance.

An instrumentalist might argue that anything observable or measurable with a device or experiment is "physical," while everything else is "unphysical." However, this perspective reduces "physicality" to a mere historical concept. Were atoms, particles, the invisible electromagnetic radiation beyond the human eye's spectrum, or galaxies and neutrinos considered "unphysical" until we developed microscopes, particle accelerators, and telescopes that rendered them "physical"?

From this perspective, nothing can ever be truly "unphysical" because it is only a matter of time before our technologies can detect it. Paradoxically, physicalists might have a valid point: Ultimately, everything is physical at its core.

While this might be a pragmatic viewpoint, it does not help clarify the questions we are exploring here. This semantic uncertainty underscores the notion that the conceptual categories we use are not as straightforward as we might think. We often operate under assumptions, implying that certain words have clear and defined meanings—which, upon closer examination, they do not.

Descartes made a somewhat clearer distinction regarding the term "unphysical." From his perspective, "unphysical" refers to something that has no extension, meaning it does not occupy space or possess a spatial location. He famously asserted that the mind—the res cogitans—is an example of such an unphysical entity.

In fact, all physical theories are fundamentally based on the concept of space. A physical theory devoid of spatial relations would fail to accurately describe the spatial-temporal reality we inhabit. Therefore, we can begin with Descartes' ideas and enhance them with contemporary physical concepts. In both quantum mechanics and general relativity, the temporal dimension plays a crucial role. Modern scientific understanding suggests that the framework of quantum field theory (QFT) cannot describe something labeled as "unphysical." Anything "unphysical" exists beyond the realms of space, time, and the fundamental laws that govern all physics as defined by QFT.[68] The central point, which we will explore in greater detail in the next section, is that QFT serves as the foundation for all interactions in the physical universe. Currently, QFT is the most comprehensive scientific theory explaining what interactions are and how they function. [69]

It should be noted that when we label something as "unphysical," placing it outside any framework of interactions as described by QFT, we are already suggesting *by definition* that it has no causal relationship with matter. Positing the mind as an "unphysical substance" means that it exists beyond a world of physical interactions. By doing so, we inherently assert that no causal connection exists between the mental and the physical from the outset! Therefore, it is nonsensical to ask how they could be causally linked and ask for a solution to the interaction problem—a problem we have created ourselves and transformed into a loaded question rooted in a category error that results in a self-referential logical fallacy.

Let us pursue this line of reasoning nonetheless while also adopting an experiential perspective to explore its implications. We will provisionally define anything that exists independently of spacetime, matter, forces, motion, energy, or any classical or quantum fields—and that the laws of physics do not govern—as "transcendent" or simply "unphysical." In short, anything beyond spacetime and causality can be described as "unphysical."

The question arises: Can we assume from the beginning that every mental event—be it an experience, a sensation, a thought, or a feeling—is an "unphysical" phenomenon? My thoughts, emotions, feelings, and sensations of pain or pleasure cannot be measured or observed in the physical world around us. In this sense, they appear to be "unphysical." However, do they fit the definition we are considering?

To explore the subtle distinction between the physical and the non-physical, we must adopt a phenomenological approach. While a definition from the third-person perspective of physics is already vague enough, it becomes even

---

[68] For the technically advanced reader: It can be shown that fundamentally all laws of physics derive from symmetry principles shaping the structure of quantum fields—that is, the principle of covariance, global and local gauge invariances, and spontaneously broken symmetries.

[69] Some contemporary efforts to unify quantum physics with general relativity into a theory of quantum gravity, which is supposed to be the "theory of everything"—that is, the ultimate theory of physical reality—are raising questions about the nature of space and time. Rather than being fundamental components of reality, they might be emergent properties arising from a deeper layer of existence that transcends spatial and temporal dimensions. If a theory of quantum gravity were to be successfully confirmed, one that views spacetime not as fundamental but as an emergent property, would the criteria used to differentiate between 'physical' and 'unphysical' substances still hold?

more ambiguous when we reflect on our experiences from an inward viewpoint.

Consider your thoughts. These are mutable mental events that change over time and are perceived within you, even though they cannot be pinpointed in a physical space. While you might not be able to separate your mind from your body,[70] it is not difficult to understand, at least on some level, what it is like to think without being influenced by physical sensations. For example, close your eyes, relax, and try to ignore any physical sensations. You can still perceive a mind generating thoughts, such as concepts of beauty, justice, or the number π, without relating them to anything embodied. Thoughts are inherently mutable and evolve over time; we experience them as a sequence of mental events. This means that while thoughts have no fundamental spatial aspects (that is, they do not require a spatial dimension), they nevertheless exist within a temporal framework. Moreover, our understanding of the causal relationships between past, present, and future thoughts remains vague. We might perceive our mental activity as being independent of space but not of time, leaving us uncertain about the nature of the causality that governs it. And, as we experience transitions from waking to sleeping states of consciousness, we recognize that our sense of time can vary significantly based on our state of awareness. Our subjective perception of time can differ greatly in waking, dreaming, and non-dreaming sleep states. Additionally, psychedelic drugs can dramatically alter our experience of time's flow.

Conversely, the mind can envision spatial relationships and contemplate immutable forms or geometric patterns that remain constant over time. For instance, one can visualize a triangle, a square, or a sphere within a three-dimensional space. Plato might have been correct in his speculation about a realm of perfect and static Forms existing in a timeless hyperuranion with spatial extension but without time. However, we can also imagine objects moving through space, separate from our sensory perceptions.

In this cognitive domain, the spatial, temporal, and causal dimensions become less clear-cut and distinct. While the mind still holds onto aspects of spacetime and causality associated with the physical world, we can perceive ourselves as disembodied thinking entities free from spatial conceptualizations in a reality characterized by a different flow of time and alternative causal relationships.

Is, therefore, the mind physical or unphysical? It appears to be something in between. Is it then something half- or semi-physical or semi-unphysical? If we define "unphysical" as something that exists beyond spacetime and causality, the dualist's claim that the mind is unphysical loses its self-evident

[70] One could entertain the controversial topic if and how people's minds function during so-called 'out-of-body experiences' and similar phenomena, but I will keep the conversation within the realm of ordinary human experience to make the rationale more accessible to everyone.

status from the perspective of an inner exploration. Or should we reconsider our definition of what "unphysical" means?

Emotions and feelings can also be understood in a similar experiential way. However, it can be more challenging to separate emotions from bodily sensations. For instance, we often feel love in the heart, while we feel hate and anger in the belly. Additionally, we associate happiness with a sense of physical elation. Despite this, I can envision myself as an undefined "bodiless ethereal blob" that experiences emotions and thoughts over time, without necessarily being confined to a body with a specific spatial presence.

The feelings and emotions associated with desires possess a unique quality that is deeply intertwined with will, force, and agency, as discussed extensively in Chapter V.4. Generally, the act of willing something is related to a desire-like psychological dimension. However, as we will explore in Chapter X.2, willing and desiring are distinct psychological forces that operate on different levels of consciousness.

This raises the question: What does it mean to will something? Can we understand will as consciousness applying itself to a work and result, even though it does not act but, rather, causes and determines actions? Can we intuitively grasp how will relates to action in a manner similar to potentiality relating to actuality or possibility to manifest reality? Will possesses causal power without necessitating a process within actual spatial or temporal dimensions. Unlike the mind, which expresses itself in time and conceptualizes space, will seems to operate outside these categories while embodying the essence of what we call causality. Will can potentially determine events if it chooses to do so. If will is a facet of consciousness, our fundamental existence might not be entirely rooted in something unphysical, particularly if we consider "unphysical" to mean beyond spacetime causality. While will might not require spacetime, it still appears to be connected to a principle of causality. Is it, therefore, unphysical, or does it retain some physical aspect? What is its true nature?

What if we further stripped ourselves of thoughts, emotions, and the will to act and think? This is much more challenging, and likely only those who can achieve advanced meditative states can provide concrete insights into what it is like to exist in a state of pure consciousness—pure awareness devoid of thoughts, perceptions, and emotions. But can there truly be pure awareness without any content to be aware of? Doesn't consciousness require some object to be conscious of? What seems like a logical impossibility to the mind might reveal itself as a natural and undeniable truth once experienced. Being pure awareness without an object of awareness is analogous to hearing without sound or seeing without light. The existence of consciousness does not necessarily depend on an object of experience. Perhaps it is consciousness rather than the mind that exists without one being conscious of anything. This could represent the ultimate unphysical transcendent "substance", standing in contrast to the physicality of spacetime and causality.

Thus, once we embark on an inner exploration, the idea that the mind is an unphysical entity, as Descartes and modern dualists posited, becomes less clear. What once seemed self-evident—that the mind is distinct from bodily existence, with spacetime and causality serving as clear boundaries between the physical and the unphysical—now raises more questions than answers. The distinction we once believed to be sharp and definitive is revealed to be more ambiguous.

This might seem like an exercise in hair-splitting philosophy, and in some ways, it is. However, one cannot overlook these questions when addressing the interaction problem of mental causation and attempting to provide a reasonable solution. The interaction problem begins with the often-unexamined assumption that a clear distinction exists between the unphysical and the physical, categorizing the mind as unphysical. It then seeks answers framed in physical and scientific terms. This fallacy results in misguided extrapolations and mostly untenable conclusions.

The introspective approach does not provide a definitive answer either, but it highlights the questionable nature of our instinctive desire to resolve questions with binary "yes" or "no" responses. People tend to categorize things into distinct camps; however, the reality is that there exists a spectrum regarding the meanings of physical versus unphysical. Inner exploration shows that many intermediary domains separate the stark distinction between the physical domain and the unphysical domain. The transition from the concrete material and physical world to what we perceive as an ethereal and unphysical plane of existence is not a sudden shift; instead, it appears to follow a gradual change from one state of being to another.

This suggests the presence of additional planes of existence that are not "physical" in the conventional sense, yet do not fit the definition of "unphysical" either. The simplistic duality of soul versus body or spirit versus matter fails to capture the complexity of reality. Between these two extremes might be realms with varying degrees of "fine-graininess" or "subtleness."

4. Why Do We Ask?

Having observed that the mind-body interaction problem is no more mysterious than the interactions between gravity and matter or photons and electrons, we now recognize that the simplistic distinction between the physical and the non-physical is also inadequate. So, once again, we must ask: Why are philosophers of mind so troubled by the concept of a mind interacting with a brain? What drives the demand for a solution to the interaction problem?

What was Descartes' original motivation? Why did he contemplate a non-extended—that is, a "trans-material" or "trans-physical"—substance? What drives humans to explore the relationship between the mind and the body in the first place?

The answer is twofold.

First, we perceive a clear distinction between our mind and conscious experiences and our physical bodily sensations. Whether consciously or not, we view things from a first-person perspective.

Second, it is human nature to wonder whether some essence of our being, or "soul," survives after death. If it does, our intuition suggests that something might continue to exist beyond the physical body. This concept implies that if such an extra-corporeal entity exists, it must be independent from our body and, at the same time, somehow interact with our body, particularly our brain, during our lifetime. Nonetheless, it assumes also a clear distinction between the physical body and a transcendent, or non-physical, soul, without considering the possibility of an intermediary state. This line of reasoning has always relied on a simplistic dual distinction, lacking any nuances or shades of grey.

However, based on the experiential exercise from the previous section, we can move away from this binary thinking. Instead, we can envision a smooth continuum between the material body and an entirely non-physical soul. It is reasonable—and perhaps necessary—to differentiate between our "gross material body" and a form of "subtle body" composed of "subtle matter."

Popular belief often envisions this subtle body as a kind of "ghost" lingering in the realm of the deceased, differing only in that, during our lifetime, it is contained within our material body. Popular literature often describes this notion as an "aura" or "life-sheath," suggesting that our body cannot live without it, yet it continues to exist after death. Similarly, the mind is not solely identified with the brain but is thought to be associated with a more subtle "mental-sheath" that also survives after death but is not the soul either.[71] Only our consciousness, or the "ultimate witness" in us, may entirely transcend spacetime and causality.[72]

Western metaphysicians might dismiss these ideas as superstitions and folklore, yet they reflect what dualists are often implicitly questioning: the nature of existence after death.

Death remains the most intimate mystery. We can measure neural activity or speak about homeostasis, but those are phenomena told from the outside. The interaction problem arises from the desire to know and understand our mortal destiny, if any. That's why we ask.

---

[71] This aligns with reports from individuals who have had near-death experiences. How can the mind remain active if one is meant to transcend both the body and the brain?

[72] This distinction might be clearer if you have read Section II.4.

5. What Is an "Interaction"?

If we refine our understanding in the light of modern science and consider the mind as a "mental body" composed of some elusive "subtle matter" that our current measurement devices—made of the ordinary "gross matter"—cannot detect, we might gain a better insight into the interaction problem. In fact, if our conscious existence unfolds in dimensions that are neither entirely physical nor entirely non-physical but instead inhabit some intermediate "subtle domains," the interaction problem takes on a new perspective. It could be resolved elegantly if we abandon the simplistic distinction between "physical" and "unphysical."

Interestingly, this approach aligns with modern physics.

To understand why and how this works, we must first examine what an interaction is from a physics perspective.

Let us clarify what an interaction between two physical substances means in the context of physics.

Contemporary physics recognizes four types of forces. These interactions include the electromagnetic and gravitational forces that we encounter in our everyday lives, as well as weak and strong nuclear interactions. Although we are not directly aware of the latter, they play a crucial role in determining the structure and stability of all atomic nuclei and, consequently, all atoms and molecules shaping our material existence.

All these interactions determine forces—or, more precisely, force fields—that are clearly defined and mathematically formalized through equations involving so-called "coupling constants." These constants are numerical values that quantify the strength of interactions, such as those between a particle and a field.

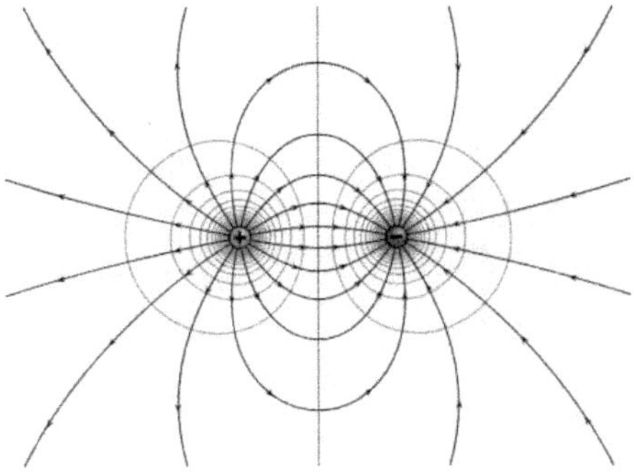

Fig. 46 In a field theory, there is no such thing as a "direct contact" between objects, only "couplings" between force fields exist (here the coupling between the fields of opposite electric charges.)

For instance, in classical physics, Newton's law of universal gravitation describes the gravitational force between two massive bodies using a simple inverse square law formula. Once their mutual distance is established, the gravitational constant "G"—the coupling constant for gravity—indicates the strength of their mutual attraction. A similar principle applies to the interaction between two electric charges.Coulomb's law also uses an inverse square law formula to describe the electric force between two charged particles. Here, once their mutual distance is fixed, Coulomb's coupling constant "ke"—the coupling constant for electric forces—determines the strength of their mutual attraction or repulsion.

At a more fundamental level reality is described by QFT which presents a model of reality in which the force between two material objects interacting at a distance is described through force-mediating fields. These fields are depicted as exchanges of force-carrier particles. For instance, the electromagnetic force is mediated by particles known as photons, which we imagine as light particles. The electric field between two interacting electrons is illustrated by a "Feynman diagram" that represents the exchange of momentum through massless "virtual photons."

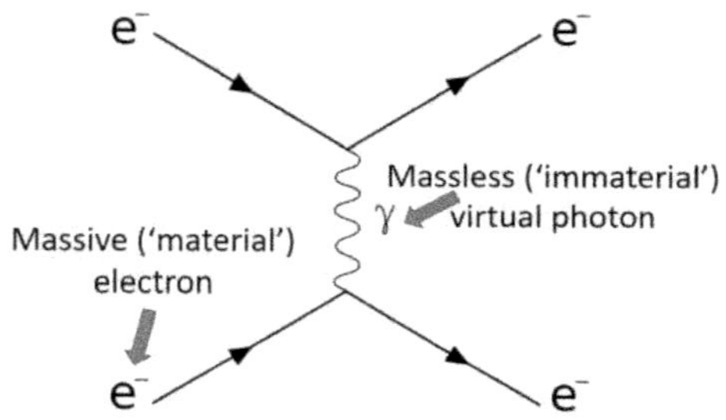

Fig. 47 Feynman diagram of the scattering between two electrons by a virtual photon.

Virtual photons are conceptualized as photons emitted by one particle and existing for very brief intervals before their absorption by another particle. They are called "virtual" because they cannot be observed in isolation; they exist only during the fleeting moments of the scattering process. Importantly, energy conservation is not violated. No long-lived energy emerges from nothing due to the interaction between "material" and "immaterial" entities. Instead, these virtual photons, as force-carriers, transfer momentum and energy from one location to another.

Similarly, strong and weak nuclear forces are mediated by their respective force fields, involving the exchange of "gluons" and "W and Z bosons." This is formalized in quantum chromodynamics (QCD), which explains the interactions between quarks (the building blocks of protons and neutrons in atomic nuclei) and gluons.

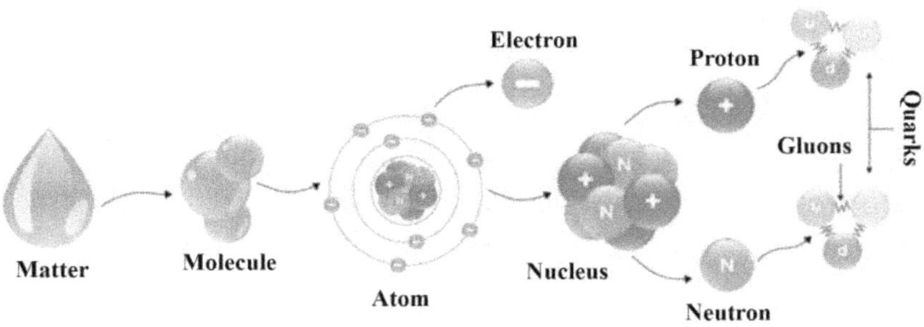

Fig. 48 Structure of matter.

That said, one should always keep in mind that this is only a naïve representation of reality. Feynman diagrams serve as convenient visual representations of particles' trajectories through spacetime (not space). However, they should not be viewed as objective depictions of real processes; instead, they are components of a more complex abstract mathematical framework used to calculate scattering processes—that is, the interactions between particles. They are only abstract shadows in our mind of an ontology that doesn't lend itself to an objective model of reality.

In fact, one can set aside the concept of particles and view fields as the sole real entities. From this perspective, what we perceive as particles are actually characterized as "excitations" of a quantum field. These excitations can be conceptualized as oscillating and traveling fields, represented by perturbations or just "bumps" in a universal and fundamental quantum field across space and time. The electron is an excitation of a universal electron field, while the photon is an excitation of a photon field. Similarly, the quarks that make up the protons and neutrons in atomic nuclei are excitations of the quark field, and so on. When two real particles—meaning non-virtual particles that are lasting excitations of the quantum field—scatter off each other, the sum of all possible virtual particle exchanges constitute what we refer to as an "interaction" or "contact" between two bodies. However, there is no such thing as a "contact" in the classical sense.

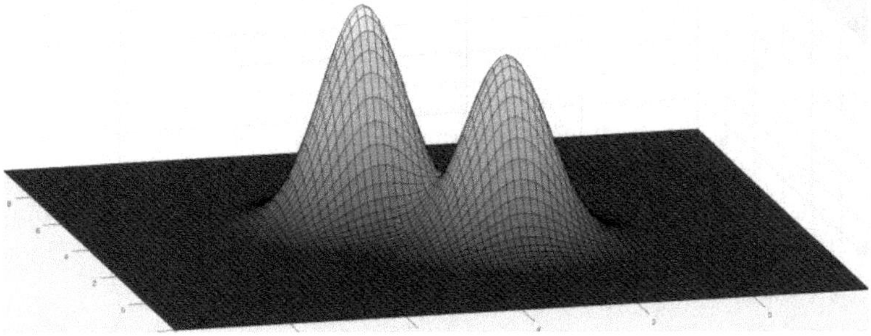

Fig. 49 Representation of two interacting particle fields.

Ultimately, what remains is a single conceptual entity: the field. At the microscopic scale of particle physics, the description of reality and the interactions between particles are more fundamental: fields and particles are two aspects of the same phenomenon. In QFT, the distinction between massive and massless particles dissolves. Massless particles, such as photons, are fields that propagate at the constant speed of light without inertia, while massive particles are fields that travel at speeds less than light and possess inertia. Despite this difference, both types are still considered fields. Even the notion of "matter" we are accustomed to dissolves. Matter is a field that manifests through localized interactions, characterized by the exchange of forces. There is no such thing like a solid, point-like particle that contacts other point-like or extended rigid objects.

At this stage, modern science can't say much more than that with regard to the fundamental nature of a physical interaction.

It's important to note that the big absentee is the force of gravity, which still lacks a coherent systematization. Since Einstein's time, physicists have sought a unified field theory that combines all four forces into a single theoretical framework of "quantum gravity." For now, QFT treats different types of particles as separate quantum fields, such as the electron field, the quarks fields, the gluons fields, etc. The ultimate unification would be to represent all particles in a unified theoretical framework with only one field. While this goal remains elusive, it is most likely that all particles—such as electrons, photons, quarks, etc.—are different "modes" of a universal quantum field. Each mode has specific "attributes," such as mass, charge, and intrinsic angular momentum. Though the standard model of particle physics treats every particle as a separate entity, modern quantum field theories tend to suggest a form of Spinozian "substance monism," proposing that at a fundamental level, there exists only one field, which we can refer to as the ultimate "substance." Deep down, everything we consider "physical" is a "mode" (or a "vibration" or a specific "state," if you prefer) of this singular substance. This is also suggested by the well-known mass-energy equivalence in special relativity, which establishes a relationship between energy and matter. Material particles with rest mass m can be "annihilated" into massless photons with energy E, according to Einstein's famous formula $E=mc^2$, and vice versa. Mass and energy seem to be two different states of the same "substance."

This is why some past and present physicists (notably Einstein) found Spinoza's worldview appealing. This worldview postulates that everything is only a "mode of a substance," with that "substance" ultimately being God. More on this will appear in sections IX.3d and X.2b.

6. How "Substances" Interact

Now that we have clarified what a physical interaction means in the context of physics, we can move beyond overly simplistic models of mind-brain interactionism.

First, observe that, within QFT, the interaction problem not only persists but actually worsens. The interaction problem revolves around the question of how two vastly different "substances," such as mind and matter, can interact. This idea assumes that the dissimilarity of substance prevents two objects from interacting. However, why don't we ask the same question about the interaction between two objects composed of the *same* substance?

In the context of QFT, the interaction between two material bodies has nothing to do with their similarities or differences. At a microscopic level, the "contact" between two objects involves electric interactions between the electric fields of the electrons in their constituent atoms. Ultimately, these interactions also entail the exchange of virtual photons, as illustrated in the Feynman diagrams shown in Fig. 47. The interaction between two material bodies is determined by the type of force to which they are sensitive, not by any similarities between them. In fact, one can make an opposite example of fields that don't couple at all, such as in the case of two photons. They traverse with each other like two water waves do. Yet, they are made of the *same* "substance," an electromagnetic field.[73] In the physical world, the sameness of substances does not imply an interaction either.

More generally, two completely distinct and independent particles associated with the excitation of different quantum fields, each with unique properties[74], can nevertheless interact. Contrary to the assumption that stands behind the interaction problem, Nature couldn't care less about the distinctness or sameness, or the independence, lack of commonality, or radical difference in the kind of properties, states, or substances. It can interact (or doesn't interact) whenever and with whatever it pleases, nonetheless.

Therefore, when people question how a mind can interact with a brain or body, they often misrepresent their own intuition by framing it as a problem of two distinct entities interacting. The more pertinent question is how something as "ghostly" or "ethereal," but not necessarily entirely beyond the physical, can engage with something so dense and corporeal as the brain. In section VI.3, we concluded that a simplistic dualistic view that separates matter and mind does not withstand scrutiny from the first-person perspective. To address the

---

[73] For extremely high intensities, this is no longer true. However, this should not concern us here, as it does not invalidate the argument.

[74] This is actually the present theoretical framework of the standard model of particle physics. One distinguishes between scalar fields, a vector field with one-half and integer spin, having properties such as mass, charge, etc.

interaction problem effectively, we must envision matter—or, more broadly, "substances"—as having varying degrees of "fine-graininess" or "subtleness."

We are already somewhat familiar with this concept. Our daily experiences show us that matter can exist in different characteristic aggregate states: solid, liquid, and vapor. Gas feels "finer" than liquid and even more so than solid, which feels like a dense and impenetrable substance. While, modern physics also knows of a state of plasma (charged and ionized matter), an intermediate state such as liquid crystal, and other states that show up in extreme conditions (superfluid, Bose–Einstein condensates, neutron-degenerate matter, and quark–gluon plasma). Particularly fascinating are superfluids, which are extremely cold fluids exhibiting quantum properties, such as zero viscosity, meaning they experience no friction or resistance to changes in shape. This frictionless quality allows them to move without losing kinetic energy, flow through pores and cracks, form vortices that rotate indefinitely, achieve other perpetual motions in closed containers, and display the "fountain effect," where they can spontaneously flow upwards and out of a container. Additionally, superfluids can interpenetrate, coexisting in the same space without mixing or significantly influencing each other's motion. Modern physics recognizes that these materials possess properties we previously deemed as almost "metaphysical." However, there is nothing "unphysical" about substances exhibiting such exotic behaviors. In fact, as we will see, physics teaches us that objects can also interpenetrate, meaning they can pass through one another with minimal disturbance and weak interactions, if any.

Ultimately, the nature of contact between substances is determined by the interactions of their constituent particles. QFT describes the interactions between particles as the "coupling" of fields mediated by electromagnetic, strong, or weak nuclear forces. When two particles influence each other, it is not because they are made of the same stuff but, rather, because they are both affected by a specific type of force. For instance, while the electron interacts with the electric field of a proton, it does not respond to the strong nuclear force of the proton's quarks. If electrons didn't "see" the presence of electric fields, they would be completely "blind" to the existence of all nuclear matter. When we don't forget how our abstract notions, such as "interaction" or a "contact" between "substances," are based on mere inter-subjective experiences that have no significance in the real world we might see the issues with mental causation from a broader and deeper perspective.

Moreover, while the reasons why certain particle fields do or do not couple with one another are complex,[75] we don't need to delve into those technical details to understand that by "subtle substance," we refer to a form of matter, energy, or force and various types of (both massive and massless) particles that interact very weakly with the building blocks of atoms and molecules, so-called "fermions."

Ultimately, the kinds of particles that constitute our (stable) material universe are reduced to only electrons and quarks. These particles, or objects composed of them, can interact with varying degrees of strength, resulting in bodies that are more or less permeable to one another. Nature not only permits this kind of varied "grossness" to "subtleness" but also exhibits it as a common occurrence.

The concept of bodies being permeable to one another, with varying degrees of permeability, is not uncommon in our everyday experience. A well-known example is radio waves. Low-frequency electromagnetic waves can pass through thick walls and even our bodies. The signals from our cell phones and radio stations are electromagnetic waves that interact only weakly with solid objects made of molecules and atoms. They interact with matter primarily when they encounter sufficiently thick layers or when the receiving material is electrically responsive and resonates with the wavelength of the signal, such as in an antenna. In other words, the intensity of the interaction—the so-called "coupling strength"—between an electromagnetic field and a matter field determines how "transparent" or "subtle" the latter is in relation to the former.

We don't need to imagine this. We know how visible light can pass through a window because the energy levels of molecules in the glass do not resonate with the electromagnetic field and, thus, do not couple with the photons.[76] This means that if the force-carrying particles couple only weakly or not at all with a material body, they are "ghostly" for that material. Recognizing that our abstract concepts like "interaction" or "contact" between "substances" stem from subjective experiences that lack real-world significance allows us to approach the issues of mental causation from a broader and deeper perspective.

These examples might seem inappropriate at first. Some might argue that radio waves aren't "bodies" made of a tangible "substance." However, this straightforward and familiar illustration demonstrates that certain entities possess evanescent and almost "spooky" properties that we typically associate with the mind in dualistic theories.

---

[75] For the technically minded: It is rooted in the fact that the field's Lagrangian must remain invariant under global and local gauge transformation and that some specific symmetry principles must be preserved. Nonetheless, the interaction strength—that is, the coupling constant—isn't determined by the theory; it remains a value that must be determined empirically.

[76] However, a silica-glass window blocks infrared and ultraviolet radiation.

Let's explore this further.

Consider the opposite example of something made of "hard stuff." Have you ever wondered why you can't walk through a wall? This seemingly naïve question is actually quite complex and requires intricate quantum field theoretical explanations. In simple terms, the answer lies in the electromagnetic forces at the atomic level. Both our bodies and the wall are composed of atoms, which consist of a dense nucleus surrounded by a cloud of electrons. When we attempt to push through a wall, this cloud of electrons in the body's atoms and molecules repel the electron clouds in the wall's atoms and molecules due to electrostatic forces. This strong resistance prevents us from passing through walls. There is no such thing as a "contact" between two solid objects as we conceive of the contact between two marbles. What we perceive as a "collision" between two material bodies is not about solid objects colliding. Rather, it is about microscopic force fields interacting. Ultimately, all these interactions boil down to the Feynman diagram in Fig. 47. The combination of this electromagnetic repulsion and other quantum mechanical effects[77] makes objects appear impenetrable—that is, "hard" or "gross" stuff.

Beyond the strength of a concrete wall and the delicacy of an electromagnetic wave, there are even tougher materials and more ephemeral substances.

For example, diamonds are among the hardest naturally occurring materials due to their carbon-based atomic structure. Each carbon atom forms strong covalent bonds with four neighboring carbon atoms, resulting in a remarkably rigid crystal structure. These chemical bonds are fundamentally driven by electromagnetic forces at the atomic level, which prevents external forces from penetrating or tearing the crystal apart.

Before coming up with the opposite example of the ultimately known most subtle entity, let us review the kind of particles and forces that make up the known physical universe in the frame of the standard model of particle physics. Fig. 50 is a compact diagram representing all the known particles and forces of the standard model of particle physics.

The fundamental components of ordinary matter are just a few types of leptons and quarks—specifically, electrons, up-quarks, and down-quarks, which together form protons and neutrons in atomic nuclei. All force-carrying particles are known as "gauge bosons."[78] Photons mediate electromagnetic fields, acting as the force carriers of light. The strong nuclear force binds

[77] It's more complicated than that. One must consider the Pauli exclusion principle, Heisenberg's uncertainty principle, and other relevant factors. However, these technical details are not essential to the point being made here.

[78] Remember, the standard model of particle physics does not include gravity. It remains uncertain how, or if, it could be integrated into a future theory of quantum gravity.

quarks within protons and neutrons through the exchange of gluons, their corresponding force-carrying particles. In contrast, W and Z bosons mediate the weak nuclear force responsible for the decay of atomic nuclei

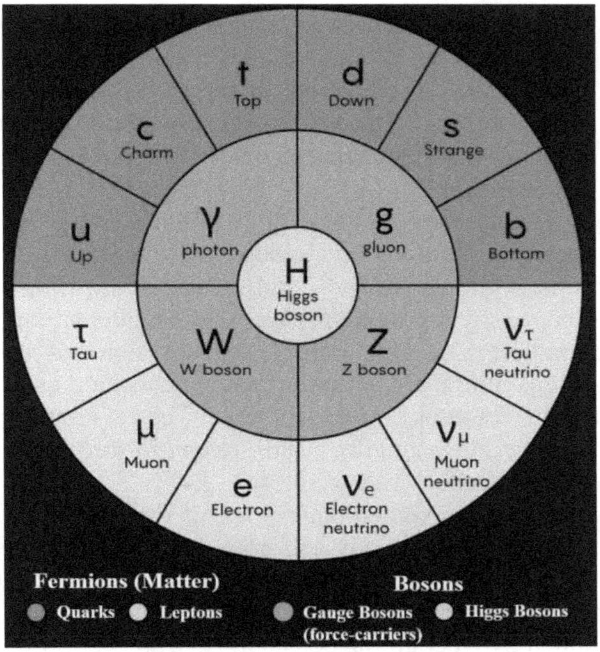

Fig. 50 The particles of the standard model of particle physics.

Unlike typical push or pull forces, the weak force primarily facilitates the transformation of one type of particle into another, leading to nuclear instability and decay. Lastly, the Higgs boson is not a particle in the traditional sense; rather, it represents a field that permeates the universe and imparts mass to other particles. The remaining particles seem to play no significant role in our current material constitution and existence, though their presence might have been essential during the early stages of the universe's formation.

The point of interest for us here is how these particles (the fermions and leptons) interact—that is, "couple"—with each other via force-carrying gauge bosons.

The strong nuclear force is termed "strong" because it is significantly more powerful than any other known force. This is due to the large coupling strength between quarks and gluons, which contributes to the extreme stability of matter, preventing it from dissolving into a chaotic gas of particles. In contrast, the weak nuclear force is labeled "weak" for the opposite reason: The coupling strength of the W and Z particles with quarks is small but sufficient to induce nuclear decay. While the coupling strength of the electromagnetic force is weaker than that of the strong force, it is still much stronger than that of gravity.

The reason we remain largely unaware of the strong nuclear force, despite its intensity, is that it operates only at extremely small distances within atomic nuclei. In contrast, light and gravity, though weaker than the strong and weak nuclear forces, act on macroscopic scales encompassing the entire universe.

Like humans, not every field likes to interact with every other field. As already pointed out, the strong force couples only with quarks and does not interact with leptons, such as electrons. In contrast, the weak force interacts with all quarks and leptons. The electromagnetic field, on the other hand, couples with anything that has a charge, including quarks, taus, muons, and electrons, but not neutrinos.

Notice how the mass of particles does not determine their interaction strength. For example, photons and gluons are massless yet mediate very strong interactions, while the W and Z gauge bosons are 150,000 times more massive than electrons, yet provide for the weakest interaction. The electron, which has the smallest mass of all particles except neutrinos, can form strong chemical bonds, enabling the creation of materials as hard as diamonds. This highlights another common misconception: The assumption that being "material" indicates the quality and strength of interactions is a misunderstanding we must eliminate.

In summary, when we contemplate how an immaterial or non-physical mind interacts with a material body and brain, we must consider the insights from modern physics; otherwise, we risk misunderstanding the nature of this interaction. In modern physics, the strength of the interaction between two fields is determined by its coupling constant. A strong coupling results in two bodies interacting intensely, behaving almost like solid, impenetrable entities. Conversely, a weak coupling leads to minimal interaction, allowing the bodies to pass through each other, potentially with some friction. If the coupling is zero, the bodies would completely ignore each other as if they were unaware of one another's existence. Ultimately, it is the coupling strength of the force-carrying particles that determines the solidity of objects and how "ghostly" or "spooky"—or, to use a more fitting term, "subtle"—the two substances appear to each other.

In conclusion, we can't ignore the most eminent particle that exemplifies this state of affairs: the neutrino. Consider the weak nuclear force; it interacts only very weakly, and its interaction range is smaller than a proton's size, making it negligible beyond that distance. This means one must get close to the quarks within protons and neutrons to interact with matter solely through the weak force. The average distance a moving particle travels through a medium—known as the "mean free path"—before engaging with a proton's quarks via the weak nuclear force is extremely large. If our bodies interacted only through the weak nuclear force, without the electromagnetic and strong interactions, we would experience almost no interaction at all with anything.[79]

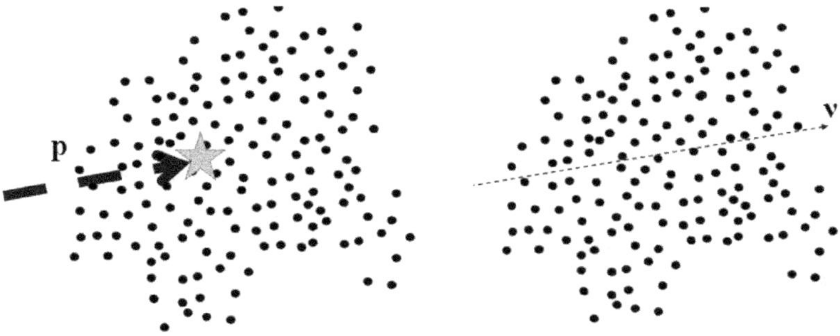

Fig. 51 A proton (left) has a much shorter mean free path before colliding with some particle than a neutrino (right).

We could pass through solid objects, planets, and stars, completely unaware of their existence and that of the entire physical universe despite being a part of it. In fact, neutrinos are produced in the core of the Sun and are able to traverse it entirely undisturbed and do the same with the Earth as well. Neutrinos are not massless; they possess an extremely small mass (about a millionth of the mass of an electron) and can never be detected directly. It's no coincidence that neutrinos are, therefore, also called "ghost particles."

However, there would be nothing "immaterial" or "unphysical" about this scenario.

How do we know neutrinos exist?

We know because, in very rare instances, a neutrino interacts with matter. Trillions of neutrinos pass through your body each second, but only one in a trillion scatters off one of your body's protons. This interaction can be detected using large and complex detectors, which indirectly reveal the existence of neutrinos by observing these scattering processes.

In these cases, particles are kicked around, seemingly violating the principle of energy conservation. We might observe a proton at rest suddenly beginning to move without any apparent cause. However, this anomaly is merely an illusion: The "ghostly" neutrino, which is invisible to us, has transferred its energy to the proton, which our detectors can observe.

The neutrino's apparent violation of the energy conservation principle should be of interest to philosophers of mind. It's possible that mental processes might seem to contradict this principle; however, this perception could be a misinterpretation. The invisible mind-substance, which brain scans cannot detect, may transfer its energy to the brain-substance that is visible

[79] Gravitational forces can be ignored because the gravitational field of the size and mass of a human body is completely negligible.

through these scans by interacting very weakly with it (with a very small coupling constant). In reality, there is no violation of energy conservation; instead, the process involves an energy transfer from a subtle mental substance to the more tangible realm of ordinary matter.

Interestingly, it is precisely the apparent violation of energy conservation that suggests the existence of other forms of "substances" or "subtle matter."[80] The energy conservation issue in the interaction problem could be addressed by recognizing the existence of previously undiscovered forces and particles. In fact, this perspective is common among modern particle physicists who are actively searching for such "anomalies."

In this way, we can resolve the energy conservation dilemma associated with the interaction problem of mental causation. Although we may observe events that appear to violate energy conservation, there is no actual contradiction. Energy is always conserved; however, the particles we currently classify as "physical" or "material" may interact with other types of particles and couple with forces that are still unknown to us. These we may label as "unphysical" or "immaterial."

7. The Subtle Matter Conjecture

This concrete example involving neutrino particles has its limitations, as neutrinos do not form bounded objects like protons and electrons, which combine to create atoms. These atoms can then bond with other atoms to form molecules and, ultimately, complex organisms like us. Neutrinos exist solely as a flux of particles and do not create bodies with complex structures. However, there are no known physical laws that prevent us from framing the conjecture of the existence of other particles that weakly interact with fermionic matter but strongly interact with unknown particles. These hypothetical particles could form complex aggregates, remaining completely invisible to us and even passing through our bodies at any moment without our awareness.

There might be other types of forces–that is, force-mediating bosons–and particles that create "planes of existence" that interact only weakly or not at all with the particles of the standard model. The diagram of Fig. 50 illustrates the particles and forces of the plane we define as "physical," but many other planes may exist, influenced by entirely different forces and particles. Our material "gross bodies," composed of ordinary fermionic matter (leptons and quarks), might be oblivious to the existence of other, more "subtle bodies" such as a "mind-body" and "life-body" governed by interactions beyond the four known ones and made of particles that couple only weakly to leptons and quarks. We

[80] Another well-known example is the detection of "dark matter" in and between galaxies. However, as of this writing, it remains uncertain whether dark matter is actually matter. To keep this brief, I will omit a discussion on this equally fascinating topic.

would be unaware of these "subtle bodies" as extended objects in space or time. However, both our material gross bodies and these subtle bodies could coexist in the same location at the same time like Matryoshka dolls placed one inside the other.

This might sound overly "ghostly," and in a way, it is. However, this is not merely a fantasy; the concept of subtle worlds aligns with the laws of physics and presents no interaction problem. Modern physical theories even suggest it is feasible. Physics currently lacks explanations for why particles have their specific properties, why their coupling constants hold particular values, and why there are precisely four known forces rather than more or fewer. It's possible that we will eventually discover evidence for fundamental forces beyond the four currently recognized, along with an entirely new "universe" of particles that interact with those forces but not (or only very weakly) with the ones we know.[81]

Physicists are actively searching for forces beyond the four currently known. As of now, none have been discovered, but we might soon realize that our universe is just one aspect of a larger reality comprising different layers of "subtler matter." One of these layers could be a mental plane that we deem "immaterial" or "unphysical" simply because we tend to think in binary terms, rather than recognizing a continuous spectrum of existence.

This subtle matter conjecture[82] could naturally support a cosmology with various "planes of existence," each linked to a corresponding "plane of consciousness." The concept of the "universal mind" or "Mind at large" in cosmopsychist or idealistic models might represent part of a universal plane of subtle mental substance. Upon the dissolution of the physical body, this mental substance could remain unaffected and continue to exist on that mental plane. However, it is neither strictly "physical" nor "unphysical"; rather, it weakly couples with the bodily reality of which we are presently aware—hence, its description as "subtle." The same could be true for a "life-plane" governed by vital forces or for higher intuitional planes of consciousness, etc.

Finally, the mind-matter interaction problem arises from our reliance on a simplistic 17th century dualism of mind and matter without a full understanding of how these interactions function in reality. By recognizing the need for a more nuanced approach that acknowledges a diverse range of planes of

[81] It is important to clarify that this discussion is unrelated to any multiverse theory or to the various "worlds" proposed by the Many Worlds Interpretation of quantum mechanics. The 'planes of existence' to which we are referring here are multiple realities that exist *within* the same universe we label as "physical."

[82] Notice that nothing stops us from formulating the opposite 'gross matter' conjecture, which suggests that *we* are the subtle bodies for beings inhabiting a world where the forces interact with their particles much more strongly than our forces do with fermionic matter. To keep this chapter focused and not overly speculative, I will refrain from elaborating on this idea.

conscious experiences, we can develop a richer cosmology, beginning in Chapter X.

8. Beyond Spacetime and Physical Causality

In the previous essays, we reformulated the relationship between the "physical" and "unphysical" by extending a mind-matter dualism into a multilayered cosmology of planes of existence or consciousness, starting from a dyadic conception of spacetime and causality. We began with an introspective investigation in section VI. 3 to clarify the ambiguous ontological status of space, time, and causality in the mental domain. In section VI.5&6, we clarified what an interaction is and how "substances" interact according to physics. Then, in section VI.7, we highlighted that the existence of "subtle worlds," where different notions of materiality (and, eventually, of space and time?) may exist, is not only a viable hypothesis but also suggested by particle physics.

Can we go even further than that?

Sections V.7-11 already provided the answer.

The relationship between quantum indeterminacy, or ontic randomness, and purposeful action guided by conscious volition, which I outlined in those sections and in the two advanced reading papers XIII.3&4, could resolve the mind-body interaction problem of mental causation, even transcending any spatio-temporal or quantum theoretical framework.

For a universal consciousness to exert any causal influence in a dynamic physical universe governed by a universal spacetime quantum field, it must be transcendent. This consciousness cannot be part of the quantum field or interfere with it directly; instead, it must "interface" with the field while remaining independent of its definitions, determinations, or influences. We have shown, in section V.10, how Schrödinger's objection has been effectively refuted: There are no concerns regarding violations of the principle of energy conservation resulting from this temporal permutation of a series of events. This is accomplished by shifting the causal principle from interactions based on physical forces, as described by the standard model of particle physics, to a conscious causal power working with the temporal ordering of events, rather than with their force-dynamics.

In the framework of libertarian quantum panpsychism, we have demonstrated that no conventional "interaction"—defined as a force exchange—is required for causal power. Instead, a change or permutation in the sequence of quantum events, governed by the laws of quantum physics and statistics, is sufficient for efficient causation. Altering the order of outcomes without affecting their overall frequency does not change the probabilities of those outcomes; it merely alters their timing of actualization.

The concept of physical causality is expanded to encompass a substance-monistic view of self-causation. We begin by asserting the existence of a single substance that possesses multiple (potentially infinite) modes and properties. The physical causation inherent in the fabric of spacetime, governed by the (four or more) fundamental physical forces, reflects a transcendent self-causation that organizes the sequence of outcomes in temporal phenomena of one single substance.

Finally, If we consider that mind, consciousness, spirit, or whatever aspect of our individuality has a subtle or non-physical nature that transcends space and time and that it “interfaces” with the world as proposed in William James's “filter theory” and Aldous Huxley's concept of “Mind at large,“ we can view our mind not merely as the brain but as something that the brain filters from a universal mentality. This universal mentality expresses itself through the stochastic nature of the quantum field. Under this framework, we can reclaim the notion of free will by remembering that, in physics, nothing prevents us from “reshuffling” the potential phenomenal events described by their probabilities in a temporal sequence.

Thereby, we have complemented and unified the subtle matter conjecture with the quantum libertarian panpsychist perspective.

9. Is the Interaction Problem a Problem at All?

In conclusion, we have explored the interaction problem of mental causation and potential resolutions free from ambiguities. Upon closer examination, particularly through the lens of modern physics, I contend that it is not a problem at all. Presenting it as a decisive argument against dualistic hypotheses, as many respected scientists do, often reveals an ideological bias rather than a robust scientific objection. While the interaction problem is a legitimate concern, it does not, contrary to popular belief, possess the argumentative strength to end the discussion. The notion that consciousness, the mind, and, I would argue, life itself are principles that transcend the “physical“—whatever that entails—remains a viable perspective. Caution is warranted when scientists or philosophers present the interaction problem or the demise of vitalism as the ultimate rebuttal to post-material interpretations without further examination; this is often indicative of unreflective thinking.

If modern science had been as skeptical toward its own theories as it is regarding more metaphysical questions, it would not have advanced beyond the science of the 16th century. I believe this argument alone is enough to silence the criticism.

The interaction problem and the notion of mental causation was condensed in a scholarly article presented in the Advanced Readings section XIII.5. It briefly takes up again the combinatorial causality discussed in sections V.10&11.

VII. Divine Action Revisited

In Chapter IV, I discussed why materialists are generally skeptical about the existence of an active God in a universe that appears to function like a clockwork machine. If everything operates under the laws of Nature and all phenomena are governed by a physical causality that adheres strictly to deterministic rules—such that each specific cause leads to only one possible effect—then all future events in the universe are uniquely predetermined by its current actuality. In this view, a Creator cannot act intervene without constantly violating those laws. If that were the case, we would be left wondering where and how this occurs, as nothing like it has ever been observed. Furthermore, why would God create a universe governed by laws only to repeatedly disrupt it?

Or, alternatively, should we adopt a deist or superdeterministic perspective and believe that God created the universe with precisely the initial conditions necessary for its history to unfold as He desired, but then ceased to intervene after the Big Bang? Everything would have been decided at the moment of the Big Bang, making the world little more than a vast encyclopedia where each moment simply turns a page, revealing what was already destined to happen without creating anything new. Were all the works of Shakespeare, the art of Leonardo da Vinci, the music of Mozart or Bach, the genius of Galileo, Newton, and Einstein, and the content of all the books that have been written by humanity already predetermined, concealed, and encoded in every detail in the primordial plasma of the Big Bang? This must be the logically inevitable—and, quite frankly, frustrating—conclusion if we adhere to a deterministic paradigm and wish to maintain a belief in God.

This assumption of a fundamentally deterministic universe has contributed to the remarkable achievements of science, technology, and the industrial revolution. For example, engineers operate under the belief that the laws of Nature are causally closed, meaning that everything that happens in the world is the result of physical phenomena governed by physical laws. They have good reasons to nurture such a belief; this understanding has enabled the construction of skyscrapers, engines, airplanes, computers, and spaceships.

However, while there is no doubt that a deterministic aspect exists within the universe, less clear is the question of whether this determinism is absolute throughout. If there had been nothing beyond the mechanical clockwork universe that Newtown described, and if science had managed to explain all aspects of life, including the function of our brains and the emergence of consciousness and volitional beings, within the framework of classical mechanics (that is, within a causally closed realm), it would become nearly impossible to argue for a reality with an open future (that is, also open to divine action or anything considered non-physical, whatever that may mean). There would be nothing but an inexorable causal chain of events that no one—not even God—could alter. This is a matter of logical principle: If everything is strictly deterministic, it is by definition closed to internal or external self-determining will. There can be no "loophole" or "backdoor" through which a transcendent divinity or an immaterial consciousness, mind, or spirit could influence a brain, let alone modify the course of material events.

Thus, if we maintain the belief in a God that is always present in the activities of Creation, we encounter an explanatory gap. If God acts intentionally to bring about changes, then there must be physical states that cannot be explained solely by deterministic physical causes. For God to intervene, this causal framework must have gaps. Science appears to contradict this possibility. Or, alternatively, we might need to reconsider our conventional notions of causality.

Fortunately for theists and unfortunately for atheists, things turned out to be more complex than initially thought. With the advent of quantum mechanics, we face a reality ruled by a principle of causality that is much more flexible and complex but also more interesting, in certain respects, than the principles of causality found in classical physics. In the quantum world, scientists have discovered that things are inherently uncertain, fuzzy, and indeterminate.

The entirety of Chapters IV-VI were dedicated to describing this state of affairs. At the microscopic scale, the world is not deterministic. There is no reason—at least not in the sense of the conventional principle of sufficient reason—for an atomic nucleus to decay at this moment or a later time. Similarly, in the double-slit experiment, there is no apparent mechanism that determines why a photon hits one pixel on the detector screen rather than another. Similarly, there is no known causal explanation for why particles in a superposition of states randomly appear in one state or the other upon measurement. Some aspects of the quantum world are simply random. However, this randomness is not epistemic (it does not appear random due to some unknown underlying process). Rather, it is ontologically indeterminate. This means it is inherently unpredictable—not due to our ignorance, but simply because that is the nature of reality.

Physicists and philosophers of science continue to speculate about the true nature of this randomness. Some hope that a future theory will reveal the hidden variables underlying this indeterminacy. Perhaps, as the story goes, this seemingly irreducible quantum randomness in the microscopic realm is caused by a hidden and still unknown process. Ultimately, even quantum mechanics might be reducible to a conventional form of deterministic causation, allowing us to recover our intuitive human understanding of a world governed by classical laws of cause and effect. However, as of this writing, over 90 years have passed since the brightest minds attempted to force Nature back to its macroscopic deterministic habits without success. This intellectual journey has been particularly marked by John Bell's famous "Bell's inequalities," which, in essence, prove that the world must be either indeterministic, nonlocal, or both, but it cannot be both local and deterministic. There remain interpretations of quantum mechanics that, in principle, could preserve determinism. These include the Many Worlds Interpretation, De Broglie-Bohm mechanics, and superdeterminism. However, while generations of physicists have proposed and studied these formulations for decades, there is still not a shred of evidence to empirically support them. Despite all efforts to find a "loophole" in quantum mechanics that could allow us to recover some form of determinism or epistemic pseudo-indeterminism, all experimental evidence supports the opposite view: Ontic uncertainty inherently governs the world. Quantum randomness appears to be quite stubborn. For almost a century, physicists have sought to close the causal openness that quantum indeterminacy seems to offer, developing a plethora of theories, interpretations, and tests in search of hidden variables. However, nothing has worked. In my view, the reason some are still trying so hard to recover our anthropomorphic understanding of causality is not scientific, but ideological: Ontic uncertainty has the kind of mystical flavor that a naturalistic and mechanistic view of reality dislikes.

Therefore, I will assume that quantum theory is a theory without hidden variables and that quantum uncertainty is ontological, not merely an epistemic limitation.

This is where some modern thinkers identify a loophole through which God could act in an otherwise causally closed universe, as presented by non-quantum science. Several scientists and theologians, such as William Pollard, Nancey Murphy, John Polkinghorne, Robert Russell, Keith Ward, Philip Clayton, and others, posit that quantum physics reveals an openness in the seemingly causally closed world of science. They argue that this fundamental indeterminacy expresses God's will, suggesting that God is always active at any moment in and through all aspects of creation.

This divine action is possible if God has a bottom-up control over quantum events, through which it exercises influence in the world. However, God may not need to be an interventionist; God does not need to modify quantum probabilities in a way that violates the principles of quantum mechanics to be causally effective. Instead, as we have seen in the second part of Chapter V, every quantum event that we regard as "random" and "uncertain" could be interpreted as the reflection of the operation of a "Schopenhauerian universal Will," which does not need to modify anything other than making the choice of which seemingly random potential phenomenon becomes actual.

By activating and actualizing one or another quantum event—that is, by the collapse of the wavefunction—God facilitates the transition from potential to actual, representing specific intentional acts. We can apply the same principle of combinatorial causality and combinatorial agency we examined in Section V.10&11, described in more detail in the Advanced Readings XIII.3&4. God realizes one of the many potentials defined by probability distributions, granting God the space to act, and thereby providing a virtually infinite range of action. God not only enacts history but also acts within it without suspending the laws of Nature. Contrary to common belief, this is not only permitted by our current understanding of the laws of physics, but is also suggested by them.

1. Does Quantum Theology Hit the Wall?

Nonetheless, naturalists still have powerful technical objections on their side. Invoking quantum effects to defend any sort of natural theology won't work for the following reasons.

The first objection is that the long-term statistical average of microscopical quantum fluctuations remains unaffected on the macroscopic scale. Any indeterminacy is "averaged out" due to the extremely large number of fluctuations and processes involved. Consider how many atoms, molecules, and cells contribute to the structure and functions of a living organism. Statistically, in our large-scale and long-term observations, all the innumerable microscopic quantum fluctuations influencing a large body make no difference because they tend to an expected mean. The pervasive indeterminism that is supposed to be the controlling hand of the Creator disappears at the macroscopic level because, statistically, it averages itself to a steady state that has no causal consequences.

To understand this concept, consider the average outcome of a fair die tossed a significant number of times.

When a fair six-sided die is tossed N times, each face (1 through 6) has an equal probability of appearing on any given toss. The average outcome after N tosses is the arithmetic mean of all observed values. According to the law of large numbers[83], as N increases, this average tends to approach the expected value of a single toss, which is 3.5 (not 3!). For small values of N, the average may fluctuate significantly due to randomness, but as N becomes very large, these fluctuations diminish, and the average outcome stabilizes at around 3.5. Fig. 52 illustrates a special case of how the average behaves after N die throws.

Initially, it "wobbles" randomly, but after a sufficient number of trials, it begins to stabilize and eventually approaches the ideal average of 3.5, which is obtained after a virtually infinite number of throws. Random variations do not sum up and are averaged out in the limit of many trials.

The second objection is that despite the randomness of quantum microphenomena, they are too small and too weak to influence the outcome of a macroscopical physical process. Quantum phenomena are simply too small to produce appreciable effects and, thereby, can't render the macroscopic universe causally open. Even the stochastic processes in our non-quantum,

[83] The law of large numbers states that as the number of trials increases, the average of the results will converge to the expected value, assuming the trials are independent and identical probability.

everyday classical world—such as the random and chaotic collisions of molecules in a gas or liquid due to thermal excitation—are significantly larger than any quantum random signal.

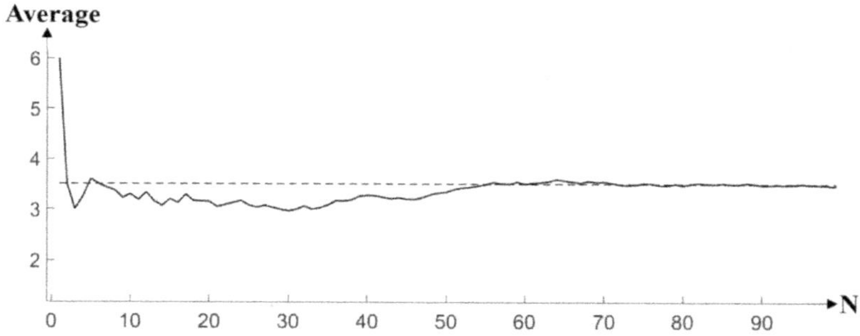

Fig. 52 The die average after N tosses.

In other words, the indeterminism introduced by the minuscule quantum random uncertainty in the realm of microscopic particles is too weak to influence the classical determinism of the relatively large-sized objects of the world.

An intuitive illustration of this appears in Fig. 53.

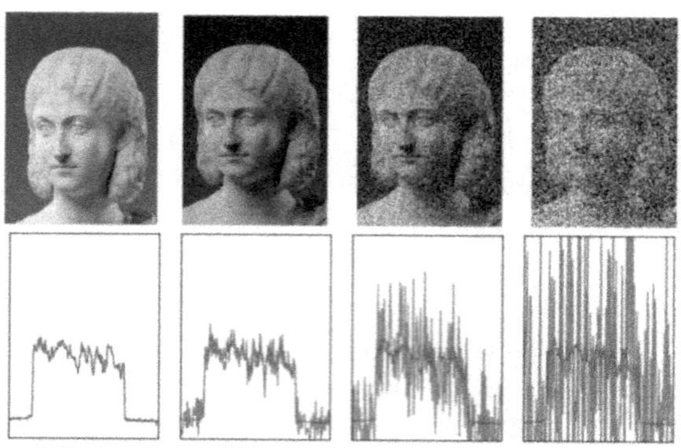

Fig. 53 Increasing noise "corrupts" an image.

From right to left, different noise signals with different amplitudes (the blue lines in the graphs at the bottom of Fig. 53) "corrupt" the signal (the red lines). The so-called "Signal to Noise Ratio" (SNR)—that is, the ratio between the amplitude of the signal and the amplitude of the overlaid noise—is, from left to right, 100, 10, 3, 1. Fig. 53 top shows how the effects of the noise "corrupt" an image. The left image shows how a noise signal that is a hundred times smaller than the signal constituting the image (SNR=100) is too small to produce discernible effects. Roughly speaking, quantum effects have an SNR of at least 10^3, meaning they are thousands of times smaller—sometimes even much smaller—than any phenomenon regulating processes such as chemical reactions in our bodies or the electrical activity of neurons in our brains. Consequently, it is difficult to imagine how these effects could influence our everyday lives.

In summary, all quantum effects are too small to have a significant impact and, even if they could, every effect would quickly be statistically "washed out" by the vast number of processes and particles involved. Therefore, quantum theology hits a wall. So the story goes.

However, while the example with die and the SNR argument are technically correct, these seemingly sound conclusions conceal a logical fallacy relevant to the philosophical argument with which we are engaging here—one that even physicists sometimes fall prey to. Let us examine more closely the two objections.

The reasoning with the die applies only to processes in which each random event does not contribute to the total uncertainty. For instance, in the case of a random walk, such as Brownian motion—also known as a "Wiener process"—the particles experience displacements due to countless random molecular collisions with surrounding particles (recall Fig. 41 or Fig. 42). These displacements contribute to the drift of the particles' positions in various directions. The drift accumulates linearly over time; each collision adds a random "kick" from all possible directions, but these directions cancel out on average. However, the uncertainty about the particle's position grows steadily. The particle's random motion does not average out. Instead, it manifests as a molecular drift that builds up over time. This is why the particle's mean displacement is zero, but its mean squared displacement (variance) increases linearly with time, as each independent random step contributes additively to the total uncertainty. To put it in simpler terms: While the noise averages out, the drift—that is, the particle's displacement—accumulates.

Fig. 54 illustrates how several trajectories of particles, all starting from the same position (Position = 0), and subjected to a noise with the same amplitude, end up in different final positions after a certain time interval, resulting in distinct "histories." Although the mean displacement remains zero, the spread of possible positions widens linearly with time.

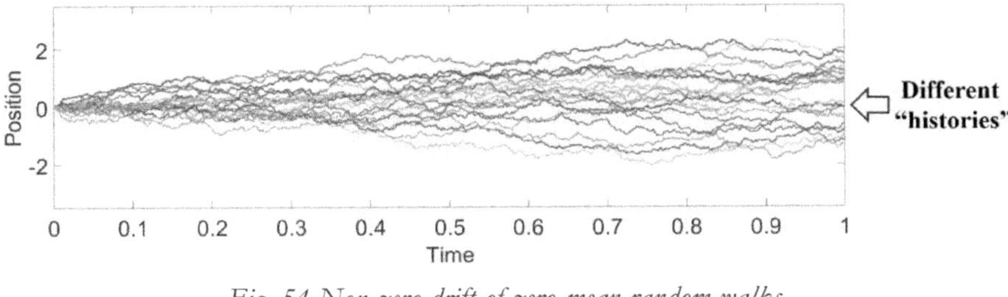

Fig. 54 Non-zero drift of zero-mean random walks.

This example illustrates that something subjected to a random process does not necessarily remain stationary, even over long time intervals or in the limit of large numbers. In the case of a random walk of a particle undergoing Brownian motion, the initial actuality—that is, what in classical mechanics is equivalent to the physical state of a system—is unique yet holds the potential for multiple future states. The analogy holds only to a certain point; classical Brownian motion is still a pseudo-indeterministic process. However, the same rationale holds for the ontic quantum indeterminacy. As we pointed out in Sections V.10&11, it is theoretically possible to determine the history of each particle by "steering" its random motion through combinatorial agency and causality without altering the zero-mean expectation value or the probability function describing the random process. There can be various potential and still non-actual outcomes: The future is “open.”

The second objection, which claims that quantum effects are too small to have observable consequences, is based on a similar fallacy. To understand why, let us connect the principle of combinatorial causality with quantum indeterminacy, which persists across all scales. This deserves a closer inspection in a dedicated section.

2. Quantum Mechanics, Chaos, and Theology: An Unexpected Alliance.

As we have amply discussed in the last three chapters, it became evident in the last century that the traditional deterministic and mechanistic Newtonian-Laplacian worldview required revision. However, this was the case not only due to the indeterminism inherent in quantum mechanics but also because of the advent of nonlinear chaos theory.

Nonlinear chaos is a fascinating field of study that delves into the behavior of complex systems that are highly sensitive to their initial conditions. Unlike linear systems, where a small change in initial conditions results in a proportionally small change in behavior, nonlinear systems can exhibit unpredictable and seemingly random behavior due to their high sensitivity to even small environmental changes. This is due to complex (self-) interactions, which can amplify the smallest perturbations over time. Chaotic systems are still deterministic in the sense that their present actuality completely determines their future actuality. They are also described by mathematical differential equations that lead to solutions—that is, the equations of motion—corresponding to unique trajectories in space and time. In this sense, they are still classical—that is, non-quantum—dynamical systems. However, due to their extreme sensitivity, even a tiny shift in the initial conditions makes long-term prediction impossible in practice. This inherent unpredictability makes chaos so omnipresent and relevant to fields as diverse as meteorology, economics, and fluid dynamics.

A well-known and evocative illustration of chaos theory is the "butterfly effect." This concept, popularized by meteorologist Edward Norton Lorenz, suggests that a very small change in one part of a system can have large-scale, unpredictable consequences elsewhere. The classic metaphor is that the flapping of a butterfly's wings in Brazil could, through a cascade of ever-amplifying atmospheric disturbances, lead to a tornado in Texas. While this might be an exaggeration, it effectively captures the core idea: Tiny changes can lead to dramatically different outcomes over time.

The butterfly effect is a direct consequence of the exponential divergence of nearby trajectories in a chaotic system. This means two points that start out infinitesimally close to each other will, over time, move apart at an exponential rate, making long-term forecasting impossible. It highlights the fundamental limits of predictability in many natural and man-made systems and serves as a powerful reminder of the interconnectedness and inherent complexity of the world around us.

The most popular example of a chaotic system is the double pendulum, where a tiny difference in its starting position (such as inclination angles, velocities, or other aspects of its physical structure) leads to a drastically different trajectory over time. Fig. 55 illustrates this high sensitivity: even if three double pendulums are released from nearly identical positions, their paths will quickly diverge and become completely unpredictable. Within a few seconds, they will begin to behave in completely different ways. Even the smallest random perturbation can induce an entirely different dynamical history.

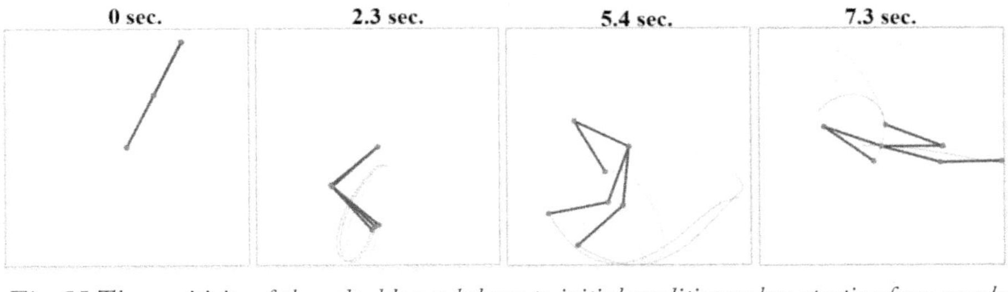

Fig. 55 The sensitivity of three double pendulums to initial conditions when starting from nearly identical positions.

Could perturbations as small as quantum uncertainties cause similar dynamical divergences? The answer is yes.

When we incorporate quantum fluctuations into the dynamical equations governing nonlinear chaotic systems, even the smallest microscopic quantum random noises can quickly be amplified to a macroscopic scale. The slightest microscopic quantum indeterminacy will become apparent over a relatively short period if the system is strongly nonlinear and chaotic.

Thus, while not all systems observed in Nature are chaotic, nonlinearity is quite pervasive in complex systems, especially in living organisms, leading us to expect that a "pervasive indefiniteness" must be present in the real world. If an ontic quantum indeterminacy is indeed amplified across all scales—from the microscopic to the meso- and macroscopic universe, including planets, stars, and galaxies—then the speculative arguments for quantum theology and divine action we made starting in Chapter IV gain a more solid scientific foundation. While we cannot claim to have all the right answers to this metaphysical question, we have demonstrated that nothing in science prevents us from positing this conjecture. Not only does science not invalidate divine action by relying on deterministic notions of causality, but it even suggests its possibility. If we consider reality from the perspective of combinatorial causality and recognize that the long-term evolution of all micro-, meso-, and macroscopic physical systems is ultimately determined by quantum indeterminism (pun intended), then divine action becomes entirely plausible.

Therefore, the objection that God's action, if it exists, can't go beyond the microscopic quantum realm is shown to be a fallacy. If reality is causally open through quantum indeterminacy not only at the microscopic scale but also across all scales, then the otherwise deterministic universe must be fundamentally undetermined. The future is not a published encyclopedia that we simply uncover by turning its pages; rather, it is open and undecided. Moreover, this recognition can provide further insights into free will and the concept of determinism itself.

Science has taken for granted that deterministic causality is the foundation of everything that exists, viewing (epistemic) indeterminism as an annoying random anomaly resulting from the complexity of the cosmos and emerging due to our ignorance of following the chain of causes and effects. However, the opposite might be true: At its core, everything is fundamentally indeterminate, and determinism is merely a large-scale, macroscopic emergent property that influences the world's evolution far less than we previously assumed. Awareness of these aspects creates entirely new philosophical and theological avenues of research.

Ultimately, we might come to realize that God's action was not impossible, nor was it confined to the microscopic realm of a seemingly clockwork universe, but rather was hidden in plain sight all along.

In the Advanced Readings Section XIII.5 titled "Quantum Indeterminacy Persists Across Scales" (Masi, 2025b), I demonstrated this phenomenon using a damped driven pendulum rather than a double pendulum, but the underlying principle remains the same. However, the paper is purely technical and does not explore philosophical aspects. (Had I included such discussions, the publisher would have rejected it immediately for obvious reasons!)

VIII. Free Will and Determinism: Integrating the Eastern and Western Perspectives

1. The Western Perspective

We have explored the concept of free will primarily from a Western and scientific perspective. While we incorporated a post-material, holistic, and spiritual vision of reality, this framework largely reflects a third-person analytical approach rooted in Western tradition. However, alternative perspectives from other schools of thought adopt a more spiritual, first-person viewpoint. These should be seen not merely as alternative interpretations but also as complementary insights to the vision we have presented.

Can these radically different perspectives be bridged? I believe it is not only possible but necessary.

An interlude that could enhance our understanding of this arc of knowledge is to reflect on some interesting observations made by the Dutch philosopher Bernardo Kastrup, who, as we will explore in a later section, is a proponent of analytic idealism. Kastrup's insights on free will can aid in the development of an integrated perspective on the relationship between matter and spirit. I would summarize his vision as follows.[84]

There is no distinction between free will and determinism because we are part of Nature, and ultimately, it is Nature that makes choices in and through us. It is Nature that determines my nature, what I am, and what I do. What Nature decides to do is both what it wants to do and what it must do. Nature wills, yet it cannot do other than will precisely what it wants due to its inherent nature. Thus, there is an identity—indeed, a strict determinism—between Nature, our nature, and our actions. In other words, you act according to who you are and could not do otherwise. If you had chosen differently, you would not be who you are.

However, my identity, character, and inherent personhood, with all their desires and tendencies, determine my actions. I cannot will otherwise without becoming something I am not. In his own words: *"Nature wills what it must will, given what it is."* Everything is as it should be because the universe could not be anything other than what it is. Even if we feel misaligned with the universe, that misalignment is part of the universe as well. Yet, Nature makes choices that could not be different because if they were, Nature would not be what it is. On the one hand, Nature determines everything, including our behavior. On the other hand, Nature is free because there is nothing beyond it

[84] I'm referring here to a YouTube conversation entitled "Why Free Will & Determinism Are The Same": https://youtu.be/y2eOtzEB8D8

that could compel it to will otherwise. Thus, the apparent incompatibility between free will and determinism fades away.

Kastrup, however, is not discussing physical causal determinism, which views the world as a collection of countless material particles governed by the laws of physics—similar to tiny marbles that, when set in motion under specific initial conditions and influenced by external forces, bounce around while blindly following abstract differential equations. In this context, the determining factor is a Will: the Will of Nature, which wills in and through us.

The decisive point arises here. When people refer to "free will," they often mean liberation from Nature (i.e., the universal consciousness or the Mind at large), allowing the individual (often identified as the little ego) to act according to personal preferences, tastes, and desires. However, because there is no true separation between "me" or "you" and Nature, we do not determine these preferences and desires ourselves. Ultimately, the desires we pursue are manifestations of Nature's "waves" flowing in and through us. Thus, even the choices we perceive as expressions of complete freedom in pursuing our cravings are still, in essence, Nature's choices.

In a nutshell, Kastrup's claim is that free will and determinism are the same.

End of summary.

While his thoughts are interesting and mostly agreeable, he seems to overlook some central aspects, treating them as secondary. If these aspects are not fully addressed, his profound insights remain on the surface.

First of all, Kastrup's standpoint assumes that Nature is not much more than a blind agency driven by an instinct lacking any self-awareness, rejecting the idea that these characteristics may be mere surface appearances beyond which our physical minds cannot perceive.

In a previous section, I began to outline a broader vision of Nature. The seemingly unconscious physical determinism that our external senses perceive in the surface world conceals a much more conscious, vast, and luminous Nature. There exists a complementarity between our subconscious, conscious, and superconscious selves and the corresponding planes of Nature's existence. In this framework, we can distinguish between an "exterior Nature" marked by the determinism of matter and life instincts and an "interior Nature" that is fully conscious and operates according to principles beyond human comprehension.

Viewing the issue from this perspective adds a new dimension to the question of free will and many other inquiries.[85] Most scientists and

[85] For instance, environmentalism could evolve from a technocratic approach to environmental protection to one that is spiritually awakened. This shift would enrich our understanding by integrating deeper insights alongside scientific and materialistic perspectives. I believe this transformation is essential for effectively tackling the challenges arising from the destructive behaviors humans are inflicting on the planet.

philosophers tend to overlook this possibility because recognizing a deeper intelligence underlying reality conflicts with the naturalistic framework prevalent in intellectual circles. Additionally, it is often dismissed even in religious contexts because it resembles a pantheistic view that equates Nature with God. We will delve deeper into this topic in the next section.

Kastrup raises an intriguing logical point: Because nothing exists beyond Nature to compel it to act differently, Nature is entirely free to will as it chooses. This leads to a natural question: Can we free ourselves from Nature? According to Western scientific and analytical philosophical logic, the answer is clearly no. If there is nothing beyond this surface reality of Nature, then how could we possibly free ourselves from it? Freeing ourselves implies moving toward something else, but toward what?

What we recognize as "my personality" is largely shaped by the forces of Nature, compelling us to act on impulses we label as "my impulses," "my desires," and "my character." According to the perspectives of philosophers like Spinoza, Schopenhauer, and more contemporary thinkers such as Kastrup, everything operates as an unconscious mechanism. Because we are part of Nature, we, too, are bound to follow this machinery indefinitely. In this pantheistic view, in which Nature and God are unified, the universal mind—or Mind at large—functions as a blind instinct devoid of aim or purpose. It simply exists and acts according to its inherent tendencies, unable to do otherwise.

This perspective raises my eyebrows, leaving me with the impression that Nature is a "dumb deity." Has God been reduced to a mere "cosmic instinct," a sort of huge "cosmic idiot'? How does this non-teleological view of a Mind at large and our existence differ from Richard Dawkins' description of the universe as "blind, pitiless, and indifferent" or Yuval Noah Harari's assertion of an "absolutely meaningless human life"? I struggle to see a significant difference. It merely replaces the mechanistic, meaningless, and unintelligent universe of the physicalist, who claims "there is only matter," with an equally mechanistic, meaningless, and unintelligent universe of the idealist, who asserts "there is only mind." Is this truly a paradigm shift?

And who are the "me," "I," or "you" that are meant to be liberated?

Rationality alone cannot answer these questions. The underlying assumption is that all aspects of our personality are fully visible to our conscious mind. However, we often overlook the fact that this external persona is merely a superficial layer, hiding a much deeper, more radiant inner self. By looking past the ordinary understanding of the mind and beginning to sense, intuit, and feel the greater reality that lies behind the veil of appearances, we may come to recognize a different truth.

The analytic mind and the strict third-person approach of science focus on an exterior bubble: the individualization of Nature's instinctual and mechanical aspects. However, if we acknowledge the existence of a universal inner Nature that functions according to broader spiritual principles, we must also recognize that this inner Nature is individualized as well. Our self-perception and conception reflect an "exterior" personality largely shaped by the determinism of matter and the life instinct. We often remain unaware of our complementary "interior" personality, which is fully conscious and operates according to principles that transcend the instincts of our exterior Nature.

Kastrup "catches the lion's tail" and begins to sense and uncover a deeper truth that most Western philosophers overlook, yet he fails to take the decisive step forward.

This is where the spiritual experience proves particularly valuable. In the following section, I will explore the topic from an Eastern perspective, enhancing the previous discussion with two chapters from Sri Aurobindo's "Essays on the Gita." I believe this Eastern viewpoint offers deeper insights with a greater explanatory power. Due to its theistic approach, it is often dismissed by Western scientists and philosophers, particularly those in the analytic tradition. However, I contend that comparing these two perspectives and integrating them into a complementary cosmology is crucial for the future of science and philosophy, as it facilitates their convergence with spirituality and beyond.

2. The Eastern Perspective

"There is no assumption of personal will in the instrument; it is seen that all is already worked out in the omniscient prescience and omnipotent effective power of the universal Divine and that the egoism of men cannot alter the workings of that Will. Therefore, the final attitude is that enjoined on Arjuna in a later chapter, "All has been already done by Me in my divine will and foresight; become only the occasion, O Arjuna, "nimitta-matram bhava savyasachin." This attitude must lead finally to an absolute union of the personal with the Divine Will and, with the growth of knowledge, bring about a faultless response of the instrument to the divine Power and Knowledge. A perfect, an absolute equality of self-surrender, the mentality a passive channel of the divine Light and Power, the active being a mightily effective instrument for its work in the world, will be the poise of this supreme union of the Transcendent, the universal and the individual." - Sri Aurobindo - Essays on the Gita, p.209.

In the previous essay, I examined the Western perspective on free will.[86] Now, I will explore an Eastern viewpoint, specifically the insights of the 20th-century Indian mystic and poet Sri Aurobindo.[87] His work "Essays on the Gita"[88] addresses the issue of free will in chapters titled "The Determinism of Nature" and "Beyond the Modes of Nature."

Sri Aurobindo's terminology is rooted in Eastern Indian mysticism, which might be unfamiliar to Western readers due to its unique language, word choices, and cultural context. However, I believe his insights can significantly enhance and enrich the Western understanding of the world, Nature, and ourselves, revealing a deeper structure of reality. To aid in comprehension, here is a brief overview of the context.

In Indian Samkhya and Yoga philosophy, which intersects with Buddhist teachings, the relationship between "Prakriti" (Nature) and "Purusha" (pure consciousness, spirit, the true being) is fundamental to grasping the philosophical and spiritual framework of the universe from an Eastern perspective. Prakriti and Purusha are viewed as two distinct and eternal principles that together account for both the universe's existence and our experience of it.

Fig. 56 Sri Aurobindo (1878-1950)

Purusha embodies conscious being, serving as the witness to and enjoyer of Nature's forms and actions. It represents the unchanging, eternal, and undifferentiated Self. As the observer of all experiences, Purusha, according to

[86] Kastrup's perspective is just one of many philosophical theories on the topic and does not represent the entirety of the Western philosophical tradition. The same applies to Sri Aurobindo's viewpoint, which should not be viewed as representative of all Eastern philosophical, spiritual, and religious practices. Both serve merely as examples for a comparative study between East and West that could lead to a synthesis.

[87] Sri Aurobindo is often labeled a "philosopher," but he did not view himself as such, especially not in the traditional Western sense. Although many of his writings employ philosophical language, his vision stemmed from profound mystical experiences rather than analytical or rational thought. His insights emerged from a state of mental silence and the suspension of intellectual activity.

[88] The Bhagavad Gita, or just the Gita, is part of the epic Mahabharata, one of the central Hindu scriptures.

Samkhya philosophy, does not engage in action.[89] Purusha exists beyond space, time, and causality, residing in a state of pure awareness.

Meanwhile, Prakriti represents the phenomenal spatial-temporal world and encompasses everything that is subject to change, including the physical plane (the world of matter studied by ordinary science), the mental plane (the realm of the mind and thoughts), and the vital plane (the domain of life energies, emotions, and feelings). Thus, both the body and the mind are manifestations of Prakriti.

Prakriti is dynamic and active, shaped by three eternal principles known as the "gunas' (inherent qualities, tendencies, attributes): "sattva' (balance, equanimity, harmony), "rajas' (activity, passion, restlessness), and "tamas' (inertia, passivity, darkness). By itself, Prakriti is unconscious and cannot perceive or experience existence. It is through its connection with Purusha that the universe comes into being and is experienced. Conversely, without Prakriti, Purusha would have nothing to witness. The process of creation, evolution, and experience begins when Purusha associates with Prakriti.

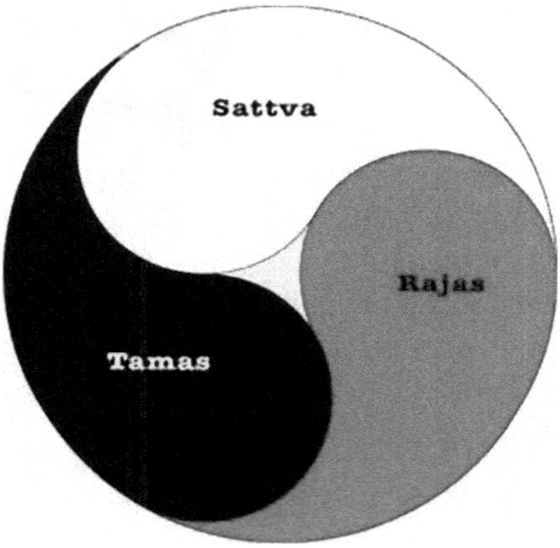

Fig. 57 The three "gunas.'

This cosmology is more elaborate than a Cartesian mind-body dualism. In our ordinary state of consciousness, the "jiva" (individual self) tends to identify with Prakriti in the form of a body and a mind due to its "avidya" (ignorance). This identification leads to "duhkha" (suffering), as the self becomes entangled in the cycles of birth, death, and rebirth. The aim of "moksha" or "mukti" (liberation), particularly in Samkhya and Yoga, is to realize the true nature of the self as Purusha and to disassociate from Prakriti. When this knowledge dawns and we are established in the "higher Self," an aspect of "Ishwara" (the supra- and intra-cosmic spirit in the cosmos), we understand that we are not

[89] However, this idea of a passive and inactive Purush is often challenged by mystical experiences that transcend traditional non-duality, as we will explore.

the body, the mind, or the vital, but the eternal, unchanging consciousness, Purusha. Once Purusha realizes that it is distinct from Prakriti and the illusion of "Maya" (the exterior "lower Nature" working in terms of separations and dualities), one reaches liberation from suffering. The ultimate aim of the spiritual journey is to recognize this distinction between Prakriti and Purusha. While Prakriti continues to exist, the realization of the separateness of Purusha allows the individual to transcend the cycles of karma and samsara and thereby achieve a state of liberation.

Sri Aurobindo makes his point as follows.

When we can live in the higher Self by the unity of works and self-knowledge, we become superior to the method of the lower workings of Prakriti. We are no longer enslaved to Nature and her gunas, but, one with the Ishwara, the master of our nature, we are able to use her without subjection to the chain of Karma, for the purposes of the Divine Will in us; for that is what the greater Self in us is, he is the Lord of her works and unaffected by the troubled stress of her reactions. The soul ignorant in Nature, on the contrary, is enslaved by that ignorance to her modes, because it is identified there, not felicitously with its true self, not with the Divine who is seated above her, but stupidly and unhappily with the ego-mind which is a subordinate factor in her operations in spite of the exaggerated figure it makes, a mere mental knot and point of reference for the play of the natural workings. To break this knot, no longer to make the ego the centre and beneficiary of our works, but to derive all from and refer all to the divine Supersoul is the way to become superior to all the restless trouble of Nature's modes. For it is to live in the supreme consciousness, of which the ego-mind is a degradation, and to act in an equal and unified Will and Force and not in the unequal play of the gunas which is a broken seeking and striving, a disturbance, an inferior Maya.

To be straightforward, the typical human state of consciousness is one of enslavement. This reality raises questions about our concepts of free will. It implies that our small egos function like puppets, bound by the rigid and mechanical determinism of Nature until we learn to embrace our higher Self.

However, it is essential to understand things in relation to the whole. It is about connecting the dots and, more precisely, uniting opposites into a Yin-Yang relationship of duality. How can we reconcile the perceived freedom of the soul with the mechanistic, deterministic, and seemingly unconscious aspects of “apara-Prakriti,“ the “lower Nature,’[90] which reflects “para-Prakriti,” the “higher Nature’ or the true “soul Nature“?

A certain absolute freedom is one aspect of the soul's relations with Nature at one pole of our complex being; a certain absolute determinism by Nature is the opposite aspect at its opposite pole; and there is also a partial and apparent, therefore an unreal eidolon of liberty which the soul receives by a contorted reflection of these two opposite truths in the developing mentality. It is the latter to which we ordinarily give, more or less inaccurately, the name of free will; but the Gita regards nothing as freedom which is not a complete liberation and mastery.

What we refer to as “free will” is only partially so. Our true inner self may sometimes provide hints or suggestions and, in rare cases, even determine our actions. However, we are largely at the mercy of Nature's whims. Our mental preferences, vital desires, appetites, and instincts—elements we identify as “my character“—are often merely the temptations of our lower Nature, leading us to actions we mistakenly think are our own.

This psychological perspective poses a challenge for Western thinkers, who often equate “free will” with the freedom to follow and indulge in our primal desires, preferences, and instincts without restraints or distinctions. However, this understanding creates an illusion of freedom; in reality, it represents a form of enslavement to our passions.

What we now call in our ordinary mentality our free will and have a certain limited justification for so calling it, yet appears to the Yogin who has climbed beyond and to whom our night is day and our day night, not free will at all, but a subjection to the modes of Nature.

We misinterpret our bondage to Nature’s movements—such as desires, subconscious thought patterns, emotions, and primitive instincts—as “freedom of choice” because we perceive the modes of Nature from a limited mentality and a state of ignorance. We like to think the impulses of Nature are our own. However, trapped in attachment to these modes, we mistakenly believe we are planning our future and making choices. In reality, these choices are

[90] Sri Aurobindo employs the terms ‘higher Nature’ and ‘lower Nature,’ while I have chosen the synonymous expressions ‘interior Nature’ and ‘exterior Nature.’ Some feel the terms ‘higher’ and ‘lower’ or ‘inferior’ and ‘superior’ have moral or judgmental implications, which was not Sri Aurobindo’s intention. He distinguished these concepts in a manner similar to how we differentiate between higher and lower programming languages, advanced and elementary mathematics, higher and lower contexts, or primary and higher education—without assigning moral value or judgment. However, I have noticed that this terminology often leads to misunderstandings and even emotional reactions. Therefore, I prefer to use ‘interior’ and ‘exterior’ in my writing except when I quote Sri Aurobindo directly.

predetermined by universal forces hidden behind the surface of life, leading our "ego-souls" to behave and act according to a different plan.

What then is this self that is bewildered by Nature, this soul that is subject to her? The answer is that we are speaking here in the common parlance of our lower or mental view of things; we are speaking of the apparent self, of the apparent soul, not of the real self, not of the true Purusha.

When one is discussing free will, it is crucial to ask: Who is supposed to be free, and free from what? Are we talking about the freedom of the ego-self or the higher Self? To use an analogy, the "I" or "me" with which we identify can be seen as a "knot" or small "vortex" within universal Nature, unable to free itself from Nature, much like a water swirl cannot extricate itself from water. We often confuse our "desire-soul" with our true soul. In failing to recognize how our true soul is intertwined with the desire soul, we misinterpret our surface personality of passions, desires, and impulses for our true personality, which stands beyond all these movements of Nature. This surface personality is shaped by desires, appetites, preferences, and tendencies largely influenced by our social conditioning and past experiences, all of which are part of Nature's movements and gunas. However, our true essence is much deeper and vaster than this flimsy phenomenal flickering.

If that is the case, the definition of free will as "the freedom to make a choice," commonly accepted by many Western philosophers, becomes highly problematic. The crucial question is: Who or what is truly free to make that choice—the lower self or the higher Self? Without recognizing this distinction, we often take pride in our freedom, unaware that we might simply be puppets controlled by forces beyond our surface awareness.

Fig. 58 The universe as the dual manifestation of Soul and Nature.

Sri Aurobindo diverges from the traditional view regarding the role of Purusha in the manifestation—the divine play or "Lila." The Purusha does not remain an inactive, passive witness indefinitely. As our true inner nature emerges and both the mind and the emotional being are transformed to align with the Light of the higher Self, this higher Self begins to influence the deterministic workings of Nature and ultimately takes the lead. Once the soul—our true identity, the "divine spark"—has matured sufficiently in its evolutionary journey, it increasingly liberates itself from the relentless material and mechanical determinism of Nature.[91]

So too with this truth of the determinism of Nature; it will be mis-seen and misused, as those misuse it who declare that a man is what his nature has made him and cannot do otherwise than as his nature compels him. It is true in a sense, but not in the sense which is attached to it,...

This is a typical understanding of free will defended by several modern scientists. For example, Francis Crick, the discoverer of the DNA helix, once commented:

"*You, your joys and your sorrows, your memories and your ambitions, your sense of personal identity and free will, are in fact no more than the behavior of a vast assembly of nerve cells and their associated molecules.*"[92]

Stephen Hawking followed the same reasoning:

"Though we feel that we can choose what we do, our understanding of the molecular basis of biology shows that biological processes are governed by the laws of physics and chemistry and therefore are as determined as the orbits of the planets." - "It seems that we are no more than biological machines and that free will is just an illusion."[93]

While evolutionary biologists and geneticist Robert Sapolsky doubles down in a podcast:

"Not only am I a free will skeptic, I don't believe there is a shred of agency that goes into any of our behavior."[94]

These were just a few examples of typical naturalist accounts. The instinctual tendency of the physical mind to dismiss the first-person perspective remains unquestioned, obscuring any distinction between material and psychological processes. The latter are often regarded as mere byproducts of the former. Meanwhile, according to Sri Aurobindo, we must ascend to higher planes of consciousness, of which sense- and physical mind-based scientific reasoning is only a superficial reflection. We should not confine

---

[91] Interestingly, many non-dual teachers deny the existence of free will. They recognize the illusory nature of the lower self, realizing the higher Self, but have yet to transform the other layers. In this way, Maya has deceived them twice!

[92] "The Astonishing Hypothesis: The Scientific Search for the Soul", 1994, ch. 1., Scribna.

[93] "The Grand Design," Stephen Hawking and Leonard Mlodinow, 2010, p.32, Bantam.

[94] Mindscape podcast, Sapolsky, 2021.

ourselves to these appearances as perceived from a narrow mental view, which misinterprets the determinism of Nature as the ultimate arbiter of existence. Instead, we must complement this perspective with a form of cognition that transcends the ordinary understanding of reality. How can we achieve this? Through self-transcendence obtained through spiritual practice.

When one has conquered one's self and attained to the calm of a perfect self-mastery and self-possession, then is the supreme self in a man founded and poised even in his outwardly conscious human being, samāhita. In other words, to master the lower self by the higher, the natural self by the spiritual is the way of man's perfection and liberation.

From this cognitive state, the context expands, revealing that Nature's determinism does not vanish but begins to make sense within a broader framework. Consciousness is not a simple binary on-and-off state; rather, it manifests in various forms that differ in kind and degree. In atoms, consciousness is nearly entirely obscured by tamas, resulting in mechanical and blind reactions with no free will. In plants, a life force emerges, yet it remains primarily reactive, exhibiting, at best, a rudimentary form of proto-sentience. In animals, the principle of rajas begins to override tamas, leading to basic forms of mind and memory, but they are still largely governed by Nature and incapable of acting otherwise. The soul acts as a passive observer, unable to influence the being's actions. Only in humans can thought fully develop, giving rise to an intelligent will (buddhi). Here, sattwa can take hold, marking a step toward freedom from the impulses of Nature.

However, as we all know, this is often not the case. Although we might not be like animals in some respects, we still grapple with natural passions, desires, mental constructs, habitual thinking, impulses, and cravings that frequently harm us rather than benefit us. We recognize that our intelligence can help us, at least to some extent, replace blind and automatic instincts with conscious, informed choices. This leads us to believe that we are free from the dictates of Nature and that we are making free choices. However, this is largely an illusion. The same human intelligence, the cherished qualities of reason and rationality that form the foundation of our scientific understanding of reality, set us apart from animals.[95] The decisive point that we should always keep in mind is that these attributes, along with our psychological traits and the ego-self to which we are so attached, are also creations of the lower Nature. [96]

The buddhi or conscious intelligent will is still an instrument of Nature and when it acts, even in the most sattwic sense, it is still Nature which acts and

[95] I am not claiming that humans are morally or ethically superior to other animal species. I am simply emphasizing the cognitive differences between us.

[96] The Buddhist mystics have a lot to tell us about this subject. However, I won't digress.

the soul which is carried on the wheel by Maya. At any rate, at least nine-tenths of our freedom of will is a palpable fiction; that will is created and determined not by its own self-existent action at a given moment, but by our past, our heredity, our training, our environment, the whole tremendous complex thing we call Karma, which is, behind us, the whole past action of Nature on us and the world converging in the individual, determining what he is, determining what his will shall be at a given moment and determining, as far as analysis can see, even its action at that moment. The ego associates itself always with its Karma and it says "I did" and "I will" and "I suffer", but if it looks at itself and sees how it was made, it is obliged to say of man as of the animal, "Nature did this in me, Nature wills in me", and if it qualifies by saying "my Nature", that only means "Nature as self-determined in this individual creature".

This is what Kastrup came to understand by an introspective move. Indeed, a powerful insight but also the point where the entire argument of the Western approach, along with the rationale discussed in the previous section, gets stuck. Nature actively wills and chooses through us, while the passive, observing Purusha remains in the background. This is not the complete picture; self-deception is always at play.

When we think that we are acting quite freely, powers are concealed behind our action which escape the most careful self-introspection; when we think that we are free from ego, the ego is there, concealed, in the mind of the saint as in that of the sinner. When our eyes are really opened on our action and its springs, we are obliged to say with the Gita "guṇā guṇeṣu vartante," it was the modes of Nature that were acting upon the modes."

Our sense of identity and selfhood, characterized by our personality, is also a mode of Nature's manifestation. Regardless of how noble our desires might be, we still act in accordance with Nature's will. Then, the identity between free will and determinism seems inescapable if we view ourselves only from a limited perspective. However, what the Western mind has not yet grasped, or does not find plausible, is that an inner higher Nature exists beyond the deterministic and mechanical exterior lower Nature in which the lower self we perceive is still embedded. To understand the workings of this inner higher Nature and distinguish it from the exterior lower Nature, a shift in consciousness is required—from the lower self of ignorance to the higher Self of knowledge. Only then can we transform our individual Nature from an enslaving deterministic chain into a liberating instrument.

In other words, freedom, highest self-mastery begin when above the natural self we see and hold the supreme Self of which the ego is an obstructing veil and a blinding shadow. And that can only be when we see the one Self in us seated above Nature and make our individual being one with it in being and consciousness and in its individual nature of action only an instrument of a supreme Will, the one Will that is really free.

Indeed, freedom exists, but it is found in the supreme Will that transcends Nature. While we identify with Nature, our true inner soul—though entirely free from it—remains a silent witness in the background. Our behaviors and choices are largely influenced by forces of which we are unaware at the surface level of consciousness. The link between this illusory sense of free will and determinism is nearly impossible to challenge. However, when we elevate our consciousness, relinquish all desires, and embrace the supreme Will, *"drawn towards it by the highest, most passionate, most stupendous, and ecstatic of all desires,"* this seemingly unavoidable connection to determinism dissolves.

Free will is an illusion and the bonds of determinism are real as long as we misinterpret the shadow we call "me" or "I" for the real self and real actor.

Nonetheless, this bond is not an unavoidable fate created by an evil demiurge. It serves a purpose and has a role in our evolutionary process.

The sense of free will, illusion or not, is a necessary machinery of the action of Nature, necessary for man during his progress, and it would be disastrous for him to lose it before he is ready for a higher truth. If it be said, as it has been said, that Nature deludes man to fulfil her behests and that the idea of a free individual will is the most powerful of these delusions, then it must also be said that the delusion is for his good and without it he could not rise to his full possibilities.

This delusion enables us to transcend ourselves. This paradoxical truth presents two opposing views. On the one hand, rational materialists assert that free will is impossible and determinism is the sole governing principle. Conversely, introspective individuals resist this notion, trusting their inner conviction that we are more than mere lumbering robots in a seemingly purposeless universe. Tension arises because both perspectives capture only part of the truth and fail to recognize their own limitations and the existence of broader horizons. Those who focus outward often misinterpret matter as the essence of the universe, while those who look inward might confuse their personality and preferences with their true selves, believing that the ego's notion of free will represents true agency. Both are still dwelling on the surface of the exterior Nature. By transcending these opposing views and embracing a third perspective, one can appreciate that both approaches are merely evolutionary tricks that Nature has orchestrated to guide our existence.

It is not wrong in thinking that there is something or someone within ourselves, within this action of our nature, who is the true centre of its action and for whom all exists; but this is not the ego, it is the Lord secret within our hearts, the divine Purusha, and the Jiva, other than ego, who is a portion of his being.

The real Self is the Will within, the true master of action. Free will exists, even in the face of Nature's determinism. However, it is not the freedom of someone or, more appropriately, "something" that is *"mastered by pain and pleasure, happiness and grief, desire and passion, attachment and disgust."* Nor is free will acquired from the ideas of the mind or the will of the identity we commonly call "I" or "me." This is where evolution plays a crucial role. As we undergo evolutionary and spiritual progress, the true self emerges, allowing us to recognize our authentic nature and the genuine source of our actions

But a time must come in our progress when we are ready to open our eyes to the real truth of our being, and then the error of our egoistic free will must fall away from us. The rejection of the idea of egoistic free will does not imply a cessation of action, because Nature is the doer and carries out her action after this machinery is dispensed with even as she did before it came into usage in the process of her evolution. In the man who has rejected it, it may even be possible for her to develop a greater action; for his mind may be more aware of all that his nature is by the self-creation of the past, more aware of the powers that environ and are working upon it to help or to hinder its growth, more aware too of the latent greater possibilities which it contains by virtue of all in it that is unexpressed, yet capable of expression; and this mind may be a freer channel for the sanction of the Purusha to the greater possibilities that it sees and a freer instrument for the response of Nature, for her resultant attempt at their development and realisation.

Sri Aurobindo's writings delve much deeper into this subject, and I encourage you to read his aforementioned essays yourself. The key point I want to emphasize in these two sections on Eastern and Western perspectives is that we can learn—through practice over time—to see these perspectives not as exclusive but as complementary to each other. Reflecting on Kastrup's thoughts on free will through Sri Aurobindo's spiritual lens serves as an example of how this approach enables us to perceive ideas from a new dimension, integrating both perspectives into a grand vision.

When we understand that the material reality we refer to as "Nature"—often perceived as a blind machine that sporadically gives rise to life in a small corner of the universe—is simply a facade for an interior Nature governed by trans-rational knowledge and self-determining power, the problematic relationship between will and determinism acquires a wider significance. If we accept, or at least entertain, the idea that our character and personality—with their preferences, desires, instincts, habits, dispositions, passions, and yearnings—are merely superficial expressions of an exterior Nature that are largely disconnected from our true identity, the question of whether the small "I" or "me" is truly free becomes preposterous. Once we recognize that what we perceive as our "identity"—the "lower self" or "ego"—is actually just a thought, a fiction, a shadow of our true soul that an inner Nature casts onto the mental plane, we start to understand who is genuinely bound by determinism and how we can attain freedom from it.

The Western analytic perspective, exemplified by Kastrup's reflections, and Sri Aurobindo's Eastern introspective approach represent two facets of an evolutionary journey. The Western approach allows us to move beyond mere intellectual reasoning into introspection, while the Eastern perspective transcends the limitations of the mind, guiding us further along our path.

IX. A First Step into Idealism

1. Is Reality an Illusion? Do We Live in a Simulation?

When I was a child, I remember someone asking the question of how we know that all we are experiencing isn't just a dream? Couldn't all our life, all the sensations we have, all the so-called "objective reality" we perceive, just be an illusion, a sort of dream with no inherent existence?

For some reason, this question never left me. As a young and naïve mind, I thought that the eventuality of our lives being nothing more than a figment, a sort of fairytale, something with no more and no less consistency than a dream, was a silly idea not worth of much attention. However, my attempt to deconstruct it and prove the opposite soon hit against very well-founded logical arguments, which left me for a long time questioning the validity of one point of view versus the other. On the one hand, it isn't easy to disprove that all of our existence is an illusion. On the other hand, it is hard to believe that nothing exists because, if so, after all, there is at least someone having that illusion. And who, then, is having that illusion? Only when I was in my twenties and encountered and studied the Indian non-dual teachings of Advaita-Vedanta could I begin to make sense of all this, putting things into context.

Here, however, I would like to avoid more or less philosophical or spiritual arguments that resort to mystical experiences and keep the line of reasoning grounded in strictly scientific and logical terms.

What recently has captivated technologists, philosophers, and the general public alike is the so-called "simulation hypothesis," which suggests that our reality might be an artificial simulation similar to a highly advanced video game. However, this notion stems more from contemporary technological hype than from a solid philosophical foundation. Its appeal is fueled by the current cultural dominance of digital technology, AI, and the speculative allure of science fiction rather than by rigorous philosophical reasoning. I believe the simulation hypothesis will ultimately be seen as a fleeting intellectual trend, reflecting the early 21st century's preoccupation with technology, AI, and digital virtual realities, rather than deepening our understanding of reality. While it resonates with today's technological zeitgeist, it is essentially a modern reinterpretation of ancient philosophical questions, such as Cartesian skepticism ("What if everything is an illusion?") and Plato's allegory of the cave, which we will discuss next. The term "simulation" only adds a contemporary gloss to these age-old inquiries without offering new insights; it merely reframes them in a way that appeals to the digital age.

Interestingly, though, taking the simulation hypothesis seriously might provide a shortcut to the core message of philosophical idealism, potentially even in a theistic context.

Firstly, a simulation itself has no causal power. Modern weather forecasts rely on computer simulations, but no one believes that these simulations cause rain to fall, the wind to blow, or the sun to shine in a computer's CPU. Similarly, while we can simulate virtual agents in a digital world, no one believes this makes them conscious, living beings.

Secondly, consider this: If we live in a simulation, who or what created the simulator? If an extradimensional alien civilization simulates us on their computers, how did they come into existence? Perhaps they were created by another simulator. But is that simulator also a simulation? Who or what, then, is the simulator of the simulator of our simulated universe? This line of questioning leads to a logical infinite regress, reminiscent of the idea of "turtles all the way down." Unless we propose that the "ultimate turtle" is a self-simulating and self-causing entity—what we commonly refer to as "God."[97]

Let's take a different approach—one I believe is much more interesting. Instead of indulging in fantasies about bizarre master simulators in cyber hyper-dimensions, we'll focus on a first-person perspective. From this vantage point, it becomes easier to understand that our perception of reality is, in a way, an illusion. Importantly, we can acknowledge this without succumbing to a solipsistic or nihilistic worldview. This perspective isn't merely a philosophical exercise or intellectual debate; it helps us detach from the overwhelming materialistic worldview that still prevails in our culture and shapes our self-conception.

To move beyond a superficial materialistic paradigm, we must first recognize how we construct reality in our minds. The strength of physicalism lies in the compelling way our minds shape what we perceive as "reality." It is difficult to accept that everything we call the "physical world" might be merely an illusion or that it doesn't truly exist. While there are valid reasons to believe in an objective reality that exists independently of our individual consciousness, minds, and subjective experiences, it is less clear that our perceptions accurately reflect reality as it exists. A critical realization is understanding the illusory nature of the world we perceive; this awareness is essential for transcending the confines of physicalism. Without it, any efforts toward achieving a post-material worldview, or even a post-material society and paradigm, will remain merely wishful thinking without the power to effect real change.

Therefore, let me elaborate on this issue, taking a particular standpoint that is purely scientific but inevitably leads to a surprising conclusion.

First, let me clarify the meaning of words and certain statements that have become slogans that don't always lead to clarity.

---

[97] Interestingly this notion of self-causation aligns with what we discussed in Sections V.7&8, and the advanced reading XIII.3.

Idealists often state that "everything is in consciousness" or that "all is mind." I prefer the former statement because, as pointed out in Section II.4, mind and consciousness aren't the same thing; there is a qualitative and decisive difference between a thought of the mind and the subject witnessing that thought. Anyway, for the time being, if you feel more comfortable with the latter statement, that's fine. It isn't so much the point I'm going to make here.

What does it mean that "everything is in consciousness'? If you have never pondered these existential questions, you might consider the idea that everything is in consciousness only a silly claim. Is a rolling rock that came down a hill and landed square on my head not really "out there'? Are the stone, the hill and, especially my headache that followed, only an illusion in my consciousness?

Some simple conceptual reflections on how we perceive the world can make clear what the real issue is.

Look at an object, say, a chair. Do you see the chair "directly'–that is, as it is in itself? Think about it.

When we look at a material object, what we really see is the reflected light of the object, not the object itself, let alone *as it is* in itself.

You might point out that, after all, if one has no visual impairment, one can safely say that one perceives a quite precise representation of the object, as if having almost direct contact with the object itself.

Let us think further.

The lens behind our pupil flips the image of the world upside down, and what reaches the retina is actually a reversed projection. It is a further modification of reality which, fortunately, isn't a real issue because our brain processes this information and interprets it as a straight image.

Let's think further because the chain of processes detaching us from reality is far from over.

We must not forget that all the 3D information of the object is converted into a 2D image on the retina. We never see the world in 3D, but the brain, again, helps us reconstruct back, or more precisely, "project back" mentally, the 2D information into a 3D mental model of reality. Furthermore, there is a transformation of a continuous mapping (the image encoded in the reflected electromagnetic light waves) into a discontinuous–that is, pixelated coarse-grained picture–due to the finite number of receptors on the retina surface.

In other words, the original information representing the objects has already undergone several modifications and transformations, and how much it retains the "reality" of the object itself (whatever that might mean) is questionable. Nonetheless, one might contend that we still have a good representation of reality as it is: We might have a bit of a coarse-grained image and lose a bit of sharpness or color, but the overall model of reality might still be a good approximation of an "objective truth.'

But do we really perceive this "model" of reality that reaches the eye retina?

The answer is negative, again: When the light impresses on the retina we still don't see a thing. At this stage, we don't see the light itself either, because the retina converts it through photoreceptor cells into electrochemical signals, and even these signals are far from having been converted into perceptible sensations. This bioelectric information on the retina is first processed further by bipolar and ganglion cells which integrate the information. Then the signal exits from each eye which is then sent to the brain via the optic nerve.

At this stage of processing, we can no longer consider the information content of the electrochemical nerve impulse as a one-to-one representation of the object we are looking at. We can't even say it is a more or less imprecise representation of the world as we considered the geometrical light projection onto the retina. At best, we could speak of "interpretations" but no longer of representations. While the image projected by a lens might be somewhat blurry and lose details, it can still be thought of as an approximate one-to-one map of the object itself. Here, however, once the light signals have been converted into chemicals and electrical impulses that travel through complicated nerve pathways in our skull, they no longer retain anything of the original geometric form, structure, colors, and other properties of the image. Everything has been encoded in neural spikes: a time series of electrochemical signals with an amplitude and frequency modulation that encode reality and no longer have anything in common with the original optical-geometric reflected "map of light rays" that triggered them.

How much of the reality of the object in itself is left at this point? And do we finally have the subjective experience of perceiving an image? Far from it.

The electric spikes traveling along the two optical nerves from the two eyes cross each other in the center of the brain at the optic chiasm. At this point, things are complicated further because the brain cuts the image from each eye into two halves and sends it toward separate hemispheres: The left (right) half of each retina is sent to the right (left) hemisphere of the brain. This contrived splitting and mixing of the retinas' information turn out to be useful for binocular vision and the above-mentioned 2D → 3D reconstruction visual process.

Do you now see something? Nope, still nothing.

Because, from the optic chiasm, the information travels through a series of structures in the brain, including the lateral geniculate nucleus in the thalamus and, especially, the primary visual cortex in the occipital lobe, where it is further processed.

Do we see something now? Still not.

All the information must finally be integrated with different areas of the brain responsible for analyzing different aspects of the visual scene, such as color, motion, and shape.

Only at this point, once all the features are "bound" together in a unique semantic whole, for an utterly mysterious reason that nobody can explain[98] are these neural spikes transformed into a unified meaningful subjective experience of a sentient being. Before that instant, it was all about the electrochemical neural activity that didn't produce any sensation.[99] However, suddenly, something binds all that convoluted information scattered throughout the brain into one single meaningful object and subjective perception that makes us say, "I see a chair."

The question is: How much of that complex and sophisticated neural processing retains anything of the reality of the object as it is in itself? The point is that there is no longer any connection, point of contact, or relationship between the object "as it is" and what shows up as a completely chaotic and indecipherable "bio-electrical storm" in the brain. What we see, perceive, and interpret of the world "out there" no longer has anything to do with that world. What we are aware of is only an interminable series of neural spikes that "represent" or "model" reality, no more and no less than the "ones and zeroes" digital digits of an encrypted and compressed image "represent" an image in a computer chip. Those spikes, bits, and bytes have nothing in common with what they stand for. They are only a projection of the brain. The question is not whether this projection retains reality as it is, it certainly doesn't, but who or what receives the projection?

This short tour into the neuroscience of our visual perception, which could also be made for all the other senses (that is, hearing, touching, smelling, or tasting), makes it clear that all what we perceive, sense, and think of the world does not reveal to us anything about the world "as it is" independently from those interpretations, figments, fabrications, or "shadows" going on in our brain (or should we speak of a mind?). In a sense, all that we call "reality" stands for reality itself, just as a shadow stands for the object casting the shadow. That shadow is a mental projection, at best, an "interpretation," but not a "model," "representation," or an "approximation" of the object itself. The shadow could be as precise as possible, but it can never, not even in principle, convey to us the real and full external structure let alone the internal structure of the projecting "entity" for which it stands. In this sense, the reality we perceive is an illusion, a metaphor, or a symbol.

---

[98] The reference is, of course, to the hard problem of consciousness. However, also the so-called "binding problem" comes into play. Nobody knows how the brain integrates all the perceptions (size, colors, textures, parts of the image, movements, etc.) into a meaningful and unique semantic whole, here, for example, the meaning 'chair.' An aspect that I will re-address in the advanced reading XIII.8: "No Qualia? No Meaning (and no AGI!)."

[99] At least not consciously. At a more subconscious or subliminal level, there seems to be some visual awareness of the environment, as experiments with people affected by blindsight seem to indicate.

The shadow-metaphor of the experience of the world is nothing new, and dates back to Plato's allegory of the cave. In this allegory, prisoners are chained inside a cave, only able to see shadows cast on the cave wall by objects passing in front of a fire behind them. These shadows represent their entire reality. When one prisoner escapes and sees the true world outside the cave, he realizes the shadows were mere illusions. Upon returning to inform the others, he finds that they don't believe him and prefer their familiar, limited worldview. The allegory illustrates how our perceptions can limit our understanding of truth and reality, and how difficult it can be to accept new knowledge that challenges our existing beliefs.

However, you don't need to be chained in a cave to experience a similar fate. All the phenomena registered by our sensory apparatus, and their transcriptions into mental images and representations, are our own "shadows."

Fig. 59 Representation of Plato's cave allegory.

Does this mean that everything is an illusion and that reality doesn't exist? Should we embrace a nihilistic philosophy?

Consider an object illuminated by a light source that casts a shadow on a wall. One might ask whether the shadow exists if the light is turned off. You would understand that once the light is turned off, the shadow—representing the phenomenon—disappears. However, unless you are a toddler no older than 18 months that still isn't aware of object permanence, you would also understand that this doesn't imply the object projecting the shadow disappears as well. It still exists, in darkness, but it is still there. We can still touch it. You know it exists even if you can't see it.

Now, suppose you mistake the shadow for the object that projects it. You aren't aware of the distinction and conflate the phenomenon produced by the object with the thing in itself. In a sense, we could say that our sensory perception of the world is like a projection—a depiction or representation—onto a screen in our mind. Then someone asks you again whether the thing we perceive still exist if we don't look at it. This question echoes Einstein's famous pondering about the strangeness of quantum physics: *"Do you really believe the moon is not there when you are not looking at it?"* The answer is that the moon is just a "shadow" in our mind, not the thing in itself as it truly is. Therefore, in a way, it isn't really there even when we are looking at it!

This is also a way to illustrate Immanuel Kant's distinction between—the phenomenon and the "thing in itself," the so-called "noumenon." The experience of the world is a projection in form of a sensed mental phenomenon on our "screen of consciousness" but this does not reveal us the inherent reality of the object itself, as it truly is (whatever that might mean.)

Therefore, this is not a nihilistic understanding of reality. It does not imply that nothing exists. There is a real world "out there," but we will never have any direct access to it and apprehend and comprehend things as they are in themselves. At least not our ordinary human consciousness.

I hope that the initial statement that "everything is in consciousness" sounds a bit less silly. After all, if everything is an illusion, who or what is having that illusion? What is the central presence that has all those perceptions, images, concepts, feelings, and emotions? There is an invariable fact that never changes throughout. There is a presence, a subject, a witness that perceives all those sensations and thoughts, no matter how unreal or delusive they are. That is consciousness.

Thus, saying that everything is in consciousness means that there is nothing we can speak, think, feel, or sense that isn't already a reflection or modification on the "screen of consciousness.'

Allow me to insist on offering another example. Another way to enhance the understanding of our sensorial abstraction of reality could be to use an approach that relies on information theory.

You might have heard of "steganography'–that is, the technique of encoding and embedding information into the same set of data. A simple example is the sentence "*Harriette is definitely done entertaining nonsense*," which hides the word "*hidden*" if we convene to read only the first letter of each word. With modern digital technologies, one can resort to similar but much more sophisticated encoding schemes to make the "real information" invisible to human senses, which see a completely different "reality."

In the two images of Fig. 60, you can see an example. The left image of the trees is what we perceive with our senses and what the brain reconstructs. However, if one removes all but the two least significant bits of each color component and further normalizes the scale of numbers therefrom obtained, the hidden image of a cat becomes visible. But the cat is not a "representation" of the trees!

In what sense is our perception of reality different?

One might object that no hidden or "alternative" images are embedded in the sensory information with which our brain works. Of course, just consider this as a metaphor. The idealist doesn't claim that there are hidden images encoded on our eye retina. Nevertheless, I hope you got the point.

Fig. 60 The image of a cat steganographically "hides" in that of a tree.

The information and semantic content of the image of the cat represents what we perceive (not represent!) of the world, while the image of the trees stands for the "world in itself" (or, if you prefer, it goes the other way around.)

And we might say that the image of the cat "encodes" or "stands for" that of the tree, but it does not "represent" it with whatever degree of precision. Nothing of the form, contours, or textures of the trees is "represented" in that of a cat. We don't sense reality as it is with "approximate representations." So-called "representations" are useful mental projections itself that serve our daily pragmatic needs but "represent" the intrinsic nature of reality no more and no less than shadows represent the object projecting the shadow, or cats represent trees. Any semantic content is reflected on a "mental screen" as well, but the associated thought is, again, a perception, an apprehension, and comprehension, still in consciousness. The common belief that science is an "approximation to truth" is untenable, because there is nothing to approximate in the first place–no more and no less than a higher resolution image of the tree will make "pop into existence" the steganographically encoded image of the cat without knowing the convened decoding scheme.

If you've made it this far and truly grasp the illusionary nature of our mental constructs of the world, you may realize that even a rock, a chair, or any piece of matter, as we perceive and conceive them, are merely abstractions—figments or shadows projected onto our screen of consciousness. With this understanding, the real issue behind the so-called "hard problem of consciousness" becomes clearer.

The difficulty in explaining how a material object like the brain gives rise to subjective experiences stems from our failure to recognize that we begin with an experiential and mental construct to explain the emergence of something that is itself experiential and mental. When we search for some contrived neuronal machinery supposedly creating conscious experience, we often overlook that our image of neurons and brains are already "shadows in the mind" that cannot explain the mind itself.

It's alike trying to understand the nature of an object, its substance, and structure, starting from its own shadow. You cannot use something to explain itself. You cannot elucidate the nature of a substance with the same substance. Similarly, you cannot grasp the essence of something by relying on the symbols that represent it. For example, you cannot comprehend what the word "wood" signifies if you attempt to characterize a forest in terms of the letters "w," "o," "o," "d."

Attempting to explain consciousness through the interaction of neuronal firings is like trying to describe the wetness of the ocean by analyzing the interactions of its water ripples. The subjective property of the former is already inherently present in the latter. We are merely trying to explain consciousness starting from waves of consciousness. It makes no sense to seek the emergence of something from the very same thing—it's already there!

If we want to go beyond a strict physicalist paradigm, we must become aware of how we construct reality in our minds. The strength of physicalism is based on the overwhelmingly convincing power of how our mind constructs, in itself, what we call "reality." Becoming aware of the illusory nature of the world we perceive is one of the most important realizations one must achieve to go beyond the grips of physicalism. This is what the idealists of the past realized.

We shall become aware of how all that we see, perceive, sense, and think is not only relative to a state of consciousness but is a figment, a phantom, a shadow of reality. It isn't enough to accept this as a belief; it must become a deep realization that we comprehend intimately.

Many great minds of the past articulated insights that predate our modern simulation hypothesis,, neurological findings, or information theory. These ideas form the foundation of philosophical idealism, which posits that reality is fundamentally mental, spiritual, or immaterial. Idealism asserts that the nature of existence and the external world is inseparable from the mind's perceptions, concepts, and consciousness. It challenges the materialist viewpoint by suggesting that the material world is either a construct of the mind or inherently linked to mental phenomena. Key forms of idealism include subjective idealism, proposed by George Berkeley, which argues that objects exist only as perceptions in the mind, and objective idealism, as presented by G.W.F. Hegel, which contends that a universal or absolute mind shapes reality. Idealism also intersects with epistemology, indicating that knowledge and truth emerge from mental constructs rather than from independent, material objects.

Philosophical idealism is only a first step toward a more integral synthesis of reality but a necessary step toward post-materialism. Intimately understanding, realizing, and comprehending this truth is the necessary basis for any form of post-materialism that goes beyond physicalism. Though this intellectual understanding does not lead to a mystic realization, it should sink into our mind and become at least an intellectual basis for how we understand the world. If we still believe that, when we look at things, we are seeing them "as they are," we will remain stuck in a naïve 19th-century scientific realism that prevents us from seeing and progressing further.

2. Is Consciousness a "Representations"?

In the philosophy of mind, another naturalistic way to explain consciousness is "representationalism." Representationalism is the view that phenomenal consciousness is representational, meaning that having a sensation, feeling, perception, or a mental state (believing, desiring, imagining, etc.) consists in representing the world in a certain way. Consciousness or a mental state is the kind of content it carries, the "aboutness" (how thoughts are about things). The phenomenal character of experience (what it's "like") is fully determined by its representational content.

For example, seeing blue feels the way it does because your visual system represents the world as containing a blue surface.. A visual perception of a red apple is a representational state that attributes redness and apple-shape to something "out there."

An interesting article that stands for this worldview is an article from Psychology Today by psychiatrist R. Lewis that supposedly explains how *"How Consciousness Emerges from Matter"* (Lewis, 2023).[100] It summarizes well the state of the art of certain modern academic consciousness studies. This is antipodal to all that I maintain about the nature of consciousness but, precisely for this reason, it is quite telling about what I call the "illusion of knowledge."

The subtitle states, "*Subjective experience is built from physically encoded internal representations*." I have already pointed out how the belief that we "represent" reality in our mind is questionable. Anyway, stating that "*internal representations of sensory perceptions of stimuli from the physical world are turned into mental phenomena*" isn't very telling. A painting physically encodes and represents bodies and objects of the world; does it, therefore, become conscious? Does the word "table," which encodes and represents with ink on a piece of paper an object of the world that we call...well... "table," awaken the paper to conscious awareness? The circumference C of a circular object of radius R is "represented" by $C=2\pi R$, with $\pi=3.14...$ So, should we believe that, at least in principle, if a certain representation is organized in the right and sufficiently complicated manner, on the canvas, on the paper, or even in a mathematical formula, a subjective experience suddenly emerges?

In a footnote, the author warns us, "*Internal representations should not be imagined as actual pictures residing somewhere in the brain. Rather, they are*

[100] Already, the title speaks volumes. That the brain produces consciousness is not in question. It is a premise requiring no further consideration.

diffusely encoded elements/traces of information stored across distributed neuronal networks, interconnected in highly complex ways."

In my view, this is only wordplay that replaces the word "representation" as a picture with that of a "configuration" as the arrangement and diffuse distribution of encoded elements across something in a particular form or combination. Think of the trees-cat ambivalence of the steganography we considered in Fig. 60. There is no "actual picture" of the trees in the cat, or vice versa; an nobody would believe this would elicit in and of itself a conscious experience.

First of all, a code is not a material entity; it is an abstract conventional agreed-upon set of symbols that already presupposes a mental agent able to read it by mapping the information (the configuration) to a semantically meaningful concept. Mind is already there. If we don't realize this, it is only because we aren't aware of how a "representation" emerges *in us*–that is, how we conflate configurations with its semantic aspect and that we call "representations," "images," and "models." You must go deeply into the first-person experience of mental processes to see that. Configurations are one thing, while their semantic value is another. The former refers to an organization of material objects in space (or a state-space); their semantic interpretation is not something inherent in them. Not the configuration itself but, rather, the meaning of the configuration becomes a mental representation, an image, and a model with meaning. These are, so to speak, "mental semantic qualia" that emerge in us and acquire meaning and significance only *after* the sensory experience of the configuration. What is "out there" are eventually states of matter, configurations of particles, and arrangements of neural networks. The states of matter and their configurations in the world could be reflected in other states and configurations and arrangements in our brains, but the semantic content does not exist in these. It is like the Rubin Gestalt face-vase figure. First, you perceive the figure–that is, the configuration–and only then does your mind decide what meaning it should present to our conscious mental awareness (the vase or the face?). One can become aware of this fallacy only if one turns their attention, through a first-person introspection, to how we perceive, cognize, and think.

Perhaps, making representations "internal" and pointing out that the brain "self-represent"' itself by cooking up "prediction model" of the environment with "neural states of resonance," will make this idea more plausible and better explain how all this "turns into mental phenomena and forms a basis for affect and emotion."

Does it? Resorting to "resonances," "feedback loops," "self-modeling," "synchrony," "phase-locking," "re-entrant circuits," or "self-referential processes" is in fashion nowadays, but why should a feedback loop make anything sentient? I regularly write software programs with feedback loops. Does my PC become self-aware when I click "run"?

Scientists rightly sympathize with "resonant feedbacks-whatever" because, indeed, these are ubiquitous in Nature, and life depends on feedback information processes based on reception, recognition/cognition, and reaction cycles (e.g., homeostasis). But constructing on these any sort of abstract philosophical speculations is an unwarranted intellectual move. The question is: Do these processes produce a conscious subjective experience, or, to the contrary, are these processes felt by an experiencing subject? By looking inside at how we perceive, understand, and react, one might well conclude that consciousness, with its apprehension of sensory qualia, the comprehension of mental images in terms of semantic values, and the will to react to the environment, uses these looping processes in Nature as a basis for making experiences–but how can these be its causal power? These recurrent processes are a medium for experience but can't explain the nature and origin of the experience itself, no more and no less than the cycles of a combustion engine are a medium for mechanical power but can't explain the nature and origin of the power itself.

So, to me, all these representational theories that are nowadays in fashion don't make much sense. Nothing in these theories substantiates their claim of having shown how consciousness emerges from matter, and they only deepen the mystery of how and why a neuronal process gives rise to subjective experience. It is an attempt to replace a seemingly immaterial entity (consciousness) with another even more supernatural process that mysteriously makes pop into existence something out of nothing. Consciousness as a representation only apparently saves us from falling into metaphysical hypotheses because, paradoxically, this comes at the cost of forcing us to make magical claims. The explanatory gap between representations (however "complex," "resonant," "external" or "internal," "predictive," you name it) and the emergence of consciousness, mind, and emotions remains abysmal. I guess these attempts will enter history as the new Ptolemaic clinging to the dying geocentric paradigm.

Anyway, this is some of today's neuro-philosophy that circulates in the best academic environments. I find it interesting to go through these theories, not only because we should always entertain ourselves with the viewpoint that stands opposite to our own–according to the good old Hegelian dialectic in which we must frame the antithesis to a thesis before summing up with a synthesis–but also because they highlight all the misconceptions and fallacies that physicalists have when they refuse to complement their limited analytic third-person perspective with first-person knowledge. Trying to dissect it is a good exercise in learning to look inward and becoming aware of how we think.

3. Platonic Forms, Ideas, and Archetypes: Ancient Wisdom for a Post-Material Paradigm?

Our society places excessive emphasis on strict linear, rationalistic, and materialistic thinking—a fact that has severely undermined our innate powers of intuition. We tend to view reality as a lifeless machine and life itself as a mere accident resulting from natural selection and random mutations in a universe that seems devoid of meaning and purpose. Intuitive thinking, subliminal perception, and spiritual insight are dismissed as remnants of emotional superstition or, at best, reduced to poetry. Only that which can be understood through analytical thought, logic, mathematics, and empirical evidence is accepted; everything else that arises from different cognitive abilities is regarded as a mirage, wishful thinking, or a chimera.

Well, I admit I have exaggerated somewhat. Fortunately, not everyone shares this worldview; indeed, many scientists do not. However, this rationalistic and materialistic perspective on life and ourselves is deeply ingrained in our culture and influences our choices at all levels, often unconsciously. I argue that this mindset arises not from a clear and rational scientific understanding of reality but, rather, from cultural conditioning rooted in a specific historical context.

Science emerged from the understanding that our senses can deceive us. For instance, while it appears that the Sun orbits the Earth, we have discovered that this perception is an illusion. This realization sparked the Copernican revolution, which ultimately propelled the advancements of Galilean and Newtonian science. However, there remains a deeper illusion that humanity must confront: beyond the deceptions of our senses, we must also challenge the illusions of our minds. The human intellect often accepts superficial appearances as the ultimate truth—something more diplomatically referred to as "objective reality." Just as our senses can mislead us, an intellect overly fixated on materiality can create delusions. Unfortunately, this understanding has yet to gain traction in contemporary scientific, academic, and, to a lesser extent, philosophical discourse. Moreover, we are often subconsciously tethered to the belief that only an analytical and rational approach to science can uncover ultimate truths. Especially when everything is expressed in mathematical language, there is a common belief that mathematization ensures rigor and scientific validity. However, the history of science demonstrates that, regardless of how rigorously a theory is represented mathematically, such rigor does not guarantee that it has anything to do with reality.[101] Upon reflection, one can see that this belief is quite anthropocentric. Ironically, it is precisely

[101] A contemporary example is superstring theory. After decades of mathematical research, it has emerged as a model of reality that exists solely within textbooks of theoretical physics and abstract mathematics.

this mindset that the scientific revolution sought to transcend, as it implicitly assumes that human reason represents the highest point of evolution and that no further cognitive development is possible.

Yet, despite the firm anchoring of materialistic thinking in people's minds, recent developments indicate a shift toward a Platonic and trans-physical understanding of reality. However, the authors of these emerging ideas often overlook the fact that their concepts are not as new and original as they believe. To avoid reinventing the wheel, it might be beneficial to briefly explore the insights of our predecessors.

How can we move from an analytic and physicalist perspective to a deeper understanding of reality, one that allows intuition to perceive beyond mere appearances, thereby enhancing our discernment? One can take many approaches to this question. An effective method is to reflect on the insights of ancient philosophers and spiritual thinkers regarding the existence of a reality beyond appearances. Are the objects and phenomena we observe in the natural world merely what they seem to our senses and minds, or do those appearances represent and signify something underlying reality? I believe contemplating these questions is not merely an intellectual exercise; it can help us rediscover alternative ways of perceiving reality and Nature that are still relevant and might soon be revisited. In this section, I will first outline the Western approach, then follow this with a discussion of a similar philosophical perspective in the Eastern tradition. [102]

a. The Western Perspective

In the third century BC, Plato developed his theory of Platonic Forms or Ideas ("eidos") as a philosophical concept to explore the relationship between the physical world and the world of ideas. He was inspired by the Pythagoreans, who believed that intelligible yet intangible structures, such as numbers, give objects their unique characteristics. The term "eidos" (εἶδος) is a Greek word meaning "that which is seen," "visible form," "appearance," or "image." For Plato, it referred to an intelligible realm of eternal and perfect "Forms," "Essences," or "Types" that exist beyond the ever-changing physical appearances. These Forms shape our perceptions of the phenomenal world and define the unique qualities of objects.[103]

[102] The perspective of indigenous cultures is missing here. I apologize for this omission, but I feel that I lack enough expertise on the subject.

[103] In contemporary terms, we can describe Plato's worldview as archetypal. The term 'archetype' derives from the Greek word 'archétypos' (ἄρχω, meaning 'to begin' or 'to start,' and τύπος, meaning 'pattern,' 'sort,' or 'type'). It refers to an original model from which copies are derived. In metaphysics, an archetype represents a fundamental transcendent principle that influences and shapes the world of human experience across various levels.

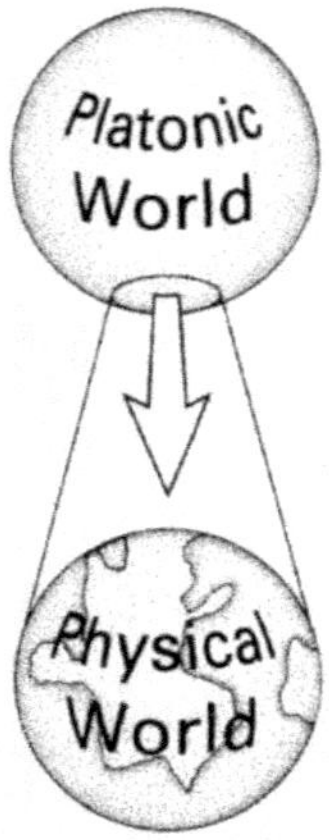

Plato proposed the existence of pure mental Forms—that is, abstract, non-material, and eternal entities that exist independently of our minds and the physical world. These Forms represent the perfect and unchanging essence of reality, transcending the sensible world while providing it with form and meaning. For instance, an object possesses a specific shape and beauty because it participates in the perfect form and beauty of something beyond itself, thus symbolizing that ideal.

The physical world consists of imperfect copies or imitations of these perfect mental entities, characterized by change, imperfection, and variability. This realm of universal Forms,[104] he called the "hyperuranion"; is beyond the material world but serves as the blueprint or ideal model for it. This includes mathematical and geometrical concepts that "in-form" our understanding of everything in the physical realm. For example, according to Plato, the perfect, unchanging, and immaterial Form of the circle exists in this realm of Forms, while any circle we encounter in the physical world is merely an imperfect copy or approximation of that ideal.

In general, all physical objects derive their qualities and characteristics from their corresponding Forms in this world of ideas. Plato believed that knowledge is a form of recollection, with our souls remembering the Forms. Through philosophical inquiry and dialectical reasoning, individuals can recollect this knowledge and come to understand the true nature of reality.

Plato's archetypal hypothesis significantly influenced the history of philosophy.

His student Aristotle proposed a less radical thesis that showcased his empirical approach to the study of Nature. Aristotle believed that universals exist, but they reside within individual things rather than in a separate realm.

[104] These universal Forms will later be called 'universals.'

This perspective is known as "hylomorphism," which asserts that form and matter are inseparable.

According to Aristotle's concept of hylomorphism, every object is composed of two essential components: matter (or substance, referred to as "hyle") and form (termed "morphe"). This metaphysical theory of substance posits that form is the essence that gives every material object and living body its organization, structure, properties, and identity. In this view, matter comes into existence only when form is present and active within it as a causative principle. Similarly, matter exists to receive its form.

Hylomorphism is also closely related to Aristotle's idea that potentiality exists for the sake of actuality. Matter represents an object's potentiality, while form represents its actuality, activating that potential. This potentiality has a dynamic character, facilitating a gradual change from one form to another, driven by an inherent impulse and purpose, i.e., the "telos," which reflects Aristotle's well-known teleological worldview.

Much later, in medieval scholastic philosophy, the 'problem of universals' became a central topic of philosophical inquiry. In the 13th century, Thomas Aquinas adopted Aristotle's hylomorphism, positing that form and matter are two distinct categories. While matter cannot exist without form, form can preexist matter; form gives matter its existence. Aquinas identified 'substantial forms' as those that determine the essential and immutable properties of material objects and 'accidental forms' as those that define mutable, non-essential properties. Building on this framework, he suggested that the human soul is a transcendent substantial form, with the body serving as its material embodiment. This reflects Aquinas' existentialism, which asserts that all things possess transcendental properties of being. At the foundation of everything lies the 'being' of things, a principle inherent in entities composed of 'quiddity' ("whatness")—their essence or nature. Consistent with Aristotelian doctrine, Aquinas also argued that all things possess final causes, serving a purpose and an aim.

Around the same time, the medieval scholastic philosopher Duns Scotus introduced the concept of "haecceity," which refers to the inherent uniqueness or individuality of an entity, characterized by its specific and irreducible qualities. This notion of "thisness"—with "haec" meaning "this"—defines what makes an individual distinct from all others.[105]

Subsequently, the late 18th and early 19th-century rise of German idealism, led by philosophers such as Kant, Fichte, Schelling, Schopenhauer, and Hegel, emphasized the significance of ideas and concepts in understanding reality.

[105] This might seem like a play on words, but it has recently been used to clarify the peculiar fact that, according to modern quantum theory, electrons are perfectly indistinguishable particles. So, how do we perceive them as separate entities? It is due to their haecceity! (Kastner, 2003)

Kant differentiated between a priori categories of knowledge, which exist independently of sensory experiences, and a posteriori categories, which rely on perceptions from our sensory apparatus. This distinction led him to establish the famous dichotomy between "noumenon" and "phenomenon"—that is, the "thing-in-itself" versus its phenomenal appearance. This concept differs significantly from Plato's perfect Ideas and Forms. In contrast to a hyperuranion beyond the sensory world where timeless ideas and forms reside, Kant's noumenon highlights the limitations of human knowledge, remaining agnostic about its true essence. Despite these differences, both philosophers agree on the existence of a reality that transcends appearances and our mental conceptualizations.

Inspired by Kant's transcendental idealism, Schelling, Schopenhauer, and Hegel each developed their own interpretation of Nature as an expression of a "World-Soul," as articulated in their major works.[106] This perspective reinvigorated natural philosophy, moving it away from the mechanistic, empiricist, and positivistic traditions that had become dominant through the Galilean and Newtonian sciences. For a time, the universe was once again viewed as a living entity unfolding according to inherent principles—at least until Darwin published "The Origin of Species" in 1859.

In the late 18th century, the convergence of Romanticism and science brought back archetypal concepts related to modified forms of Platonism and Thomism, particularly through Wolfgang von Goethe. His work "The Metamorphosis of the Plant" introduced the idea of the "Urpflanze" or primal archetypal plant. Goethe envisioned this archetype not as a static Platonic ideal but, rather, as a dynamic whole that evolves in both space and time. He never depicted it as a fixed image; instead, he presented it as a Gestalt in constant metamorphosis.

Similarly, he developed a broader natural philosophy centered on the "Urphänomen"—the primal archetypal phenomenon—especially evident in his theory of colors. The Urpflanze and Urphänomen cannot be understood through logical reasoning or inference; they must be grasped through extensive observation of plant growth and various natural phenomena. Goethe's archetype is not an immutable, timeless universal but, rather, an inherent aspect of organisms as a holistic "time-form." It represents a single unified form that unfolds over time, embodying qualities that are not immediately discernible to the senses or the rational mind yet manifesting a recurring motif within the organism as a dynamic whole.

---

[106] Kant's "Critique of Pure Reason," Schopenhauer's "The World as Will and Representation," Schelling's "On the World-Soul," and Hegel's "The Phenomenology of Spirit."

Fig. 61 The "anima mundi," or "World-Soul," inspired mystics and philosophers.

A creative unity is characterized by constant change and a dynamic flow of forms driven by an intentional primal force. Through repeated observations of various types, employing a disciplined approach of "delicate empiricism," we can identify the common invariances that all organisms share. Only then can the mind—or, more accurately, an intuitive act of cognition that goes beyond the mind—grasp the archetype within, thereby diminishing the distinction between subject and object.

"There is a delicate empiricism that makes itself utterly identical with the object, thereby becoming true theory. But this enhancement of our mental powers belongs to a highly evolved age" (Goethe 1995, 307).

Schelling's natural philosophy similarly emphasized the organism as the driving force behind the activities of Nature rather than a mere mechanism. He took this idea further by positing that will is the most fundamental element of reality, characterizing Nature itself as inherently willing and meaningful (Pitkänen, 2019). In this context, Schelling addressed the objection regarding whether the sense of unity in life and the purposiveness of natural products—seemingly guided by an "Idea"—is an arbitrary or necessary conjunction of our concepts.

"Whenever you conjoin things which are separated in space in a single aggregate, you act quite freely; the unity which you bestow on them you

transfer to them simply from your thoughts; there is no reason residing in the things themselves which required you to think of them as one. But when you think of each plant as an individual, in which everything concurs together for one purpose, you must seek the reason for that in the thing outside you; you feel yourself constrained in your judgment; you must therefore confess that ***the unity with which you think is not merely logical (in your thoughts), but real (actually outside you)***"[my emphasis] (Schelling [1797] 1988, 32).

Plant photographs capturing the sensual beauty of nature from "Urformen der Kunst" (Primal forms of Art) (1928) by Karl Blossfeldt (1865–1932).

Thus, Nature and life exhibit a fundamental unity and purpose, characterized by an intrinsic organization that cannot be dismissed as mere illusion or mental trickery. Organisms display a foundational archetypal phenomenon, revealing an internal relationship of forms that is inconceivable without a pre-linguistic archetype underlying all their variations and expressions. Although both Goethe and Schelling rejected the notion of teleology or divine determination, they viewed organisms not as passive entities shaped by their environment but, rather, as participatory, agential, and purposeful unities within the natural history of a World-Soul, whose transformation is driven by internal forces.

In the 20th century, the British philosopher Alfred North Whitehead developed his "process philosophy," which shares some important similarities with Platonism.

Whitehead considered a system of eternal objects that resemble Plato's Forms in a metaphysical reality beyond mere physical existence and that provide a structure and function to reality as pure potentials that can be realized in the world. His was a participation theory that saw reality as composed of "actual occasions" participating in eternal objects. Whitehead's universe is structured by an underlying rational order that emerges dynamically through the creative advance of the universe. Whitehead's eternal objects are immanent in the world's processes. They do not exist separately but, rather, are realized through actual occasions. In Whitehead's metaphysics, agency emerges due to these actual occasions—that is, through the "prehension" of possibilities and by its creative actualization. "Actual occasions of experience" are actualized in processes, with eternal objects existing only as pure potentials.

The difference, however, is that while Plato's Forms are unchanging, timeless ideals, Whitehead's process philosophy sees reality as fundamentally in flux. Instead of a realm of fixed Forms, Whitehead's universe is an ongoing creative process of becoming. The underlying Nature's creativity is a guiding aim that he calls the "divine lure." He did not impose fixed structures but, rather, conceived of the world as a system developing toward higher forms of order, interacting with the creative advance of reality.

Whitehead's process philosophy can be seen as a dynamic reinterpretation of Platonism, replacing Plato's static hierarchy of Being with a fluid, evolving cosmos. While both recognize the importance of ideal structures in shaping reality, Whitehead integrates them into a world of processes of evolutionary becoming rather than being.

Finally, when discussing archetypes, one can't omit Jung's perspective. At the beginning of the 20th century, Jung defined archetypes as fundamental, universal symbols and patterns found in the collective unconscious. This concept parallels Plato's ideal Forms, which are unchanging and transcendent entities shaping our perceptions and experiences. However, while Plato's Forms pertain to abstract ideals, Jung's archetypes relate to psychological qualities that embody universal, innate patterns or images within the collective unconscious of all humans. These archetypes significantly influence our thoughts, feelings, and behaviors. Just as Plato's Forms represent ultimate reality and the foundation of knowledge, Jung's archetypes serve as essential concepts that illuminate recurring themes, characters, and symbols in various cultures, literature, and art.

Jung also engaged in a fascinating exchange of ideas with physicist Wolfgang Pauli regarding the potential connections between psychological archetypes and quantum physics. The "Pauli-Jung Conjecture," speculated that archetypes extend beyond mere mental constructs, suggesting that meaningful coincidences, or synchronicity, might reflect deeper connections between psychology and physics. The conjecture posited the existence of an unknown third realm, beyond both the mind and matter, which gives rise to both psychic phenomena (archetypes) and physical laws (quantum mechanics).

This established the foundation for the mind-body theory of "dual aspect monism," which posits that there is an underlying reality that is neither purely mental nor purely physical. Instead, the mental and physical domains are viewed as aspects or manifestations of this unified reality, in which they do not exist as separate entities. This perspective serves as a springboard for what we will later describe as a multi-modal integral cosmology.

In summary, my question is: Should we dismiss old metaphysical speculations as outdated ideas worthy only of history books?

While many believe this to be the case, I disagree. One need not adopt a strictly Platonist perspective—an approach I find questionable and limited—or revert to some form of Aristotelian geocentrism. There is no need to fully embrace their theories. However, we might consider whether these diverse worldviews were merely philosophical abstractions or, instead, if they reflected a deeper truth captured by intuitive minds.

As we will explore in the next section, similar ideas exist in Eastern cultures. Is this merely a coincidence, or could it be a trans-cultural exchange of knowledge of which we have lost trace? Alternatively, to frame it in Jungian terms, might this lost wisdom represent a shared spiritual knowledge within the collective unconscious that emerges across different times and cultures?

b. The Eastern Perspective

Platonist and archetypal concepts are typically linked to the Western philosophical tradition. However, similar ideas exist in Eastern cultures, addressing themes of abstract principles, universal archetypes, and the nature of reality. The key distinction lies in the context in which these philosophies developed. Eastern philosophies evolved within a more mystic, spiritual, and religious framework, contrasting with the analytical approach predominant in Europe. Eastern thought emphasizes a holistic understanding of reality rather than focusing solely on its individual components. In this context, mystics, sages, yogis, and rishis perceive and articulate their understanding of the world through transcendental, first-person experiences, relying less on analytical reasoning, though some use rigorous philosophical language to express their insights. Nonetheless, the concept of an underlying order that transcends appearances and shapes the phenomenal world pervades Eastern traditions as well.

The Platonic notion of 'eidos,' or that of Jung's archetype, representing an intelligible yet intangible realm of forms and transcendent first principles that the phenomenal world reflects, resonates with several aspects of Eastern metaphysical theories of reality. Similarly, the modified Aristotelian interpretation, which posits that immaterial forms determine the potential of physical substances, and the noumenal ideas of Kant, are not foreign to the ancient spiritual intuitions of the Asian continent. While there are differences and nuances between these philosophies, they are not significantly more pronounced than those found within Western perspectives.

Let us briefly survey some of these concepts from a bird's eye perspective.

In Taoism, the impersonal, formless, and immutable Tao (or Dao) represents the eternal principle that underlies all of reality and existence. It is the source of everything and operates universally. While it does not align perfectly with any Western philosophical category, it shares similarities with

Plato's Forms as a formless essence that gives rise to form. However, unlike Plato's mental realm of perfect Ideas and Forms, the Tao is transcendent and foundational to all existence.

In Chinese philosophy, the principle of Yin and Yang describes the duality, complementarity, balance, and interconnection of opposite forces. Common polarities in Nature include male-female, light-darkness, and receptive-active dualities. This principle harmonizes all universal processes and extends to the living realm and the human psyche. Rather than being an object or an immaterial entity like a Platonic Form or archetype, it represents an inherent principle within the cosmos that generates the diversity and multiplicity of all things and processes.

Fig. 62 The Ying Yang symbol of Chinese cosmology.

In Buddhism, the concept of Śūnyatā, often translated as "emptiness" or "vacuity," carries various meanings. However, in most schools, it signifies an ultimate undifferentiated reality from which all distinctions arise. Notably, the Mādhyamaka teachings of the Indian Buddhist philosopher Nāgārjuna argue that all things and phenomena (dharmas) are "empty" in the sense that they lack inherent true existence. Everything emerges from a relational interconnectedness with other objects, which means each thing has no identity or essence apart from its constantly changing status relative to others.[107]

This philosophical stance might align with Kant's distinction between noumenon and phenomenon; however, a key difference lies in the absence of relativity in Kant's framework. For Nāgārjuna, the "emptiness" of things is knowable, unlike Kant's unknowable "thing-in-itself." Furthermore, not only sensory perceptions but also mental events and ideas are considered "empty phenomena." Our thoughts and ideas are just as real or unreal as any physical phenomenon. Despite these differences, a common thread remains: An ultimate truth exists beyond mere phenomenal appearances.

Central to our discussion is Hindu spirituality and philosophy, which asserts the existence of an all-pervading, transcendent, and supreme reality known as

[107] This idea resonates with the linguistic contextuality proposed by Swiss linguist and philosopher Ferdinand de Saussure, suggesting that language reflects a deeper truth about reality.

"Brahman." This essence is characterized as "Satcitananda," and it encompasses Existence, Consciousness, and Bliss. Brahman represents the ultimate, undifferentiated, formless, and unchanging infinite reality that underlies and supports all manifestations. Its fundamental nature embodies unity amidst diversity and multiplicity, yet it remains the "One without a second," as described in Advaita Vedanta.

Brahman transcends the Platonic realm, surpassing even the mental world. It is a universal consciousness that shares similarities with, but also diverges from, the World-Soul of German idealism. The Upanishads and Advaita Vedanta highlight that our individual selves, or Atman, are identical to the impersonal Brahman that encompasses all selves, famously expressed in the phrase "Tat Tvam Asi," meaning "Thou art that."

Fig. 63 The symbol of the sacred sound Om (or Aum), is said to be the essence of the supreme Brahman, underlying the Cosmos.

Brahman is transcendent, formless, and devoid of attributes or qualities ("nirguna Brahman") while simultaneously manifesting in the universe with infinite forms, attributes, and qualities ("saguna Brahman"). Our ordinary experience of the world is viewed as an illusion of separation and duality—a cosmic illusion referred to as "Maya." However, those who attain liberation ("moksha" or "mukti") can transcend this illusion and recognize that Maya itself is merely another manifestation of the same Brahman.

In the Advaita Vedanta tradition, the ordinary human, or the 'non-liberated' individual, lives in a state of illusion likened to the metaphor of a snake and a rope. When we mistake a rope for a poisonous snake, we live in unnecessary fear. Only by awakening to the realization that there is no snake but merely a rope can we free ourselves from this fear and anxiety. Here, the rope represents the ultimate reality, Brahman, while the snake symbolizes our superficial and false perceptions of the world, known as Maya. This metaphor echoes Plato's cave allegory, illustrating how we become entangled in the play of appearances and fail to recognize their true nature. We confuse the illusory world with true reality—a situation that also applies to the realm of names and forms (nama and rupa). This illusion dissipates once we realize the true nature of reality.

For the Vedic rishi, this understanding was not merely a metaphysical speculation or poetic imagination; it was a lived experience. This insight emerged through spiritual practices like yoga and asceticism, which involved detaching from superficial sensory perceptions and exploring reality through meditation, contemplation, and self-realization to attain higher states of consciousness and cognition. While there are both similarities and differences between the Vedic ontology and various forms of Western archetypal Platonism or Kant's transcendental idealism, they share a common transcendent and metaphysical nature. Both assert the existence of a reality beyond appearances, beyond the physical and material world.

Moreover, it is hard not to draw parallels with the metaphysics of the Neoplatonists, particularly those of Plotinus. He argues that by detaching ourselves from sensory perceptions, we can connect with supra-sensible realms. This leads us to understand the three "Hypostasis," or the fundamental substances that underlie all of Reality: the One (the Good, God, that which transcends being), the rational intellect (the "nous," or mind), and the irrational soul (which is driven by desire, entangled in matter, and capable of sensations).

Analogies can be drawn between Spinozian pantheism, which identifies God with Nature as a manifestation of various "modes of substance," and the idealism of the German philosophers discussed earlier. This duality of World and Will, or World and Soul, is well represented in the Vedic vision, particularly in its later forms found in Advaita Vedanta and Samkhya philosophy, which we explored in the section on free will. Here, we see a distinction between the static and active aspects of Brahman: 'Purusha' represents the soul, while 'Prakriti' refers to Nature or Maya, the cosmic illusion. Purusha embodies the static, stable, unmoving, unchanging, and eternal soul of the infinite Brahman, the One Spirit underlying the universe. In contrast, Prakriti is the dynamic play of Nature, the active aspect of Brahman, which realizes and experiences itself through its powers, will in action, and expressions of force and form in both spatial and temporal manifestations.

The duality of Purusha and Prakriti might resemble various philosophical concepts, such as Spinoza's identity of God and Nature, Cartesian mind-body dualism, and the Platonic and Aristotelian distinctions between form and matter. It also echoes Kant's differentiation between a priori and a posteriori categories. However, this notion transcends mere dualistic cosmology; the universe encompasses much more than just Soul and Nature.

A clear East-West divergence is evident in the cosmological symbolism of the Hindu pantheon. Each deity represents a specific quality or universal property. In contrast, Platonism typically views archetypes as impersonal, devoid of the personal traits associated with gods or deities. However, these distinctions are largely contextual and can seem arbitrary. Qualitatively, there is little difference between Hindu deities and the supernatural beings that Abrahamic religions refer to as 'angels.'

The primary distinction between the Vedic tradition and Western thought is that the Vedic perspective goes beyond a simple matter-soul duality. It introduces several "planes of consciousness" or "planes of existence" that exist in between. The Vedic and Vedantic framework is based on three fundamental principles of our ordinary experience: matter, life, and mind, each with both universal and individual aspects.

First, there is the physical universe, which is referred to as the "physical plane of matter," alongside the physical body, or "sheath" of any living organism, known as the "annamaya-kosha" or "food-sheath." Next is the cosmic life-plane, along with the individual life force present in every living being, called the "pranamaya-kosha" or "life-sheath." This is followed by the cosmic mind-plane and our individual mind, represented by the "manomaya-kosha" or "mind-sheath."

The fourth plane is the superconscious principle of knowledge, known as "vijnana," which signifies free or spiritual intelligence—a concept that Sri Aurobindo referred to as the "supermind." The fifth and highest level is Satcitananda, or Brahman, which embodies the triune aspects of existence, consciousness, and bliss.

For the soul to ascend to its Source, it must navigate these intermediate levels of consciousness, undergoing a gradual evolution in the process.

This subdivision in planes should not be taken literally; it arises from an arbitrary mental distinction along a continuum of states of consciousness. This framework aids our intellect, which operates on the mental plane but cannot fully grasp the nature of these superconscious planes. Exact categorizations are unnecessary. We will explore this aspect in more detail later.

For now, it is crucial to understand that we can comprehend these spiritual cosmologies only by broadening our perspective beyond a simplistic two-modal mind-body Cartesian dualism or Platonic matter vs. ideal Form metaphysics. Instead, we could adopt a multi-modal spiritual cosmology that recognizes various levels of consciousness and distinct planes of subtle substances. These levels separate a "beyond" from gross matter, which, taken together, reflect the same divinity that has forgotten itself.

This was merely a superficial overview of Eastern philosophies regarding reality, mind, consciousness, and phenomena. Nonetheless, these worldviews collectively reflect a shared human desire to explore whether and how our minds might be misled by appearances. This is the same theme I discussed in relation to Western philosophies in the previous section. When we honestly examine our perceptions and understandings of the world, we can't help but conclude that they are largely projections, shadows, or masks rather than true representations of reality. This is what all cultures throughout history have recognized: the existence of something that transcends mere appearances.

A scientifically-minded realism might lead us to believe that these ideas are merely cultural products rather than first-person perspectives that reveal something about reality. However, the universality of certain concepts across cultures and throughout time makes this viewpoint difficult to sustain. Reflecting on these questions and choosing a path to understanding reality based on our intellectual interests and spiritual inclinations is not merely an abstract philosophical exercise; it can actually expand our awareness of dimensions we often ignore or refuse to acknowledge. Contemporary materialism, which profoundly shapes our culture and influences our perception of the world, life, and ourselves, has roots in a strictly physicalist and rationalistic mindset. If we limit ourselves to dissecting, rationalizing, and reducing reality to mere blind and unconscious processes, we will perceive only that narrow reality, thus leaving our intuition, imagination, and deeper spiritual insights underdeveloped. Conversely, by exercising our intuition and developing a cognitive ability to see beyond the surface, we may open ourselves to dimensions of reality that could guide us toward a post-material future.

c. The Faculties of Consciousness: An East-West Perspective

I would like to provide a practical illustration of how Eastern philosophies—and, more broadly, other non-Western philosophies—can inspire us to expand our mental horizons. By adopting a more integrated approach to knowledge, science, philosophy, and culture, we can enrich our understanding. I will explore this with a specific example, arguing that the five senses might be fundamental rather than emergent.

According to contemporary theories of evolutionary biology, the five primary senses—sight, hearing, touch, taste, and smell—evolved over hundreds of millions of years as adaptations that helped organisms survive and reproduce in their environments. Their evolutionary origins can be traced back to simple sensory systems in early life forms, which gradually became more complex through natural selection.

Touch is the most ancient sense, as it appeared in the earliest single-celled organisms. Primitive life forms, such as bacteria, used mechanoreceptors to detect pressure, temperature, and chemical gradients in their environment. We know that single cells already reach out and touch each other. They respond with ion channels that open on contact, making intercellular communication come alive. Mechanosensitive channels reside not only on the cell's membrane but also in lysosomes, which are organelles inside the cell. The instinct of "reaching out and touching" something or someone seems primordial and was present from the very beginnings of life. As multicellular organisms evolved, specialized nerve cells developed to process tactile information. This led to the complex somatosensory systems found in modern animals.

Meanwhile, the ability to detect chemicals in the environment was crucial for early life to find food, avoid toxins, and communicate. Single-celled organisms had protein receptors that detected specific molecules—an ability inherited by more complex life. In vertebrates, taste buds evolved to detect nutrients and harmful substances, while olfactory receptors became specialized for detecting airborne chemicals. Smell evolved as an extension of taste, allowing terrestrial organisms to detect volatile chemicals from a distance and thereby boost their chances of survival.

Vision originated from simple light-sensitive cells in early organisms; it helped them differentiate light from darkness. The evolution of the eye began with ancient light-sensitive cells around 600 million years ago. It followed a gradual process, with pinhole eyes evolving into lens-based eyes that allowed for sharp images. Over time, these cells formed clusters, leading to rudimentary eyes in ancient animals like trilobites. Ultimately, it culminated in the sophisticated visual systems found in modern animals, including humans. Different organisms developed specialized vision systems, such as color vision in primates and ultraviolet vision in birds.

Hearing is another complex bio-physical process; it converts sound waves into nerve impulses. The evolution of hearing started 250 million years ago along with mammalian evolution, evolving from the faculty to detect vibrations in the environment. Early aquatic animals had simple vibration-sensitive cells that later formed the lateral line system in fish. As vertebrates moved onto land, specialized structures like the eardrum and inner ear bones developed to detect airborne sound waves. Mammals evolved a highly refined auditory system, including three middle ear bones that amplify sound. The sensory cells responsible for hearing, known as "hair cells," are located in the cochlea. These hair cells transform mechanical vibrations into electrical signals through a process called "mechanotransduction." The cochlea contains a basilar membrane that vibrates in response to sound waves, with different regions reacting to different frequencies. These vibrations are then transmitted to the brain via the auditory nerves. Mammals display a remarkable variety of hearing abilities. For instance, elephants communicate using infrasonic sounds, while bats navigate their surroundings through echolocation calls at ultrasonic frequencies.

According to Darwinian theory, these senses provided significant survival advantages, as they allowed animals to detect food, predators, mates, and environmental changes. Over evolutionary time, natural selection favored species with increasingly complex and specialized sensory systems. The result was the wide diversity of sensory abilities seen in the animal kingdom today.

This, in a nutshell, is how modern science describes the nature and origin of our five senses. In other words, science suggests that these senses arose from a lengthy bottom-up process of biochemical and biological evolution.

On the other hand, in the Indian Vedic tradition, people thought the other way around. According to the Aitareya Upanishad, the Self or Atman—the Self-Existing Being—was alone in the beginning. This Being encompassed all the "modes of consciousness," which were unexpressed and existed only in potential. Not until the primordial Purusha (the eternal Spirit, the consciousness of the cosmic being) projected itself into the manifestation of matter, space, and time did these modes of consciousness emerge as "faculties of consciousness," as his Word, Breath, Sight, and Hearing.

In other words, while the Veda had no concept of evolution, at least not in the way we understand Darwinian evolution, it did view our physical senses with its sensory perceptions as something imitative of more fundamental faculties of a universal consciousness that have always existed. These faculties of consciousness pre-existed life, corresponded to higher cognitive capacities, and, during the creation of the universe, were cast down and "involved" in matter in the form of physical sensorial senses. The Upanishads specifically envisioned seeing, hearing, and touch as the primary faculties of consciousness, while taste and smell were considered derivatives of these. Rather than being the end result of a complex biological process, these faculties were considered the starting point and inherent qualities of the creative Spirit that brought this universe into existence and projected them in physical embodiments in sense organs as its shadowy and limited sensorial forms, with which we are familiar. However, the true inner senses of seeing, hearing, and touching are not merely the physical sensations we experience in our ordinary state of consciousness; instead, our ordinary sensations echo the Spirit's true Seeing, true Hearing, and true Touching. They serve as a means of knowledge that connects us only indirectly to the spiritual truths of things and that can be apprehended and comprehended only from a higher level of consciousness.

The rishis, or Vedic seers, categorized the faculties of consciousness into two groups: Seeing and Hearing as perceptive faculties and Mind and Word as their active counterparts. Mind and Seeing are associated with Forms (rūpa), reflecting the aspect of Power, while Word and Hearing are related to Names (nāma), representing the aspect of Knowledge.

When the rishis referred to "Names," they meant more than verbal labels. "Names" signify powers, qualities, and characters (features and traits that distinguish one thing from another) caught up by our consciousness. These "Names" are latent and inherent in the nameless and timeless Absolute but manifest in the temporal world as the qualities and powers of its earthly forms. Similarly, "Forms" encompass not only signs, words, and symbols but also sound waves, ideas, mental concepts, and even feelings or emotions, all of which are manifestations of consciousness.

Thus, Names serve as expressions of Knowledge, while Forms function as expressions of Power. Together, they constitute the phenomenon of Consciousness in manifestation. Through these aspects, Brahman can engage with its creation.

In the post-Vedic period, Sāṃkhya and Yoga viewed the mind (manas) as one of the faculties of consciousness, alongside seeing and hearing, as understood by Vedic and Vedantic seers. The mind was not seen as the product of an evolutionary process; instead, it was regarded as a fundamental aspect of the universe that existed prior to the emergence of life.[108]

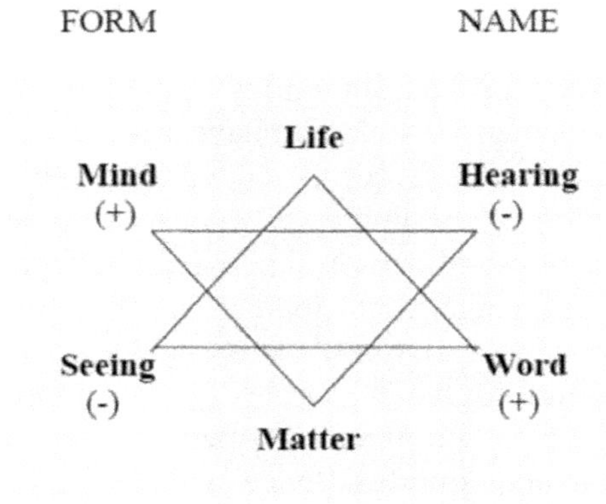

Fig. 64 Credit: "On the Origins of Faculties of Consciousness" (V. Yatsenko, 2016; 2021)

The original theory of knowledge in the Vedic tradition is far more complex.[109] However, let's consider whether this intuitive perspective is really so far off from our current scientific understanding.

When I first came across this idea, my Western rational and scientific instincts prompted me to dismiss it as just a religious parable—a convenient symbolic fiction from a pre-scientific culture. I believed that the modern theory of Darwinian evolution—based on natural selection and random mutations—is sufficient to explain how life arose and that no other assumptions were necessary. However, over time, my dissatisfaction with the explanatory power of science regarding the nature and origin of consciousness, the mind, and the seemingly teleological character of evolution grew. I found these explanations increasingly, if not superficial, at least incomplete. Moreover, upon closer examination, I realized how one can view, conceive, and understand the scientific and the Vedic standpoints not as mutually exclusive but as complementary.

While there's no reason to doubt the validity of evolution, the key question is whether the mechanisms that evolutionary biology has identified are sufficient to account for the complexity of life. Darwin questioned how

[108] One might wonder if we have additional senses beyond the ones we recognize. Perhaps there are subliminal, subconscious, or even extrasensory perceptions of which we're not consciously aware. In the Veda, the faculties of consciousness extend beyond the five senses and the mind to include speech, feeling, and other forms of supra-rational cognition or its expressions, as we will see later. For the sake of brevity, in this section, I will focus solely on the former.

[109] For an in-depth reading I suggest Vladimir Yatsenko's works (Yatsenko, 2016; 2012).

complex organs like the eye could evolve through selection alone, especially because they appeared in different species independently. Modern evolutionary biology posits that mechanistic processes of selection can explain the development of complex sensory organs. Creationists and supporters of Intelligent Design question this, but I found no compelling argument to doubt that natural processes have led to the rich diversity and complexity of life and its organs. That being said, while these processes undoubtedly contribute, it remains unclear whether they are sufficient on their own to account for the emergence of such complex sensory organs. The idea that adaptation, random mutations, inheritance, and the "sifting" process of natural selection are alone sufficient to gradually assemble such complex systems is only an assumption, a working hypothesis, and not a scientific fact.

Consider a blind and deaf scientist who lacked sensitivity to touch and who, nonetheless, possessed all knowledge of chemistry, biology, and physics 600 million years ago. Not only would such a scientist have been unable to predict the evolution of photoreceptive, mechanoreceptive, and sound-receptive cells into fully formed eyes, ears, and sensitive skin, but the very concepts of sight, hearing, and touch would have seemed completely abstract and devoid of meaning.

Yet, it is remarkable how these senses developed almost immediately with the emergence of life, not limited to just one or a few species, but appearing universally across many forms. It was as if evolution felt the need to reinvent the senses multiple times from scratch, starting from an already present and inherent archetypal idea. The emergence of sensory organs seems to reflect a general rule, an inherent power, and fundamental principles of life rather than emerging from processes ruled by mere mechanistic coincidences.

Furthermore, why should the complex biochemical and bioelectrical processes underlying our sensory perceptions lead to sensations and subjective experiences, ultimately creating the rich, phenomenal world of a sentient being? Of course, here I'm referring (again) to the hard problem of consciousness discussed in Section II.1. Proposing that consciousness and its faculties are fundamental (the Vedic perspective) rather than emergent (the biological perspective) might provide a more coherent theoretical framework with more explanatory power.

Was this also true for the mind, cognition—that is, manas? If so, one should expect brains to develop independently from species to species. In fact, it turns out that birds and mammals did not inherit the neural pathways responsible for intelligence from a common ancestor; instead, these pathways evolved independently. Modern biology interprets this as "convergent evolution"—that is, the independent evolution of similar traits or features in distantly related species, often as an adaptation to similar environmental pressures or ecological niches.

However, we could also adopt an alternative (or complementary) hypothesis suggesting that intelligence might be an inherent aspect of life rather than merely a product of random evolutionary processes. Nature "e-volves" the senses and intelligence from within, whenever and wherever it finds an opportunity to do so, by adaptation, speciation, and natural selection. It is not the other way around.

Moreover, despite all progress, the process that shapes the form of an organism, tissue, or organ—known as "morphogenesis"[110]—remains one of the most enigmatic aspects of biology and which cannot be explained solely through genetics and evolutionary biology. We are still far from grasping how a severed salamander limb can regenerate into an identical new limb, how tadpoles can have their heads scrambled yet still rearrange and develop into normal frogs, or how planarian flatworms can be manipulated to grow the heads of different flatworm species without any genetic changes. New ideas and bold hypotheses are needed.

There are signs that a deeper intuitive approach to the facts of Nature is experiencing a comeback. An example, I would like to mention is American biologist Michael Levin's theory (Levin, 2025) of the relationship between morphogenesis and a "Platonic space"—that is, a latent space similar to Whitehead's domain of potential forms and actual occasions of experience or pre-existing possibilities that have not yet been actualized.

Levin challenges the conventional paradigm that living beings are solely products of genetics and the environment, and proposes a framework in which evolution favors forms that exploit mathematical and computational truths. He argues that the relationship between mind and brain is similar to the relationship between mathematical patterns and morphogenetic outcomes, suggesting that cognitive patterns "ingress" from a Platonic space. The theory hypothesizes that instances of embodied cognition ingress from a Platonic space containing both low-agency patterns (like facts about triangles) and higher-agency ones (like mental properties). At least some patterns arise from mathematical causes, independent of physical or historical explanations, and are beyond genetics and the environmental constraints, which do not reduce to facts of physics but, instead, exhibit emergent goal-directedness and problem-solving capabilities. According to Levin, these patterns are derived not solely from genetic information but also from a problem-solving process guided by bioelectric pattern memories structured by a Platonic space of forms.

Whether Levin's approach will succeed or remain a mere speculative line of research, I don't know, but I appreciate that he was bold enough to cross the red line of naturalism and see if and how this could lead us further.

What I strongly sense is that all these natural processes might reflect a deeper creative aspect of Nature, one expressing an even more profound non-

[110] In biology, morphogenesis defines the origin and development of morphological traits.

natural law of self-expression that transcends them all. By broadening our understanding of life to include, but not be limited to, material, biological, and morphological aspects, we can view the cognitive abilities of all living organisms as manifestations of the emerging faculties of a universal consciousness. From this perspective, many previously unexplained evolutionary processes make more sense.

The Vedic perspective that saw hearing, touching, seeing, name, and form as manifestations of a deeper truth, or Plato's realm of perfect Forms, might be not just outdated and abstract philosophical fantasies but also intuitions of a deeper truth inherent in reality. I argue that, despite being based on mystical experiences or on the intuitive minds that perceived archetypal realities expressing them with idealist philosophies, these viewpoints can, nonetheless, help us transcend a naïve realism. They encourage the intuitive mind to detach from the physical mindset of the naturalist, who believes only in what physical appearances suggest. This shift is desperately needed in our current society, which remains trapped by materialistic thinking and has lost touch with its inner soul.

d. Substance Monism and Creative Evolution

In his work "Ethics," Baruch Spinoza presented a pantheistic cosmology in which Nature and God are fundamentally the same, with both being expressions of a single infinite "substance." All things are considered "modes" of this eternal substance, which possesses an infinite number of "attributes"—each reflecting an eternal aspect of Nature. Modes represent the "states of substance" or "affections," meaning they are modifications of the substance itself.[111] Attributes, on the other hand, are the infinite "essences" of the substance, representing what we understand about it through our human perspective.[112] In this framework, the one existing substance manifests itself in an infinite variety of modes—changing states of substance—each characterized by attributes that define its eternal essences.

If this sounds confusing, don't worry. Even scholars disagree on the correct interpretation of Spinoza's ontological categories. However, we don't need to delve into abstract metaphysical details to focus on what matters most for our purposes: Spinoza's "substance monism."

Substance monism is the belief that only one ultimate substance exists. According to this doctrine, no other substance exists; everything we perceive

---

[111] "*By mode I understand the states of a substance, or that which is in another through which it is also conceived.*" B. Spinoza, Ethics, Part 1, V.

[112] "*By attribute, I mean that which the intellect perceives as constituting the essence of substance.*" B. Spinoza, Ethics, Part 1, IV.

and conceive, including all of Nature, is simply different forms, qualities, or manifestations of this singular substance, often referred to as 'God.'

It naturally follows that mind and matter are not separate entities, as suggested by Descartes's dualism; instead, they are two of the infinitely possible attributes of an underlying substance.

In contrast to the beliefs of Abrahamic religions, there is no true distinction between God and Nature, nor is there a substantive separation between God and humanity. We are expressions of the same divinity that seeks to understand itself through Nature.

It is important to note that, within this framework, this deity does not act through design, purpose, or free will; rather, it acts out of necessity and cannot do otherwise.

Nearly a century and a half later, Schopenhauer presented similar ideas. In his work "The World as Will," he described a blind mechanism driven by an unconscious Will, lacking both purpose and knowledge.

Some contemporary philosophers have embraced this idea. For instance, philosopher Bernardo Kastrup developed a theory of analytical idealism, envisioning the universe as a "Mind at large" (a term borrowed from Aldous Huxley). This idea suggests a universal consciousness that fragments itself through a process of self-dissociation into various "dissociated boundaries" manifesting as individual minds, bodies, and individuals and that he called "alters"—a phenomenon equivalent to "cosmic dissociative identity disorder." Ultimately, we are these individuated alters projected by the universal consciousness, each possessing unique personalities, characters, and traits.

In this view, evolution is perceived as a process of self-discovery, leading to increasingly self-aware beings like us, emerging from a Mind at large that is not inherently self-aware and operates solely on instincts, urges, and unintelligent drives. It remains fundamentally a quasi-mechanistic universal consciousness, or at best, a blind and instinctive Mind that lacks self-awareness, free will, self-knowledge, and comprehension of both its transcendent and immanent aspects and processes.

This viewpoint is appealing to some, as it appears to reconcile modern naturalistic evolutionary theory—which focuses solely on the unconscious or subconscious deterministic aspects of the world—with a non-physicalist understanding of Nature. However, a closer examination reveals that it only scratches the surface and leads to inconsistencies that require resolution. If we stop here, we are left with a paradox. While I appreciate that Spinoza's substance monism, Schopenhauer's concept of the World as Will, and Kastrup's analytic idealism are significant steps in the right direction, they ultimately conclude too hastily, thus bypassing critical processes and leaving a substantial explanatory gap.

How could evolution transform seemingly insentient, unconscious, aimless matter into living organisms that are surprisingly sentient, conscious, goal-driven, and purposeful?

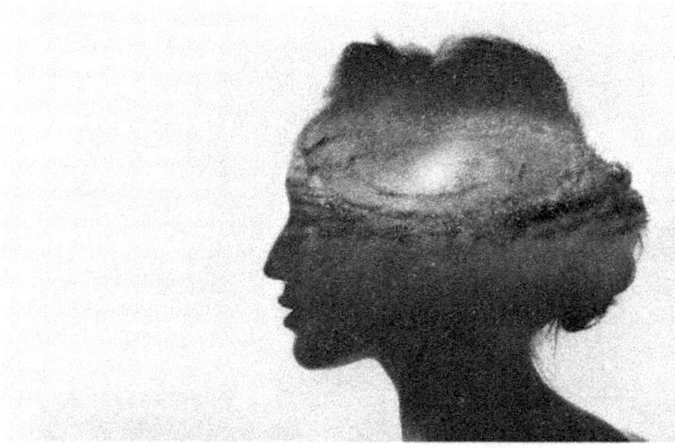

Are we self-dissociated subjects of a "universal Subject"?

Because evolution is widely accepted as a scientific fact, I contend that if this universal or cosmic consciousness—whether referred to as Mind at large, God, Nature, or any other name—is truly blind, subconscious, aimless, and devoid of knowledge, then it is incomprehensible how it could be so remarkably creative. This supposedly ignorant and almost comatose entity was nonetheless creative enough to give rise to an incredibly complex and diverse universe, which does not reflect the work of a mindless intelligence. Moreover, it has led to the emergence of relatively intelligent and purposefully driven beings like us. I fail to see any logical mechanism through which this could occur without an inherent higher form of Intelligence and Will at play. The idea that countless blind instincts could combine to create a self-aware, intelligent being is even less plausible than believing that randomly tossing 10,000 Lego pieces into a pot would result in an Eiffel Tower being assembled automatically or "instinctively."

The point is that before the formation of the first metabolic unit (a biological cell), this perspective suggests that there was no intelligence or knowledge at play, only a primitive and instinctive basal consciousness operating on a spatially and temporally localized scale. At that stage, there existed only an instinctive Mind at large without specific centers of self-awareness. The question arises: How can mere instinctive impulses give rise to something as complex, efficient, and self-reproducing as a living cell? How could a bottom-up process of abiogenesis, which theorizes the origin of primordial life from non-living matter, result in the emergence of the first biological cell only by meddling with macromolecules in a primordial soup? A cell is far more complex than the most advanced supercomputers, airplanes, rockets, or spaceships, which do not self-assemble starting from mere instincts, let alone reproduce.

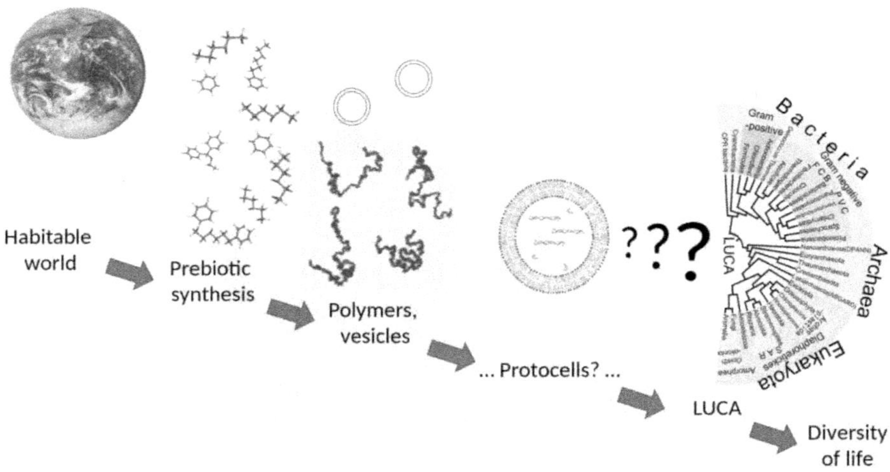

Fig. 65 Abiogenesis: The origin of sentient and mental life from insentient, unconscious, and lifeless matter?

It seems implausible that this could occur for a biological cell, even after 13.8 billion years of "instinctual development." I struggle to understand how the assembly of countless mechanistic instincts, regardless of the time involved, could lead to anything creative or cognitively advanced, aside from yet another form of blind "super-instinct."[113]

The issue at hand concerns not only the origin of life but also its entire evolution, from the first eukaryotic cell to modern humans. To me, it is scarcely credible that a purely instinctive and blind will, devoid of any intelligence, could almost magically initiate a creative evolutionary process of such remarkable complexity, resulting in an astonishing array of life forms.

The metaphysician might argue that this initial "instinctive divinity" could simply "will" things until the "willed" things are perfected, refining them over hundreds of millions of years through trial and error. This perspective aligns

[113] This conceptual issue is similar the "combination problem," of panpsychism—or more precisely, micropsychism.—which addresses the challenge of explaining how simple, individual forms of consciousness, attributed to fundamental particles or basic components of matter, combine to create the rich, unified consciousness we experience as humans. Specifically, how do these basic conscious experiences, differing greatly in scale and complexity, merge to form more complex and integrated states of consciousness, such as human awareness? What mechanisms allow any combination of these basic experiences to yield a coherent and unified consciousness?

with modern evolutionary theories of natural selection and random genetic mutations.[114]

We could suggest that a transcendent yet primitive Mind at large, or basic universal consciousness, learns through experience by acting on its instincts and engaging with the universal interplay of matter and energy across space and time. While this learning process seems plausible, it cannot originate solely from blind instinct. After all, how can a transcendent instinct evolve into a higher transcendent intelligence simply by immersing itself in growing material complexity? If one looks closer at this rationale, things do not add up.

Of course, materialists often tell us a very different story. They believe in a goalless and undirected Darwinian evolution that supposedly explains everything through natural selection, random mutations, inheritance, variations, speciation, and various other evolutionary mechanisms. If something does not fit within this naturalistic framework, they can still invoke the concept of "evolutionary advantage" to account for the emergence of any phenomenon, even the most contradictory ones. While I have no reason to doubt the reality of Darwinian evolution and its mechanisms, I find it to be equally insufficient with regard to addressing deeper existential and philosophical questions about life, consciousness, and the absence of aim, purpose, or design in Nature.

While, the underlying (Spinozist or Schopenhauerian) assumption is that higher cognitive functions are neither inherent nor transcendent but, rather, a later emergent aspect of evolution. This perspective aligns with modern scientific, non-teleological, and naturalistic views. However, as discussed in Chapter III, this stance is ultimately untenable. It relies on speculative extrapolations that are no more convincing—arguably even less so—than other metaphysical theories. We are far from demonstrating that evolution, based solely on natural selection, variation, and inheritance, can adequately explain the emergence and complexity of life and cognition. While these mechanisms account for certain aspects of the development of living organisms, extending this reasoning to the rise of increasingly intelligent organisms and the entirety of our natural history remains speculative. We might wish to believe this is true, but no evidence exists to support its feasibility. Darwinian evolution, in

[114] Something along the lines of the theory of 'cosmological natural selection,' proposed by American physicist Lee Smolin, which draws an analogy between biological natural selection and the evolution of the universe as a whole.

its current naturalistic form, claims an explanatory power that it does not possess.[115]

After all, if those who prefer a post-material perspective on reality, the world, life, and evolution feel compelled to adopt a naturalist narrative that reduces God to a mere machine in order to create a coherent theoretical framework, what is the point of embracing idealism or any trans-physical view of life in the first place?

Regardless of whether one identifies as an idealist or a Darwinist, the fundamental question of how an ordered universe—and even more so, complex conscious life with its higher cognitive dimensions—came into existence remains unanswered. Also, Eastern philosophies like Advaita Vedanta, which assert that all is Brahman and that everything else is the illusion of Maya, provide little insight into how this illusion originated or what life and evolution fundamentally are.

Ultimately, it makes sense only if a bottom-up Darwinian evolutionary process, which operates without intelligence, is complemented by a top-down causation originating from a higher intelligence that is inherent in matter but remains unexpressed in its unevolved form. The circle will be complete only if we balance the physical constraints imposed by evolution and the environment from the outside with an internal impulse that guides development from within.

e. The World as Real-Idea

I prefer the vision of the Indian mystic and poet Sri Aurobindo, who claimed to have attained a higher state of consciousness, which he called the 'Supermind.' From this state, he developed an integral perspective that unites idealism and evolution, aligning with science while avoiding any assumptions of a dreamy, instinctive, and amnesic divinity.

Aurobindo's vision, like many Eastern philosophies, centers on the concept of a transcendent and ultimate Absolute. However, his cosmology presents a multidimensional structure that encompasses more than just mind, body, and soul. In this framework, the Supermind represents the highest plane of creative agency, with gnosis underlying all universal dynamics. It acts as a mediator between the unmanifest and the manifest, as well as between the transcendent

[115] Implementing even the simplest life processes in computer programs to create increasingly intelligent artificial life is not overly difficult. However, attempts to demonstrate that this leads to increasingly complex and intelligent agents have seen limited success. While it's possible to simulate cellular automata competing for resources or to create relatively complex structures using genetic algorithms, the overall results, after decades of research, have been disappointing. These digital creatures fail to evolve into more intelligent cognitive agents. A key issue is that every evolutionary algorithm requires human intervention to function effectively. In other words, the bottom-up simulated evolution that begins with unintelligent virtual agents works only when guided by the top-down oversight of an already intelligent programmer.

and the immanent, embodying what we recognize as "intelligence" but beyond human intelligence.

The term "super" highlights its nature as a trans-rational gnosis, encompassing knowledge and power that surpasses the analytical mind—the "scientific reason" of the Age of Enlightenment, which is merely a diminished reflection of this deeper cognition. The Supermind expresses itself as a force of consciousness and an immanent will that organizes the forms, qualities, powers, and characteristics of things eliciting our lower forms of mental perception.

While the mind operates through dualities, oppositions, and separations, leading to fragmentation, the Supermind perceives all seemingly distinct objects and their relationships in a comprehensive and holistic manner. It views them not as separate entities but, rather, as within an eternal unity, fully aware of the multiplicity and diversity present without falling into divisive oppositions.

In Aurobindo's words:

"The fundamental nature of this Supermind is that all its knowledge is originally a knowledge by identity and oneness and even when it makes numberless apparent divisions and discriminating modifications in itself, still all the knowledge that operates in its workings, even in these divisions, is founded upon and sustained and lit and guided by this perfect knowledge by identity and oneness. The Spirit is one everywhere and it knows all things as itself and in itself, so sees them always and therefore knows them intimately, completely, in their reality as well as their appearance, in their truth, their law, the entire spirit and sense and figure of their nature and their workings. When it sees anything as an object of knowledge, it yet sees it as itself and in itself, and not as a thing other than or divided from it about which therefore it would at first be ignorant of the nature, constitution and workings and have to learn about them, as the mind is at first ignorant of its object and has to learn about it because the mind is separated from its object and regards and senses and meets it as something other than itself and external to its own being."

The Synthesis of Yoga, Pt. IV, Ch. XIX, p. 786.

This perspective transcends traditional Western and Eastern philosophical idealism. It envisions the world as more than an image or figment of a universal Mind at large. Instead, it expands its cosmology beyond a silent and unchanging Brahman. This view understands the universe as an expression of a supra-conscious Being that manifests itself through what the mind interprets as "names and forms." By "names," Sri Aurobindo referred not only to the words used to describe objects but also to the powers, qualities, and characters (features and traits that distinguish one thing from another) caught up by our consciousness. In this context, all "names" are inherently present in the nameless and timeless Absolute, but they are articulated by a supreme

Intelligence in the temporal realm as the qualities, powers, and characters of its forms.

Unlike Spinoza's concept of Nature or Schopenhauer's notion of the World as Will (both of which function through necessity, instinct, and blind will without self-awareness), the Supermind represents the pinnacle of cognition, wisdom, and knowledge within a universal consciousness. This supramental consciousness operates utilizing its inherent powers, forces, and will. It shapes the universe through a 'Consciousness-Force' and 'Consciousness-Will,' influencing processes, structures, patterns, and, ultimately, living forms in accordance with a divine Intelligence. This Intelligence acts through a creative power that reflects an original Idea, which Sri Aurobindo referred to as the 'Real-Idea' or 'Seed-Idea.'

The transcendent Real-Idea exists beyond matter, space, and time, yet it projects itself into the realm of space and time as an expressive power of conscious force. It embodies an Idea of oneness that initiates a dual and divided representation of itself, thus giving rise to a world of dualities and divisions. This primeval, perfect, and undifferentiated Knowledge, or Consciousness—often referred to as "Gnosis"—manifests in various mutable names and forms. These forms represent what the mind interprets through symbolic images of this Seed-Idea. At the human level of understanding, intuition captures these as archetypes, symbols, and signs, each carrying meaning and significance related to all processes and phenomena.

This vision posits that the Real-Idea embodies a gnostic power where Spirit, Truth, and Knowledge represent three aspects of the same Reality. It is a self-creative form of Knowledge and "Truth-Consciousness" that organizes—through design, quantity, and number—a foundation for manifesting the qualities and properties of its own Idea. These ideas take shape in the forms we perceive as a living, material world within the spatio-temporal realm. Life, minds, and bodies are partial expressions of this original transcendent Idea, which our ordinary consciousness only vaguely senses. Philosophers refer to this as archetypes, Platonic ideals, or universals. It represents the vestige of significance, meaning, design, and purpose some poets, philosophers, and seers intuit in Nature, a reality that materialists often dismiss as an illusion or anthropomorphic misconception. Thus, everything we perceive in the world, from non-living matter like rocks, mountains, and rivers to living entities such as plants, animals, and ourselves, are "symbols," "signs," or "significant forms" of this ultimate Real-Idea, which, despite being veiled, resides within them.

This power of Knowledge, while ultimately of transcendent nature and beyond manifestation, expresses itself in the manifestation of matter, force, space, and time by plunging itself into the world of duality and separation and determining its processes, which, however, retain and manifest something of this Idea. Evolution is the outer reflection and appearance of this process expressed within the limitations of a universe rooted in spacetime and causality. It is a process of self-finding a Spirit-Idea already existent beyond mind and body that, in matter, is expressed in "coarse-grained" images, meanings, and semantic representations of itself. A self-finding process of a Spirit struggling to emerge from self-oblivion and ignorance through a trial-and-error drama, which we discover in the form of the Darwinian paradigm of natural selection and random mutations. The Spirit of a supreme Consciousness, Knowledge, and Gnosis projects an image of itself in the self-forgetful exterior Nature and strives for increasingly self-conscious representations of itself in both life and mind by emerging in and through matter.

Life serves as the basis by which this conscious force and its Idea create, sustain, and dissolve individual forms, acting within them. The Real-Idea shapes the relationships among all things, which can be interpreted through a higher intuitive cognition as occult signs and codes, though they are seldom recognized at a mental level (like, for instance, in the DNA code with its genetic information). This is what modern biology intuits, particularly through the lens of biosemiotics, in the vast array of communication and signification present within and between living systems: Meaning in Nature is pervasive.

The Real-Idea manifests through the power of Knowledge and Will, drawing from infinite potentialities. It "rolls out" or "e-volves" what is internally contained as a "seed" within the oblivion of form. This emergence is a self-development, an outgrowth of itself, occurring through a gradual, sometimes abrupt, eonian process. The Real-Idea is perpetually present; it is a hidden knowledge and an indwelling spirit that exists within minerals, plants, and animals, all responding to its diminished will-force.

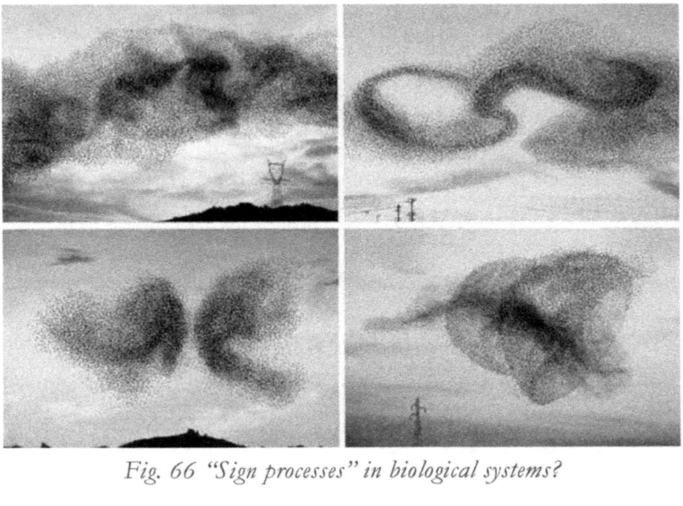

Fig. 66 "Sign processes" in biological systems?

The Real-Idea shapes life by groupings and associations of elementary constituents, building significant forms, types, and species. These are the external signs of a conscious force playing with its own Idea and creating all types in all possible combinations.

The manifestation we observe is merely a pale reflection of the underlying Idea. This Idea is fundamental to all cosmic existence; it influences the forces of creation and the evolutionary processes that sustain it through an inherent presence in every form. The undifferentiated yet multifaceted Real-Idea operates through a variety of Idea-Forces. Similarly, our own complex, diverse, and multidimensional psychology serves as a 'symbol' of the Idea, capable of expressing itself through countless characters and individual personalities.

As we intuitively recognize that the world around us is not merely a blind, mechanical system but, rather, conceals a deeper meaning, aim, and purpose (suggesting the development of an overarching project or Idea), we begin to sense that our limited minds can still intuitively grasp that transcendent Real-Idea. This realization prompts us to question whether meaning and significance are solely constructs of the human mind or if, instead, they are inherently present in Nature itself, extending beyond human understanding.

"In all material things reside a mute and involved Real-Idea, a substantial and self-effective intuition, an eyeless exact perception, an automatic intelligence working out its unexpressed and unthought conceptions, a blindly seeing sureness of sight, a dumb infallible sureness of suppressed feeling coated in insensibility, which effectuate all that has to be effected." (The Life Divine, Bk.II, Ch. X, p. 570)

The automatic and seemingly blind mechanisms of Nature are not the original starting point; rather, they are a terminal superficial expression of a divine intelligence that shapes forms and powers. The interactions of all elements in the universe are not simply a series of unconscious events; instead, they reflect a deeper knowledge in which what we call the 'laws of Nature' embody a Real-Idea that drives everything toward its necessary outcome. The intelligence we observe in the world is not a mere illusion; it arises from a hidden intelligence operating beneath our ignorance.

An evolution occurs whereby forms, qualities, and quantities are shaped by a Real-Idea from the inside out. All forms and properties of the natural world are already present within an eternal divine archetype on other trans-material planes of existence. Every formation that emerges from Nature is an expression of a preexisting Idea that our ordinary human minds perceive as names and forms.

This sheds light on how a conscious, goal-driven, and willful life could arise from unconscious, goalless, and unwilling matter. Viewing the world, the universe, Nature, life, and evolution through the lens of Sri Aurobindo's Real-

Idea reveals clarity and meaning in what once seemed mysterious. This perspective embraces a subtler, self-aware Will and Spirit that subconsciously aligns with an original super-conscious Knowledge and Idea structuring all existence. This understanding might sound very mystical, but it is not at odds with modern science.

All processes that appear "random" or governed by "chance" to our ordinary minds are actually local expressions of a deeper order and wisdom that encompasses a global perspective. They seem like blind chaos to us for the reasons outlined in Sections V.9-12.

Natural selection, random mutations, and other evolutionary processes become visible manifestations—or, more accurately, "signs"—of a self-limited and inconscient power. In its oblivion, this power resorts to mechanical and seemingly blind processes in matter and life to achieve its goal of expression through matter. At our current stage of evolution, life is perceived as a symbolic representation of a will and idea within matter, originating from a unitary Seed-Idea that transcends the polarities of mind, life, and matter.

The world and its natural phenomena serve as symbols whose deeper significance is revealed only at a gnostic level of consciousness. It is shaped by elemental and inherent influences, and its true power lies in a gnosis that transcends mere "mind" or "Mind at large." It is far from a blind instinct or an unconscious, aimless necessity, as Spinoza or Schopenhauer would suggest.

Thereby, one can view evolution as a two-way process. It was preceded by an involution of the Spirit into form—essentially, a *"plunge of Light into its own shadow."* [116] Spinoza and Schopenhauer , were not entirely incorrect; they recognized the inherent Will in Nature manifesting in various forms. However, because they perceived only the shadow and overlooked the light from which it originates, they misinterpreted the final stage as merely the beginning. The blind instinct of an amoeba and the reflex actions of an insect are not the origins of the evolutionary journey toward higher forms of cognition. Instead, they represent an expression of veiled Knowledge and infallible Vision that concealed itself. This knowledge, having become lost in the darkness of unconscious matter, seeks to rediscover itself through a trial-and-error process.

From this standpoint, a materialistic understanding of evolution can be harmonized with idealism in a post-materialistic synthesis that transcends both.

[116] The Life Divine, Bk.I, Ch. XVIII, p.173.

X. An Integral Cosmology of the Conscious Universe

1. Materialism Failed, but Idealism isn't Enough

Modern scientific research tries to explain the emergence of consciousness, mind, and life in embodied biological organisms, with their mental properties, cognition, behavior, sense of selfhood, and subjective feelings, as well as their purposeful, willful, and intentional action, via an exclusive naturalistic, mechanistic, and materialistic approach. Science compares living entities starting from mindless, lifeless, conscious-less, purposeless, and intention-less phenomena like self-organizing units driven by homeostasis, genetic mutations, natural selection, and adaptation. Another layer of sophistication arises from the perspective of nonlinear dynamical systems theories, which regard an organism as a system utilizing its bodily and physical/energetic resources to solve survival and persistence-related problems, autonomously adapting to an ever-changing environment. According to this worldview, the different forms of biological agency stem from a complex interaction with the environment. The sense of self arises from a homeostatic equilibrium with it, and an individuated system self-realizes itself as an adaptive, non-linear, complex feedback control system within the framework of evolutionary and developmental biology. The modern scientific approach aims to explain all life with its different forms of biological agency in purely mechanistic and functional terms in analogy with artificial intelligence, robotics, cybernetics, autopoiesis, neural mechanisms, and other extended and more complex types of processes and machineries supporting agent behavior.

Recently, several complex-systemic theoretical frameworks have evolved, marking a paradigmatic shift from a reductionist bottom-up approach to a more holistic top-down perspective that emphasizes the whole organism rather than its individual parts. These frameworks have transitioned from a strictly machine-like computational view to a broader understanding of non-computable forms of cognition, possibly including quantum cognition. Additionally, they recognize organisms as active participants in evolution rather than passive entities.

We have already mentioned some of these theories.

Enactivism in biology highlights the active role of organisms in shaping their experiences and environments through embodied interactions. It challenges traditional views that separate organisms from their surroundings and instead proposes that cognition emerges from the dynamic interplay between an organism's biological structures, sensory-motor activities, and environment. Enactivism sees life as a continuous process of self-organization in which organisms maintain their identity by interacting with their surroundings in ways that ensure their viability. This perspective integrates insights from biology, cognitive science, and philosophy, emphasizing how living systems both influence and are influenced by their ecological contexts.

Among other approaches worth mentioning is the pioneering work of American biologist Michael Levin, who explores the role of bioelectricity in regulating development, regeneration, and cognition in living organisms.

Meanwhile, theories of consciousness have proliferated, with one notable example being Integrated Information Theory (IIT). This framework posits that consciousness arises from the integration of information within a system. Another example of a theory of consciousness is the "Global Workspace Theory" (GWT), a cognitive framework that explains consciousness as the result of widespread information sharing throughout the brain.

On a more biological and cellular basis, we can mention the theory of "Cellular Basis of Consciousness" (CBC) by the American cognitive psychologist Arthur S. Reber. It posits that consciousness originates at the cellular level, with individual cells, particularly neurons, contributing to the foundation of conscious experience.

I found Evan Thompson's theory of "biopsychism" to be of particular interest. It proposes how life and mind are deeply intertwined, suggesting that all living organisms, even single cells, possess a form of basic, embodied cognition or subjectivity.

However, both Reber's and Thompson's approaches remain rooted in a naturalistic worldview that will not lead us further in answering the deeper philosophical questions about life, the nature of the mind, and consciousness.

Overall, these theories are interesting and offer insights into certain aspects of our cognitive dimensions. However, they fall short of addressing the true nature of consciousness, cognition, and the emergence of sentience. A common thread among these modern theories is their effort to eliminate any non-naturalistic and non-physical principles that could explain life, mind, and biological agency. The (often unacknowledged) ultimate goal is to eradicate any non-material and spiritual notions, with the hope that they will no longer influence scientific explanations.[117]

That being said, I believe it is time to move forward and seriously consider non-naturalistic approaches—those that are not grounded in religious beliefs or abstract philosophical speculations but, rather, that embrace a worldview anticipating a post-material paradigm aligned with science.

Fortunately, contemporary philosophical thought has expanded beyond a strictly biological and materialistic perspective. We have already observed a resurgence of interest in panpsychism, whether in its micropsychist form or its cosmopsychist form. Additionally, other theories of universal consciousness

[117] Levin's ideas might be an exception, as he posits the existence of a Platonic space. I mentioned this briefly in Section IX.3a.

that align more closely with the integral framework we will present deserve mention.

Max Velmans' Reflexive Monism is a philosophical theory that aims to reconcile materialism and dualism. It suggests that the universe serves as a map illustrating how the mind interacts with the physical world through a reflexive act of perception. This theory emphasizes our instinctive and often unconscious tendency to project our conscious experiences into the external world. In analogy to our discussion in the chapter on idealism, the reflexive model posits that the contents of consciousness seem to exist "out there" because our minds project these experiences into what we perceive as "3D space," which appears to extend beyond our physical bodies. The reflexive nature of this process lies in the fact that once our visual system processes stimuli, the physical processes originating from the world are perceived as external, or "reflexed," outside the brain. However, in reality, everything we perceive occurs within our skulls. Similar reflexive loops take place within the body. Reflexive Monism ultimately acknowledges that all observations can be deemed "objective" only in the context of being "intersubjective.'

One of the most influential theories of universal consciousness is the analytic idealism proposed by Dutch philosopher Bernardo Kastrup. We have already explored some of his views on free will, but his most notable model of reality is a form of modern neo-idealism. Kastrup's analytic idealism asserts that consciousness is the fundamental substrate of reality, with the physical world emerging as a representation within a Mind at large rather than existing independently. Drawing on philosophical idealism and insights from quantum mechanics, he suggests that individual minds are dissociated segments of a universal consciousness, referred to as 'alters.' Analytic idealism seeks to integrate philosophical reasoning with empirical evidence, thus offering a compelling alternative view of the nature of reality and the role of consciousness within it.

American cognitive psychologist and popular science author Donald Hoffman proposes an "Interface Theory of Perception" (ITP). Our perceptions do not reflect objective reality but instead function as an evolutionary interface—a simplified, user-friendly representation that helps us survive and reproduce. Much like a computer desktop hides complex code behind icons, our senses present an adaptive 'virtual reality' that emphasizes fitness, not truth. This is a form of idealism analogous to that which I introduced with the neuroscience of our visual perception in Section IX.1 and that was asking whether reality is an illusion. According to Hoffman, organisms that perceive reality accurately would be outcompeted by those that see only what they need in order to thrive. This suggests that what we experience is more about utility than accuracy.

Moreover, Hoffman and his colleagues developed a physical theory with a micropsychist flavor that models the interactions of conscious agents along the lines of the scattering processes in particle physics, thus providing a geometric framework to capture their relationships and experiences beyond space and time. It is a mathematical approach that supports the idea that reality, as we perceive it, is a simplified, adaptive construct rather than an accurate depiction of the world.

One intriguing metaphysical speculation is Steve Taylor's theory of "panspiritism," which suggests that spirit, or a fundamental essence of consciousness, exists in all things, both living and non-living. Taylor argues that consciousness is not limited to biological organisms but, instead, is a universal characteristic of reality, manifesting in varying degrees across different entities. In contrast to panpsychism, which ascribes basic forms of awareness to matter, panspiritism highlights a spiritual dimension, proposing that an underlying, transcendent flow of Spirit infuses the material world. Taylor's perspective aims to bridge scientific and spiritual worldviews, suggesting that the universe is not simply physical but is animated by a pervasive spiritual essence connecting all existence, reminiscent of the animistic spirituality found in indigenous cultures.

These are just a few of the most notable theories of consciousness and cognition among the many that are currently popular. It is impossible to cover them all in detail here. I mention them to highlight how, in contrast to just a few decades ago, this field of metaphysical speculation has become vibrant and active. Paradoxically, in a society heavily influenced by materialistic and reductionist thought, a spiritual, post-material understanding of life and the universe has experienced a revival. Velmans' concept of a self-knowing universe and Kastrup's intuition that a cosmic dissociative process underlies subjective individuation—suggesting that our experiential physical-mind contents are a 'redux' of something with a cosmic dimension—align with the mystic traditions of Eastern philosophies.

Panpsychism, reflexive monism, analytic idealism, the interface theory of perception, panspiritism, and other spiritually based cosmologies view consciousness as a fundamental primitive rather than an emergent property, as materialism suggests. These perspectives represent a significant advancement over the prevailing reductive forms of naturalism.

However, I'm going to argue that they remain overly simplistic models of reality. We must refine and expand them into a more comprehensive and integrated cosmology.

For instance, Velmans' approach is largely a modern reinterpretation of Kant's idealism. In contrast, Kastrup's monistic idealism, which posits that everything is mental, becomes trapped in a simplistic theoretical framework that fails to adequately represent the complexity of reality. A significant obstacle arises from the conflation of mind and consciousness—a linguistic confusion I regard as a serious fallacy that hinders deeper understanding, as I previously discussed in Section II.4. Additionally, the ontological status of concepts such as feelings, emotions, desires, and cravings isn't addressed. What exactly are these? How should we distinguish them from mental states? I believe this lack of clarity is a critical flaw that obstructs our progress and understanding. It resembles Alice in Talbot's 'Flatland,' who attempts to understand the properties of a three-dimensional spherical surface based solely on the circles she observes on her two-dimensional plane.

The exclusively idealist approach is fundamentally flawed. Our perception of the physical world as mental phenomena on our 'screen of perceptions' raises the question: Does this mean that everything must be mental? The fact that we perceive everything through the lens of the mind does not mean that everything is inherently mental.

Additionally, I previously noted that this perspective lacks a coherent evolutionary framework. How can Mind at large create the first metabolic system from raw, insentient matter? Is fundamental consciousness merely a state of "mindless" pure being and passive existence devoid of intelligence? If that is the case, the evolutionary aspect of life remains an unsolved mystery unless one falls back on a purely mechanistic neo-Darwinian view of evolution which, ultimately, is just as speculative regarding the deeper existential questions surrounding the emergence of life, mind, and consciousness. These theories focused too heavily on solving the hard problem of consciousness, neglecting other first-person qualitative aspects and the evolutionary emergentist dimension of life and consciousness. As we will see, these elements are naturally harmonized within an integral cosmology.

Finally, Hoffman's approach is also too restrictive. While it is reasonable to conclude that our perceptions do not accurately reflect objective reality and that they function as an evolutionary interface, this observation does not support an idealist philosophical stance. Although it makes sense that natural selection would not favor a truthful perception of reality, this is not the fundamental reason we can't perceive reality as it truly is. The inaccessibility of the noumenon is principled, no matter what evolution does.

Thus, these new metaphysical trends in modern Western philosophy of mind are a positive development representing a significant advancement over the restrictive view of physicalism. However, they are only a first step toward transcending the meaningless, purposeless, and spiritually empty universe that a 19th-century kind of materialism seeks to force on us as the "official" scientific view.

Before exploring the nature and origin of consciousness from an evolutionary perspective, it is helpful to first consider some ancient mystical teachings that may provide insights. One does not need to adopt a specific religious belief to appreciate certain elements that can guide us toward a direction or working hypothesis, that emphasize the multidimensional nature of an original and all-pervasive universal consciousness. This perspective could serve as a foundational basis for developing a theoretical framework that aligns more closely with modern scientific understanding.

2. From a Coarse-Grained Idealism to a Multi-Modal Conscious Universe

To illustrate how we can move beyond simplistic forms of idealism, let me once again resort to the metaphor of epicycles.

In resisting Copernican heliocentrism, people resorted to intricate epicycles to maintain the appearance of an Earth-centered universe. As a means of explaining the unusual paths of the planets in the celestial sphere, they introduced epicycles—circles rotating around other circles.

In contemporary science, a similar tendency emerges. Matter is placed at the center of the universe, with life considered to 'orbit' around it. From life arises cognition as an epiphenomenon, which acts like an epicycle, adding complexity to the explanation. Unfortunately, the narrative doesn't stop there; we must also account for the emergence of consciousness, leading science to introduce yet another epicycle around the mind. This results in a "matter-centric" model of reality that resembles attitudes prevalent in the early days of science.

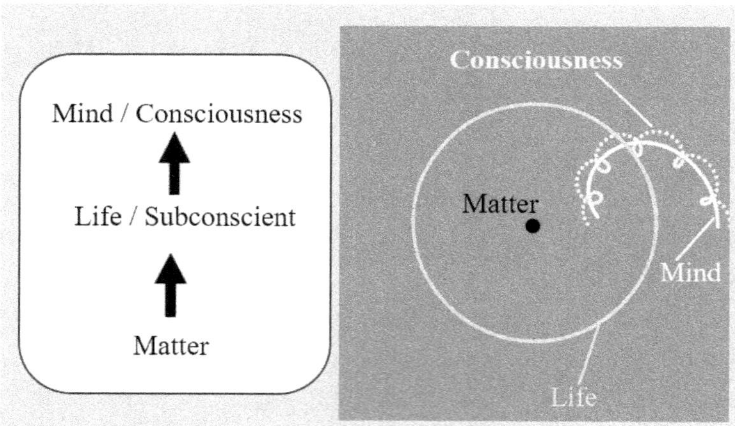

Fig. 67 How a bottom-up "matter-centric" physicalism posits matter at the center of the universe and accounts for life, mind, and consciousness.

Additionally, idealistic approaches that suggest the existence of a cosmic mind or universal consciousness often lack a coherent evolutionary perspective. More importantly, they tend to rely on overly simplistic, "coarse-grained" theoretical frameworks, which fail to capture the complexity of the world. By "coarse-grained," I refer to the fact that idealism remains a low-dimensional representation of a multidimensional reality. While we might sympathize with idealistic views, we must recognize that reality is seldom black and white, and the properties of the universe are rarely reducible to on-off or up-down dichotomies.

This undermines not only modern idealism but also much of Western philosophy of mind, which, as its name suggests, seeks to reduce everything to the mental. Philosophy of mind engages in this reductive approach because philosophy is a mental activity that naturally views everything through a mental lens. We must move beyond this singular perspective and explore a more nuanced and richer understanding of reality that offers greater explanatory power. This involves pursuing an evolutionary, multi-modal spiritual cosmology rooted in first-person experience.

To understand how our worldview is strictly dependent on our way of seeing, we must first deconstruct some deeply rooted misconceptions and learn to use an introspective approach to reality that can be just as insightful as, and sometimes even more revealing than, a strictly third-person analytic perspective.

a. An Inquiry into the Structure of Reality from a First-Person Perspective

Science examines reality, the material universe, the nature of life, the mind, and consciousness from a third-person perspective. This external viewpoint treats life as a machine. Thus, it overlooks our internal experiences of feelings, perceptions, and sensations, which are often deemed deceptive or illusory. In its pursuit, science investigates the brain's functions while abstracting the subjective aspects of the mind. It questions the origins of internal dimensions such as consciousness, thoughts, desires, feelings, and will and analyzes them from an external standpoint. This leads to the puzzling inquiry into how these elements could emerge in our inner worlds, seemingly out of nothing and almost magically.

Thus, let us take a different approach and explore not just through the lens of analytical and philosophical thought but also by observing, perceiving, and experiencing ourselves from a phenomenological perspective.

To facilitate this, I invite you to participate in a psychological exercise aimed at uncovering what we perceive, sense, think, and subjectively experience. Initially, this might seem like unnecessary academic hair splitting, as we tend to regard all aspects of our inner experiential world as simply 'mental' without further distinction. This tendency has contributed to significant confusion in our understanding of reality. However, by clarifying the quality and nature of the phenomena within our psychological dimension, we can significantly enhance our conception of the universe. I encourage you to engage in this exercise seriously, even if only once.

The exercise can be summarized with a simple question: "What do you perceive in this moment, and how do these perceptions differ qualitatively from one another?'

Due to our conditioning by the dominant materialistic culture, most of us will likely first identify our bodily-sensory experiences, such as hearing sounds, seeing objects, and smelling scents. The initial type of perception we recognize, which philosophers refer to as "qualia," is material and physical in nature. We can categorize these as sensations and experiences specific to the bodily or "physical plane" of consciousness. By "plane of consciousness"? I mean a phenomenal "layer," "level," or "stratum" proper to some particular kind of qualitative and mutable experience. The "physical plane" is the layer of all the phenomena we perceive through our physical sensorial senses.

Let's characterize it as follows.[118]

The Physical Plane

All bodily sensory experiences (seeing, hearing, touching, tasting, smelling, bodily sensation, etc.)

Most people stop here. They often refer to the American philosopher Thomas Nagel's definition of consciousness as "what it is like to be" in a particular psycho-physical state. Typically, discussions focus on sensory experiences, such as seeing a red tomato (the subjective experience of perceiving the red qualia) or feeling pain when placing one's hand in a fire. However, do these strictly physical sensations encompass the full range of our conscious experiences?

Consider emotions. When we feel fear, anxiety, love, hate, sadness, depression, joy, happiness, or grief, are these simply physical sensations? Certainly, they often correlate, sometimes intensely, with physical experiences. It's no coincidence that we feel an emotion in our bellies or say that we have a "gut feeling" or "knots in our stomach." These phrases highlight the link between emotions and the physical sensations they induce. But what is the cause, and what is the effect?

By looking inward and adopting a first-person perspective, you may recognize that sudden and intense emotions can arise without an external physical stimulus. These emotions can be triggered by a vivid memory or a simple thought, resulting in strong feelings. While emotions can elicit physical sensations, they originate from within and are not inherently physical. Emotions such as joy, anger, sadness, fear, and excitement are not necessarily prompted by sensory organs responding to external stimuli, unlike sight, sound, touch, smell, or taste. Instead, they encompass a broad spectrum of unique qualities that extend beyond the physical sensorial domain. Although emotions are intertwined with the body, they have a distinct character that differs from any physical sensation related to the five senses. When one is looking through the lens of lived experience, it becomes clear that emotional experiences are fundamentally different from physical experiences driven by

[118] The following distinctions will be my personal categorization and are not intended to be rigorous or exhaustive. You don't have to agree with this conceptual arrangement. The main point I want to convey is that mind-matter dualism alone is insufficient. There are also emotional, mental, subconscious, and intuitional dimensions that are inherently connected to a willpower and that are neither physical nor mental.

external material conditions. Rather than trying to understand this intellectually, take a step back and observe your emotions as a witness, without using analysis.[119]

Additionally, one often overlooked aspect is the role of emotions in driving action—such as desires, passions, and excitement. The term 'emotion' derives from the Latin 'emovere,' which means 'to put something into motion' or 'to stir up.' Emotions result from a will, agency, intention, or volition within a living being, granting it an active role. While all complex life forms experience emotions, the intensity and sophistication of these emotions can vary significantly, with different degrees and kinds across all living organisms.[120] The urge to act and the desire to achieve goals, along with the need to escape pain in favor of pleasure, reflect a specific and fundamental aspect of life. This intentionality and volition drive us from within.

Thus, emotions, feelings, moods, and the sensations of will, agency, intentionality, and volition emerge from within us rather than from our perceptions of external phenomena or objects. They are not experienced as material objects in the outside world; instead, they manifest as qualitative phenomena in us, though they can occasionally have a profoundly material concrete impact.

We should distinguish these emotional sensations and life experiences as belonging to what we might call the 'life-plane.'

This lack of distinction has also led to the exclusion of vitalism from biology by modern science. Science has lost the capacity for introspection and dismisses, without rational justification, a concept it never falsified, as discussed in Section III.

From now on, when I use the term "life," I refer to a broader meaning that transcends mere material organic form. Instead, it encompasses emotional, desire-driven, and volitional dimensions. Let me summarize this as follows.

The Life-Plane

Emotions: *Sensations/perceptions/impulses of desire, passions, sentiments, affections, attraction or repulsion, polar perceptions of joy or anger, happiness or sadness, courage or fear, love or hate, etc.*

---

[119] One might argue that there is a distinction between an emotion and a feeling, as well as between a feeling and a mood. While the list below acknowledges and outlines these differences and nuances in separate categories, they are not particularly relevant to our current discussion. For the sake of simplicity, let us consider both feelings and moods as part of the broader 'emotional' or 'vital' domain of our inner existence.

[120] Do bacteria, plants, or jellyfish possess some form of elementary emotions? As unlikely as this might sound, I wouldn't dismiss the possibility outright.

Feelings: *Thought-sensations about emotions, mental designation of (more or less subtle or intense) emotions (e.g., frustration, sadness, sympathy, compassion, shame, guilt, etc.)*

Moods: *Psychological states of mind about a feeling/emotion (e.g., contentment, excitement, serenity, indifference, etc.)*

Will, intentionality, and agency: *Volition, want, disposition, force, action.*

Moreover, we are not just material and emotional beings; we are also mental creatures. By "mental," I mean the faculty responsible for our thinking, with all its thoughts and intellectual activities. The form of cognition so cherished by the Age of Enlightenment—that is, the faculty within us that enables reason, rationality, logic, etc. However, the mind is not merely an analytic faculty. Additionally, animals can have some form of more or less complex mental abilities. Non-human animal behavior clearly shows that it also has inner representations of the environment, distinguishes between other harmless life forms or predators, has a memory, possesses at least some rudimentary associative logic, can anticipate, etc. Animals also have a "semantic awareness" that allows them to understand, apprehend, and comprehend things. To the surprise of scientists, nowadays, we know that even plants and unicellular organisms possess some form of rudimentary "basal cognition" or "minimal cognition". Even the simplest forms of life—namely, bacteria—have been shown to be able to sense the environment, actively move within it, target food, avoid toxic substances, and meaningfully change their swimming direction with a collective intelligence for problem-solving abilities to achieve a common goal.

The mind, in its broader and more nuanced sense, represents a continuous spectrum rather than a simple on-off cognitive faculty, and it exists throughout the living realm. Although we do not expect cats or dogs to program computers or drive cars, we recognize that they possess their own form of intelligence, exhibiting rudimentary yet sometimes surprising mental abilities. While we might not refer to the minds of bacteria or plants in the same way, we can acknowledge the presence of a subconscious proto-intelligence within them. The concept of the mind appears to have a universal quality, differing in degree and kind across various forms of life.

Can we then consider the mind as something inherently present in all reality, existing prior to brain functions?

Before we address this question, let's once again turn to introspection. Is the mind the source of experience, or is it merely something that is experienced? Are you the mind, or is the mind another layer of experience? Avoid answering this question intellectually; instead, take a step back and observe your thoughts. You'll soon realize that thoughts come and go, appearing and disappearing against a backdrop that perceives them as phenomena or fleeting events. Thoughts are perceived; they are not the perceiving entity. In a sense, we can describe thoughts as "mental qualia" that arise for someone. They are experiences and perceptions, but they are neither emotions nor physical sensations. Thoughts belong to a distinct category of perception. We understand this not through rational or philosophical reasoning but through lived personal experience. It isn't merely an opinion; it is a truth that, while difficult to prove experimentally, you know to be accurate from introspection. Thus, it makes sense to differentiate between the physical and life planes.

Now, let me list the qualities specific to the mind.

The Mental Plane

Thoughts, reason, rational intellect, analytic & logical thinking, deductive & inductive reasoning. The mind structures sensory perceptions and translates them into representations, ideas, concepts, and meanings based on symbols, forms, shapes, and numbers. It interprets sense perceptions, analyzes data, judges, reflects on concepts, decides, discerns, selects, discriminates, and makes logical inferences. It is the intellectual instrument that allows the homo sapiens to think rationally and logically by an analytic, deductive and deductive thought process. The mind has 'semantic awareness,' i.e. it apprehends and comprehends mental states as concepts, ideas, images, imaginations, etc.

People often refer to emotions as a "state of mind," but are they truly the same? Is an emotion, such as fear, anger, hate, or love, merely an abstract thought? Conversely, can abstract concepts like the number $\pi = 3.14...$ or the notion of "existence" be considered emotions? These questions can be answered only through direct subjective experience—a fact that encourages us to explore our inner selves from a first-person perspective.

While thoughts and emotions are intertwined and interdependent, they exist on different planes of perception.

Now that we have distinguished between matter, life, and mind, a crucial question arises: Who or what is experiencing them? Who is the witness to these physical, mental, and emotional qualia? When we take a step back as observers, we notice that physical sensations, emotions, and thoughts are ever-changing phenomena. These sensations, emotional experiences, and fleeting thoughts arise and fade before a witness that remains constant, much like a film projected on a screen. This unchanging yet sentient aspect is what we refer to as "consciousness." Consciousness pertains to our unique subjective experience—the witness that feels its own existence and truly knows all phenomena. This can also be described as "witness consciousness."

Although words fall short when it comes to fully capturing the essence of consciousness (as every definition inherently relies on mental categories, sensory experiences, or emotional phenomena perceived by consciousness itself), I will still attempt to provide some descriptors.

Phenomenal Consciousness[121]

Witnessing sense of existence, presence, and sentience, the knower who knows oneself, a self-existent reality, pure 'self-aware beingness' aware of the phenomenal world. It exists before any mental, emotional, or physical perceptions. It is that in which all experience appears and is known as pure and undifferentiated irreducible awareness and sense of existence which witnesses phenomena but remains unaffected by it. It is a changeless, silent, space-less, featureless, and immobile witness that stands as a pure and undisturbed background behind all our activities.

In our framework, consciousness is not a 'plane' but, rather, the essence within us that observes the various planes *of* consciousness. The 'planes' we refer to represent an inner psychological dimension characterized by specific yet changeable qualitative phenomena such as physical perceptions, emotions, and thoughts associated with the physical, life, and mental planes, respectively. These phenomena arise and fade over time. In contrast, consciousness itself remains the unchanging 'witness' or "spectator" that becomes aware of these phenomena. It is important to recognize that this is not merely an intellectual distinction; it arises from a profound inner discernment gained through an introspective first-person exploration.

Yet, the physical, life, and mental planes do not encompass the entirety of the witness consciousness's experiential spectrum. There are still a few domains we perceive to exist but to which we cannot directly relate.

One such domain is the subconscient plane, which has been extensively explored in modern psychology. This psychological dimension, largely beyond our awareness, significantly influences our thoughts and emotions. It manifests in dreams, shapes our thought patterns, and can suddenly evoke intense feelings. It underlies trauma, unexplained suffering, and pain stemming from past experiences, dysfunctional family dynamics, and education. Within us, there is a part that records, notes, and shapes our instinctual reactions, thoughts, and behaviors. We recognize this not through philosophical speculation but through our everyday experiences. We understand that something submerged and infra-rational within us, though only vaguely acknowledged, profoundly influences our lives. This domain can be characterized as follows.

The Subconscient Plane

A concealed and unexpressed consciousness below our surface consciousness. It is mechanical in its nature due to persistent, habitual, and repeated movements and thoughtless reactivity, behind the instinctive reflexes

[121] The term 'consciousness' has taken on various meanings depending on the discipline and context. In this book, I will always refer to 'phenomenal consciousness' simply as 'consciousness,' as characterized above, unless stated otherwise.

of the mind, life, and body. It is the source of our conditionings, habits, uncontrolled reflexes, religious and cultural dogmas, or educational imprints. It receives impressions and stores them in a subconscious memory which can emerge as instinctive mental formations or surge in dreams. It is uncontrolled by the mind but responsible for 'sending up' the automatic and old imprinted responses to the mind, body, nerves, and cells. It can support good or bad habitual actions and determine our character traits. It is the evolutionary basis with no waking conscious coherent thought, will, feeling, or organized reactions.

However, we must make an important distinction. There exists a vast suprarational domain that eludes our ordinary waking consciousness as well, yet contrasting sharply with the qualities of the subconscient. At some point, all of us have experienced sudden intuitions, moments of deep inspiration, creative bursts, and effortless insights. Where do these cognitive "revelations" come from? Are they from the subconscient? If you reflect on how you felt during these moments of wisdom, creativity, and clarity and compare those feelings to experiences rooted in the subconscious, the difference becomes apparent.

While both originate beyond our conscious awareness, the moments of harmony and wisdom have a distinctly different quality from the obscure, often painful, and reactive nature of subconscious experiences. There also exists a notable contrast between analytical intelligence and profound wisdom, which encompasses 'vision.' The flashes of knowledge, wisdom, and inspiration that artists, musicians, and scientists often describe are not products of the subconscious; they are "superconscious" phenomena. These insights possess a completely different character and quality than those arising from the subconscient.

Often, we overlook this distinction because we tend to confuse superconscious intuitions with the mechanical and reactive responses of the subconscious. However, recognizing this difference doesn't require any abstract philosophical theory. Just take a moment to compare your feelings during a brilliant idea or a moment of clarity with those you experience when sudden fears and anxieties arise from the subconscious. In both cases, the origins lie beyond our immediate awareness, yet they possess distinctly opposing qualities and sources. Therefore, I will differentiate between the superconscious and the subconscious as follows.

The Superconscient Plane

Intuition, transrational cognition, creative, inspirational, extra sensorial, and subliminal perception. Direct perception of truth or fact, without any reasoning process, lightning-flash of cognition, the proverbial "aha-moment," intimate perception of a truth, beyond analytic conception, a vision of things that carry with itself a sense of certitude that the mind can't explain, a higher form of discrimination, an awareness of knowledge the intellect can't understand or perceives only vaguely.

We must clearly differentiate not only between the mind and consciousness but also between the subconscient and the superconscient. We do this not merely as an intellectual exercise but also because these distinctions represent genuine lived experiences and observable psychological phenomena.

b. Towards a Multi-Modal Cosmology

Thus, starting from an experiential perspective, we can clearly see that our personality is not singular; it possesses a complex psychological structure. We are not just a body and a mind, nor is our being merely a byproduct of bodily functions. Instead, we embody a multi-faceted existence that includes physical, mental, emotional, subconscious, superconscious, and conscious dimensions.

Could this perspective hint at an alternative cosmology, one that transcends the conventional mind-body dualism and physicalist paradigm? Might these various planes of consciousness within us reveal insights about the universe as a whole?

Let us explore the implications of placing consciousness, with all its inherent qualities, at the center of existence rather than matter. Instead of assuming that everything else is simply an epiphenomenon of physical processes, we could consider the reverse. We might not seek to explain the emergence and nature of life, mind, sentience, cognition, supercognition, will, agency, purposefulness, and goal-directedness as products of material mechanisms. Instead, we could posit that these qualities are fundamental modes and attributes of consciousness, which serves as the foundational essence of the universe. In this framework, consciousness becomes the primary element, with life, mind, matter, the subconscious, and the superconscious representing different 'planes of consciousness'—or layers of reality.

Yet, how can we reconcile the distinction between the manifest mutable planes of existence and a transcendent immutable center of consciousness?

It is natural to consider a form of substance monism, similar to what we explored earlier in Section IX.3d, or when we were speculating about the subtle matter conjecture in Section V.VI6. The manifest is seen as a mutable aspect of the transcendent.

From a Spinozian perspective, the planes of consciousness represent the spatio-temporal modes and attributes through which the same consciousness expresses itself. Consciousness is not merely a mode or attribute of something; instead, everything is an expression of a specific attribute of a mode of consciousness. Ultimately, our emotions, thoughts, and physical sensations are different "modes," "vibrations," or "excitations" of the same "substance" we call consciousness. By adopting an integral perspective, we can view everything that appears distinct as an aspect of the indistinct, with every differentiation being simply a modification of the undifferentiated.

This perspective regards the mind as a veiled or involved aspect of consciousness, life as a veiled or involved aspect of the mind, and matter as a veiled or involved aspect of life. None of these are distinct or separate from consciousness; instead, they represent different "modes," "states," "excitations," "forms," or "phases" of the same underlying consciousness. As an analogy, consider the various states of the same substance, such as water vapor, liquid water, and ice. These three phases have unique physical properties and can take on different shapes and forms, yet they are all manifestations of the same substance: H_2O molecules.

SOLID LIQUID GAS

Fig. 68 Solid, liquid, and gaseous phase of the same substance: water.

By adopting 'consciousness-centrism' instead of materialism, we can position consciousness at the center of the universe. In this framework, consciousness precedes its own self-modification, which we refer to as 'mind'—that is, mind emerges from consciousness. Meanwhile, mind precedes life—that is, life emerges from mind. Finally, life precedes the most involved form of consciousness, which is matter—that is, matter arises from life.

How so? This sounds absurd and turns everything upside down, doesn't it? How could consciousness, mind, and life exist before matter? How can life exist without matter? This doesn't make sense, does it?

It sounds illogical because we always assume consciousness, life, and mind are only localized phenomena emerging in cells, organs, and bodies rambling on some minuscule planet in an immense dead and meaningless universe. However, if we look upon individual consciousness, mind, and life, and the super- or sub-conscient, not as being confined in a biological bubble but, rather, as universal principles that are active, or veiled and involved, throughout the universe (even in a rock on Mars), thereby conceiving of them as universal planes permeating all reality, then things might begin to make sense. There is not just the "Mind at large" or the cosmic mind of the idealist but also the cosmic life, the cosmic sub- and super-conscient, along with the cosmic physical—namely, what we ordinarily consider to be the physical universe studied by science. The physical universe is only one of several universal planes. While our mind is an individuation of the universal plane of mind, life in a living cell or organism is an individuation of the universal plane of life, and the material body is an individuation on the physical plane of matter.

This is a spiritual framework in which consciousness, life, and mind do not emerge from or are produced by matter through complex mechanisms. Instead, consciousness is seen as the fundamental essence that shines through or manifests in and through the mind. The mind, in turn, shines through or emerges in and through life, while life shines through or emerges in and through matter.

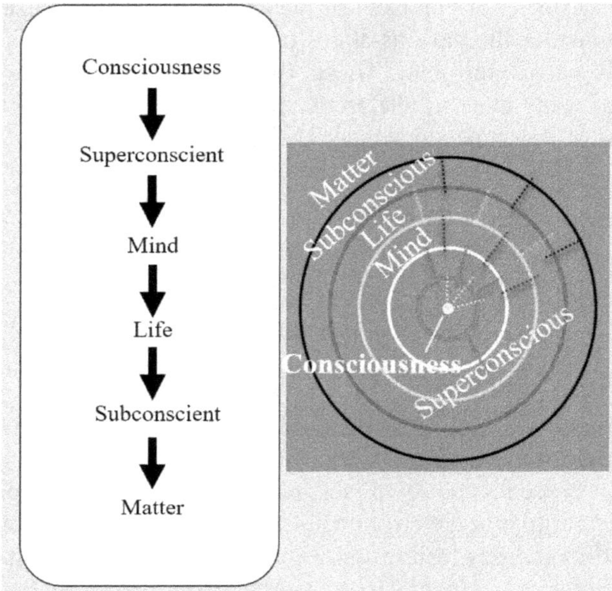

Fig. 69 The bottom-up and top-down integral substance monist posits consciousness as the fundamental primitive. Matter, life, mind, the super- and subconscient planes of consciousness as "modes" of existence of the very same "substance": Consciousness.

This raises an important question: If the mind is a localized aspect of a universal mind, what role does the brain play in this context?

Here, William James 's "transmissive hypothesis" or "filter theory of consciousness," which we discussed in Section II.5 and in the first advanced reading XIII.1, becomes particularly relevant. The brain is a physical organ that exists only within the physical realm, yet it does not belong to the other subtle planes of existence. Its main purpose is to connect the gross physical plane of matter with what lies beyond it. It functions as a 'filter' and connecting 'interface' between the physical plane and the other life, mind, and superconscious planes. Its primary role is more about transmission than generation. The brain does not create the mind or consciousness; rather, it channels and transmits them from the subtler superconscious, mental, and life planes to the denser physical plane. It receives and transmits information to and from the universal planes. To put it in Aldous Huxley's terms, the brain can be viewed as a reducing valve, interfacing with what he referred to as the 'Mind at large.'

This metaphysical model aligns with scientific understanding, as it presents a holistic view and avoids reducing all reality to a one-dimensional physical existence.

The brain is indeed responsible for cognitive functions, but I believe these functions are tied primarily to processing physical facts and perceptions from the physical senses rather than abstract reasoning and intuitive understanding. The brain acts as a 'physical/mechanical mind' focused mostly on tangible, physical aspects of reality. The brain could be seen as a 'pre-processing' unit that follows (not precedes) more abstract and semantic thinking to physical information. Additionally, the brain oversees crucial physiological functions, including heartbeat regulation, digestion, thermoregulation, homeostasis, and the processing of motor-sensory information and pattern recognition proper to the physical plane. However, it seems to be much less responsible for higher cognitive functions or any superconscious activities. Instead, the brain is deeply rooted in the subconscious realm. It takes time to rewire, habituate, and reprogram its functions. Consequently, the brain tends to reflect a mechanistic and repetitive style of cognition rather than a comprehensive and intuitive one. This aligns with the paradigm that sees both matter and the subconscious representing the mechanical aspects or 'modes' of the superconscious planes.

The distinction between the various planes or 'modes' of universal consciousness should not be seen as a strict separation regarded as a system of rigid classifications with sharp boundaries. Rather, it represents a continuous spectrum that begins with universal Consciousness and progresses through various modes and states, including the superconscient, mind, life, subconscient, and matter. It's like a color spectrum. While there are clear differences between colors like red, yellow, green, and blue, the exact points where one color begins and another ends are largely conventional. Nonetheless, this does not diminish their reality.

Finally, we developed a framework that is more nuanced than current theories of idealism, surpassing the simplistic mind-body dualism of traditional dualists and the one-dimensional mind-brain identity theory of material monists. This new framework possesses its own complexity, yet it offers a richer spiritual perspective and has greater explanatory power. It is grounded not just in theoretical speculation but also in our direct experiences of what we see, perceive, feel, and ultimately know about ourselves.

The objection I frequently confront is: Why should we even differentiate between multimodal planes of reality? Are these distinctions merely semantic categorizations? If we embrace substance monism, why not limit ourselves to the one substance and simply call it "mind" or "consciousness"?

To some extent, this is true. From a broader perspective, everything can be seen as an 'excitation' within consciousness, ultimately pointing to the same underlying reality. The distinctions we make between mind, consciousness, life, and the superconscious might seem like an unnecessary proliferation of concepts. Why not embrace the principle of parsimony and apply Occam's razor to eliminate all but the most fundamental element, whether that be consciousness or mind?

My answer is that we shouldn't eliminate everything based solely on the desire for simplicity or the principles of parsimony, as reality is often much more complex than our simplistic understanding of it. The notion of what is supposed to be “simple” or “parsimonious” is highly subjective. To truly grasp the world, we need a certain level of distinction and nuance. Without this, our perception of the inner and outer universe becomes overly simplistic, like a pixelated image that lacks clarity. Just as we wouldn't advocate for a black-and-white representation of the world for the sake of parsimony, we shouldn't apply the same approach to constructing a cosmology. Recognizing and understanding the various 'excitations of consciousness' can enhance our comprehension of the world, both spiritually and scientifically.

Using a metaphor, we might say the semantic distinction between water vapor, ice, and liquid water is negligible because the ontology of H_2O molecules remains the same. While this is true, neglecting the fundamental differences in their physical states can lead to confusion when one is trying to understand a world composed of oceans, rivers, clouds, glaciers, and polar ice caps. By cutting out these distinctions with a razor, we risk conflating one state with another and failing to grasp the complexities of the world.

Most contemporary thinkers throw these all together without distinction, reducing them to a materialist, one-dimensional physicalist ontology or, at best, to a mind-matter dualism. This approach is often justified by the principle of parsimony—Occam’s razor—which is said to take precedence. However, conflating different entities and treating them as equivalent or synonymous may simplify our worldview, but it has never resulted in conceptual clarity or new discoveries. On the contrary, it creates confusion and leads to false premises.

The tendency to reduce everything to a single fundamental principle, often ignoring its diverse modes and states, is a mental instinct inherent in the analytic mind. This inclination leads to a belief in unity through uniformity, which fosters an uncritical acceptance of material monism while simultaneously rejecting the idea of substance monism as incompatible or unnecessary. However, as we will explore, embracing the multiform aspects of the one undifferentiated Consciousness—operating at various levels of reality—provides a more powerful framework for understanding fundamental questions about our true nature, psychology, life, the principles of evolution, science, the universe, and, ultimately, ourselves.

At this point, the key question isn't whether you should believe in or adopt this model of reality. My aim is not to promote another belief system or engage in abstract metaphysical speculation. Rather, I wish to offer insights into our nature through introspective investigation and to explore where that leads us.

How we perceive reality influences what we observe and ultimately discover. If traditional third-person science has not provided answers to fundamental existential questions, we should consider whether a science that embraces subjective experience will lead us in a more fruitful direction. This framework is not intended to be the ultimate theory of everything; instead, it serves as a working hypothesis—a temporary conjecture that invites testing. [122]

In the next section, I will demonstrate how this 'spiritual cosmology' offers a fresh perspective on evolution, potentially illuminating many aspects of life and evolution that remain otherwise unexplained.

3. Evolution, Involution and Spiritual Emergentism

We expanded the idealist worldview from a coarse-grained mind-body dichotomy to a multi-modal cosmology that includes various "planes" of conscious existence. This shift was based not on abstract analytical deduction but, rather, on introspective experiential observation. I will refer to this perspective—which encompasses different layers of existence, ranging from matter to the superconscious, each possessing its own ontological reality—as "integral cosmology."

In this paradigm, the universe is not merely a Mind, as some limited forms of idealism suggest. Instead, mind, matter, life, the subconscient, and the superconscient represent different "modes," "concretizations," or universal "planes" of the same essence: consciousness. This cosmology bears some resemblance to Spinoza's pantheistic monism, which posits that Nature and God are identical, with mind and matter as merely two "attributes" of the same "substance." However, the key distinction is that integral cosmology is not pantheistic (where Nature is God) but, rather, panentheistic (where Nature exists within God, yet God transcends Nature). It broadens the vision beyond a simple God-Nature duality.

In Sections IX.3d&e, and X.1, we examined how traditional forms of idealism fail to satisfactorily address questions regarding the nature of evolution. These perspectives often envision a universal Mind, a Mind at large, or, as Spinoza suggests, a One-Substance presumed to operate as a blind instinct devoid of intelligence and free will. This concept of a self-unaware universal consciousness was also adopted by Schopenhauer, who interpreted the "World as Will and Representation." Contemporary idealists, such as Bernardo Kastrup, continue to defend this viewpoint.

---

[122] Occam's razor is a philosophical principle that suggests the simplest explanation, with the fewest assumptions, is usually the best one. It encourages choosing the most straightforward solution that adequately explains the phenomenon, thereby avoiding unnecessary complexity. This principle is widely used in philosophy to favor clarity and efficiency. Contrary to popular belief, Occam's razor is not a scientific principle. Additionally, the history of science demonstrates that it has not always been the correct choice. Nature can be complex.

On that occasion, I clarified my perspective by noting that the origin of life and evolution are unlikely to be fully explained within such a one-dimensional idealistic framework. There must be an inherent higher intelligence expressed as a Real-Idea in Nature, one that is more self-aware, to account for the complexity and creativity we observe in Nature.

The lingering sociological question is: What drives the shift from the religious concept of an almighty God to the indifferent and mute divinity of the philosopher? I believe this shift stems from a coarse-grained metaphysical perspective in philosophy and science that struggles to move beyond mind-body dualism or, worse, a simplistic material monism.

Given the superficial mechanistic and seemingly deterministic nature of the universe, if we equate Nature with God without recognizing any 'planes of consciousness' in between, we must perceive God as mechanistic as well. If Nature, with all its phenomena, is viewed as the conscious universe or merely an appearance of it, with little beyond those appearances, we are led to the logical conclusion that a creation governed by blind processes must reflect an equally mechanistic creator, one that is not much more than its creation. If Nature behaves like a mechanical, automatic, and unfeeling entity and is thought to embody the properties of a higher Self, then that Self cannot be a superconscious and omniscient being. Instead, it becomes a rather disappointing, dumb, and subconscious divinity.

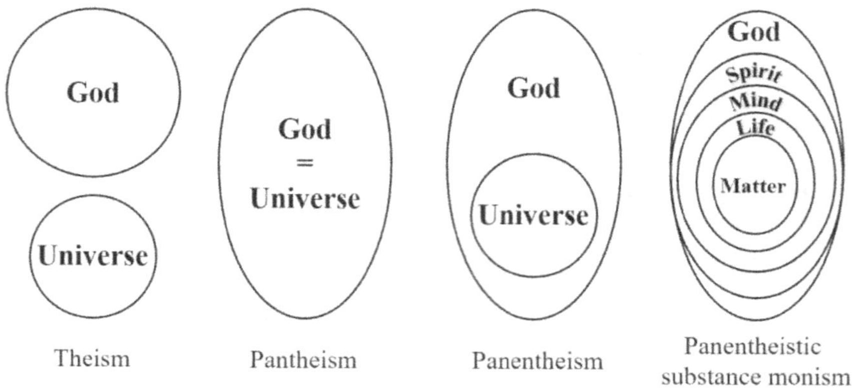

Fig. 70 Diagrammatic representation of theism, pantheism and panentheism.

The issue resolves itself when we move beyond a black-and-white view of the universe and a purely pantheistic perspective. We must recognize not just the physical realm of Nature and a semi-conscious universal mind but also a spectrum of planes of consciousness that extends beyond these. The instinctive and unintelligent aspects of Nature represent the lowest rung on the ladder of superconsciousness. In a panentheistic framework, what we perceive as the automatic, unconscious, and insentient qualities of the physical plane we call the 'exterior Nature' are the final manifestations of a process in which a supremely creative 'interior Nature' forgets itself. This higher Nature reflects the superconscious, all-aware, and almighty Godhead.

If we consider Nature as a seemingly unaware and ignorant external layer of consciousness, which is a projection of higher, more aware planes of existence (such as the life and mental planes), then we begin to reconcile apparent contradictions. These internal planes, in turn, are expressions of a supreme superconscious plane that transcends yet encompasses them all.

From this perspective, the notion that Nature and evolution resemble a machine, with a creator possessing opposite attributes, becomes clearer. What might appear to be an irreconcilable contradiction from a coarse dualistic or pantheistic viewpoint—prompting philosophers to reduce God to the level of Nature—reveals a different understanding through the lens of a multi-modal panentheism. This approach acknowledges the multiple planes of consciousness descending from the superconscious to the unconscious realm. In this light, we can discern that unconscious Darwinian evolution reflects the workings of a concealed superconscious knowledge. In other words, ignorance is only one of the many masks that the Omniscient wears.

Can this perspective enhance our understanding of evolution?

The term "evolution" originates from the Latin word "evolvere," which means to "unroll," "open out," or "uncover" (with "e" signifying "out" and "volvere" meaning "to roll"?. Interestingly, this contrasts with the modern Darwinian definition of "descent with modification," which defines biological evolution. The term "evolution" suggests a deeper wisdom—that it is an emergent process revealing something inherently present within the unfolding. Similarly, the concept of "development" implies that something "de-velops" by emerging from an envelope, thus hinting at an already fully formed essence that evolution uncovers. It seems to suggest that evolution unveils something that is already fully developed but concealed or hidden. However, this essence may not be merely material, molecular, or genetic. We can view evolution as the manifestation—the "unrolling"—of universal Spirit through matter. It is not matter that produces consciousness; rather, it is a universal Consciousness that emerges within and through matter, taking on the form of individuated conscious entities. The distinction between the universal and the individual reflects an inherent polarity.

Conversely, this suggests that before the evolutionary process that unfolds the individual spirit within matter, a reverse process must have occurred. The emergence of consciousness, mind, will, and agency in the physical realm reflects powers that a process of "involution" previously obscured. Initially, the universal spirit immersed itself in a state of fragmentation through a movement of self-forgetfulness, becoming matter—that is, it "involved" itself in matter.

Viewing the physical manifestation as an involution of the Godhead into the material offers a different perspective. It represents a divine superconsciousness that has willingly descended into oblivion by "projecting" and "sacrificing" itself into its own shadows, which we refer to as "matter," "life," and "mind." In this process, it conceals itself from its own awareness, becoming a self-masked yet still divine consciousness that remains latent and dormant within the depths of inconscient matter.

The true significance of this process lies beyond intellectual understanding, though analogies, metaphors, or imagery can help us intuitively grasp its meaning. For instance, we might say that the One, in its desire to become the many, sought self-awareness by manifesting itself in multiplicity. In this act, supreme consciousness fragmented itself, creating myriad individual identities that are no longer aware of their original Source-Consciousness. Consequently, the Infinite fell into its opposite, leading to a loss of the original bliss found in unity. Now, it is engaged in an evolutionary journey to reclaim that unity through a reverse movement.

Matter is a projection and formulation, a 'shadow' of the same Consciousness in manifestation. Despite this, it remains an expression of original divinity. However, it behaves as a mechanical and inconscient substance within a gross material universe, existing in what can be termed 'cosmic ignorance.' From this state, the spirit gradually re-emerges through an evolutionary principle.

Matter represents a form of veiled life; life is a form of veiled mind, and mind is a form of veiled super-consciousness. It is the One that has become self-involved in the depths of a descending ladder of consciousness, transforming into the many. It was this involutionary process that gave rise to the planes of consciousness and the polarity between the universal and the individual manifests across all planes of existence. The universal spirit does not merely break into material particles; rather, it begins a process of individualization that occurs on every level of existence. Each individual reflects and embodies all universal planes through its unique expression. This means for every plane of existence, there is a corresponding "part of being." The individualized material bodies, which emerge through the evolutionary process starting with the formation of the first cell and progressing to the development of multicellular organisms, represent the individualization in the physical plane. Our subconscious psychological nature reflects the cosmic subconscious in its individualized form. Furthermore, our inner emotional being, which we previously distinguished from our mental faculties and conscious awareness, represents the individualization of a universal life-plane. The universal cosmic mind, or the Mind at large, manifests in our physical existence as individualized minds. Our mental activities and thoughts are shaped by our unique identities, which are situated in space and time yet arise from the broader universal mind. This principle also applies to our intuitions, inspirations, and sudden bursts of genius. These moments represent localized expressions of higher forms of knowledge that descend from the universal superconscious.

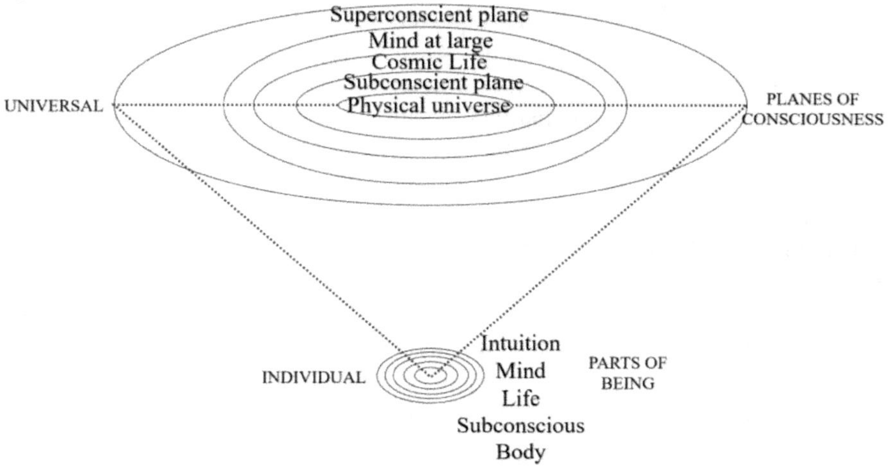

Fig. 71 Diagrammatic representation of the relationship between the universal and individual in integral cosmology.

Although this perspective differs from the scientific understanding of evolution, it is important to note that this spiritual viewpoint, which we can call "spiritual emergentism," does not contradict any scientific findings. The integral vision does not reject the traditional Darwinian model; rather, it complements this model with a dual process of emergence. On the one hand, there is an involution of a superconscious spirit descending from superconscious realms into unconscious ones, which then develops through a bottom-up process as an unconscious spirit in matter. This unfolds from a nearly unconscious state into progressively more complex states of consciousness, which the analytical physical mind interprets as Darwinian evolution. On the other hand, this process is also guided by the influence of the same superconscious spirit from above, operating within the individual living organism as well.

This perspective aligns with the modern scientific understanding of evolution in that it suggests matter possesses a will and a drive to bring forth something from the inconscient. This process begins slowly and without awareness, relying on trial and error through natural selection, random mutations, etc.; where speciation, with its enormous morphological variety, reflects the infinite multiplicity inherent in the original Unity. Sooner or later, it must reveal itself.

While evolutionary biologists attempt to explain the primordial impulses of agency, goal-directedness, volition, and intent that appear in all forms of life through complex physical, computational, or functional processes, their efforts always fall short. These traits, which emerge from matter or as propensities of mechanical processes, can be traced back to the earliest single-celled organisms. In the context of spiritual emergentism, however, these behavioral and psychological aspects are viewed as fundamental elements of consciousness itself. The 'life impulses,' such as the instinct for survival, are merely superficial manifestations of a deeper, original impulse. This impulse is not a product of blind, mechanical biophysical processes but, rather, arises from a deeper desire to exist that is inherent in consciousness. Because consciousness and existence are one and the same.

When the Spirit's concealment is nearly complete, the consciousness embedded in matter becomes forgetful of itself and is largely dominated by mechanical forces. Consciousness emerges through a slow and arduous bottom-up process, shaped by eons of Darwinian evolution driven by natural selection, mutations, and habitual behaviors. However, once consciousness begins to unfold from matter, it gradually assumes a leading role.

Simultaneously, there is a top-down influence. The superconscious realm, while not intervening directly, indirectly guides and influences the evolutionary process from above. The morphological development of species on the physical plane represents only a superficial aspect of a deeper spiritual force at work—one that acts from the superconscious levels above and manifests as “seed-power” within the physical and subconscious below. This spiritual process reveals the Real-Idea in Nature, as discussed in Section IX.3e.

Thus, there is no contradiction or irreconcilable viewpoint between matter and spirit, science and spirituality, or spiritual emergentism and the mechanistic materialism of Darwinian evolution. The latter is simply one aspect of the former and does not encompass the entirety of biological evolution. Once one recognizes this trans-physical-spiritual dynamic in creation, contradictions fade away. There is no insurmountable duality separating seemingly unconscious matter from the omniscient spirit. Additionally, there is no conflict between the non-teleological, mechanistic view of evolution and teleological interpretations that account for final causes. Both represent different aspects of the same process, which can be harmonized within this integral vision.

For example, Spinoza's instinctive God that couldn’t do otherwise, as well as Schopenhauer's undirected "World as Will," which also influenced Kastrup's analytic idealism, represent a cosmic subconscient life-plane. They depict a specific facet of an all-knowing and all-powerful divinity that, within this realm, manifests as an instinctive universal will, repeating habitual behaviors almost mechanically. However, this cosmic subconscient life-plane is merely a small segment of a vast spectrum of being in the layers of an involved consciousness.

The mind is also not a creation resulting from an aggregate of matter. Rather, it is already involved in it and re-emerges—that is, evolves—out of matter during the evolutionary process once the physical aggregate is ready to hold it. Mind exists as a universal plane already present and pervasive throughout the cosmos before any evolutionary processes began. However, it is a derivative and subordinate aspect of the higher levels of consciousness.

This suggests the universe came into being through a creative process in which matter is the last manifestation, not the initial foundation of existence. Evolution represents the unfolding of the Spirit within matter, with matter acting as a densification or condensation of the same self-involved and self-limited Spirit. From this viewpoint, a universal Consciousness contracts and confines itself into what we refer to as "matter." Matter serves as a self-demarcation or exclusive self-concentration through the differentiation of the One. Just as a seed, cell, or egg contains the potential for the full organism into which it will eventually develop, matter holds within it all the potentialities of the Spirit. However, at this stage of evolution, matter's conscious aspect remains largely inert and only subconsciously active, characterized mainly by its mechanical and deterministic nature.

The forces driving this evolutionary process originate from below and are shaped by material processes governed by the laws of physics and established Darwinian principles such as natural selection, genetic drift through seemingly (apparently) random mutations, environmental influences, adaptation, and the survival of the fittest. Simultaneously, other forces arise from above and guide the process from within through an inherent consciousness. This guidance varies in directness depending on the planes of existence and maintains a connection to an original superconsciousness.

In stone, which represents the most concrete form of the spirit on the physical plane, the planes of life, mind, and superconsciousness are present but remain unexpressed. In plants, the universal life-plane emerges as the first individualized expression of life, although the mind is still largely dormant, beginning to manifest only as a basal cognition. In non-human animals, a more developed brain enhances cognitive and sensory capabilities, thus allowing for further development of both the life and mental planes in the individual organism. In humans, fully developed analytical reasoning—an inherent aspect of the cosmic mind—emerges on the physical plane, while initial glimpses of transrational intuition from the superconscious begin to appear. It is reasonable to predict that humanity's next evolutionary step will be to transcend itself toward a superconscious existence that fully integrates and harmonizes all universal planes within the individual.

The aim of evolution can now be understood as the progressive self-discovery of the universal superconscious spirit through a virtually infinite variety of material forms. These forms increasingly possess the capacity to embody the spirit's powers, qualities, and knowledge. Evolution unfolds as an epic journey in which every event, manifestation, and creature represents a 'Thought,' an 'Image,' or a 'Projection' of a powerful 'Idea' within a unified essence that embraces infinite diversity.

From this perspective, the Christian belief that humans are made in God's image, standing in contrast to the notion of humanity's fall from a perfect state of grace and obedience to one of disobedience and corruption as described in the book of Genesis, begins to make sense. This contrast reflects a collective, albeit distant, memory of a partial truth. From an integral perspective, the individualized spirit emerges from a 'fallen' state, shaped by the complexities of our existence, while simultaneously receiving influence from the universal Spirit above.

This raises important questions about our nature: What does this integral cosmology reveal about us? What is our true essence? Are we merely lost "fragments" or "dissociated alters" of the universal Spirit? What, ultimately, defines our identity?

4. The Evolution of the Soul

The previous sections presented a spiritual and emergentist cosmology that transcends traditional dualism and philosophical idealism. The universe cannot be reduced to merely a Mind; we are not simply a body and a mind, and reality encompasses much more than the 'mental.' From the standpoint of substance monism, the universe is ultimately a single, undivided consciousness. However, this consciousness manifests in various 'modes,' 'excitations,' or 'states of being,' which are not merely different mental states but represent qualitatively distinct experiential modes that constitute the foundation of the universe. Matter is just one of these states, while mind and life are rooted in deeper, submerged subconscious planes of existence.

Most importantly, we recognize the existence of a superconscious universal plane, which we are largely unaware of but which is fundamentally different in quality and character from our subconscious. As individualized entities, we are projections of this universal Consciousness. We embody and reflect these universal planes and their qualities within our limited forms—specifically, as body, life, mind, subconscious, and superconscious individualities. These aspects are projections of the One undivided Self in form of physical, vital, mental, subconscious, and superconscious universal planes. The distinction between the universal and the individual arises because the Spirit has undergone an "involution" by an "exclusive concentration"[123] into different modes of its existence; meanwhile, through a process of spiritual emergence within matter, it is reemerging in what we refer to as "evolution."

---

[123] This "exclusive concentration" is analogous to B. Kastrup's notion of the self-dissociation process of Mind at large into individual "dissociated boundaries" or "alters," representing minds, bodies, and individuals, and that he described as equivalent to a "cosmic dissociative identity disorder." Integral cosmology extends this phenomenon to encompass everything that appears differentiated from the undifferentiated One. Molecules, atoms, and particles also represent exclusive concentrations of the One into itself.

When we simultaneously recognize the evolutionary, multilayered dichotomy between the universal and the individual, a clearer and more coherent vision emerges. This perspective offers greater explanatory power regarding many of the mysteries with which both idealists and physicalists still grapple.

The key questions then arise: What is the nature of this individuality, and what role does it play in the universal manifestation? Why did a universal Consciousness or Being fragment itself into countless semi-conscious or subconscious entities, such as living beings like us? What led to the emergence of this universal vs. individual distinction? Most interestingly, what is our place and destiny as individualized beings in relation to the universal Being?

A productive way to explore this aspect is to take the introspective approach seriously. We have already recognized an experiential fact that does not depend on philosophical abstractions: The ordinary human mind can distinguish itself from consciousness. This witness-consciousness serves as a passive observer, remaining detached as an unperturbed testimony to all the phenomena it encounters. This is what we consider the essence of our true being.

However, if we listen carefully, we realize there is more beneath the surface.

Within us is a feeling centered in the heart (rather than in the head) that reflects superconscious intuitions and emotions as well. We often express this as "the mind may say one thing, but the heart knows otherwise." There is an innate understanding of what is good or bad for us, even when our analytic mind resists this wisdom. This inner nature sends us quiet intimations and possesses deeper knowledge despite external circumstances suggesting otherwise. It is this "inner voice" that inspires our sense of purpose and life mission. It embodies our instinct for beauty, seeking expression through art and music, and upholds a profound sense of goodness even in a world often governed by the survival of the fittest.

The average person is only vaguely aware of it and often perceives it as the popular "inner voice" in our heart that softly whispers messages to our conscious mind. Many traditions describe this heart center in various ways; for instance, Sufism refers to it as the "five levels of the heart."

The Trappist monk, mystic, and author Thomas Merton called it the "point vierge" —the virgin point— describing it as a *"center of our being, a point of nothingness which is untouched by sin and by illusion, a point of pure truth, a point or spark which belongs entirely to God... which is inaccessible to the fantasies of our own mind or the brutalities of our own will"*.

Our true innermost individuality, or "divine spark," possesses qualities primarily of a higher emotional nature, including love, beauty, empathy, compassion, an innate perception of truth that transcends mental reasoning, a sense of goodness, harmony, emotional balance, sweetness, peace, and a deep connection with its Source. While these qualities reside within us, they are often obscured by the distractions of our untransformed mental, emotional, and physical activities. For instance, genuine empathy originates not from the mental or life-plane but from the soul, which experiences a profound sense of identity and connection with others and all living beings beyond superficial separations and personalities.[124]

If this remains elusive, we can turn to the insights of Don Salmon, who offers practical steps to help us reach this inner presence.

"It requires only an open heart, not a fully silent mind or years of philosophic study. One can start simply looking at a flower, or an animal, a baby, or anything that sparks a sense of wonder, beauty, and deep noble goodness. Then take attention away from the object and dive deep into the feeling (not vital, emotional feeling). Then expand attention, allow attention to widen, widen until it embraces all of experience. Then let that inner feeling expand throughout the field of attention.

Do this for even 10-15 seconds at a time on and off throughout the day. More and more, recall this sense of beauty and wonder, expanding attention, looking at the sky, recalling the ocean, the infinity of outer space, and allowing that combined sense of deep wondrous beauty with the awesome boundlessness of space. Then allow "something else" to take over. And be patient.

It may be days, weeks, months or even years. But there is a call, a yearning deep in everyone's heart, that we all can contact. This call is said by the Sufis (Islamic mystics) to be what Allah was referring to when He declared that, "I was a hidden treasure [hidden in the depths of our hearts] and sought to be known." It is what Paul was pointing us toward when he spoke of all of creation yearning for the return of Christ.

This yearning is common to all human beings, no matter how much it may appear to be hidden. When we can begin to intuit the warmth and radiance of this loving call, this profound yearning in the depths of our being, we will begin to discover it more and more manifest not just within, but in the beauty and goodness we begin to perceive in everyone and everything around us."

The more or less conscious perception of our inner nature has driven many religions to adopt various concepts of the soul as a central element for understanding human existence and spirituality. The soul is often viewed as a sacred, immortal aspect of human identity, intimately connected to God. In Christian belief, after death, the soul faces judgment. Depending on one's faith, actions, and relationship with God, the soul will either enter eternal communion with Him (heaven) or be separated from Him (hell). Some

[124] This explains why some individuals exhibit a strong sense of empathy toward their own family, ethnic group, or nation but show little to no empathy for others. Similarly, they might feel a deep connection with animals while lacking empathy for humans or vice versa. This phenomenon reflects an emotional 'life-empathy' rather than a 'soul-empathy.'

Christian traditions also embrace the concept of purgatory, a temporary state of purification for souls destined for heaven. Individuals can be forgiven, and their souls can be saved through a notion of salvation deemed necessary for eternal life. Notably, the Christian tradition, as well as every Abrahamic religion, emphasizes the spiritual development of the soul during earthly life. This development involves practices such as prayer, worship, sacraments, and acts of charity, all aimed at nurturing and strengthening one's relationship with the Divine and transforming the soul.

Traditional Indian philosophy, particularly Vedanta, introduces the concepts of the “Self” and “Atman” as reflections of the eternal and immutable Brahman. This non-dual philosophy encourages individuals to transcend ordinary states of consciousness through psychological, mental, and, ultimately, physical practices, such as yoga and meditation. The journey culminates in the realization of the nonexistence of the ego and the recognition that it is a mere illusion, a mental figment of reality that is part of the phenomenal world. Over time, the individualized mind stops identifying itself as separate and becomes aware of the universal Mind. One sees the ego as an individual manifestation of a singular mental reality. Our personality, with its traits, tendencies, appetites, habits, and moods, is merely a “knot” in the mental, emotional, and physical realms—a mask we identify with and become attached to, though it does not represent our true nature.

This is why non-dual masters are cautious in discussing the concept of a “soul.” When they do, they do not refer to the personality that identifies with an “I” and its associated feelings of selfhood. Instead, these traits are seen as external manifestations that do not define our innermost being. Our true essence transcends these individualizations and remains constant, described as an “atemporal beingness”—a state beyond verbal expression, sometimes referred to as “nothing,” “emptiness,” or the Buddhist concept of “sunya.” Once this realization occurs, it leads to “liberation,” “moksha,” or “nirvana”—freedom from suffering and the cessation of the mind's activities.

However, the general religious, theological, or philosophical context is still missing something: the evolutionary aspect of the soul. The concept of the soul, when accepted, is often viewed as a static, passive, and unchanging entity that embodies our immaterial and spiritual individuality, persisting even after death. In many Eastern traditions, the soul continues its journey through a cycle of reincarnations but is not seen as a spiritual individuality capable of change and evolution during this process. Instead, it is regarded as an eternal witness (the Purusha) that does not actively influence or shape our lives. The idea of an evolutionary soul is notably absent.

Thus, the concept of 'soul' being discussed here should not be conflated with what we refer to as 'character,' 'personality,' or 'ego.' While the soul is characterized by a divine emotional quality, our surface emotions are shaped primarily by external vital impulses, habitual behaviors, physical and emotional preferences, mental tendencies, and unexamined biases and beliefs. These elements represent only the untransformed aspects of our personality, influenced by education, environment, socio-cultural factors, and genetics. The 'ego' is merely a shadow that affects our mind and superficial self rather than a true reflection of our innermost nature.

The metaphysical assumption of a changeless soul has influenced modern Western psychology. Most psychological schools adopt a secular approach, often ignoring the hypothesis of an immaterial soul-entity. Even among the few psychological theories and practices that consider spiritual perspectives, such as Ken Wilber's transpersonal psychology or Roberto Assagioli's psychosynthesis, the soul is typically regarded as having little or no impact on our psyche and life. Rarely is it viewed as possessing an evolutionary nature or purpose. In contemporary psychology, philosophy, and metaphysics, only a few exceptions exist, but even they are constrained by this a priori assumption, which, like many rationalistic and naturalistic doctrines, has significantly hindered progress. [125]

How can we understand the evolution of the individual differently?

If we aim to move beyond the typical physicalist perspective that attributes evolution solely to physical forms, thereby neglecting our inner spiritual dimension, we must explore the meaning and purpose of evolution. Is it merely a blind, mechanistic journey through one era after another? In this regard, proponents of Darwinian materialism hold a distinct advantage. They argue that evolution lacks meaning, sense, purpose, or aim; it is merely a series of coincidental morphological changes driven by natural selection and reproduction. Naturalism does not concern itself with questions about the existence of purpose, let alone "souls."

However, those willing to look beyond the confines of an anthropomorphic materialistic science—and, similarly, anthropomorphic religious doctrines that conceive of a transcendent individual entity—must provide a more compelling explanation of the relationship between this supposedly static and unchanging soul and the continuous evolutionary changes in the natural world. Is life merely a one-time experience after which we are judged and sent to heaven or hell? Or do Eastern philosophies suggest that nothing truly evolves and, rather, everything is a cosmic illusion, or Maya, with our only purpose being to escape it forever? If that is the case, why did this cosmic drama begin in the first place?

What if we envision an evolutionary spiritual cosmology in which the inmost entity that transcends our body, mind, and emotions is not merely a static and eternal presence but, rather, the dynamic essence of each individual that evolves from life to life through the process of transmigration? Furthermore, what if biological Darwinian evolution on the physical plane reflects the evolution of the soul that inhabits it?

---

[125] For an interesting review of this aspect, see Teklinski (2018).

This perspective flips conventional thinking on its head. First, it introduces the concept of an evolutionary soul, thereby challenging traditional metaphysical ideas, as well as secular science and psychology. Second, it shifts the focus from viewing evolution solely through physical forces, such as genetics, the environment, or biochemical processes, to placing the evolution of the soul at the center of both material and morphological development. When we examine life, evolution, our minds, and our psychological dimensions through a soul-centered lens of evolutionary spiritual emergentism, many aspects of reality that once relied on convoluted and ultimately unconvincing theories or simplistic religious dogmas begin to make more sense. This perspective opens us up to new ways of understanding the world and ourselves—ways that we previously could not have imagined.

After all, the concept of an evolving soul aligns with our everyday experiences. Life experiences are fundamental in shaping who we are as individuals. Each experience, whether positive or negative, presents an opportunity for learning and growth. Through challenges, successes, failures, relationships, and encounters, we gain new knowledge, skills, and perspectives that enhance our personal development. Facing adversity fosters resilience, while overcoming obstacles teaches us to adapt, develop coping mechanisms, and emerge stronger. Our life experiences help us uncover our identity, values, and aspirations. By exploring various paths, interests, and relationships, we gain insights into our strengths, weaknesses, passions, and purpose. Ultimately, these experiences shape our character, perspective, and life trajectory, thus acting as catalysts for our inner evolution.

As Eckhart Tolle uses to say:

"Every being is a spark of the Divine, or God. Look into the eyes of the dog and sense that innermost core. When you are present, you can sense the spirit, the one consciousness, in every creature and love it as yourself."

This isn't abstract metaphysical speculation or rocket science. We already understand this, yet we resist the idea of viewing it from a universal perspective that encompasses all life. Instead, we tend to limit our conception of evolution to the superficial morphological traits of organisms and their biology. Then, we wonder why and how we got here.

When we are present and sense the spirit—the one consciousness—in every creature and love it as ourselves, something deep within us naturally connects with qualities such as love, beauty, truth, and peace. Our individual and evolutionary self reflects these because they emanate from the universal impersonal superconscious plane, where all these qualities naturally reside. One could argue that the true qualities and powers of universal Consciousness extend beyond mere perception and subjective experiences. They encompass love, bliss, superconscious knowledge, will, power, and force, among others.

On the physical, mental, and life-plane, these qualities manifest as human love veiled by egoistic desires. However, these superficial manifestations do not define our true nature; rather, they are expressions of our emotional and mental inclinations, closely intertwined with bodily necessities and our nervous and subconscious identities.

Our true nature is an individual essence that understands us more profoundly than the surface personality with which we typically identify. This innermost individuality emerges throughout the evolutionary process, which does not occur within a single lifetime. During its life experiences and the journey of transmigrations, this soul entity grows, evolving from one lifetime to the next. From the perspective of this integral evolutionary cosmology, death is neither a malady nor an inevitable accident. Instead, it is an essential part of the evolutionary process that enables the soul to have diverse experiences across different forms and lives. Death is not annihilation, as materialists believe, nor is it a final judgment, as some religious doctrines suggest. It is also not a reset button that forces us to repeat the same cycle, as some mystics claim. Rather, death is a transition that facilitates further transformation and evolution.

Thus, biological evolution, complemented by a view that conceives of an evolution of the soul, offers a completely new perspective.

It is important to clarify that referring to an "evolution of consciousness" is technically inaccurate, as the ultimate universal Consciousness simply exists. This Consciousness is transcendent and beyond space and time; thus, it doesn't evolve. It does not need to, as it is already complete and, in and of itself, does not require evolution, even while being inherently present and active across all levels of its creation. Instead, it is the individual soul—the "inner flame"—that evolves. This individualized aspect of the superconscious and universal Godhead undergoes a process of rebirth, growing in consciousness from one life to the next. Such an individualized "soul-evolution" manifests in the physical realm as the transformation of species: from inert matter to plants, from plants to animals, from animals to humans, and now from humans to a future superconscious being. It represents the emergence of Spirit within matter through the increasing complexity of organic forms, which become progressively self-aware and capable of containing and expressing these higher powers.

It is reasonable, even inevitable, to believe that humanity is gradually ascending to the next evolutionary status, slowly yet surely advancing to a new form of inner development and knowledge that will transform our perspective of the world. This aligns perfectly with the current naturalistic understanding of evolution, life, and the cosmos. Unless humanity faces self-destruction or a catastrophic event, such as a massive asteroid impact that wipes out all life on Earth, it is inevitable that Homo sapiens and our way of understanding the world will be surpassed. The question is not whether this will happen but, rather, when and how it will occur.

However it is, as long as science continues to ban the soul and other, more subtle domains of existence as anathema, positing it a priori as non existent, not because of a scientific rationale but because of an axiomatic and almost unquestioned dogma, it will hardly be able to go beyond a superficial understanding of life, reality, and the universe. It will forever be able to answer only the question of how things work and never why they are, let alone what they are.

In the next section, we will see some examples of how this cosmology has more explanatory power than the conventional forms of idealism or reductive naturalism.

5. The Spectrum of Life and the Qualities of Consciousness

A multidimensional vision has been presented, envisioning a panentheistic ontology in which reality encompasses not only the material, subconscious, and mental realms but also the superconscious, with each individuality representing a soul in evolution.

In Section III.1, we explored the question of "what is life" and discovered that life cannot be confined to material functionality or characterized solely by genetic, chemical, or morphological traits. By turning inward and embracing first-person observation, we recognize a fundamental truth: Life also includes a rich psychological dimension, expressed through feelings, desires, emotions, passions, instincts, and motivations that drive intentionality, volition, and goal-oriented behavior. Additionally, life encompasses a cognitive dimension involving mental activities such as thoughts, ideas, and imagination. A strictly materialist biological perspective often overlooks these aspects, yet they are tangible and real for every living being. Ultimately, they are rooted in various forms of sentience that we refer to as "consciousness."

When you ask biologists about the origins of all these subjective aspects of life, they often steer the conversation toward topics like reproduction, adaptation, genotype and phenotype, homeostasis, and metabolism. However, this redirection does not address the underlying issue of where the variety of psychological qualities of life originated from.

The crucial question is: What causes this blind spot? It stems not from rational reasoning but from a (more or less subconscious) reaction that disregards psychological dimensions. If they acknowledged these factors, they would have to force themselves to shine a critical light on the foundational naturalistic assumptions that guide them. Metaphysics and teleology could infiltrate their work like a dangerous virus; to some extent, this is already occurring behind the scenes. The biologist J. B. S. Haldane famously remarked that teleology—the belief in purpose within Nature—*"is like a mistress to a biologist: he cannot live without her but he's unwilling to be seen with her in public."*

The latest sidestep to this dilemma is known as 'teleonomy.' This doctrine asserts that while the purposefulness and goal-directedness of structures and functions in living organisms should not be dismissed, they should be regarded as merely apparent rather than genuinely real. In my opinion, this perspective will prove to be a temporary compromise between rational thought and a deeper intuition. Such an attitude will fail to reveal the true nature of life's processes and will not answer the fundamental questions.

Why do all living beings act as if they possess agency, will, and purpose? What drives the will to act, reproduce, grow, expand, and seek knowledge? From where do the creative impulses of novelty and the desire for increasing complexity and diversity in the evolution of life originate? In what way are emotions, feelings, and instincts easier to explain than the elusive problem of consciousness? How is it that even single-celled organisms display primitive forms of cognition before any brain functions develop? These characteristics appear to be intrinsic to life, beginning with the earliest bacteria.

It is now evident that even single-celled organisms possess a level of sensing and information processing that challenges conventional explanations. Cells and plants have a form of “basal” or “minimal” cognition previously thought to be exclusive to organisms with brains or nervous systems. Additionally, collective intelligence influences the behavior of bacteria and cell colonies, giving the impression that they operate with a unified, collective mind. This basal and collective cognitive behavior in response to environmental stimuli shapes their evolutionary path, which continues to defy naturalistic explanations.

Of course, we might like to believe that, despite centuries of almost no or scarce progress in accounting for these life phenomena, we have only to wait another century of progress and research, and then we will get there and dispel the mystery. These phenomena may be explained away by complicated selection, adaptation, metabolic, or other mechanistic processes. However, since the vast majority of scientists have adhered to this (unwarranted) naturalistic premise for generations without making any progress on these deeper existential questions, it may be time for at least a small minority of scientists to explore alternative avenues.

Instead of proposing a blind, bottom-up material process that supposedly leads to the emergence of intrinsic psychological properties of life, could we explore the idea that these properties are already inherent in Existence? What if we consider life not as a creation but as a creative power already embedded in matter—something fundamental that, once favorable conditions are met, inevitably will manifest? What if we posit the existence of intentional principles, such as a “desire-force” and a “will-force,” as intrinsic to life from the beginning? Although this might sound absurd, why not treat it as a working hypothesis, a temporary conjecture, or a serious speculation, and see where it leads us?

Nothing in science precludes us from seriously considering the hypothesis that consciousness, desire, sentience, mind, goal-directedness, will, and similar psychological phenomena are not emergent properties of matter and its physical interactions but, instead, are fundamental elements that drive the organization and function of living matter itself.

This perspective might be reminiscent of vitalism, and in some respects, it is. As I discussed in Chapter III and its advanced reading in the appendix XIII.2, science has largely dismissed the vitalist option, not because it has been proven wrong or for any rational or observational reasons but, rather, due to a commitment to a specific and narrow metaphysical framework. However, it is not a reversal of historical vitalist theories that can bring us further. We should rather ask the right questions, starting from a different perspective and worldview, before seeking the right answers.

How might the current integral cosmology broaden our understanding of life in comparison to theories such as panpsychism, cosmopsychism, analytic idealism, or other contemporary concepts of universal consciousness?

From the perspective of integral cosmology, universal consciousness, along with its superconscious plane, is not merely a passive observer, whether transcendent or immanent. Consciousness encompasses much more than experiencing something that exists. It surpasses Thomas Nagel's definition of "what it is like to be." The Indian tradition describes ultimate reality as a transcendent consciousness characterized by Sat-Cit-Ananda (Existence-Consciousness-Bliss), with no distinction existing among the three; they are a unified triad, akin to the Christian Trinity of Father, Son, and Holy Spirit.

More broadly, consciousness embodies, contains, and transcends infinite qualities, properties, and attributes. It is not only sentience but also will; it is not just experience but also existence. Consciousness cannot be reduced to creative power alone; it is also a force. It encompasses will and bliss, as well as knowledge, love, joy, beauty, empathy, compassion, and many other qualities that transcend the terminology of our everyday experience. It embodies all these qualities simultaneously, without distinction, while manifesting different forms of what we refer to as "phenomenal consciousness" within the experiential dimensions of a spatio-temporal realm.

Universal Consciousness is not simply a silent, immovable, and inert self-awareness, nor is it merely the sum of these qualities. Instead, these qualities reflect different aspects of the same essence, existing simultaneously and without distinction. The distinctions we make—such as between existence and bliss, knowledge and will, or joy and power—emerge only when consciousness manifests across various planes, differentiating itself through a process of exclusive concentration from the universal to the individual. However, none of these qualities are absent on any plane; they are intertwined and expressed only partially in various forms on each level.

For instance, when supreme Knowledge aligns with the mental plane, it manifests as "cognition," which in humans reflects a dim and pale version of the superconscious Gnosis known as "reason." Similarly, when this supreme Knowledge and will connect with the subconscient, they give rise to "basal cognition" in cells or plants and what we refer to as "instinct" in all living beings. When supreme unconditional Love engages with the life-plane, it expresses itself as "human love," which remains largely conditioned and egoistic. When the supreme aspects of Love and Will interact with the life and physical planes, they express as "emotions," "desires," "feelings," and "passions." When Love and Unity merge with the duality of the physical plane, rooted in the subconscious, they manifest as sexual attraction, a drive to reproduce, and various bodily instincts based on attraction or repulsion, as well as the craving for physical pleasure. When supreme Will operates through the life-plane, it manifests as "agency," "intentionality," "volition," and "goal-oriented behavior." When the supreme creative Power engages with the physical plane, it becomes what we call a "physical force or power." When the supreme quality of "delight of existence" interacts with both the physical and life planes, it appears as the instinct for survival and egoistic self-preservation. And so on.

In other words, all the properties and qualities of the world, along with the traits of our psychological inner lives, are ultimately mere reflections or "projections" into manifestation of the supreme qualities that exist on an originating Source-plane. Specifically, life instincts arise as a derivation of consciousness, not the other way around. However, materialists perceive only a single physical plane and mistakenly attempt to explain life, consciousness, and the mind by starting from the physical plane only. Science, in its current naturalistic approach, confuses the endpoint with the starting point, which leads to confusion about paradoxes, the hard problems of consciousness, and the origins of inexplicable agency and goal-oriented behavior in living beings.

Contemporary idealistic philosophies and cosmologies of universal consciousness share similar shortcomings with material monism. Although they advance beyond a purely materialist viewpoint, they remain constrained by a coarse ontology. Reducing everything to mind alone, without distinguishing between life, matter, the superconscious, and the subconscious, leads to a superficial understanding of the nature of reality. This perspective tends to attribute the qualities of the subconscious plane, the physical plane, and the life-plane as the fundamental nature of the Mind at large, ultimately grounding everything in primal desires, basic emotions, instinctual drives, rudimentary cognition, blind agency, intentionality, and volition, as if these were the fundamental primitives that create life by a bottom-up process. These elements are mistakenly viewed as the foundational building blocks of life. Ironically, idealists find themselves in the same reductionist trap as the material sciences they sought to transcend. Instead, the integral cosmology approaches these phenomena from both a bottom-up perspective and a top-down one.

Existence, Being Consciousness, Sentience, Bliss, Love, Oneness, Unity, Knowledge, Will, Power, Beauty, Joy, Creativity

↓

Mind plane

Knowledge→ reason, duality, intelligence, curiosity, creativity, desire for knowledge and beauty
Will → intentionality, volition, attention
Love → morality, ethics

↓

Life plane

Love → human love, emotions, desires, feelings, moods, passions
Will → agency, pleasure seeking goal-oriented aim, purposeful behavior
Existence → instinct of survival

↓

Physical and Subconscious planes

Love→ sexual attraction, a desire to reproduce, bodily instincts, cravings for pleasure
Knowledge → basal cognition, sensory perception
Creative Power→ physical force

Fig. 72 All qualities and properties as "projections" of the One-Consciousness.

From an integral perspective, we recognize that the first-person approach has inverted the logical order of evolution. It is not merely a physical, environmental, or genetic process imposed on a purely physical organism; rather, evolution also flows from the inside out. Organisms are not just passive vessels shaped by their genetic code and passively adapting to environmental influences; they are also active agents that partially determine their own destinies and even the environments they inhabit. Each living entity possesses unique qualities, embodying a "divine spark" of the same supreme consciousness. This internal impulse contributes to the process of evolution as well. Such an understanding has long been an intuitive truth for several natural philosophers, particularly German romantics like Goethe and Schelling because, contrary to the established mechanistic naturalism of their times, the introspective approach wasn't alien to them.[126]

Conversely, the involution of infinite Consciousness not only obscured its original powers and qualities but also led it to conceal itself within what appears to be its opposite. From a state of superconsciousness, it descended into unconsciousness and insentience; existence morphed into non-existence, knowledge transformed into ignorance, love turned into hate, bliss became pain, and creativity shifted to passive inertia. Nevertheless, beneath these appearances, it persists as a supreme secret intelligence, working toward its reunification through the evolutionary process, by an evolution of the species that is only apparently directed by outward factors.

A universal consciousness has immersed itself in opposing qualities to gain self-awareness. To understand attributes like bliss, unity, and love, one must also experience their opposites: pain, duality, and hate. In integral cosmology, this interplay of duality is the fundamental reason for the coexistence of these contrasting aspects, with which all religions and spiritual traditions have grappled.

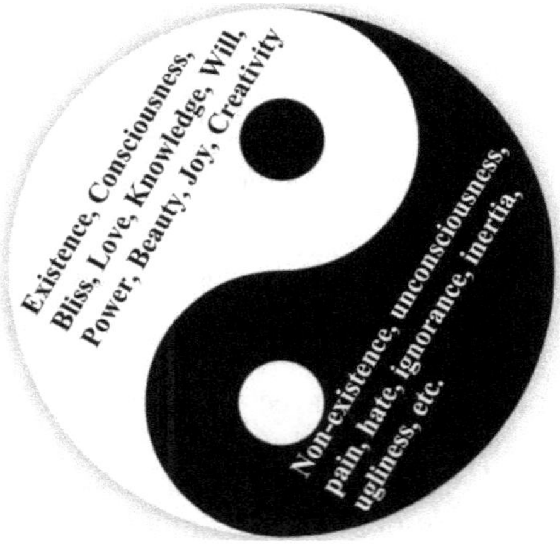

Fig. 73 The polarity of opposites in universal consciousness.
The Hermetic wisdom of "As Above, So Below."

It is within this Ying-Yang qualitative polarity of opposites inherent in universal consciousness, and which reflects both the evolutionary and involutionary process, that we can reframe our understanding of the origin and nature of the problem of evil.

[126] French biologist Jean-Baptiste Lamarck suggested a similar idea in its secular form that was dismissed due to a lack of evidence—though one might argue that no evidence could be found without actively searching for it. Interestingly, modern biology is now beginning to suspect that this idea might have merit, as seen in the so-called "niche construction process," even if it remains only a surface-level observation.

6. The Problem of Evil

The problem of evil presents one of the most enduring challenges in philosophy and theology. It questions how suffering and moral wrongdoing can coexist with an all-powerful, all-knowing, and wholly benevolent God. If such a deity exists, why do natural disasters, human cruelty, and innocent suffering exist? Moreover, what kind of God creates souls only to send them into a world filled with suffering? If God is all-powerful and all-loving, why does it allow a reality in which survival of the fittest prevails and organisms prey on one another?

This portrayal of God seems to indicate either a limitation in power or a troubling indifference. The resulting image is not that of a benevolent divinity but, rather, that of a distant architect who either permits or actively engineers suffering as part of a larger plan, leaving humanity to face the fallout.

This dilemma has prompted various theological and philosophical responses. Some abandon traditional theism, while others seek to reinterpret God's nature, suggesting limitations to the divine or a more dynamic relationship with creation. The problem continues to be a central point of debate, prompting thinkers throughout history to explore the complexities of faith, morality, and the nature of existence.

For example, the free will theodicy[127] posits that evil exists because God has given humans genuine free will—the capacity to choose between good and evil. This freedom is essential for meaningful relationships, moral responsibility, and authentic virtue. Without the option to make wrong choices, true freedom and love would be unattainable, as coerced goodness lacks moral value. While free will does lead to moral evils, such as cruelty and injustice, it is seen as a necessary condition for a greater good: a world where individuals can freely choose love, righteousness, and personal growth. God allows evil not out of a desire for it but because its removal would necessitate taking away humanity's freedom, thus rendering a morally significant world impossible.

However, if suffering is the price of free will, one might wonder why an omnipotent God could not have designed beings with the freedom to choose but also with an innate resistance to evil. Other questions are legitimate. What free choice or wrongdoing could an innocent victim of aggression possibly be responsible for? How can we explain the suffering of a newborn child afflicted with a painful disease or dying from hunger?

Personally, the free will argument doesn't convince me. However, it's also true that there can be many varied answers. Everyone must find their own.

Nonetheless, I believe that, while I'm not claiming to have a definitive answer, an integral cosmology framework might provide us with a clearer

[127] Theodicy is the branch of theology that defends God's goodness and justice in the face of the existence of evil.

understanding of the subject and bring us closer to some truths. In the following, I highlight a few key aspects that exemplify how to approach this subject within the framework of universal consciousness.

The moral question

If we are God's children, why does He send us into a world filled with tears and suffering? God is believed to exist in a transcendent heaven, enjoying perfect spiritual bliss and delight, yet He places us in a realm of pain. This seems inconsistent with the nature of an all-loving and compassionate Father or Mother. It resembles a demon more than a divinity.

From the perspective of a theology centered on universal consciousness, the enigma disappears at the outset. There is no separation between the Source and its "sparks" or souls. We are emanations of the One, without a second, and our soul is not distinct from the One-Soul. The perceived duality and separation between us and God are illusions, devoid of deeper reality. There is no separate God who punishes His children; our suffering is His suffering, our joys are His joys, and our afflictions are His afflictions. We exist in God, and God exists in us. This means that God experiences every pain, misery, and agony through us. No one and nothing has cast us into a realm of struggles and imperfections. We are manifestations of the same divinity that has willingly individualized and subjected itself to this experience in a world filled with ignorance and materiality. Once we relinquish the idea of an external and separate God who judges us detached from our everyday reality, the moral dilemma disappears. This standpoint is not so alien to the biblical tradition, from which we learn that *"God created mankind in his own image"* (Genesis 1:27) and *"In our suffering, God suffers with us"* (Isaiah 63:9).

But what about babies suffering and dying, children victim of sexual abuse, or mistreated animals? This doesn't appear to represent a teaching moment for them.

To put it in the words of my friend and psychologist Don Salmon:

"Take the perspective that the baby, the abused child and the mistreated animal, are all formations of a transcendent and universal consciousness, part of which remains transcendent and looks on its own forms, much as Shakespeare looks on Romeo, Julius Caesar and other of his characters. They have no existence outside of Shakespeare's consciousness, so it is not the baby or the animal that is learning lessons - it is Shakespeare himself.

Perhaps a book is not a clear enough analogy. Imagine Shakespeare with the capacity to project his own consciousness as a 3D holographic world, and he is the sole character appearing as a vast multiplicity of characters (much as you or I might be in a conscious dream, in which we know we're dreaming).

Do you see how the perspective might be radically different?

And remember, from an evolutionary perspective - in which, say, I (Don) am the Shakespeare creating the 3D holographic world - I have chosen to dive in, forget myself, for the sake of the adventure of rediscovering myself. I have chosen this adventure, fully recognizing that in what will appear to be aeons, the various characters I identify with will be experiencing what to us is known as "suffering."

I have consciously chosen this, for the sake of my infinite Self.

Does this change your perspective at all? (please keep in mind, this is a very rough analogy)

As far as whether "you" or "I" are one with the Mind of God, I hope you realize that every mystic teaching the world over, including those underlying the major religious traditions, East and West, as well as various indigenous traditions, teaches us that we are living in Ignorance of our true nature. This Ignorance, which Sri Aurobindo refers to as the result of an "exclusive concentration," keeps us identified with the various characters appearing through countless lifetimes who mistake their mind-body individuality as utterly separate from the Mind of God.

This is the value of an evolutionary perspective. This perspective can give us at least a hint of this adventure of consciousness in which we are all participating—an adventure in which the Infinite "I Am" has chosen to hide in material, plant, animal, human form in ignorance in order to: "grow up" – and to ultimately realize that the limited "I am" has never truly been separate from the Mind of God (and perhaps, to grow even beyond the human form)."

This doesn't mean we shouldn't strive to minimize suffering. It is not a license for excusing or normalizing evil, and it may offer little comfort to those who are abused or dying from cancer or Alzheimer's. However, it can help us find some understanding in an otherwise senseless, purposeless, and pitiless universe—which senselessness doesn't make sense either.

The evolutionary function of pain

One might wonder why pain and suffering exist at all. Why would a perfect and complete divinity choose to subject itself (through us!) to a world filled with hardships? First, let's examine this from a sociological perspective.

Sooner or later, we learn that life's challenges shape our character and skills. The proverb *"every cloud has a silver lining"* reflects this wisdom. Pain and suffering serve as profound catalysts for learning and growth throughout our lives. During these difficult experiences, we are compelled to confront our limitations, tap into inner reserves of strength we never knew existed, and develop resilience in the face of adversity. By navigating hardship, we gain invaluable insights. These trials can reshape our perspectives, clarify our values, and ultimately mold us into more insightful and resilient individuals, thus allowing us to emerge from the darkness with newfound wisdom and strength.

From this spiritual perspective, the deeper motivation for the existence of pain lies in the spirit's need to awaken from the inertia and forgetfulness of the material world. This spirit must experience shock and disgust, sometimes even horror and torture, to break free from its lethargy. Without pain and struggle, our inner spirit—our evolutionary soul—would lack the incentive to grow and would remain trapped in the rigid form of a rock forever. At best, we might not have gone much beyond the amoeba. Pain serves an essential evolutionary purpose: the growth of the soul during its journey of transmigration.

This is, in essence, the notion of karma, which is central to several Eastern religions and philosophies. It has nothing to do with judgment or a punishing God. It posits a principle of cause and effect that governs our present and future lives. It suggests that every action[128], whether physical, mental, emotional, or verbal, creates a corresponding energy or imprint that will eventually return to the individual. Positive actions, characterized by compassion, kindness, and integrity, are believed to generate positive consequences, while negative actions rooted in selfishness, harm, or ill-will lead to negative outcomes. This isn't necessarily a system of immediate reward or punishment; rather, it is a natural law that emphasizes personal responsibility for one's choices and their reverberations throughout the cycle of life, death, and rebirth. It is also a necessary condition for the evolution of the soul.

While the notion of reincarnation is absent in Abrahamitic religions, the notion of karma isn't foreign. Instead, it assumes different names and nuances and is confined to the frame of one single life. For example, here also exists the idea that sins—that is, all transgression against divine law—committed with willful intentions always have consequences.

Why does suffering exist in the first place?

If God, Brahman, Allah, or whatever name we use for Him is omniscient, all-powerful, infinite, and perfect, why did He create a finite world filled with imperfection, weaknesses, ignorance, and suffering? Why not simply create a heaven where we could live in bliss without affliction and death? Why would a divine Being want to have this painful experience of throwing itself into the opposite of itself and creating duality and separation, leading to suffering and evil?

As we have previously discussed, integral cosmology suggests that evolution was preceded by an involution of the universal Spirit, which immersed itself in its opposite. Pain, ignorance, and unconsciousness are part of the polarity of opposites within universal consciousness. If God embodies bliss, knowledge, and existence, then He must have also experienced pain and ignorance, plunging itself into a state of apparent non-existence.

One could argue that the original deity, representing supreme knowledge, bliss, power, and pure existence, is truly all-powerful precisely because it impelled itself to experience the opposite of its true nature. A divinity that confined itself solely to its essence, denying the experience of its opposite, would be a limited entity rather than infinite Consciousness. If universal

[128] The word 'karma' has its origin in the root 'kri,' meaning 'action' or 'work.'

consciousness is infinite, it must encompass infinite experiences, including those we perceive as "bad" and painful. Otherwise, it would not be truly infinite and complete.

Moreover, the individuated souls that emanate from God forget their divine origin. This identification with a lower state of consciousness is what we may refer to as "evil." The concept aligns with religious mythologies that depict demons as fallen angels, reflecting a human intuition of a deeper truth. Ultimately, all evil appearing in the form of demons, or of criminals and sinners, is also a "mask of God" with which God tricks itself into oblivion to surge in the evolutionary process.

What is "evil"?

Is evil truly evil? Our human concepts of "good" and "bad," "moral" and "immoral," "ethical" and "unethical," and "right" and "wrong" are shaped by our limited consciousness and a human understanding of reality. However, the perspective of a divine Gnosis, what we label as "evil" and "bad" might be perceived very differently. We must recognize that ignorance might cloud our view; the mind cannot fully comprehend the workings of a higher consciousness, and we cannot determine what is genuinely good or bad for ourselves or for the collective development of humanity. A heavenly Mother or Father certainly knows better what is best for us.

We might question what lessons can be learned from the deaths caused by earthquakes, hurricanes, or volcanic eruptions. Are these not merely natural phenomena that impact us randomly? However, from a cosmic perspective that transcends the notion of a vengeful God, even such disasters might hold meaning within a framework of spiritual emergentism. These events could represent a divine process through which the universal consciousness shapes us—that is, itself—and fosters the growth of its emanated souls. Perhaps even at the core of pain lies the delight of being and existence itself.

Be that as it may, adopting an integral cosmology—in which universal consciousness arises through spiritual emergentism—might allow us to move beyond constricting religious beliefs, or soulless materialistic worldviews, that once served a purpose but that have become overly simplistic with regard to understanding our earthly condition. Embracing a multidimensional view of existence that transcends dogmas and scriptures while taking spiritual insights seriously could enable us to look beyond and see further.

7. Concluding Remarks on the Integral Cosmology of the Conscious Universe

Here are some final thoughts on free will and some concluding remarks regarding the integral cosmology of the conscious universe.

Notice that in this integral context, the question of free will can be reformulated, thus providing further insight into our discussion in Section

VIII.2. When we adopt an integral perspective, the longstanding question of free will takes on a new significance, revealing itself as fundamentally flawed. The question lacks clarity about who is supposed to be free and from what. Whose will are we discussing? Are we referring to the freedom to indulge in the lower instincts of our outer self, the ego, or the mentalized "desire-soul" of the life-plane? Or are we considering the will and freedom of our true soul, our innermost being? When we mention "will," do we mean the mind's volition that clings to the appetites of its egoistic individualization? Or are we talking about the Will of universal consciousness, of which our real soul is merely a spark? Or, rather, are we addressing something in between?

These nuances are often overlooked in traditional philosophical debates, which tend to compress all aspects of existence into simplistic monistic or dualistic frameworks. In contrast, the integral paradigm offers a new meaning and clearer interpretation of these questions.

We often perceive an exclusive binary dilemma regarding free will: Are we free or not? This perception arises from our lack of self-awareness and our ignorance of the complexity of our existence. However, from an integral perspective, we can resolve the free-will dilemma by reconciling these opposing views. Determinists are partially correct in asserting that physical laws govern us, while libertarians and compatibilists also have valid points in their claims about free will. However, each side has its shortcomings as well. By expanding our understanding beyond simplistic material monistic or dualistic frameworks and acknowledging the complex nature of the universe, we can clarify who is truly free and from what. This comprehensive perspective allows us to appreciate the richness of our being across all planes of consciousness and ultimately leads to a coherent, contradiction-free answer to the question of free will.

The will, my will, your will, can be a subconscious will, a will of desire, a will of the mind, or that of a superconscious Will. Only when all the former are aligned with the latter can we speak of authentic free will. And then, at that point, our will coincides with the supreme Will and all inconsistencies disappear.[129]

Overall, the key takeaway is that the integral cosmology presents a new way of conceiving of the universe—a broader perspective and a higher vantage point from which to view life. This isn't a mere philosophical musing because, ultimately, how we conceive and, thereby, perceive the world, rather than what we see, shapes our thoughts and actions. For example, we often think of the

[129] Yet, the mind raises the question: Are we merely puppets of a universal superconsciousness? Does freeing ourselves from the constraints of Nature make us slaves to God? To the ordinary mind, that doesn't seem like a favorable trade-off. However, there is no dilemma, as the will of the individual aligns with the divine Will. Or does it? I will leave you with this uncertainty…

Renaissance as a period focused on art. However, it was actually a revolution in the way people perceived the world, shifting reliance from religious authorities to a quest for knowledge through observation and inquiry. It is time to strive for another similar shift that enhances our awareness of the world and our understanding of ourselves.

The failure of science in addressing the true nature of the psychological powers and qualities we observe in life becomes clear. Science, with its narrow focus on the physical plane, attempts to explain phenomena using an increasingly complex naturalistic framework. However, these phenomena originate from other planes of consciousness. While the external physical plane reflects some qualities of these internal planes, it does not generate them. You can't get milk from cheese. Efforts to explain or explain or even replicate these qualities solely through physical means will ultimately fail or, at best, result in a hollow imitation that will inevitably reveal its emptiness.

Our cultural and psychological challenge in accepting this perspective lies in the observation that, on Earth, basal instincts emerged first and were followed later by complex cognitive functions. We have not considered that both might be expressions of something that was initially implanted as a seed through a top-down involution and that is now unfolding through a bottom-up evolutionary process. Science will never truly understand the essence of life by attempting to derive consciousness, life, and mind from their primitive forms, just as one cannot grasp the true image of an object by starting with its shadow.

After all, this understanding of the world is neither new nor particularly original; it appears throughout the spiritual history of humanity. A similar message resonates in Gnosticism, Kabbalah, Hermeticism, Neoplatonism, and various Eastern philosophies. My aim is to reflect this understanding from a contemporary perspective, particularly through the cosmology of Sri Aurobindo, which I believe aligns most closely with modernity. Despite their differences and nuances, the foundational principles and forces shaping life and Nature, as outlined in these mystical traditions, developed independently across cultures and have remained consistent throughout history. They have not lost their relevance due to science; rather, science has succumbed to a narrow form of scientism, failing to recognize what it should have considered more seriously long ago.

The article in the Advanced Readings section XIII.7, titled "The Integral Cosmology of Sri Aurobindo: An Introduction from the Perspective of Consciousness Studies" (Masi, 2023c), was originally intended as a general overview of Sri Aurobindo's integral cosmology. However, I later realized it could be more effectively linked to the Western philosophy of mind, as this field stands to gain the most from his mystical insights. Consequently, this long essay serves as a detailed summary of the current chapter, enriched by perspectives from contemporary consciousness studies.

XI. The Dead End of Materialism: The Case Study of Artificial Intelligence

This chapter marks a departure from the scientific, philosophical, and spiritual themes I have explored thus far. In the preceding chapters, I maintained a theoretical approach without delving into how a post-material worldview might be applied in practical life. Understanding universal consciousness, recognizing the Logos underlying all reality, viewing evolution as the emergence of spirit in matter, and advocating for a perspective that transcends the limitations of naturalism are crucial first steps. However, if this new understanding does not connect with our daily lives, it risks becoming mere abstract and sterile metaphysical speculation. Therefore, it is time to present specific examples or at least reflections that demonstrate how a theoretical framework beyond the established paradigm can lead to a different understanding of science and suggest alternative approaches to the same subjects.

Many examples could be provided, but I will focus on a specific case study that has sparked considerable debate in contemporary research: the widespread discussion about artificial intelligence (AI) or the speculations about the impending artificial general intelligence (AGI) revolution that is expected to shape our future. I chose this topic because AI is a prominent subject in the media and is presently one of the most heavily funded areas of research. In an accompanying advanced reading XIII.8, I will also address the subject of linguistics, as it is closely related to AI in ways I will explain. One cannot overlook this connection when adopting a more comprehensive perspective that considers all aspects from a post-material viewpoint.

Following the success of ChatGPT and the remarkable capabilities of large language models (LLMs), discussions about the impending arrival of AGI—human-like intelligence—have surged. We have witnessed several breakthroughs, with some experts predicting the emergence of AGI within the next few years. Given that machines have already demonstrated the ability to pass medical, law, and business school exams, it seems plausible that we might soon encounter the first humanoid intelligence capable of interacting with us and potentially working alongside us. This advancement raises concerns about mass unemployment, and some take seriously the possibility of these entities becoming more intelligent than humans. In a worst-case scenario, this could lead to a situation reminiscent of a Terminator movie, in which machines take control and dominate the human race.

Allow me to clarify why this perspective is misguided.

These overly optimistic beliefs—or pessimistic, fear-based projections—are not due primarily to flawed reasoning or a lack of technical understanding of intelligence. Instead, they arise from a psychological inability to understand ourselves fully and from an excessively externalized mindset that has neglected the importance of adopting a first-person perspective and viewing matters from within.

In fact, the question of whether machines will soon achieve consciousness is crucial—not because I believe they will but because the answer shapes and is shaped by our self-perception. My argument is that AI, or any form of AGI, will remain an empty (black) box without real semantic understanding unless we discover a way to make machines conscious.

Before we delve into the technical details of this argument, here's a brief overview from the metaphysical perspective.

If the mind and its intelligence are not solely the result of computations performed by a material entity, such as the brain, and if William James 's 'filter theory of consciousness'—suggesting that our thoughts and mental phenomena are filtered from a non-material universal mental plane, which cannot be fully explained by physical processes on the physical plane only—is accurate, then no machine can achieve consciousness or true mental capability. If the human mind is not merely a computation executed by physical processes, then any computer simulating the brain or any sophisticated AI will not attain human-like intelligence, let alone consciousness. This is true unless the computer can somehow connect with the mental plane, becoming a "filter" similar to biological brains. If the mind is inherently not completely physical, attempting to create a mind using a material machine is like trying to generate rain, wind, and sunshine from a computer weather simulation. If something's essential nature is X, it cannot be reproduced with Y unless one discovers a way to 'filter' X into Y. This is sheer logic.

Modern AI could be considered 'intelligent' or as possessing some human-like cognitive abilities. However, it still lacks genuine human-like intelligence and the capacity to emulate animal abilities. There is an essential quality, a sense of agency and cognitive aliveness, that only living organisms possess. Most importantly, even the most advanced AI contains no indications of consciousness or self-awareness. Currently, it shows no signs of spontaneous will, agency, volition, or true autonomy characterized by self-directed, goal-oriented behavior. Additionally, computers lack creativity, original thinking, and reflective or self-reflective skills. They struggle with generalization and do not possess the imagination necessary for inventing new concepts or making discoveries on their own. Moreover, and this will be a key focus of my discussion, AI lacks a deeper understanding, common sense, and semantic awareness of the world that all living beings have to some degree. Ultimately, even the most sophisticated AI does not truly understand anything. Its ability to convincingly emulate certain skills is a result of modern computers being powerful enough to process the vast knowledge humans have accumulated throughout history and replicate it through imitation and educated guesses. This is why AI can pass PhD-level exams in molecular biology yet struggle to replicate the agency of a mouse.

A fundamental principle and a significant issue regarding the pursuit of building human-like and conscious AGI are largely overlooked by IT experts, AI researchers, and many philosophers of mind. This neglect arises from their strict adherence to a physicalist framework.

Beyond the technical aspects, what matters more in this context is our perspective. By viewing the issue through a different lens, we can gain a broader understanding and deeper insights. This narrow focus might lead to a considerable waste of human and financial resources in the pursuit of something that is not achievable, even in principle. Meanwhile, a first-person perspective can illuminate aspects that science focused solely on problem-solving from a third-person viewpoint often misses.

Therefore, I would like to express a counterpoint to the prevailing belief that we are on the verge of creating unconscious but true AGI. The emergence of machines with human-like intelligence might soon challenge my prediction. I am taking a risk by basing my prediction on a non-materialistic understanding of the mind, and if it is proven wrong, one could perceive this as a victory for naturalism.

1. Matter, Life, Mind, and Artificial Intelligence

As a teenager, I enjoyed watching the 1980s TV series 'Knight Rider,' featuring David Hasselhoff as a lone crimefighter who battled evil with the help of KITT, a virtually indestructible and artificially intelligent supercar. Around the same time, I began writing my first software programs in BASIC on the historic Apple II Plus PC. It felt like the dream of building self-driving cars powered by AI was just around the corner. I remember being firmly convinced that within twenty years—around the turn of the millennium—KITT would become a reality. I was confident that with advancements in computing power, a deeper understanding of brain structure and function, and increased research funding, we would soon be able to replicate human intelligence in machines.

Looking back, I can clearly see that my optimism was rooted in the naïve wishful thinking of a young mind unaware of the simplistic nature of my understanding of the brain, the mind, and the childlike worldview that shaped my thoughts. Much less clear to me is why, today, respected scientists continue to hold those same naïve beliefs and simplistic views of the world.

There is a significant ongoing discussion about the future of AI technology and the potential emergence of human-like artificial general intelligence (AGI). The recent advancements in large language models (LLMs), particularly the Generative Pre-trained Transformer (GPT) developed by OpenAI, especially in its ChatGPT version, have revitalized the conversation around the arrival of an AGI era.

The natural language chatbot interface is remarkably human-like and impressive. It effectively emulates various cognitive tasks previously thought to be exclusive to humans. I won't delve into the specific capabilities of these LLMs or how closely they resemble the functions of an intelligent, if not conscious, agent, as I assume that, by now, most people recognize their power and potential applications.

Instead, I would like to explore whether this represents a genuine advancement toward a humanoid mind and possibly the first hint of artificial sentience or if it is merely an illusion created by a system that mimics human thought processes.

I support the latter hypothesis for several reasons.

First, it's important to clarify what an LLM truly does. Although it is built on deep learning neural networks that attempt to replicate certain functions and structural properties of a biological brain, its learning process and sources of information are fundamentally different. The functional organization of an LLM is not a direct emulation of biological brains; instead, it relies on a vast corpus of human-generated training data. The model processes an immense amount of text, images, and videos sourced from the internet, books, and other documents. Without this extensive pool of pre-existing written knowledge, an LLM would be ineffective. Unlike a brain that receives input from sensory organs, an LLM cannot train itself through mere interfaces like microphones or cameras. As of now, it predominantly consumes textual information; to a lesser extent, generative AI works on images and videos. It depends on the existence and insights of cognitive agents—namely, humans—to replicate human-like cognition. Furthermore, after an initial phase of unsupervised learning, fine-tuning requires human intervention, as humans must guide the model on what is considered right or wrong during the training process.

Most importantly, it lacks a conceptual understanding of truth and falsehood. For instance, it does not deduce from mathematical logic that 2+2=4; it only provides this answer because it has been trained to recognize it as the correct or most probable response. If you train it to believe that 2+2=5, it will accept that as true without question.

It's essential to remember that ChatGPT or the like does not arrive at solutions through inference, logical reasoning, or analytical thinking. Instead, they guess what the best answer might be. When given an input, they predict the most likely output by identifying patterns and relationships based on hundreds of billions of parameters. This is why some people refer to them as 'stochastic parrots'; they generate responses drawn from a vast pool of knowledge found on the Internet and other documents, but their outputs lack semantic understanding or any true perception of meaning.

The question at hand is whether we can truly classify a machine that merely emulates intelligence—essentially, a machine adept at 'playing the game'—as a genuine AGI system. Many people believe we can; I do not.

To illustrate my point, I could reference the concept of a 'philosophical zombie' or 'p-zombie.' This idea describes beings that behave as if they are sentient and capable of subjective conscious experiences yet lack any such awareness. These entities can mimic feelings, emotions, pain, and pleasure and can even replicate human intelligence, but nobody is 'in there' experiencing or thinking anything. I contend that modern AI systems exhibit similar characteristics. While they can emulate experiences, they do not truly feel them. This distinction is significant. If you consider the first-person perspective, you can almost physically sense the difference. Here comes the decisive point: Then, you can reflect on how this distinction might manifest in the form of behavioral differences.

No evidence suggests that these systems possess will, agency, desire, or any urge to act. They do not have instincts for survival or a drive for pleasurable experiences. If a chatbot does not receive input, it will remain a silent and passive black box indefinitely.

This raises questions about the origins of instincts in living organisms. What are desires? Are they merely computations? Taking an introspective approach, do you perceive your desires as calculations?

Cognition and mind might be fundamental rather than emergent.

In Nature, we observe cognition that exists independently of brains, yet we do not find life that is disembodied from cognition. No evidence supports the idea that life and mind can be separated, such as in the pursuit of creating a general intelligence in computers. Everything suggests that mind and life are deeply intertwined; they cannot exist apart. While we see instances of cognition without a brain, we never find cognition without life. Phenomena like chemical reactions, tornadoes, hurricanes, ocean currents, and volcanic eruptions do not possess cognition. It appears that mind cannot exist without life and vice versa.

So, where does this pervasive yet rarely questioned belief in a mind existing independently from life come from? Why is this concept, which lacks observational support and contradicts our evidence, so deeply ingrained in our scientific perspective?

This belief arises from the idea that, nonetheless, we can recreate a limited form of 'physical intelligence' based on the material processes occurring within a computer. We think that by doing so, we can extend this understanding from the ground up to encompass all forms of living and non-living cognition. This perspective suggests that there is no inherent mind or life; instead, there are merely complex dynamics of material particles governed by differential equations—nothing more. It represents a materialistic, physicalist worldview that views matter, life, and mind as different large-scale emergent phenomena resulting from biochemical reactions. However, as we have explored within the framework of integral cosmology, the physical mind is actually the outcome of a top-down involution of a supermind rather than a bottom-up evolution. If this is correct, attempting to recreate the latter by starting from the former is a naïve and futile endeavor.

I would like to adopt the perspective of philosopher Mazviita Chirimuuta, who, on the basis of this rationale, questions our assumptions about future AI and its potential evolution into AGI. She advocates for a "bio-centric" view, arguing that life and intelligence are inherently linked; without an existing form of life, there can be no mind. This perspective suggests that matter, life, and mind represent three distinct fundamental realities. Therefore, it might never be possible to recreate AGI using a purely material digital computer, even in principle, unless we can find a way to connect a machine on the physical plane with a mind on the mental plane. In this context, a worm possesses greater general intelligence than a chatbot. This insight aligns with what we have previously discussed: Matter is a veiled form of life, life is a veiled form of mind, and mind is a veiled form of consciousness. Ultimately, these are three expressions of the same essence, or, as Spinoza would say, three "modes of existence" of the same substance. If this is the case, AGI might remain nothing more than wishful thinking, a fantasy of science fiction.

You might argue that we have already created machines with artificial intelligence. In this sense, these machines can be seen as artificial minds with cognitive abilities that only partially align with human cognition. The critical question then arises: What is cognition, and how does artificial cognition differ from human cognition?

2. What Is Cognition?

Cognition is a fundamental concept in the cognitive sciences, referring to the processes by which humans and other living organisms acquire, process, store, and utilize information.

The point is that a question that once kept only a handful of cognitive scientists and philosophers busy has gained renewed attention in contemporary science thanks to recent developments in AI and new findings in biology, with potential practical implications for AI, biology, and cognitive science.

The advanced intellectual capabilities of modern AI systems, such as LLMs like ChatGPT, raise the question of whether these systems truly 'cognize' as humans do or if they simply mimic human behavior through statistical predictions. Researchers studying cognition seek to understand how the human mind operates, how intelligence develops, and how artificial systems can replicate mental functions. Their work offers valuable insights into both natural and artificial intelligence.

Meanwhile, in recent decades, biology has found that even plants and single-celled organisms exhibit forms of "proto-cognitive" abilities. Cognition, or at least what is often called "basal" or "minimal" cognition, does not require a brain. Remarkably, tiny unicellular creatures can sense their environment, navigate it, target food, avoid toxic substances, change their swimming direction, and even solve problems collaboratively to achieve a common goal. Discoveries have shown that organisms such as bacteria, plants, mushrooms, and slime molds—despite lacking a brain—can engage in cognitive tasks previously thought possible only in more complex organisms.

Defining cognition is challenging, much like defining concepts such as "life," "consciousness," "mind," "thoughts," and "agency." The underlying question is: Why are these concepts, which seem self-evident and integral to our essence and daily experience, so difficult to define?

Contemplating questions that blur the boundaries between biology and psychology, artificial and natural intelligence, or even physics and metaphysics can be an interesting exercise. This reflection reveals our often unacknowledged premises and assumptions. Such discussions are essential for understanding *how* we think.

Before we continue, you might find it helpful to attempt to define cognition on your own. You will soon discover that what appears to be a simple idea is, in fact, difficult to articulate in precise terms. Give it a try...

Experts offer a variety of definitions of cognition, encompassing concepts such as information processing and storage, problem-solving, goal-directedness, decision-making, planning, the ability to interact with and adapt to the environment, etc. These definitions also include mental processes like perception, memory, attention, reasoning, and language comprehension. Very broadly speaking, scientists tend to think of cognition as a faculty that uses information to solve problems and guide adaptive behavior.

The existence of many definitions highlights the uncertainty and controversy surrounding the concept. My issue with these approaches is that they attempt to define cognition solely in Darwinian evolutionary terms, portraying it as an adaptive function that improves an organism's survival and reproductive success. These definitions, grounded in an exclusive third-person perspective, fail to capture the essence of cognition and conflict with our first-person understanding of it. I believe my cognition extends beyond mere adaptive and problem-solving functions. For example, when I watch a sunset, listen to music, or immerse myself in a poem, I do not perceive these experiences as being purposeful only for acquiring data or adapting to my environment. Cognition maintains a qualitative aspect that can't be reduced to mere computation.

Therefore, allow me to take a different approach.

Sometimes, starting with exploring the etymology of words can provide valuable insights. According to the Online Etymology Dictionary, the term "cognition" originates from a combination of "com," meaning "together," and "gnoscere," which means "to know" (derived from the Proto-Indo-European root "gno," also meaning "to know"). The root "-scere" is linked to the Latin word "scire," meaning "to know," which is the basis for the term "science" as a form of knowing. Interestingly, it is also associated with "sciss," meaning "to cut off" or "to incise," as seen in the word "scissor." Thus, cognition can be understood as the process of "knowing together what is cut into pieces," thereby reflecting the act of integrating and comprehending information into a semantic whole.

The notion that cognition involves the integration of information is a common theme in modern literature. Interestingly, people thousands of years ago appear to have intuitively grasped this understanding.

For example, when you look at an apple, you perceive its shape, size, position, orientation, various colors, background, and, eventually, motion in space. You do not recognize these features as separate elements; instead, they come together in a nearly instantaneous cognitive act, leading you to simply state, "There is an apple." A vast amount of information is seamlessly integrated into a single, unified conscious experience.

Another well-known example of information integration is illustrated in the figure below. At first glance, the image might seem like a collection of random, meaningless patches. However, it won't take long before a coherent scene almost suddenly takes shape in your mind. This leads to a sudden shift in perception, in which a meaningful whole emerges.

Fig. 74 The Dalmatian dog.

This might appear to us as an obvious and entirely normal cognitive process that requires no explanation. However, the mechanism behind this "integration" or "binding" remains unclear.

First, while we know the brain processes various features in distinct areas, we don't know how it manages to combine these features into a single, unified mental object of perception. No "center of integration" has been found. Numerous theories attempt to explain how feature integration occurs at the neurological level, but so far, no conclusive answers exist.

Second, implementing such an integration or "binding" in a computational process proves to be an extremely challenging task. It has taken decades of research to develop neural networks that can reliably perform pattern recognition and categorize it meaningfully. Even today, modern neural networks have not achieved human-level generalization. While this does not mean they will never reach that level, it highlights how a cognitive activity we often consider trivial is, in fact, quite complex.

This issue is commonly referred to as the "binding problem."

Furthermore, this binding is not confined to visual experiences; it is a pervasive process throughout all levels of cognition. Feature integration is a fundamental aspect of all our cognitive processes, functioning across a temporal dimension as well. For instance, consider how we combine a series of still frames to perceive fluid motion in a movie. Similar integration can be shown to take place at the level of auditory and tactile perception.

Another example is the meaning of a sentence extending beyond the individual words. The emphasis we place on certain words situates the sentence within a specific context that can change. Even with the same words and sentence structure, entirely different semantic meanings can emerge based on the emphasis. Consider these sentences, where the words are emphasized differently.

"I **never** said she stole my money." (I never said it at all.)
"I never **said** she stole my money." (I might've implied it, but didn't say it.)
"I never said **she** stole my money." (Someone else might have.)
"I never said she **stole** my money." (Maybe she borrowed or found it.)
"I never said she stole **my** money." (She stole someone else's money.)
"I never said she stole my **money**." (She stole something else, not money.)

Therefore, the emergence of meaning in our mind is the integration of the context-dependent apprehension of diverse features, followed by an almost instantaneous cognitive act of comprehension. Once we bind, integrate, and coalesce all the elements, we perceive a single, coherent, meaningful whole despite the presence of numerous words, subjects, objects, predicates, types of emphases, etc. This integration of information is at the very basis of our meaning-making.

In hindsight, it is not surprising that creating a theory of semantics that enables AI software to grasp meaning turned out to be an extraordinarily challenging task. Such a theory cannot simply combine individual words; it must also take into account the entire context, which encompasses the subject's broader experiential world. Next, we will explore how modern AI achieves a form of cognition that resembles human-like meaning-making.

Meanwhile, note how the meaning we derive from our sensory perceptions is not inherent in the external objects themselves. Physical information does not uniquely dictate semantic information.

Moreover, from the same set of physical information, one can create two distinct interpretations. Popular examples of this phenomenon include the Necker cube and the Rubin vase. In these cases, you can alternate between different interpretations: Your mind can switch between two orientations of the cube or between viewing the vase and viewing the two faces.

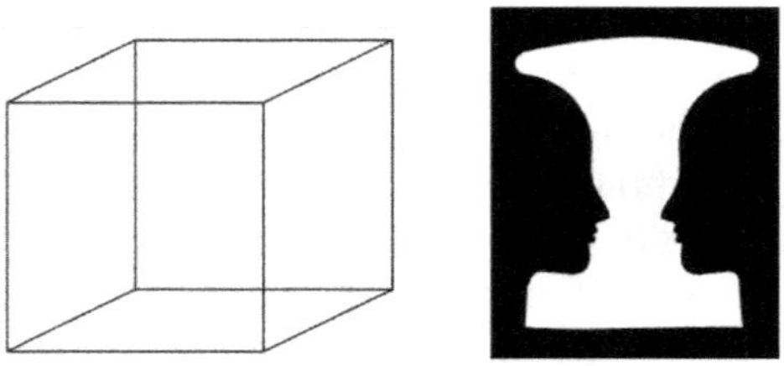

Fig. 75 The Necker cube and the Rubin vase.

This shows that while the concept of meaning arising from context and a system of relations holds some truth, it is only partially accurate. Meaning is indeed relational and context-dependent, but this does not tell the entire story. Completely different meanings can emerge from exactly the same context and physical information content. One figure, representation, pattern, or configuration can lead to multiple and mutually exclusive semantic interpretations. There is no reason to assume that our everyday sensory perception and brain activation patterns operate differently.

This should further clarify that, as we have already seen in Section IX.1, our mind does not create a "representation" or a "map" of the external world based on sensory perceptions. Instead, it constructs an interpretation—essentially, a fabricated story.

Some people suggest that our brain generates internal, more or less approximate representations of the environment. However, exactly what "representation" means is unclear. Which of the two cubes more accurately "represents" reality? Is the vase the correct "interpretation"? Or is it the two faces?

What we cognize is not an "approximation to truth," let alone a "vision of truth." We do not understand the world as it truly is. Instead, we select one of many possible interpretations or models based on appearances, which nonetheless helps us navigate life. Since the time of Plato, philosophical idealism has recognized this idea. Kant argued that we can know only the phenomenon, not the noumenon. Some modern philosophers and scientists have taken up this idea.[130]

Thus, cognition extends far beyond merely collecting data. Sensory information alone does not equate to cognition; for instance, a hard drive can store all physical information about an environment without any actual understanding. Cognition occurs only when information is integrated, comprehended, and experienced as a meaningful whole.

[130] A popular example is cognitive psychologist Donald Hoffman who proposes an "Interface Theory of Perception" (ITP) discussed in Section X.1.

However, modern AI, such as LLMs, can be quite effective in simulating "understanding." A conversation with ChatGPT demonstrates how these machines appear to possess a semantic understanding of the world. In what ways does the "cognition" of an LLM differ from that of a human being?

That's a complex story that we can't fully explore here. However, we can examine some examples and compare machine cognition with human cognition.

For example, a neural network for pattern recognition works by mimicking the way the human brain processes data. It consists of layers of interconnected nodes, or neurons, organized into an input layer, one or more hidden layers, and an output layer. Each neuron receives input data, processes it using a weighted sum, applies an activation function, and passes the result to the next layer. Over time, the network learns to extract and recognize complex patterns from raw data—whether it's recognizing faces in images, detecting speech, or identifying anomalies in data. Fig. 76 illustrates how a multilayered neural network processes the pixels of an image of a handwritten "2" and integrates them to activate a cell labeled as "2." In a sense, this is information integration, a binding that collapses all data into a symbol.[131]

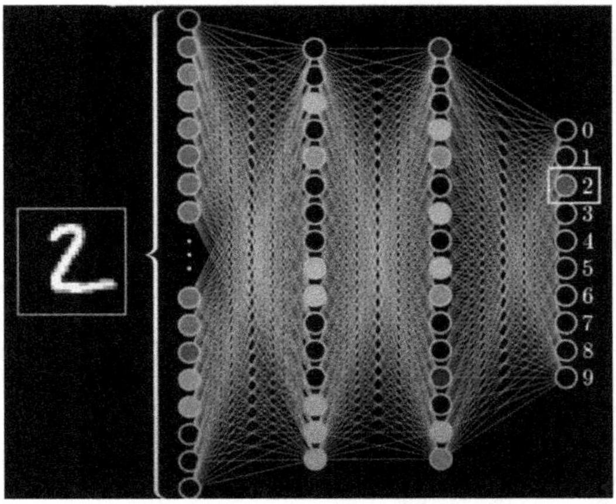

Fig. 76 A neural network recognizing the handwritten number "2".

Can this be called "pattern recognition"? In a sense, yes. 'Cognizing'? Well, only a machine-like cognition. Is this real semantic "understanding"? I doubt it.

This information processing can be expanded to a multidimensional integration. For a further illustration of 'binding,' consider how a neural

[131] Notice how this is a powerful method to compress information. Presumably, a similar mechanism works with the genome in the DNA molecule.

network categorizes both an object and its features within a multidimensional vector space.

Fig. 77 Credit: Greff et al. (2000)

For example, these are the elements of one vector distinguishing between the entry “apple” and “pear”:

$$Apple = \begin{bmatrix} 1 \\ 0 \end{bmatrix}; \quad Pear = \begin{bmatrix} 0 \\ 1 \end{bmatrix}$$

While the elements of the second vector designate its color as “red,” “green,” or “yellow”:

$$Red = \begin{bmatrix} 1 \\ 0 \\ 0 \end{bmatrix}; \quad Green = \begin{bmatrix} 0 \\ 1 \\ 0 \end{bmatrix}; \quad Yellow = \begin{bmatrix} 0 \\ 0 \\ 1 \end{bmatrix}$$

How can we integrate the information that an apple is green, or the pear is yellow? We can represent them with 3x2 matrixes, where the columns distinguish between an apple and a pear, while the rows stand for its colors, as follows:

$$Green\ apple = \begin{bmatrix} 0 & 0 \\ 1 & 0 \\ 0 & 0 \end{bmatrix}; \quad Yellow\ pear = \begin{bmatrix} 0 & 0 \\ 0 & 0 \\ 0 & 1 \end{bmatrix}$$

While the statement “a red apple and a yellow pear” (case c of Fig. 77) could be synthesized as:

$$Red\ apple + yellow\ pear = \begin{bmatrix} 1 & 0 \\ 0 & 0 \\ 0 & 1 \end{bmatrix}$$

Notice how colorful objects, concrete items, and whatever kind of concepts can be encoded in a language-independent abstract space. For instance, these matrices represent red apples and yellow pears without relying on specific language. They can be part of a vast abstract digital space that reflects the world.

Obviously, this is an extremely simplified explanation of what occurs in a computer. An LLM can contain millions of neurons, each connected by synaptic weights totaling hundreds of billions of parameters, with the outcomes represented by multidimensional vectors, matrixes, tensors, etc. However, this overview provides a foundational understanding of what it means to "bind" or "integrate information" in a machine.

Does the machine "cognize"?

It certainly doesn't align with the human understanding of "cognition." When we look at a green apple or a yellow pear, we don't perceive vectors or matrices.

However, one might argue that the machine does possess a form of cognition. We shouldn't assume that human-like cognition is the only possible one. In fact, representing reality as a multidimensional vector space enables AI to engage in meaningful conversations with us.

Fair enough. Yet, without succumbing to anthropocentrism and placing our cognition at the center of the universe, we must further explore the differences between human and machine cognition.

The challenge lies in understanding how the multiplication of vectors and matrices, represented as digits in a memory chip, translates into the rich experience of perceiving the "greenness" of an apple, feeling its shape and size, and even tasting it. A memory cell labeled "apple" or "pear" is merely a symbol encoded in the physical state of a circuit. There exists a subjective and experiential dimension that a symbol alone cannot encapsulate, regardless of the complexity of the computations leading to its identification.

This issue is not exclusive to machines; it also applies to humans. If you have never tasted an apple, no amount of description—be it through language, science, mathematics, computation, or neurological explanation—can truly convey what that experience is like.

I asked ChatGPT to describe the taste of an apple. Its answer:

An apple's taste depends on its variety, but generally, it offers a balance of sweetness and tartness. A crisp, fresh apple bursts with juicy flavor—some are more honeyed and mellow, like a Fuji or Gala, while others, like a Granny Smith, have a sharp, tangy bite. The texture adds to the experience: firm and crunchy at first, then turning into a smooth, refreshing juiciness. There's also a subtle floral or citrus-like undertone in some apples, making each bite both familiar and exciting.

That's an amazingly vivid description of the subjective experience of what it feels like to taste apples. But does ChatGPT truly "know" what it is like to eat an apple? Obviously, this can't be the case. It could come up with this description only because it has crawled the entire internet, indexing countless websites that describe the flavor of apples.[132] It can mimic the human experience of tasting an apple only because it relies on texts written by people sharing their own experiences. Without this information, neither ChatGPT nor humans who lack firsthand experience can truly "understand" what it is like to taste an apple, let alone "cognize" that an apple "offers a balance of sweetness and tartness."

This highlights what philosophers of mind refer to as the "symbol grounding problem." This problem addresses the challenge of understanding how symbols in computational systems—whether they are words, numbers, vectors, matrices, streams of bits, signals, or more complex representations—can convey meaning without being grounded in sensory experiences or real-world interactions. A fundamental gap exists between a simple textual description of the world and the meaningful mental state that arises from it. Words and sentences do not truly represent ideas; instead, they evoke them. In fact, modern neuroscience and linguistics are finally realizing that language is a low-dimensional reduction into labels of what constitutes our rich cognitive phenomenal experience, and while it has provided humans with significant advantages over animal intelligence, it has also diminished certain cognitive abilities. A single word can't capture the essence of a beautiful sunset. While poetry can convey the deeper feelings it evokes, only those who have directly experienced it, or at least can experience it in the first place, can fully understand its meaning across all psychological dimensions.

This also tells us something about language itself. The nature and origin of language must be found in what goes beyond symbols, tokens, or bits. We will investigate this in the next section and, for those that are looking for a deeper dive, in the advanced reading XIII.8. However, you don't need to be a mystic; simply practicing meditation shows that, without language, there is greater mental clarity, not less. Language is the final rung—not the starting point—on the ladder of conscious expression.[133]

If meaning is not found in the arrangement of symbols, then who or what generates that meaning?

The same can be said for how the brain represents information. Neuroscience suggests that everything we perceive through our senses is mapped and coded as neuronal representations. All the stimuli we encounter and the thoughts we generate are stored in a configuration space as synaptic weights within a neural network.

[132] Google "sweetness and tartness," or "apples subtle floral or citrus-like undertone," or "honeyed and mellow, like a Fuji or Gala," and you will find tons of websites describing the taste of apples.

[133] It is beyond my understanding that even great minds like Jean-Paul Sartre believed language creates consciousness, rather than the other way around.

This is certainly true, but it can't be the whole story.

For example, the figure below on the left illustrates another "Gestalt switch," depicting the well-known ambiguous image of an old woman and a young woman. A neural network may recognize one or the other figure, with the choice depending on its synaptic weights which, however, can't be determined solely by "the world out there." The physical information is exactly the same in both cases. The decision may be influenced by the context, our current mental or emotional states, or other factors. However, we also have the ability to make a conscious choice to switch our perspective. A volitional "will-factor" enables us to see things in one way or another.

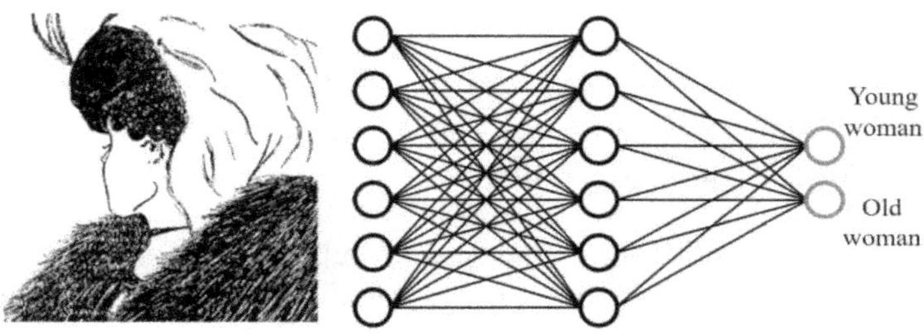

Fig. 78 The young/ old woman ambiguous figure and its recognition by a neural network.

These Gestalt figures highlight a broader truth: Our perception of the world is not a direct reflection of reality, nor is it merely an interpretation. Instead, it serves as a symbolic transcript—essentially, a sign or token that represents something else but holds no intrinsic meaning until the mind ascribes one to it. We don't "cognize" reality as it is; we "re-cognize" it by re-constructing arbitrary (and often false) appearances to meet our needs and evolutionary demands. Ultimately, we recognize nothing beyond a narrative of illusory appearances, even though they are permanent and quite useful.

Again, there is something inherently different between machine cognition and human cognition. What the neural network "sees" is neither a young woman nor an old one. After all, its "cognition" reduces only to the on-off activation state of two cells.

In other words, while the physical state of a neural network is a much more complex configuration than the symbols in a typical text or the patches of a figure, it does not contain meaning on its own any more than the text or the figure do. A neural configuration inherently contains no meaning. The notion of a neural representation is, in and of itself, insufficient to explain how meaning is instantiated—no more and no less than the vectors and matrices tell us something about how our mind instantiates the semantic awareness of a subject experiencing the sight of a green apple and a yellow pear. Or, no more and no less how an association of symbols, such as written words on a piece of paper, elicits concepts with meaning. Patterns of neural activity do *not* have causal power solely by virtue of their meaning. The meaning is *not* inherent in the configuration of the circuitry. The claim that neural patterns have causal power only because they represent or reflect something about the world or our internal states is a common fallacy.

Therefore, something else must intervene to "collapse" a neural pattern into one semantic whole attended by a conscious experience of a "perception of meaning." In that instant of binding and integration of information in between, something special suddenly results in the materialization of the meaning of things lived as a first-person subjective experience. There is that fleeting moment when an initially meaningless but very concrete sensory experience materializes all at once in a compound conceptual object of cognition.

If the meaning of things is not found in the external world and, thereby, not in the arrangement of neurons in our brains, then who or what generates meaning? A gap exists between the physical information that computers process and the semantic information that we, as sentient beings, experience.[134]

Not so unexpectedly, the issue boils down to the "hard problem of consciousness" discussed in Section II.1. We still do not understand how the unconscious, insentient, experience-less, and subject-less functions of neurons in the brain give rise to a conscious, sentient being that has experiences. There is an explanatory gap between information processing mechanisms and an account for conscious experience.

Here, we encounter a similar "hard problem of semantics." We do not understand how the unconscious, insentient symbols or neuronal patterns in the brain give rise to a conscious, sentient subject that has a "semantic experience." This is because meaning is inherently connected to consciousness. Integrating symbols or any complex mechanistic processes into a coherent semantic whole inevitably requires conscious experience. The cognition that living sentient beings exhibit cannot exist without conscious experience. To cognize is to grasp and integrate sensory data into a coherent mental object, which allows us to perceive and comprehend meaning. Understanding is rooted in consciousness; without consciousness, true understanding is impossible. Just as there can be no seeing without the experience of light, and no hearing without the experience of sound, meaning is derived from the experience of a mental phenomenon. Meaning-making is an expression of consciousness rather than a mechanical process that generates conscious experience as a fortuitous byproduct. This explains why we often find ourselves in a cognitive state in which we understand something deeply yet struggle to articulate it in words. Words, tokens, bits, and bytes fail to capture the perceptual experience because they are merely labels that refer to an experience; they do not embody and ground it.

---

[134] Physical information is a concept that can be precisely defined and quantified in physics. In contrast, semantic information remains elusive; no one has yet been able to define it clearly, let alone quantify it using a formula.

Let's take a closer look at the type of human-like cognition that involves a first-person perspective. By stepping back from the urge to rationalize your thoughts and instead observing how your mind operates, you will see things in a new light.

This observation reveals that the process of integration or binding involves a conscious act that diminishes the distinction between the subject and the object of perception. In other words, it reflects a will to identify with the sensed object by re-cognizing, re-capturing, or re-assembling it within oneself to become one with it. However, the mere integration of information into a representational unit does not inherently result in conscious experience or semantic awareness. Through this "re-combination," my singular sense of self can connect with this integrated unit and unify with it by a "seizing of significance." It is a "thought-conception" that by a sight takes together the contacted object into the subject. As Swiss linguist Ferdinand de Saussure once recognized, in written text, the syntactic relational structure of the words and their context (the signifiers) coalesce into an evoked semantic unit (the signified), which the mind then internalizes.

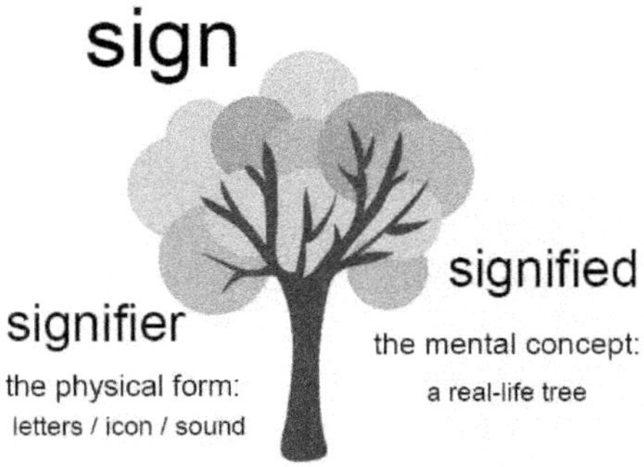

Fig. 79 A signifier (symbol, letter, word, etc.) is not the signified (the mental experience of what the signifier stands for.)

Thus, cognition in living organisms is a faculty of consciousness, not the reverse. Contrary to the naturalist's assumption, in living organisms, consciousness does not arise from cognitive processes; rather, it is cognition that requires consciousness. Meanwhile, cognition devoid of conscious experience is characteristic of machines.

What about cognition in non-human animals? While a clear difference in intellectual abilities exists between humans and, for example, dogs or cats, I believe animals also engage in meaning-making processes based on their lived experiences. I do not subscribe to the Cartesian view that regards animals as mere automatons.

However, as we move down the ladder of biological complexity, the situation becomes less certain. The question arises: Do bacteria "cognize"?

They are unicellular organisms, so their behavior is governed by simple stimulus-response mechanisms that inform them about their surroundings. However, it is difficult to believe they maintain any conscious awareness of the environment. There is no entity interpreting the meaning of the signals they process; they lack the semantic understanding that humans, and likely other more complex life forms, have.

What about plants or other organisms without a brain or with only a few neurons? In this case, biologists typically view the behavior of plants, jellyfish, and insects as a manifestation of machine-like basal cognition. This perspective is grounded in biochemistry, genetics, physical processes based on thermodynamics, homeostasis, bioelectric signaling, or more recent biological concepts like complex system dynamics, autopoiesis, free energy principles, etc. It reflects a naturalistic understanding of life as a machine—a viewpoint that has been ingrained in our thinking for about four centuries.[135] A perspective that is not inherently incorrect but that can become misleading when regarded as the only valid viewpoint.

This worldview might be the fallacy behind decades of debate about how to define cognition. It might take a considerable amount of time, possibly generations, for us to learn to appropriately contextualize the mechanistic aspect of Nature.

As a matter of fact, we can envision simple organisms exhibiting basic forms of cognition alongside simple, basal experiences. Rather than explaining cognition through a reductionist, bottom-up approach[136] that views it as a mere product of unconscious, insentient, machine-like problem-solving processes, we should consider the hypothesis that even the primitive sensing and instinctual reflexes of a paramecium coexist with a fundamental, proto-conscious form of sentience. This perspective could provide greater explanatory power. There might be not only an elemental form of sentience but also a basic form of meaning-making—what we might call "proto-semantics"—that connects entirely unconscious and insentient mechanical processes with the simplest instances of conscious and sentient signification.

---

[135] As modern findings increasingly challenge this worldview, many biologists tend to identify themselves as "naturalists" who reject the reductive machine metaphor. However, when pressed, they often resort to descriptions of life that rely on processes, such as "complex system dynamics," "self-referential feedback loops," "enactive embodied cognition," "cellular automata," "Turing non-computable information processing," or whatever concepts ultimately reiterate the same mechanistic perspective in disguise.

[136] Modern findings increasingly challenge the exclusive bottom-up understanding of life. In response, many propose theoretical frameworks rooted in the opposite paradigm of a top-down causation. While this perspective is not inherently incorrect, it is, again, the same mechanistic perspective in disguise. I don't believe it will provide the key to explaining the emergence of cognition, consciousness, agency, and life that they hope for.

Viewing things from this perspective might help us bridge the gap between a mechanistic understanding of life, which constructs complex organisms from a bottom-up assembly of insentient agents acquiring knowledge solely from physical information, and that of a sentient living organism, in which "proto-knowledge" is inherently rooted in "proto-semantics." The deeper "understanding from within" that characterizes human meaning-making might be involved in the fundamental basis of all cognitive processes, even down to the bacteria, rather than an emergent function. Knowledge could be a fundamental aspect and power of consciousness rather than merely the outcome of intricate, blind, mechanistic physical interactions.

As described in Chapter X, seeing the universe as a living entity could enhance our understanding. Matter may be viewed as a concealed form of life, while life itself could represent a hidden aspect of mind. Using the same reasoning, we might conjecture that the mind acts as a hidden expression of a supermind that has yet to be fully revealed on Earth but that is latent and waiting for the right moment to manifest. Also, the vital principle in life might be not mere superstition but, rather, a reality that science overlooks and vehemently rejects. Life, consciousness, and cognition could be three facets of the same phenomenon, as I described in the advanced reading XIII.2.

In other words, we do not try to explain consciousness, cognition, and the mind by starting from microscopic, unconscious, and non-mental processes; instead, we view consciousness as fundamental, with (1) cognition serving to integrate conscious experiences and (2) the mind synthesizing these cognitive acts into what we commonly refer to as a "thought." While we should avoid conflating mind and consciousness, we can loosen the strict distinctions between consciousness, cognition, and mind, considering them as points along a continuous spectrum rather than as entirely separate categories.

In conclusion, I would like to address one of the potential objections to the viewpoint that connects meaning-making with consciousness.

Many of our cognitive processes occur without our conscious awareness. If this were not the case, an overwhelming amount of information would inundate us, making life nearly impossible. There is a significant distinction between what we perceive as conscious thought and unconscious cognition.

For example, think of how we can navigate a complex environment—whether walking or driving—almost on "autopilot," mentally absorbed in unrelated thoughts while still registering what is around us through peripheral vision. This suggests that we can perform meaningful tasks and comprehend our surroundings even when we are not fully consciously aware of every detail. It appears that human-like cognition, which implies a true semantic understanding of the environment, does not require conscious awareness.

Moreover, experimental psychology demonstrates that sophisticated cognitive abilities seem to exist without consciousness. For instance, individuals can process semantic content and focus on objects even when masking techniques prevent the information from reaching their access consciousness. In cases of blindsight, individuals who are blind due to lesions in the primary visual cortex, can respond to visual stimuli they do not report perceiving,. Remarkably, they can often correctly identify objects that they believe they do not see.

Does this show that understanding is possible in the absence of consciousness?

I believe this conclusion is premature. Our understanding of consciousness is largely based on a superficial first-person experience. What we refer to as "unconscious" or "lack of attention" might actually represent another form of conscious awareness and attention—not unconsciousness but, rather, a subliminal, non-metacognitive (that is, a non-reportable) type of awareness. We often confuse non- or less ego-centered states of consciousness with being "unconscious" simply because we cannot relate them to our ordinary ego-centered state of awareness. In these states, we might not be genuinely unconscious; instead, we might simply lack mnemonic access to those experiences.

What is it like to be in a conscious state without memory? Are we truly "unconscious" during dreamless sleep? Are we unconscious during anesthesia? Are we unaware while in a hypnotic trance or sleepwalking?

We do not have definitive answers, but in Section II.2 I have shown how phenomenal consciousness—that is, some form of sentience—can be present but fade away once we return to our ordinary state of consciousness. Different layers of consciousness exist, but their content might not be transferred among them. Consciousness-mediated semantic cognition might still function in these altered states. Therefore, we should avoid making hasty conclusions, as this remains an open question.

In any event, the main point of this section is that asking questions about our psychological dimension while trying to detach from a first-person perspective is misguided. The nature of cognition cannot be fully understood from a third-person perspective alone. An investigation limited to the third-person approach forces us to posit unconscious, mindless, and purposeless processes as foundational, ultimately leaving us puzzled about how these processes could give rise to conscious, living, and purposeful beings like ourselves.

That's why, after decades of research and contrived speculations, cognitive sciences and biology have yet to provide satisfying answers to the deeper philosophical questions surrounding the nature of consciousness, cognition, life, and mind. Ultimately, the process of discovery relies not just on hard facts but even more on our beliefs and thought processes.

3. The Problem with Machine Intelligence

Viewing consciousness as the foundation of cognition and semantic awareness instead of starting with the unconscious mechanistic processes that supposedly lead to it is a shift that has significant implications for AI and the prospects of AGI. Human understanding arises from conscious experience and cannot be reduced to a mere insentient algorithm. Likewise, a human-like AGI must, by definition, rely on a form of cognition that surpasses simple best guesses, next-token predictions, or high-dimensional vector manipulations. If true semantic understanding requires conscious experience, then any future AGI will also need to possess consciousness.

This does not imply that AGI is impossible or that truly intelligent machines will never exist. However, it suggests that consciousness might be a necessary component for machines to achieve AGI. While the advent of LLMs has significantly advanced machine intelligence, there has been no progress in making these machines conscious. ChatGPT and similar systems are undeniably impressive, but closer examination leaves one doubtful that they possess semantic awareness. They still lack the generality of human intelligence.

One of the fundamental errors we make when we compare ourselves to machine intelligence is similar to the mistake we make when we question the essence of life. As we have seen in Section III.1, biology attempts to explain life without considering its psychological aspects, which include feelings, emotions, desires, instincts, subjective experiences, sentience, and consciousness. Similarly, AI seeks to replicate human intelligence while overlooking these psychological dimensions and reducing our cognition to mere computational, logical, and rational abilities. However, the human mind encompasses much more than that.

Comparing AI to living intelligence highlights a crucial difference: It is not just a matter of varying levels of intelligence but, rather, the lack of fundamental life-qualities. Even the most sophisticated AI systems lack agency, volition, intentionality, desire, autonomy, and goal-directedness. It isn't clear how much creativity and originality a machine has. For example, there is ongoing debate about whether generative AI can exhibit any creative or original impulse beyond human instructions and the data it has processed.

Ultimately, without input, a chatbot remains a passive entity, taking no action on its own. There is not even a beginning, a spark of agency, and a desire to reach a goal.

Moreover, while humans are fallible (often more so than machines), a fundamental difference exists in how we make mistakes. Humans can become aware of their fallacies by self-reflection, then reconsider their actions, deliberate, and review their belief systems. We can recognize our errors and restart the process to arrive at different conclusions. In contrast, once a machine provides an answer, it never doubts or reconsiders that answer unless a human prompts it to do so. This lack of active self-reflection stems from the same root as its cognitive limitations: Without consciousness, there can be no genuine agency, will, intentionality, or semantic awareness of one's own conclusions. All of these are inherent faculties of consciousness and cannot be separated from it.

Assuming that these psychological cognitive qualities are inherent to and an expression of consciousness itself, as discussed in Section X.5 on the faculties of consciousness, things begin to make more sense. Let's explore the differences between human and machine intelligence and examine why this distinction is significant.

A notable contrast between human and machine intelligence is that machines often excel in areas where humans struggle, and conversely, humans can easily handle tasks that confound machines.

For instance[137], generative AI systems, such as LLMs, frequently fail to grasp negation in questions (the 'not-questions'). When asked to avoid depicting an elephant in a room, these models typically generate images that include an elephant, as illustrated in Fig. 80. The reason for this inconsistency is not entirely clear; modern LLMs are complex enough that their inner workings remain hidden. They operate as "black boxes," which means we can observe the outputs based on specific inputs, but we cannot fully understand how they arrive at their conclusions.

Fig. 80 ChatGPT prompt: "Create a picture of an empty room with no elephant in it. Absolutely no elephant anywhere in the room."

What we do know is that LLMs do not truly comprehend their tasks; instead, they attempt to predict the most likely answer based on their training. Because they are trained on vast datasets in which the phrase "elephant in the room" is common, they conclude that the presence of both "elephant" and "room" necessitates that they depict an elephant and disregard the negation.

[137] The AI failures mentioned might have been resolved by the time you read this. I use them here only to illustrate a general point.

Fig. 81 provides another example from Google DeepMind's Gemini generative AI. When prompted to illustrate someone writing with their left hand, it often produces the opposite scenario.

Fig. 81 Gemini prompt: "Generate an image of a person writing with their left hand."

The issue is that the system does not simply "refuse" to create images of someone writing with their left hand; it fundamentally does not comprehend the request. Why is this the case? The likely answer: The system operates on probabilistic guesses rather than semantic understanding. It has been trained on a vast number of images, the majority of which depict right-handed individuals. Consequently, it concludes that the most likely response is an image of a right-handed person. It does not grasp the question; it merely predicts the statistically most probable answer.

Consider another interesting example. Despite being trained on extensive textual datasets, the system demonstrates a lack of understanding regarding the relationship between names and their quantities. For instance, if asked how many US states have names that begin with the letter K, they might respond that there are four: Kansas, Kentucky, Kansas, and Kentucky. This issue reflects both a lack of semantic awareness about the concept of 'states' and an inability to grasp the notion of quantity.

These examples illustrate not only that AI can make mistakes—something we all recognize, as humans do too—but also how the absurdity of some responses highlights AI's lack of deeper understanding. Simply having data and recognizing relationships is insufficient. In reality, meaning-making is an incredibly complex cognitive process, and something fundamental seems to be missing.

To support the assertion that meaning transcends mere symbols, raw data, and symbolic information processing, American philosopher John Searle introduced the "Chinese room argument" in 1980. Through this thought experiment, Searle aimed to demonstrate that a computer program, operating on a classical algorithm (as opposed to other forms of computation like deep learning neural networks or quantum computation), cannot understand meaning, even in principle. For a more detailed description see my book "Spirit Calls Nature" (Masi, 2025a).

Searle's thought experiment was reinforced by the "symbol grounding problem," which Stevan Harnad discussed in 1990. This problem highlights the challenge of understanding where symbols—such as words, numbers, bits, and signals—derive their meaning. There exists a gap between symbols and their computation on the one hand and our semantic awareness, intuition, understanding, and knowledge of their meaning on the other. The formal framework of any computational model cannot accurately capture semantic understanding. This issue can also be framed within the more specific context of neural coding.

As noted in the previous section on the nature of cognition, a binary string of ones and zeros doesn't "have" or "possess" meaning in and of itself until a conscious mind interprets it. Meaning is bestowed upon symbols by an observer; without this conscious engagement, bits, bytes, letters, or symbols are devoid of significance. For example, a string of bits in a computer is merely a series of electrical signals in a microchip. Similarly, the string 1010 could be represented by two white stones flanked by two grey stones in a dusty desert, yet it would hold no meaning in that context. Unless a semantic agent interprets it as semantic information, it remains simply a "displacement" or "puncture" in matter—nothing more than physical information.

Based on these arguments and other implications, Searle introduced the distinction between "strong AI" and "weak AI" (with the former now referred to as AGI). Strong AI is a hypothetical concept in which machines genuinely understand meaning as humans do. In contrast, weak AI refers to machines that only simulate human intelligence, relying on computations similar to a Chinese room and using smart inferences or probabilistic guesses without any true understanding of their actions.

Searle's Chinese argument and Harnad's symbol grounding problem have received numerous critical responses and remain matters of contention. However, they highlight that the algorithmic processing of symbols alone cannot fully explain the human mind. There exists an explanatory gap between syntactic rules and the semantics as humans perceive them. The perception of meaning cannot be reduced to mere information processing. While symbols, like the letters in the book you are reading, can evoke meaning, this occurs only in the presence of a conscious being capable of understanding. Meaning is a subjective experiential phenomenon *within us*, not an intrinsic property of symbols or external material objects, and it cannot emerge from computation alone.

The prevailing belief is that achieving AGI requires simply providing these systems with more computing power, more neurons, trillions of parameters, and larger datasets. This includes expanding their capabilities from text to images and videos, as well as integrating them into robots that can navigate their environment and gather sensory information about the world. The current goal is to enhance the AI of computers, which rely primarily on textual data, to incorporate all the sensory perceptions that complex living organisms, like humans, possess. Essentially, machine intelligence must be "embodied" so that it can explore the environment through various sensory modalities and create a "world-model." AI research has shown that machines' semantic maps are often inadequate because they are trained predominantly on text rather than through a focus on the external world. The next step is to implement genuine "world knowledge"—the conceptual and relational understanding of objects, the external environment, and the interactions between the perceiving agent and the world it inhabits.

Many believe achieving human-like AGI hinges on the concept of "embodiment," which is thought to lead to the creation of truly intelligent machines with genuine understanding. Once we fully implement this embodiment, we might finally develop AGI in humanoid form.

However, while this approach might advance our progress toward more intelligent machines, I am skeptical that it will be sufficient for achieving AGI. Although this idea seems plausible, it ultimately leads to an infinite regress.

Consider this: When you observe an object, such as a chair, your mind does not perceive it as an isolated entity. Even if you visualize it floating in zero-gravity space, you still understand it within a vast context of logical and conceptual relationships stored in your memory, enriched by past experiences. Behind the mental image of a simple chair lies a complex web of concrete and abstract sensory and conceptual impressions. A chair is a thing for sitting; it has four legs to prevent tipping, and it is made of a material—such as wood—that comes from trees. Trees are plants that belong to forests, which are intricate ecological systems composed of many trees. These ecological systems are part of the biosphere, which, in turn, is part of the planet's surface.

Moreover, who is 'someone'? To grasp what a person does with a chair (i.e., how humans use chairs), you must first understand what a human being is. This latter concept encompasses a fantastically complex set of material and psychological ideas.

One can continue this line of reasoning without truly resolving it, which leads to an endless semantic universe.

How does the human mind escape this infinite regression?

If you ask ChatGPT, it might seem to provide insights akin to our own semantic awareness. However, its responses are derived not from personal experiences but, rather, from the encoded experiences of others, represented through signs, symbols, letters, and sentences. Its knowledge is based on patterns and information derived from its training data. When you input "chair," it relies on a vast range of textual data from the Internet, translating words into numerical representations—high-dimensional vectors that capture contextual relationships among words. There is no concept of a chair in the human sense, only numerical patterns associated with chairs, including their typical features, uses, materials, and contexts. LLMs lacks visual experience, perception, imagination, and consciousness; it operates solely on learned patterns in the data. At best, its processing involves syntax, not semantics.

Moreover, as previously noted, chatbots remain entirely passive tools. They possess no agency, will, or desire to create something new. There is no indication that AI systems have any internal drive to act or create autonomously or that they possess intentions unless they are explicitly directed to do so. Without input, they remain passive black boxes. While these AI applications are sophisticated and effective at processing symbols, there is no evidence that they will develop conscious self-awareness. There is no reason to believe that within them exists any sentient being capable of understanding or having the motivation to create.

4. No Consciousness? No Meaning and No AGI!

Thus, despite its remarkable capabilities, modern AI does not truly understand anything. This might seem like an overly negative assessment of what LLMs can achieve. When we interact with a chatbot, it often gives the impression of genuine understanding. In a sense, we might argue that modern computers have a good grasp of the data they process, especially in terms of language. However, it is more accurate to view this as an imitation of deep understanding rather than as the "semantic awareness" to which I refer when I discuss meaning-making.

Scientists are gradually realizing that it isn't that simple and that there might be fundamental flaws in our understanding of what intelligence and reasoning are about. They are beginning to sense that the gap between syntax and semantics is likely much wider than previously thought. Even with enormous data sets and immense computing power, a computer will not gain a true understanding of the real world.

There is a gap between human and machine intelligence. Shannon Vallor summed this up: *"We are more than efficient mathematical optimizers and probable next-token generators."*

Nonetheless, many AI and machine learning researchers hesitate to fully acknowledge this. Few are willing to confront the elephant in the room: the possibility that human-like intelligence, characterized by semantic awareness, might not be naturalizable and could never be fully replicated in a machine—unless that machine becomes conscious.

We don't need to delve into complex metaphysical debates about the mind-body dualism. There is a straightforward experiential truth that everyone acknowledges yet many overlook. From a first-person perspective, it becomes clear that our semantics-based cognition is inherently linked to conscious experience—always.

You cannot truly understand colors, sounds, tastes, smells, temperature, touch, or sight without having had a conscious experience of these sensations[138]. For instance, one can explain everything about the H_2O molecule, chemistry, and physics, but true comprehension of wetness comes only from experiencing water firsthand. Without a conscious subject to have qualitative experiences (referred to as 'qualia'), all textual descriptions are insufficient. A machine can only grasp the abstract relationships of symbols, letters, words, sentences, numbers, and mathematical functions; it cannot truly understand them.

This limitation might explain why self-driving cars, which rely solely on visual data, are easily misled and unlikely to achieve level V autonomy; they do not understand what they are seeing. They perceive the world only as numbers, vectors, and matrices.

That's why building self-driving cars has proven to be far more challenging than initially anticipated; as of this writing, their applications remain limited to specific driving contexts. For years, we were led to believe the bright future of self-driving cars was just around the corner, but expectations have since begun to wane. The reason for this is shown in Fig. 82, which illustrates what a self-driving car 'sees' when it encounters a vehicle with an advertisement on the back.

Fig. 82 What a self-driving car's camera sees when it looks at a car with an advert on the back.

The AI's ability to classify objects and people is remarkable; it can recognize and differentiate between cars, bicycles, and pedestrians within its environment, even in bad lighting conditions. Yet, it cannot discern between real and fictional patterns. This is because classification does not equate to

[138] This does not mean that congenitally blind people cannot understand discussions about light, colors, or other visual sensations. They certainly can. However, their comprehension is indirect, as they rely on information derived from their interactions with others and their experiences through other senses. They "see" through sound, touch, taste, and smell, as well as through more complex spatio-temporal conceptualizations that are also qualitative subjective experiences. This enables them to semantically represent and understand visual concepts, despite their inability to directly perceive them, because other forms of conscious experience support their understanding. The same applies to congenitally deaf individuals.

understanding. The AI categorizes and sorts information without truly comprehending it, as there is no 'ghost in the machine' with semantic awareness of what it perceives, processes, or analyzes. There exists a strong and inseparable connection between feeling and knowing, sensing and understanding, perceiving and comprehending. These two aspects of cognition cannot be disconnected or treated separately, as current research attempts to do. Without consciousness, understanding is not possible, not even in theory. When you grasp the meaning or semantics of something, you experience an inner flash of realization, a spark of knowledge, often referred to as the 'aha moment'—an experience that an unconscious machine will never have.

The same holds true for any narrative surrounding AGI. There is a direct correlation between general intelligence and consciousness. My argument is that AGI cannot exist, even in theory, without consciousness. True intelligence necessitates semantic understanding, which is fundamentally tied to sentient experience. Simply adding trillions of neurons, increasing the number of parameters, flooding an AI system with more data, or enhancing its computational power will not suffice. Without consciousness, AGI cannot be realized. This becomes self-evident from a first-person perspective.

Other implications arise from the fact that, sooner or later, the machine makes an extremely stupid mistake that reveals its lack of understanding and, thereby, shows that it has no subjective experience. Notably, this serves as a compelling argument against the existence of so-called "philosophical zombies," or "p-zombies" for short. A p-zombie is a hypothetical entity that is physically indistinguishable from a normal human but that lacks inner life and conscious experiences, or qualia, while behaving exactly like a conscious human. However, because a p-zombie lacks the sentience necessary for meaning-making, it will make glaring mistakes, similar to other non-conscious AI systems, thus revealing its true nature. Therefore, it is reasonable to conclude that beings that mimic human behavior without subjective experiences probably do not exist; otherwise, we would already be able to differentiate them from conscious humans or non-human animals.

A meaning-making test could provide a more robust and convincing measure of a computer's consciousness and intelligence. This approach might even introduce a new standard for evaluating intelligence, thereby potentially replacing the well-known "Turing test." The goal would be to create a test that determines whether a machine has genuine semantic understanding and deep comprehension similar to that of a human. If it does, this would imply a form of "intelligence" that suggests the presence of experience as well.

Consequently, one could argue that if we are to achieve AGI, we must first develop conscious machines. However, science is far from understanding how consciousness emerges in organic intelligence and how it is connected to agency and life.

The Advanced Reading in Section XIII.8, titled "No Consciousness? No Meaning (and No AGI!)," provides a more detailed analysis of the relationship between human and machine intelligence—a topic that has been a longstanding area of discussion. Initially, I hadn't intended to delve deeper into this subject. However, I have noticed that the connection between meaning and sentience is largely overlooked. I argue that semantic awareness is unlikely to exist without consciousness, which has significant implications for the future of AGI. The article summarizes my previous sections but from a more rigorous academic perspective.

5. Consciousness, Meaning and Language from an Eastern Perspective

"Ultimately, language is both an enhancement and a constraint, a double-edged sword that structures thought while simultaneously obscuring other modes of being." Jeremy Skipper

In the previous sections, we explored the interconnections between consciousness, meaning, and language. We also examined how an introspective approach can uncover insights that a solely external, third-person perspective might overlook, particularly regarding our understanding and aspirations in creating machines with human-like cognitive abilities.

Despite extensive research, the origin and nature of human language remain contentious topics among linguists, anthropologists, and philosophers. The emergence of human language is a mystery. Did it develop gradually from primitive communication systems, or did it arise suddenly as a unique cognitive leap? Why is human language so much more complex than animal communication? What cognitive natural or social pressures drove its development?

While the world's languages exhibit vast differences, they also share fundamental structural similarities, suggesting an innate human capacity for language. This apparent paradox intrigues linguists: How can consistent rules govern such diversity? Notably, the American linguist Noam Chomsky posits a genetically determined "language faculty" in the human brain based on a theory of "Universal Grammar." Others argue that language emerges primarily through social interaction and cognitive development and does not require a specific biological blueprint.

Nevertheless, the exact nature of this mental framework and its interaction with culture, experience, and environment remains elusive, leaving the origins and nature of language an open question. Linguists have not achieved a clear consensus. The lack of agreement on these foundational issues illustrates just how elusive and multifaceted the study of language remains, with each theory offering compelling insights but falling short of a definitive explanation.

However, this analysis was largely grounded in Western scientific and philosophical naturalistic frameworks.

While philosophers like Ludwig Wittgenstein often dismiss metaphysical interpretations as "nonsensical," we should consider whether the Eastern tradition offers valuable insights. If consciousness and meaning cannot be reduced to mere computational processes, how can we engage with them and explore their relationship from a different perspective? What might a spiritual viewpoint reveal about our cognitive activities in meaning-making and language? Is there a possibility of a new approach that connects the origins of language and consciousness beyond naturalistic explanations?

In an era dominated by LLMs driven by extensive textual data, questions surrounding the origins and nature of language have received renewed attention and are as compelling as those concerning consciousness. However, the Western tradition tends to overlook the relationship between consciousness and language. In particular, it ignores the potential of a first-person perspective.

I found it particularly interesting to explore the nature and origin of language through the perspectives of the Indian mystics and philosophers Abhinavagupta (AD 950-1016) and Sri Aurobindo (1872-1950).

Eastern philosophies within the Vedic tradition emphasize that all speech originates from a transcendent sound referred to as the "Word." This concept is also reflected in Christian theology, as seen in the Bible: *"In the beginning was the Word, and the Word was with God, and the Word was God"* (John 1:1).

Abhinavagupta's grammatical cosmology is based on Vedic ontology, positing that consciousness and language are inherently linked. He describes the genesis of words as a process within consciousness, whereby mystic sounds become concrete human language. He explains that at the origin of existence, consciousness expresses itself through sound, and this primeval vibration generates all forms of speech. He delineates four levels of consciousness corresponding to different stages of language manifestation. Each level reflects a deeper understanding of the relationship between sound and meaning, emphasizing that true meaning arises from a transcendent source.

According to Abhinavagupta, the four levels of speech are as follows.

Parā Vāc: This is the highest non-physical and transcendent level. In it, consciousness is pure and undifferentiated. It's like the ultimate source of all sounds and meanings, existing in a state of potential without any physical form or separation between sound and meaning.

Paśyantī Vāc: At this level, the will to know and the visualization of thoughts begin. It's where ideas and thoughts start to take shape, but they are still not spoken or fully formed. It's like seeing a flash of intuition or a visual image of a word before it becomes a concrete thought.

Madhyamā Vāc: This is the mediating level, where thoughts and words start to become more distinct and structured. It's the stage where the mind begins to organize and differentiate between subject and object, and words take on a temporal sequence. It's like the internal speech we hear in our minds before we speak.

Vaikharī Vāc: This is the final level. In it, thoughts are expressed as physical sounds through the speech organs. Only at this stage does the non-physical idea that dwelt on other planes of existence finally manifest. It's the stage of spoken language and written words, where ideas are communicated in a concrete form. It's like the actual words we speak and write, and it's the level at which modern AI, such as LLMs, operate.

Meanwhile, Sri Aurobindo builds upon Abhinavagupta's insights, focusing on the psychological aspects of language development. He recognizes that words stem from "mind-impressions" elicited by "seed-sounds" with psychological qualities. Language initially emerged from emotional and sensory expressions, then evolved into a structured system of symbols. Seed-sounds carry inherent meanings and psychological qualities, suggesting that human speech reflects these deeper vibrations. He also posits that the original language, spoken in an 'age of truth,' was rooted in a perfect relationship between sound and meaning—a relationship that has since been lost in modern languages.

However, for Sri Aurobindo, the origin and nature of speech run deeper than that.

As we discussed in Section IX.3e, the universe itself is the expression of a creative transcendental Real-Idea, and which is foundational for all meanings, signs, words, and human language. The universe is a manifestation of the Real-Idea, a gnostic power that shapes existence through its inherent Knowledge and Will. Accordingly, all forms in Nature that we apprehend are symbols of this ultimate Idea, which transcends the material realm. This perspective allows for a synthesis of Eastern and Western philosophies, suggesting that the universe operates as a language expressing deeper meanings and truths.

The integration of Abhinavagupta's and Sri Aurobindo's spiritual experiences provides a comprehensive framework for understanding language's origin and nature. Their insights challenge conventional linguistic theories and suggest that semantics (not syntax) is not merely a human construct but is also a reflection of a deeper cosmic reality. This perspective opens avenues for reconciling materialistic views with spiritual insights, thus emphasizing that meaning is inherent in all existence and that language reflects this deeper truth.

It is not surprising that LLMs are highly effective at capturing the world based solely on language. This is because the world can be understood as a symbolic and linguistic expression of the divinity. LLMs excel because they reflect the same relationships found in language that exist in the world.

Language encodes the relationships among various entities—objects, phenomena, and physical events—through linguistic representations such as words, symbols, and tokens.

This suggests a post-material understanding of language that needs further exploration. However, this should not replace Western scientific perspectives; rather, it should integrate them.

I had never paid much attention to the origin and nature of language until I encountered the frameworks of Abhinavagupta and Sri Aurobindo. In my opinion, these perspectives offer a distinct yet complementary approach to Western analytical methods. Furthermore, they provide valuable insights into the interconnectedness of intelligence, semantics, AI, AGI, and consciousness, demonstrating how Eastern and Western cultures can integrate and support one another. This realization prompted me to write the article "The Nature and Origin of Speech in Abhinavagupta and Sri Aurobindo," which constitutes the Advanced Reading of Section XIII.9.

6. Why All This Matters

Many might think that knowing the answers to these questions is not practically important, as the primary concern is simply whether the self-driving car can safely take me to my destination. However, this reluctance to address deeper issues has misled numerous investors into pursuing an enterprise that remains far from meeting expectations and, as of this writing, is currently stalled. This doesn't mean the game is over; however, a bit more wisdom could have led to better investment decisions.

Consequently, there is a fundamental and principled argument against the emergence of AGI. Unfortunately, IT scientists often dismiss philosophical considerations and are only now beginning to confront insights that were understood over four decades ago.

The root cause of this excessive faith in a humanoid-like coming revolution and bright future lies in humanity's lack of self-understanding. We consistently externalize our focus and are overly inclined to objectify reality—a fact that hinders our ability to perceive and understand our own cognition. Had we not adhered to a mechanistic view of the mind and life, we might have been less naive and could have directed our resources toward more promising avenues while also gaining a deeper awareness of consciousness, cognition, and the essence of life itself.

In cognitive sciences and biology, scientists face a similar challenge from the opposite perspective. Consciousness, mind, life, cognition, and agency are already present, leading us to question how this is possible. Recently, there has been a slight shift away from a strictly reductionist worldview, though it remains rooted in a materialistic understanding of reality. Some attempt to incorporate free will into their view of life while still adhering to a deterministic perspective (as discussed in Section V.6 on compatibilism).

Others are willing to abandon the machine metaphor in biology, recognizing its inadequacy, yet they still cling to blind mechanistic processes as the fundamental basis of life. The theory of enactivism acknowledges that traditional naturalism is outdated and, thus, seeks to introduce a 'new naturalism,' but it struggles to explain the emergence of experience within it. Ultimately, we persist in trying out more of the same and would like to have it both ways.

I believe, ultimately, AI will reveal more about us—our consciousness, our true inner nature, and our psychological and spiritual dimensions—than it will clarify what consciousness actually is or how it supposedly emerges from a biological and neuronal process. We might reach a point where we recognize that, regardless of how advanced and sophisticated AI becomes, it will always lack the essence that defines our humanity.

If this is the case, we will be compelled to acknowledge that mind and consciousness are not merely physical epiphenomena, which might lead us to embrace a post-material perspective. Once this understanding becomes widely accepted, we can collectively break free from the constraints of a mechanistic and material view of reality, especially concerning life and our own existence. When this shift occurs, the world will have the potential to change for the better. Sometimes, the path to truth requires exhausting all alternatives. This long and challenging journey might ultimately lead to a profound revolution—one that I believe will be far more interesting and fulfilling than any transhumanistic AGI fantasy.

Our choices are influenced not just by the facts around us but also by our worldview and, particularly, our self-perception. If you embrace the idea that Earth is not the center of the universe, your decisions will likely reflect a broader mindset. Conversely, if you remain trapped in a geocentric perspective, you might struggle to see beyond a narrow horizon, thereby restricting your potential and growth. The matter-centric paradigm has already caused significant harm.

Humanity currently finds itself navigating the modern hype surrounding AI. If we view ourselves merely as 1.4 kg of wet brain tissue, we will likely adopt an approach to mental illness that remains fixated on a brain-centric perspective. This narrow view might prevent us from moving beyond purely pharmaceutical treatments. If we dismiss the 'hard problem of consciousness' as a trivial philosophical issue or a weekend rumination, we risk perpetuating a society that cannot evolve from a materialistic, industrial framework rooted in economic exploitation and injustice.

This perspective has already led us into a complex dilemma that hinders the progress of medical science. A gene-centric and neuro-centric view has fostered a neuroscience delusion that threatens to negatively impact how future generations treat mental illnesses. You might question the relevance of science to mystical experiences or view philosophical inquiries—such as whether reality is an illusion or whether free will exists—as trivial. However, the answers we provide to these questions reflect a mindset that will shape our future.

Moreover, our tendency to conflate mind and consciousness, as well as our engagement with the complexities of the mind-body interaction and the relationship between determinism and free will, are philosophical considerations that fundamentally influence our worldview. The framework we choose—whether materialistic or post-materialistic—will ultimately guide the direction humanity takes in the coming decades.

We stand at the cusp of a technological revolution that will define our future. Placing our understanding in the right context will steer us toward one path, while failure to do so might lead us astray.

XII. Conclusion

1. The Lack of Perspectivism is Killing Us

I will always vividly remember a key experience I had as a physics teacher in a German high school. You might think that physics lessons are unaffected by political and sociological debates, but that isn't the case. For example, when it comes to the topic of the peaceful use of nuclear energy, you quickly learn that teaching a dry and technical lesson in front of a blackboard about the structure and functioning of a nuclear power plant—intended to convey the so-called "objective knowledge" to young minds—won't get you very far. Especially if you are in a German Waldorf school which is markedly environmentally oriented, it doesn't take long until a debate for and against the use of nuclear energy, will heat up.[139] Needless to say, my students were all aligned with an anti-nuclear policy, unanimously depicting nuclear energy as a demonic force.

However, when I asked for their arguments, almost no one was able to present a coherent line of reasoning to support any thesis beyond simply referencing Hiroshima, Nagasaki, Chernobyl, and Fukushima. They did not even mention common concerns, such as radioactive waste or radiation exposure. Instead, they were essentially repeating typical narratives they had encountered randomly on social media or from their parents, without having thought them through or conducted any research.

"Great," I thought, "after all, that's what schools and teachers are supposed to do: not impose political views, but create an environment for study and reflection that offers fact-based knowledge, providing a foundation for more thoughtful debate." I invited my students to conduct research and present a logically and scientifically coherent argument based on facts that support their thesis.

However, I had one requirement: they needed to return with a piece of paper divided into two sections, one section listing arguments against nuclear power—as they would like—but also the other section containing an equal number of arguments in favor. In a sense, without mentioning it explicitly, I was asking them to engage with the classic Hegelian dialectical method: thesis, antithesis, synthesis. Only in the synthesis should they express a final judgment, not before.

This sparked a rebellion in the classroom. "Why should we seek arguments in favor of something we don't agree with?" "Aha...You're obviously biased, aren't you, Mr. Masi?" Within 24 hours, the entire school was questioning

[139] This must also be considered in light of Germany's decision to phase out nuclear energy since 2011, which has sparked significant national political skirmishes that continue to this day.

whether I was being paid by the nuclear lobby (… and you are thinking that too, aren't you? Haha!). Nobody believed me when I later explained that I am actually opposed to nuclear energy; in a different context, I would have voiced strong opposition.

After all, the story had a happy ending. Much later, after graduation, some of my students thanked me for introducing them to this type of research approach. But it took me a while to realize that I was the one who learned a real lesson.

Are you able to consider multiple perspectives simultaneously?

This one-sided and uncritical thinking is also widely prevalent among adults. For example, when my school board colleagues discussed how to foster critical thinking in young minds, their verdict was unanimous: we should direct them to reliable sources.

This makes no sense to me. Who decides what "reliable" means, and based on what criteria? Ask a MAGA Trump supporter what constitutes a "reliable source" of information. Is it Fox News? None of my colleagues appeared to be interested in promoting a perspectival understanding of reality. I suspect this is because it implies that you allow students to form their own opinions, which can be dangerous.

This attitude reflects a more or less unaware belief with which we have been raised since childhood: there are good guys who know the truth, and we have to distinguish them from the bad guys. Once we've done that, we can simply look to the "good guys" to tell us what "truth" really is. This is evidenced also by the fact that we easily tend to believe in authority, shutting down opposing views by citing an expert or authoritative figure's opinion rather than doing our own research, collecting evidence, fact check, and especially diversifying our sources with contrasting and, eventually, opposite views.

This isn't exactly news. It's clear that about one third of society (my subjective estimate) has been left behind and lacks the ability to think critically. This group is easily manipulated and tends to believe whatever "truth" the recognized "authorities" or leaders present to them. They take things at face value, without scrutinizing or questioning the narrative, and blindly follow a Pavlovian stimulus-response conditioning. This behavior has always been the default mode for the masses, but it has come to the forefront more recently due to the rise of social media. A typical strategy adopted by the fossil fuel industry involves bribing an authority (corruptible scientists, academics, or anyone with a PhD) to fabricate seemingly plausible arguments against human-made climate change.

However, while influencers, think tanks, policy advisors, academics (researchers, professors, institutions), cultural figures (artists, writers, celebrities), media outlets (journalists, editors, commentators), advocacy groups, and political strategists fervently discuss how to awaken (or keep asleep) that segment of society living under a rock—with political and electoral strategies or by luring them into another materialistic dream of a capitalistic renaissance—nobody addresses the root causes deeply anchored in our institutional and parental mono-perspectival educational system.

This issue is not confined to the uneducated; it is also glaringly evident among highly educated academics. Here's another personal anecdote.

In the past three to four years of my intellectual journey, I have been publishing on highly contentious topics that involve philosophy, physics, cognitive science, and metaphysics simultaneously, and, thereby, inevitably ask the reader to adopt a multiperspectival view as well. I could only do so as an independent scholar without academic affiliation because these subjects, for obvious reasons, don't receive institutional funding. About 50% of the dialogue with referees focuses not on the content of the essays but on an endless struggle to clarify their misunderstandings or misreadings that are rooted in a mere physical, mere philosophical, or mere psychological viewpoint. Philosophers accept the philosophical aspects but reject the physics; physicists accept the physics but dismiss the philosophical interpretations; cognitive scientists ask you to ignore both and rewrite everything in their jargon; while, most of the time (though not always), they typically reject the metaphysical components because they are convinced physicalists. The clash between the material, religious, and spiritual mind is fundamentally the same. The materialist views the world solely through the lens of scientific realism, the religious individual through scripture, doctrine, or dogma, and the spiritual person through a mystical, experiential perspective. It is a clash between characters who are willing to embrace only the objective, only the abstract mentality, or only the subjective, yet refuse to acknowledge that the same reality can be viewed from all these perspectives simultaneously.[140] For example, when I adopt the stance of substance monism, I ask people to consider three perspectives simultaneously: that of the philosopher, like Spinoza; that of the physicist, who can speculate about a universal quantum field; and that of the phenomenalist, who argues that ultimately everything boils down to experience. Can you do that?

There is an absence of perspectivism that prevents individuals to view an argument, a line of reasoning, and even everyday facts from different angles. We have been raised and educated to see the world through a particular lens, often shaped by our studies in school and college—be it the scientific, humanistic, or artistic viewpoint. We tend to embrace only a physical reductionist, or only a philosophical, or only psychological, or only sociological perspective, but seeing integrally is perceived as an impossibility,

---

[140] Even though, deep down, they are all subjective. What we call "material" and all abstract philosophical conceptualizations are ultimately subjective phenomena—"shadowy projections" onto our screen of consciousness.

an oxymoron. And, we not only struggle to see the same thing from these various standpoints but often refuse to do so. Like my middle school students, who were both unable and unwilling to compare and share different perspectives.

Generation after generation, we enthusiastically promote multi-, trans-, and interdisciplinarity. However, our efforts often fall short in an increasingly hyperspecialized and compartmentalized academic landscape that faces interconnected challenges requiring coordination, openness to diverse approaches, and mechanisms for genuine dialogue between fields. We once believed this would lead to true integration, but it rarely goes beyond fragmented and cumulative contributions. Educational institutions largely consist of siloed academic departments with rigid funding structures. While the noble intention of having different disciplines work side by side, interact, and exchange methods, concepts, and frameworks is laudable, it does not automatically foster perspectivism. Without a multiperspectival approach, further progress remains elusive or becomes sterile.

The crux of the matter is that we (often unconsciously) still believe it revolves around information, knowledge, data, abstract concepts, or learning new things. We tend to think that once we know the true facts, we will become wise and knowledgeable. However, this has little to do with the ability to see things from multiple perspectives simultaneously. Seeing differently is not (or not necessarily) about knowing more.

This is also true for very theoretical, intellectual, and philosophical questions. To get a better understanding, you might want to see some concrete examples. As we have seen, AI scientists often fantasize about the impending revolution of conscious machines. They are regarded as the true experts on the topic. However, most of them are software engineers who view cognition and life as mere Turing machines or simulations running on a computer but lack a deep understanding of human psychology, have no philosophical background, and are unable to adopt a subjective experiential first-person perspective to explore how their own minds work and what their consciousness is—they essentially do not know themselves. The notion of considering all three perspectives coherently does not occur to them. Even when it does, it is often dismissed as unnecessary or too challenging. While I do not claim to have achieved true perspectivism, I at least strive to do so.

This intellectual inability to recognize perspectivism is just the tip of the iceberg. Beyond rational and analytical discourse, being multiperspectival also involves the capacity to feel, sense, and perceive from multiple standpoints simultaneously. Ultimately, it is about entering the minds of others and understanding how they see and feel from their perspective. It focuses on learning how they think and feel, rather than merely what they think and feel.

The typical objection is that this is impossible because one cannot truly "know" what it feels like to perceive or experience something unless one directly experiences that feeling or perception. For example, congenitally blind or deaf individuals will never "know" what it is like to see the blue sky or hear a symphony. They may understand the concepts intellectually, through analogy, language, or other sensory replacements, but the so-called "qualia"—the raw subjective experience of vision or sound—remains inaccessible to them. This is highlighted by Frank Jackson's "Mary's Room" thought experiment, which imagines Mary, a neuroscientist who knows everything about perception and the brain. However, since she lives in a black-and-white world, she cannot truly understand what colors are or what they mean. This thought experiment illustrates the gap between so-called "objective knowledge" and subjective, first-person experience. How can we embrace a perspectivism of qualia? We can't!

The objection is technically correct but has evolved into an argument that justifies a kind of multiperspectival laziness. While I may never truly understand what you have gone through, as your experiences are uniquely yours, I can still feel sympathy, compassion, and closeness to you and your suffering. I may relate to it, even if only indirectly, through my own experiences, even if they are not the same. Yet we tend to dismiss emotional perspectivism because it isn't an exact, quantifiable science. This has historical roots in early 20th-century psychology, which largely overlooked empathy, particularly in Freud's psychoanalysis and later even more so in Skinner's behaviorism. And now we complain about the lack of empathy in our society, while others even label it as a weakness.

However, perspectivism offers much more than an emotional connection to others' suffering. It also enhances our aesthetic senses and, most importantly, deepens our understanding of the inherent unity in diversity that makes us a human family. To become multiperspectival means developing the ability to appreciate perspectives that may be alien to our culture, education, and geographic background. We will never achieve unity in diversity and multiplicity if we view the world—and especially our fellow humans—from a mono-perspectival standpoint. An ideal of human unity will never be achieved through political revolutions or economic reforms. War will continue to exist as long as there is a sense of separation between nations. This sense of "otherness" is rooted in the inability—and often the unwillingness—to see reality from multiple perspectives.

This inability or unwillingness to see and feel the beauty of diversity from many frames of refence is the root of all forms of intolerance. Racism cannot exist in a multiperspectival mind, as such a mindset dissolves any sense of separation. Within a framework of perspectivism, ethnicity, skin color, culture, language, and religion are not separate and independent civilization traits but different styles with which the one-spirit expresses itself. The only way to overcome war and racism is by learning to live and work together in a multicultural environment that allows us to appreciate others' perspectives, rather than isolating ourselves on an ideological island where we view the world from a single angle. The same applies to sexual preferences. You may not understand a way of loving that differs from your own, but you can recognize that your way is not the only "natural" form of love.

However, the opposite of intolerance cannot be tolerance. "Tolerating" still implies that we must accept something we dislike. As long as we believe we must tolerate rather than appreciate one another, we remain in a state of separation and division. An attitude observed in some religiously "tolerant" individuals who advocate for interfaith dialogue while privately asserting that their own religion is the "superior" one and viewing all other religions as, at best, auxiliary or supplemental; and then thanking God for being born Christian, particularly Catholic. That is a "tolerance" that endures a diversity perceived as adversity and which, again, is ultimately rooted in a refusal to see things through a different lens.

The true opposite of intolerance is recognizing that the "otherness" of others is a unique form of flourishing that stems from the same source as my own "otherness." It is the realization of unity in diversity, where variety and multiplicity are seen as strengths rather than weaknesses. This state of mind perceives diversity not as a threat, but as a desirable and enriching evolutionary force that fosters growth. It has nothing to do with morality or ethics; it reflects an inherent quality of life. One cannot grasp this without a multiperspectival and integral grand vision of the whole. Only through perspectivism can we apprehend and comprehend the true essence, meaning, and power of the unity that is many.

This perspectivism isn't just a different way of seeing. Seeing differently is great, but who or what determines that one perspective is better than another? Perspectivism values the simultaneity of standpoints rather than attempting to determine which one is superior. I don't think there is a perfect, all-encompassing viewpoint that can be declared the ultimate one, like the "tolerant" interfaith preacher believes. It isn't even about discovering a final truth. We know all too well how the word "truth" can be misused to serve its opposite. However, perspectivism can bring us closer to the truth.

This may sound philosophical and abstract, but the lack of perspectivism is a root cause of misunderstandings and misjudgments that shape our society and contribute to human suffering. Historically, societies characterized by isolationism, mono-culturalism, and a homogeneous worldview could exist—and even coexist peacefully—until recently due to a lack of significant physical and cultural contact. However, in today's globalized and interconnected world, which is subject to massive migratory flows, this mono-perspectivism is becoming increasingly unsustainable and even self-destructive. It is killing us.

Nonetheless, the ability to simultaneously see multiple perspectives is not a genetic trait or a gift from God. It is a skill that can be practiced and learned.

It's time to focus not on what we learn, but on how we learn. New political or economic orders won't solve this problem; education is the key.

2. Towards a Soul-Centred Post-Material Worldview

The Copernican revolution famously repositioned the Earth in a peripheral role, placing the Sun at the center of the known universe. This conceptual shift was not confirmed by any evidence; it was a hypothesis, only a conjecture, but it nonetheless enabled humanity to move from cultural stagnation to a vibrant cultural and scientific renaissance. Although the Ptolemaic worldview, which maintained that the Earth was the center of everything, held some influence for a time, it ultimately had to yield to the advancements of the opposing paradigm. Where would humanity be today if it had stubbornly clung to an anthropocentric belief system that refused to reconsider its stance?

In the mid-19th century, Charles Darwin catalyzed a profound transformation in science. The belief that humans occupy a central place in creation was shattered again, revealing that humans, like all other species, are products of an evolutionary process. This shift from anthropocentrism to an 'evolution-centered' perspective on life marked a significant change in understanding the world and ourselves. Unlike Copernicus, Darwin based his theories on empirical evidence. He was particularly influenced by the fossil record, which illustrated a progression of life forms from simpler to more complex over geological time. Additionally, he observed human practices of selective breeding, which showed that deliberate selection for specific traits could lead to substantial changes in populations over generations. He recognized that similar processes could occur in Nature through natural selection. However, was Darwin's change in perspective driven solely by new empirical data, or did he also acknowledge the need for an ideological shift to accommodate this evidence? What would humanity's understanding of life be today if we still adhered to a strictly creationist viewpoint?

Less known is the "second Copernican revolution" that occurred in the mid-1930s and that shifted our perspective from a Sun-centered view of the universe to a galaxy-centered one. The Sun is not the center of the universe; rather, it is just one of the billions of stars in the Milky Way. No significant advancements in astronomy would have been possible without embracing this second paradigmatic shift.

Around the same time, Einstein's theory of relativity and quantum theory emerged and fundamentally altered our understanding of physical reality. In relativity, the notion of a preferred reference frame—such as one fixed to the Earth or the Sun, where the laws of physics could be considered invariant—was replaced by a conception of reality that has no preferred position or specific state of motion. Contrary to its name and popular belief, relativity emphasizes the invariant aspects of reality. The laws of physics remain invariant across all frames of reference and states of motion, and the speed of light is constant for all observers. Ultimately, it was not just a groundbreaking discovery; it was Einstein's shift from one worldview to another that paved the way for further progress.

On the other side of Einstein's 20th-century scientific revolution stood Max Planck, who introduced a crazy idea: Energy is not emitted and absorbed continuously; rather, this process happens in the form of tiny "packets" or "quanta." Later, Einstein himself demonstrated that light could be understood as a stream of particles, now called "photons," rather than waves. Werner Heisenberg further contributed to this shift in perspective with his famous uncertainty principle, which asserts that everything is inherently "fuzzy," "unsharp," or "indeterminate."

The universe was no longer a precise clockwork, which our analytical minds had cherished for so long. Instead, we were compelled to accept yet another "Copernican shift," moving from a world centered on certainties, where everything could, at least in principle, be predicted, to an inherently unpredictable reality. This transformation was not driven solely by a series of discoveries; rather, it was the lack of progress and the presence of anomalies that prompted these thinkers to reevaluate their assumptions of how reality works.

The advent of quantum physics paved the way for significant advancements in solid-state physics, which form the foundation of microelectronics and ignited the digital revolution beginning in the 1950s. One might wonder where humanity would be today had it stubbornly clung to a classical and deterministic view of the universe.

Other examples from the history of science and human cultural evolution illustrate this concept. The key aspect shared by all these "paradigm shifts," as Thomas Kuhn termed them, is not merely a discovery, observation, or anomaly. Rather, it is a fundamental change in mindset that breaks free from the constraints of previous belief systems. It involves a transition from an anthropocentric worldview to a more universal perspective that encompasses a broader understanding of the world and our place within it.

What has followed since the mid-20th century? Have we experienced any new paradigm shifts, groundbreaking discoveries, or innovative ways of perceiving reality? Thus far, the answer is no. This raises an important question: Do we need one?

Some might argue that, given the remarkable successes of science and its countless applications, a new paradigm shift is unnecessary. Everything seems to be functioning well, and science is more productive than ever.

The question of whether science is in robust health is debatable. I can present strong arguments suggesting that, despite appearances, modern science is stagnating. It has entered a phase of incremental progress that requires significantly larger financial and human resources than before to achieve the promised advancements. I believe that, in particular, the so-called big science projects are facing a deep crisis. For instance, despite the tremendous technological advancements in neuroscientific tools, the actual progress and practical applications of neuroscience are vastly overrated. This has resulted in limited practical applications in psychiatry and medicine—far fewer than anticipated.

Allopathic medicine remains firmly rooted in a mechanistic and materialistic understanding of life, our bodies, and health. This perspective inevitably shapes how medical sciences are practiced and how related research and development are funded.

Another prominent example of this situation is the so-called "war on cancer," which has been ongoing for nearly 50 years. Despite some significant advances, the overall results have fallen short of expectations, and cancer remains one of the most feared causes of death. While the mapping of the human genome and decades of intensive research and development in genetics have led to groundbreaking discoveries, these have been primarily foundational rather than practical. The anticipated "personalized medicine" and therapeutic applications have yet to come to fruition. The common response is that the complexities are greater than anticipated, necessitating another 10, 20, or even 30 years of research, along with substantial funding, before we achieve meaningful progress. How long must we refrain from asking critical questions?

However, there might be a "life force" that influences our personality and well-being. If this is the case, it becomes evident how limited and superficial a purely physicalist and mechanistic worldview of ourselves and our bodies can be. Our bodies are more than just complex machines; they are more than a collection of organs, bones, and a brain. The reductionist physicalist approach, which confines our understanding of physical health and well-being to genes, cells, and chemistry while neglecting broader perspectives, has contributed to stagnation and a lack of progress in many areas of medicine. The popular belief in significant leaps in medical science often dissolves under closer scrutiny and critical analysis, revealing itself to be far more limited than conventionally assumed.

This is not to say that these sciences lack value. Medicine, rooted in pharmacology and surgery, has proven useful and will continue to be important. It has contributed to increased life expectancy and has made several deadly illnesses curable. However, a candid evaluation of past research projects reveals that the oft-touted successes and progress have been largely overstated. This slow and disappointing advancement has yet to gain collective awareness within the professional community and is often denied. The sooner we acknowledge that many aspects of modern medical research resemble a "no-progress quest" rather than a steady march forward, the better it will be for everyone. We continue to deceive ourselves by calling for more funding for research and development, more technology, and even more detailed microscopic and functional analyses. We ask for another two decades to create even more precise maps that promise to lead us to a bright future, only to find, as usual, that these maps offer little insight into the actual territory. Eventually, we will need to confront the reality that we are merely scratching the surface of a much deeper, non-material reality that lies beneath.

The rapid advancement of digital technologies and the Internet has obscured our view of the more modest progress in other fields. Much of modern science is overhyped, exaggerated, and amplified by social media. Every day, we hear about groundbreaking discoveries and inventions touted as world-changing. However, if you closely examine these so-called scientific breakthroughs, you'll find that most fade into obscurity after a brief moment of fame. Science and technology are unlikely to deliver on the expectations we often hold, and stagnation is evident across many areas.

Even if you disagree with this critical view of contemporary science, the reality is that the world is facing multiple crises: environmental, geopolitical, mental health, and financial. These issues suggest that something is fundamentally wrong, thus leaving a pervasive sense that something is missing.

Should we reconsider our perspective?

Perhaps it's time to move away from a strictly matter-centered and reductionistic view of Nature—and of ourselves—and adopt a broader, spirit or 'soul-centered' perspective. Not by viewing it as a dogma, but rather as a working hypothesis, and exploring where it takes us.

Focusing primarily on the physical properties and interactions of matter might not be sufficient for fully understanding reality. While this approach has significantly advanced scientific knowledge and technological progress, it has inherent limitations in grasping the complexity of reality and the nature of consciousness. Materialism struggles to explain subjective experiences, such as consciousness, emotions, and thoughts, solely through physical processes. The hard problem of consciousness remains a mystery, often dismissed by those who do not understand what the problem is in the first place.

Reductionism posits that complex systems can be fully understood by breaking them down into their constituent parts and analyzing their interactions. However, emergent phenomena demonstrate that the whole can be greater than the sum of its parts—a notion that challenges the reductionist viewpoint. Properties like consciousness, life, mind, and self-organization arise from complex interactions that cannot be predicted by studying individual components alone. One cannot truly understand a forest by examining it solely as an aggregation of microscopic cells devoid of life and mind.

A matter-centered science tends to focus on measurable and quantifiable aspects of reality, such as mass, energy, and physical or chemical properties. However, it overlooks the qualitative dimensions of existence, including consciousness, experience, will, agency, life's desires, goal-directedness, and, particularly, meaning, purpose, and our existential place in the universe.

We often assume that the complex interactions of neurons can explain our entire being. There is a hope that mental health issues will eventually be resolved through drugs that alter our brain chemistry. This belief suggests that mapping every single cell, molecule, and atom, along with their interactions, will provide a complete understanding of the territory.

Yet, even tiny cells are much more than complex genetic automatons; they have minds of their own. Similarly, our bodies are more than mere machines; they embody a unique consciousness. Beyond our physical form lies a subtle layer of bodily existence. The healing process not only relies on individual mental and subconscious self-suggestions but also might be influenced by external suggestions originating from a collective and universal realm. Life might extend beyond mere material and biochemical processes to potentially involve other dimensions.

As long as we dismiss these possibilities and cling to a matter-centered perspective, we risk ignoring or denying the existence of a spiritual or "psychic" dimension. This denial might lead to painfully slow progress or even permanent delays in our understanding. Once we address an illness, infection, or genetic disorder, we might find ourselves facing new physical or mental challenges that were previously unknown or unexpected. This is the nature of consciousness evolution: True progress requires inner growth as well.

For at least a century, various forms of scientism have ingrained in our collective consciousness the belief that reason, along with a strictly materialistic and reductionist approach to science, constitutes the ultimate means of acquiring knowledge. This perspective insists that everyone must align with what the analytical mind champions. Ultimately, it reflects an unaware form of anthropocentrism, as it suggests that human reason is the sole pathway to discovering truth.

In the sciences, with their emphasis on analytical thinking, we should strive to look beyond mere mental constructs, which are ultimately anthropocentric, and learn to perceive the Spirit at work behind appearances and which the physical mind alone can't recognize. By doing so, we can view all our questions in a new light, recognizing that the world's processes—along with seemingly mysterious phenomena such as mind, life, and consciousness—are not merely emergent properties of matter but, rather, are manifestations within and through it. The real question is not what actions we should take, but how we should see, feel, and perceive Nature and the World. What assumptions should we be willing to explore? If we open ourselves to the unspeakable, we might take a more significant step forward than if we had clung to outdated paradigms offering only superficial conceptual shifts that fail to capture deeper realities. It is essential to allow ourselves to perceive, feel, apprehend, and comprehend something beyond our ordinary sensory and analytical awareness and to see where that leads us.

Therefore, it is time to transition away from a materialistic—matter-centered—understanding of reality and toward a post-material, soul-centered vision. Both science and the humanities need a spirit-centered Copernican paradigm shift rather than a continued reliance on an increasingly outdated materialistic perspective that remains blind to anything that cannot be quantified or mapped.

The so-called "Age of Reason" increasingly longs for an "Age of Wisdom."

3. Endnote and Further Reading

I'm not sure if I successfully conveyed an alternative to conventional wisdom through my somewhat random essays on a post-material paradigm, which examines existence, consciousness, and the nature of reality from a different perspective. Despite my love for science, I have never fully embraced the exclusivist rational and scientific worldview that often shies away from exploring the deeper existential questions of life and Nature. This self-censoring attitude has always struck me as flawed, superficial, and limited. More than half a century later, when that eight-year-old boy, looked up at the sky and was wondering what that strange cloud might be, he still feels compelled to express a different way of thinking, seeing, and feeling.

Therefore, if you've read this far and are now at this endnote, I hope you've found something of value and wish to explore it further.

If so, as already mentioned, my suggestion is to take a look at my more in-depth analysis, which you will find in the book "Spirit calls Nature" (single vol. 3.ed)(Masi, 2025a). Other topics that were not developed here and that you will find are, for example, an overview of the "higher mind seeing," as I used to call it, of past philosophers whose ideas are rediscovered in modern times–from the ancient Greeks, to Plotinus and Meister Eckert, to modernity with Henri Bergson, Jean Gebser, and Owen Barfield. Another aspect that is frequently missing when people talk about science and spirituality is the phenomenology of natural philosophy. In particular, I focused on Goethe's way of seeing the wholeness of Nature and natural science. A third part is an extensive description of the metaphysical framework of Sri Aurobindo, which, I believe, is to date the most integral vision aligned with modern science, and that lays down a basis for a synthesis of knowledge. It further explores his vision, revisiting some aspects discussed in this volume. I also suggest taking a look at the reading suggestions in the final pages, as they can provide further insights into the connections between science, philosophy, and spirituality.

If, instead, you are more interested in the conceptual foundations of quantum mechanics beyond a simplistic popular guide, but desire something comprehensible like a high school-level vademecum, or something for an undergraduate or graduate philosopher who doesn't want to go beyond high school math, you might want to check out my two-volume book series, *"Quantum Physics: An Overview of a Weird World"* (Masi, 2019 a&b). It is a comprehensive guide that will furnish you with all the basics necessary to deal with the deeper philosophical questions of quantum mechanics. The first volume starts with the origins of quantum physics throughout the 20th century. The second volume goes beyond classical quantum physics to describe recent discoveries in the 21st century. Once you have gone through especially the first volume, you will be in a much better position to appreciate the conceptual subtleties that distinguish the quantum world from the classical world to better appreciate the philosophical implications that have been elaborated in Chapters IV-VII of the present volume.

However it is, as I pointed out in the beginning, here again, I would like to emphasize that what really matters is not so much that we converge on the technical details but rather that the reader can sink into the mentality and intuitional view that goes beyond the materialistic conceptualizations of reality. I hope this perspectivist approach will trigger some to look upon the world and life, guided by both reason and intuition, beyond the present paradigm and participate in the construction of a post-material science, society, and way of life. The aim is an integration that offers a bridge transcending the limitations of each domain, fostering a more comprehensive understanding of the human experience and the depths of our own being by cultivating awe and wonder. We must seek wisdom from diverse sources and foster a sense of interconnectedness that transcends cultural, religious, and disciplinary boundaries, finally going beyond all of them.

I hope that this book has served as a catalyst, sparking further reflections and inspiring further inquiry. This journey marks the beginning of an ongoing exploration. May this exploration foster a deeper appreciation of the intricate dance of science, philosophy, and spirituality inspiring us all to embrace a more holistic and post-material vision of our place in the grand tapestry of existence.

With humility and curiosity,
Marco Masi

XIII. Advanced Readings

The present chapter contains several of my published articles in peer-reviewed academic journals. I created a separate chapter of advanced readings to provide a deeper theoretical and technical background for the more experienced reader, while keeping the main chapters accessible to the general reader. This structure allows those who wish to explore the subject in greater depth to do so without overburdening the discussion with complex details or specialized terminology. I hope that, by isolating the more advanced material, the core chapters remain clear, focused, and easier to follow, ensuring that the broader audience can grasp the essential ideas without being distracted or discouraged by excessive complexity.

1. An Evidence-Based Critical Review of the Mind-Brain Identity Theory[141]

Abstract

In the philosophy of mind, neuroscience, and psychology, the causal relationship between phenomenal consciousness, mentation, and brain states has always been a matter of debate. On the one hand, material monism posits consciousness and mind as pure brain epiphenomena. One of its most stringent lines of reasoning relies on a 'loss-of-function lesion premise,' according to which, since brain lesions and neurochemical modifications lead to cognitive impairment and/or altered states of consciousness, there is no reason to doubt the mind-brain identity. On the other hand, dualism or idealism (in one form or another) regard consciousness and mind as something other than the sole product of cerebral activity pointing at the ineffable, undefinable, and seemingly unphysical nature of our subjective qualitative experiences and its related mental dimension. Here, several neuroscientific findings are reviewed that question the idea that posits phenomenal experience as an emergent property of brain activity, and argue that the premise of material monism is based on a logical correlation-causation fallacy. While these (mostly ignored) findings, if considered separately from each other, could, in principle, be recast into a physicalist paradigm, once viewed from an integral perspective, they substantiate equally well an ontology that posits mind and consciousness as a primal phenomenon.

1. Introduction

Since the times of René Descartes in the 17th century, the mind–body problem has been one of the central debates in the philosophy of mind, psychology, and neuroscience. The conventional Cartesian dualism is no longer considered tenable but other forms of dualism, or theoretical frameworks of philosophical idealism, or more generally, non-physicalist ontologies, state that mind and consciousness cannot be explained as a mere result of neural processes.

Dualism is opposed by an identity theory, which, instead, considers mind processes as identical to brain processes, and consciousness as nothing other than an emergent epiphenomenon arising from the collective interaction of the neuronal activity. Sentience, with all its subjective dimensions of experiences,

[141] The present article was published in *Frontiers in Psychology*, Sec. Consciousness Research, vol. 14. DOI: **https://doi.org/10.3389/fpsyg.2023.1150605**

feelings, and thoughts, is a physical process determined only by the laws of physics. Qualia–the subjective, phenomenal, and mental experiences we can access only introspectively, such as the perception of color, or that of pain and pleasure–are physical brain states, while any speculation concerning an immaterial mind or consciousness is considered an unnecessary hypothesis.

Dualists and monists have different schools of thought but, despite the variety of opinions, it is fair to say that most scientists and philosophers consider themselves to be material monists. For example, according to a survey (Bourget and Chalmers, 2020) 51.9% of philosophers declare themselves 'physicalists' vs. 32.1% as non-physicalists, and 15.9% as 'other'. On the other hand, exceptional human experiences occur frequently in both the general population and in scientists and engineers (Wahbeh et al., 2018).

However, there is a growing awareness that a mere functional investigation will not answer questions of a more philosophical nature. The belief that the progress of modern neurosciences would soon shed light on David Chalmer's notorious 'hard problem of consciousness' (Chalmers, 1995) has turned out to be too optimistic. This is because, unlike other physical processes, in which both causes and effects can be observed from a third-person perspective, in consciousness studies, one is confronted with a cause–the brain activity–that one can still analyze from a third-person perspective that, however, apparently produces an effect we call 'conscious experience,' or just 'sentience,' which can be apprehended only from a first-person perspective. This 'perspectival asymmetry' makes consciousness in its subjective and experiential dimension stand out as a phenomenon alien to any attempt at conceptual causal and ontological scientific reduction. Inside a naturalistic framework, the origin and ontology of the phenomenal subjective conscious experience remain unclear.

While most arguments were based on a physicalist line of reasoning (for a review, (see Seth and Bayne, 2022)), and also other post-materialistic models of consciousness that are not exclusively based on brain activity exist (for a review and discussion (see Wahbeh et al., 2022)), here it is shown that there are also strictly neuroscientific facts that have not received sufficient appreciation and that give us good reasons to look upon the physicalist assumptions with a more critical eye. Non-neurocentric paradigms of consciousness that posit mind and consciousness as a fundamental primitive, rather than matter, remain a viable option. No particular dualistic, panpsychist, Eastern philosophical, or metaphysical scheme is favored. Rather, a variety of findings, especially when seen jointly and in their relationship to each other, could suggest other possible ways of interpreting the neuroscientific findings, and that this might even have more explanatory power in terms of an underlying post-material ontology.

A preliminary note of conceptual and terminological clarity is necessary. In psychology, or the philosophy of mind, and neurological sciences, the words 'consciousness,' 'mind', and 'self-awareness' are defined and used with different significances, sometimes with overlapping or conflating semantics. In fact, for historical reasons, the mind-brain identity theory used the terms 'mind' and 'consciousness' somewhat interchangeably (Smart, 2022). Here, however, 'consciousness' will relate to phenomenal consciousness–that is, Nagel's famous 'what-it-is-like' states (Nagel, 1974) underlying our subjective qualitative experiences, 'qualia,' that what makes us sentient of perceptions, feelings, sensations, pleasures or pains, and self-aware as a unified subject. Phenomenal consciousness is not to be confused with 'mind' which, at least in the present context, relates to the cognitive functions of thought, memory, intelligence, ideas, concepts, and meanings. The two are to be kept distinct in the sense that the mind's thoughts come and go, while the conscious experiencing subject is permanent. I deem this distinction necessary because the question relating to the physicality of the spectrum of all our psychological dimensions, as we are going to see later, may not have a unique answer. For example, one can argue for the unphysical nature of phenomenal consciousness but maintain that memory is in the brain, or that low-level cognition (e.g., sensory perception modalities) are neuronal epiphenomena, while other high-level functions (decision-making, agency, reasoning, and planning) are not.

Having made this distinction, in the following, I will first examine more closely the logical framework that sustains a mechanistic conception by pointing out some conventional neurological causation-correlation fallacies.

et us first question some basic assumptions. Does the physical change of a brain state leading to cognitive impairment or altered states of consciousness provide a necessary and sufficient logical proof that mind and consciousness are an emergent cerebral phenomenon?

After all, it is undeniable that there is a direct relation between the physical state of our brains and our subjective experiences (e.g., Aguinaga et al., 2018), (Vollenweider and Preller, 2020), (Davis et al., 2008). Dopamine is a neurotransmitter molecule that enables biochemical transmission among neurons and that is responsible for the effects of a drug like cocaine. We know that psychedelic drugs can lead to intense subjective effects. It is a well-known fact that brain damage can lead to severe cognitive impairments. If Broca's area, a left cerebral hemisphere area, is lesioned, one loses the ability to speak (interestingly, though, not the ability to comprehend language). Someone being anesthetized using anesthetic drugs (seemingly) 'loses' consciousness. And nowadays, we have a number of sophisticated brain scan technologies making it clear, beyond any reasonable doubt, that for every conscious experience, there exists a neural correlate in our brains.

Thus, apparently, a neuroscience that is based on brain chemistry and loss-of-function lesion studies leaves no place for any form of non-material monistic approach. Mental states and conscious self-awareness seem to emerge from matter; there is no distinction. Our personalities, identities, moods, and states of consciousness seem to depend on the biophysical state of our brains.

And yet, few further critical thoughts should make it clear that such a correlation is not a sufficiency criterion. One must secure one's theoretical framework from a possible logical fallacy believing that correlation implies causation. The fact that two events are always coincidental or always happen shortly, one after the other, does not imply that the first event caused the second event to happen. If event B always follows event A, we are not entitled to conclude that A is the cause of B. These sorts of logical fallacies are known as 'post-hoc fallacies'.

Nevertheless, the necessity and sufficiency that the explanation of our qualitative experiential dimension is to be chiefly found in neural circuits remains a rarely questioned belief [with few exceptions, e.g., in the field of behavioral processes (Gomez-Marin, 2017)]. There is a general tendency to believe that causal mechanistic explanations based on neural lower-level properties are better than higher-level behavioral accounts. For example, Krakauer et al. pointed out that neuroscientists (and, I would add, too many psychologists and most analytical philosophers of mind) frequently use language to hide more than to reveal, by assuming that a neural causal efficacy equals understanding–that is, charging it with an explanatory power it does not have. The result that "neural activity X is necessary and sufficient for behavior Y to occur" allows a causal claim often added by a further explanatory sentence that rearticulates the same causal result employing 'filter verbs' (such as "produces, "generates," "enables," etc.) and that, however, masks the faulty logic to cause a metaphysical position to pass as empirical data (Krakauer et al., 2017).

But, what are the alternatives to the mind–body identification that could be in line with the above correlation between mental states and physical neural correlates of consciousness?

In fact, the metaphor most idealists prefer is the 'filter theory of consciousness,' which dates back to an original idea of William James, who stated: "My thesis is now this: that, when we think of the law that thought is a function of the brain, we are not required to think of productive function only; we are entitled also to consider permissive or transmissive function. And the ordinary psycho-physiologist leaves this out of his account" (emphasis in the original text) (James, 1898).

James thought of the brain and thought in the frame of a 'bidirectional transducer theory' using the analogy of the prism separating white light into respective colored beams. If a broken prism fails in its function to 'reveal' the colored light beams, this should not lure us into the logical correlation-causation fallacy that the prism 'produces' colored light. The material and structural modification of the optical medium modifies the refractive gradient that 'transduces' light with a different chromatic dispersion but does not 'create' it. A prism is just an object with a transmissive function; it does not 'generate' anything.

Aldous Huxley expressed a similar idea and proposed that the brain is a 'reducing valve' of what he called a 'Mind at large,' a universal or cosmic Mind comprising all of reality with all ideas and all thoughts. According to Huxley, our mind filters reality under normal conditions because, otherwise, we would be overwhelmed by the knowledge of this universal Mind. Psychedelic drugs can remove the filter and bring us into contact with the Mind at large, leading to the experiences that several mystics describe. In his words: "To make survival possible biologically, Mind at large has to be funneled through the reducing valve of the brain and nervous system" (Huxley, 1954). For Huxley, the brain was a material 'connecting device,' an 'interface' or 'relay station.' In this view, human mind is a localization of a universe-wide Mind projected into our brains. The brain filters and suppresses this universal Mind but does not 'produce' it.

An understanding of the mind-brain relationship reminiscent of Eastern philosophies, and that maintains similar views, is neatly summarized by the Indian mystic and poet Sri Aurobindo: "Our physical organism no more causes or explains thought and consciousness than the construction of an engine causes or explains the motive-power of steam or electricity. The force is anterior, not the physical instrument" (Aurobindo, 1919).

From these perspectives, mind uses the brain as an instrument, as an interface of expression. Mind and consciousness are constrained and interdependent from the brain but aren't generated by the instrument itself.

Notice that this standpoint is not entirely alien to our ordinary understanding of how a digital computer works. Knowing everything about its hardware, and recreating its exact physical structure in every detail, would not lead us to a machine that makes anything meaningful or useful. Software–that is, a running code written by an intelligent external agent–is needed. Here, also, a computer is only an instrument, a means of expression for a cognitive entity, not its origin or source. In fact, studying a microprocessor with the same criteria employed by modern neuroscience, trying to reverse-engineer its functions by analyzing local field potentials, or selectively lesioning its units by correlating this with its behavior, would turn out to be a quite difficult task: We would still have a long way to go to explain how it works and figure out the whole running code, which is the real 'agent' causing the behavior of the machine (Jonas and Kording, 2017).

Thus, neural correlates of consciousness, or loss-of-function lesion-based studies, do not constitute a sufficient logical foundation for a mind-brain identity theory. We have the right to maintain the contrary hypothesis: Consciousness, mental states, and emotional states are more or less 'funneled through' depending on the physical state of a brain. The brain could equally well be seen as a physical substrate through which these conscious states manifest without leading to any inconsistency with current scientific knowledge. How current neuroscience not only fails to falsify this hypothesis but maybe even suggestive of this claim, is the purpose of the next section, with a review of old and new neuroscientific findings that are asking for clarification if one wants to save the mind-consciousness-brain identity theory. Another part will review the evidence for the neural correlates of memory. A brief section will focus on the emergent fields in the study of plant and cellular 'basal cognition.' A discussion and concluding remarks will follow.

2. From (lack of) evidence to interpretation

2.1. The search for the 'seat of consciousness'

Crick and Koch once postulated that the claustrum, a sheet-like neuronal structure hidden beneath the inner surface of the neocortex, might give rise to "integrated conscious percepts"–that is, act like the "seat of consciousness" (Crick and Koch, 2005). Modern neuroscience, however, indicates that the claustrum behaves more like a neuronal information router than an organ responsible for a specific function (Madden et al., 2022). To date, there is no evidence, not even indirect or circumstantial, of a single brain region, area, organ, anatomical feature, or modern Cartesian pineal gland that takes charge of this mysterious job of 'producing' or 'generating' consciousness. Most of the brain is busy processing sensory inputs, motor tasks, and automatic and sub- or unconscious physiological regulations (such as the heartbeat, breathing, the control of blood pressure and temperature, motor control, etc.) that do not lead to qualitative experiences. Neural activity alone cannot be a sufficient condition to lead to phenomenal consciousness. The vast majority of brain activity is unconscious–that is, non-conscious cognitive processes (e.g., mnemonic, perceptual, mental or linguistic tasks) and physiological processes (e.g., cardiac, hormonal, thermal regulation, etc.) taking place outside of our conscious awareness. This raises the question: What distinguishes a neural process that leads to a conscious experience from that which does not?

For example, the cerebellum is almost exclusively dedicated to motor control functions, and its impairment leads to equilibrium and movement disorders. However, it does not affect one's state of consciousness. Its role in 'generating' experience seems to be marginal, if any. There are also rare cases of people who live without a cerebellum ('cerebellar agenesis') and have only mild or moderate motor deficits or other types of disorders (Feng et al., 2015). This is a fact that seemingly confirms the brain's proverbial neuro-plasticity, which we will see next through other extraordinary examples.

It may be worth recalling that the neuronal architecture in our bodies is not confined to the brain–that is, it goes far beyond our heads, through the brain stem, and down through the spinal cord. The central nervous system is made up of the brain and the spinal cord. The latter is responsible for the transmission of nerve signals from and to the motor cortex; as is well known, injury to it can result in paralysis. But, again, no cognitive deficit or state of consciousness is altered by impairments of the spinal cord. This leaves only one option: If there is a 'seat of consciousness,' it must be identified somewhere in the cerebral cortex or subcortical areas of the brain (Fig. 1).

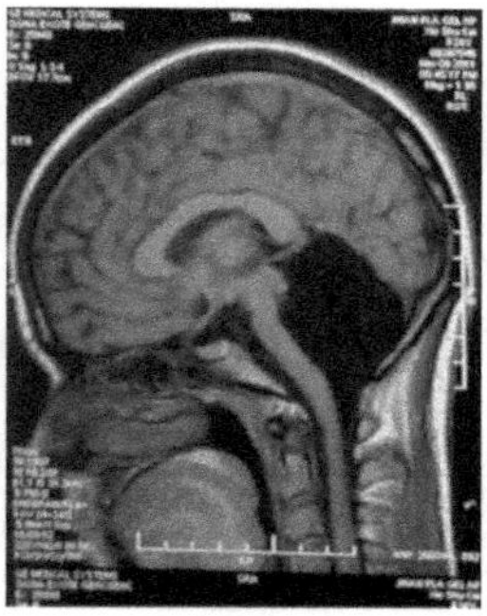

Fig. 1 Case of cerebellar agenesis: Living (and walking) without the cerebellum. Credit: Feng & al. (2015).

Another interesting example of how the correlation-causation fallacy conditions scientific and popular understanding of the mind–body problem can be illustrated by an interesting experimental finding that showed how stimulation of the thalamus arouses macaques from stable anesthesia (Redinbaugh et al., 2020). The awake, sleeping, and anesthetized states could be aroused with the stimulation of the central lateral thalamus. The straightforward conclusion seemed clear. The ultimate origin and switch 'modulating' consciousness was discovered. If your consciousness 'depends' on the state of your thalamus, which is 'switched' on and off with the touch of a button, then the thalamus must be the 'seat of consciousness.' Is this an unavoidable conclusion?

First of all, observing from a third-person perspective the absence of an external physiological signature as evidence for a lack of internal first-person sentience is yet another correlation-causation fallacy that has too frequently led to unwarranted conclusions. For example, that anesthesia induces an unconscious state with the patient having no subjective experience is far from obvious. We simply do not know if it really induces a completely unconscious state or a conscious but non-metacognitive no-report state that makes one unable to recall past experiences once one is back in the waking state. The former assumption is, unfortunately, taken in most cases as the standard scientific approach. Whereas, indications suggest that anesthetic-induced unresponsiveness does not induce complete disconnectedness (Radek et al., 2018; Turku, 2018). Interestingly in this regard is also the so-called twilight anesthesia, an anesthetic technique that sedates patients only mildly and induces amnesia but no loss of consciousness (Scheinin et al., 2020). During this 'twilight state,' patients are responsive and can be asked to perform some tasks that they will not be able to recollect after the surgery. This case alone shows that the inability to recall events during sedation is no proof of unconsciousness.

Moreover, there is now a non-negligible amount of scientific literature, presenting empiric evidence on parasomnia (sleepwalking), hypnosis, non-REM sleep, and subjects in a vegetative state, that some form of conscious awareness is also present in all these non-responsive states of consciousness (e.g., Owen and Coleman, 2006; Oudiette et al., 2009; Cruse et al., 2011; Siclari et al., 2018; Mackenzie, 2019). Arguing and extrapolating from the lack of superficial physical cues and mnemonic retention to a verdict that declares someone to be 'unconscious'–that is, as having no subjective phenomenal experience–is, at least from the philosophical perspective, again betraying a logical correlation-causation fallacy.

But even if we assume that there is no internal experience when we are anesthetized, the relevant question remains: Do these sorts of experimental findings confirm that the thalamus is the 'seat of consciousness'? Is it a sort of modern replacement for Descartes' pineal gland in its mechanistic-material monist version?

The thalamus is responsible for sensory information processing. It is known that its main job is to function as a relay and feedback station between sensory brain areas and the cerebral cortex. For example, it functions as a hub between the optical nerves that transport the visual information coming from our retinas to the visual cortex. Even if one remained conscious by turning down the functionality of the thalamus, one would no longer see anything because the neural pathways between the retina and the visual cortex are interrupted. From that, however, nobody would conclude that the thalamus is the seat of the visual experience for which the visual cortex is responsible, as we know that it is a 'hub,' a 'transducer' or a 'filter.' From this perspective, the thalamus' function is to 'integrate' the information flow of the several brain areas; if this is disrupted, it leads to a 'loss' of consciousness.

Thus, these findings do not tell us much about the generation of conscious experience. However, if there is not one single 'seat of consciousness,' could it be that the combination and activity of some or all of the different brain areas do 'produce' the subjective experience? Considerable attention in this direction has been focused on theories such as the 'Integrated Information Theory' (IIT) (Oizumi et al., 2014; Tononi, 2015) and the 'Global Workspace Theory' (GWT) (Baars, 1988), according to which the amount and integration of information and the momentarily active and accessible memory determine the level of consciousness leading to a conscious entity. A process of integrating the information and the memory coming from all the brain areas may be the efficient cause of our experiential richness. In fact, we have sufficient evidence that compels us to abandon this simplistic view of a compartmentalized brain, with modern neuroscience thinking more in terms of network science, in which several brain regions are highly interconnected and interdependent. No brain region does only one thing, and no neurons supposedly have only one function. Most neurons have several functions, not a single purpose. It turns out that whenever we hear a sound, have a visual experience, have feelings or emotions, or perform a motoric task, the whole brain is involved. Even such an apparently highly specialized brain region as the primary visual cortex carries out information processes related to hearing, touch, and movement (Merabet et al., 2008; Liang et al., 2013). The reason why we nevertheless tend to associate specific brain regions with specific cognitive, sensorial, or motoric functions is that brain scans show only a temporal snapshot of the brain's most intense activity. We are seeing only a few 'tips of the iceberg' and missing the overall activity in the noise. When studies are conducted using less noisy but much more expensive and complicated detection methods, most of the brain's activity becomes visible (Gonzalez-Castillo et al., 2012). Therefore, it would seem plausible that if consciousness arises from the activity of a complex aggregation of neurons, at least some brain areas must work together in a unified whole via thalamic activity.

However, how far these conjectures align with reality is questionable.

Because a natural question could be that of asking if and how a subjective feeling of selfhood changes if someone were to split your brain into two parts? Would you feel somewhat less conscious and less 'yourself'? As is well-known, this is a very real surgical procedure performed since the 1940s: the corpus callosotomy (although used only rarely nowadays). It is performed to treat the worst cases of epilepsy (patients having up to 30 seizures a day) that did not respond to medical treatment. In this procedure, the corpus callosum, the nerve tract connecting the left and right brain hemispheres, is severed (in part or, in some cases, entirely), thereby avoiding the spread of epileptic activity between the two halves of the brain (Figure 2). Its natural function is to ensure communication between the two cerebral cortexes of the two hemispheres to integrate and coordinate motor, sensory, and cognitive functions, such as moving left and right limbs, the visual integration of the left and right sight, etc. Because most of the brain's activity is distributed throughout both hemispheres, with no indication of one or the other part being responsible for generating our sense of 'self,' one must wonder how the patients who have gone through such an acute surgical intervention feel. Do their split brains 'generate' a dual consciousness and split personality?

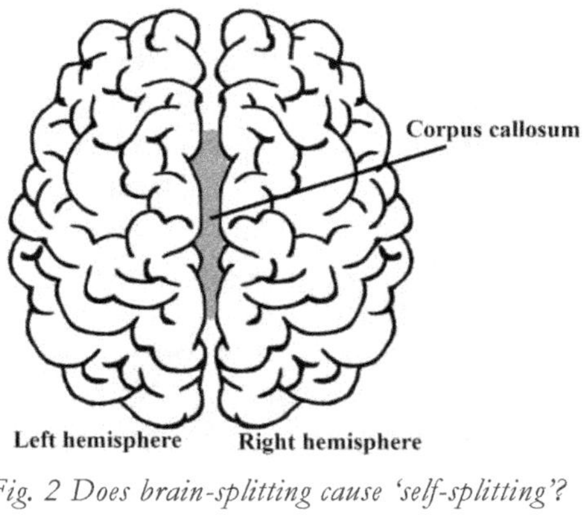

Fig. 2 Does brain-splitting cause 'self-splitting'?

Disagreement exists about whether in these patients a subject unity is present or if they display any signs of multiple first-person perspectives (De

Haan et al., 2020). They deny being a different person from what they were before surgery, and close relatives who knew the split-brain patients before and after surgery do not notice any personality change (Bogen et al., 1965; Sperry, 1968, 1984; Pinto et al., 2017),

Of course, there can be more or less severe drawbacks. In some cases, the so-called 'alien-hand syndrome' can take over, in which one hand appears to have a mind of its own. This occasionally happens when the two hemispheres' representations of reality come into conflict and one wants to override the other. In these instances, decision-making and volition between the two hemispheres clash. An example is the patient's struggle to overcome an antagonistic behavior, such as knowing what cloth they want to wear, while one of their hands takes control and reaches out for another cloth they do not want at all. However, this should not be confused with two personalities competing against each other (as in the case of dissociative identity disorders), as split-brain patients identify with only one body and perceive their disobedient limb as being subjected to annoying motoric misbehavior; they do not report any sensation of some other internal personality taking control. The brain–or, more precisely, our two brains–tell us two different 'stories.' Split-brain patients seem to identify with one of the stories–that is, consciously access one of its interpretations–and keep the other in a subconscious or subliminal awareness, what the American cognitive neuroscientist Michael Gazzaniga used to call the 'left-brain interpreter.'

Recent investigations also question the canonical textbook findings (Pinto et al., 2017, 2020). While it is confirmed that a corpus callosotomy splits the visual perception of the environment in two, several patients can nevertheless see them both and report it to the outside world–that is, they can access their language centers. Moreover, there is no evidence for memory loss (Forsdyke, 2015).

In my view, confusion surrounding split-brain psychology arises only if we conflate the 'unity of mind' with a 'unity of consciousness' and sense of selfhood. If we do not confuse mental states as being the origin or efficient cause of consciousness, then any apparent paradox dissipates. Split-brainers may have two (eventually even conflicting) hemispheric and motor-sensory mental states (something not entirely unusual in healthy subjects) but even if one argues and provides evidence for a 'two-minds' model, that would not imply a split sense of identity or self-awareness. One can consciously and subliminally be aware of a plurality of experiences, yet retain the experience of singularity. There can be several experiences and representations generated in a brain, with or without a representational unity, which, nevertheless, belongs to and is experienced by one subject [for a more detailed analysis of this point (see De Haan et al., 2021)]. A 'split subjective identity' resulting from split-brain in the sense of a symptomatology similar to what we know from dissociative identity disorder characterized by the disruption of identity in two distinct personalities, differing not just in sensory-motor functioning or depersonalization disorders, but also each with two psychological behaviors, characters, affects, social preferences, and experienced as alternating 'possessions' with cognitive discontinuities and different memories of autobiographical information, as observed by others and reported by the (alternating) subjects themselves, is not observed.

So, if our subjective and conscious experience is generated by the integrated activity of the whole brain, why does not such a radical bisection lead to any modification of our state of awareness? Given the severing of the corpus callosum of a brain, one would expect a loss or at least a diminishing of conscious awareness because there would be a loss of working memory and information integration. However, nothing like this happens. The 'unity of consciousness' remains unaffected and, thereby, unexplained.

To save the paradigm, those who endorse the view that in such brain condition consciousness can no longer be 'integrated', point out that in not all documented cases was a complete transection of the corpus callosum performed. The truth, however, is that in several cases, the complete sectioning was performed and even confirmed by MRI imaging or radiological means (Gazzaniga, 1985).

Yet, one may still point out that a complete transection still leaves some residual subcortical structures intact, which allows for some communication between the two hemispheres, potentially maintaining the 'self' of the patients.

To further substantiate the contrary hypothesis, one could mention cases in which there is no second hemisphere to communicate with in the first place. To treat epilepsy, the most extreme surgical intervention is to remove an entire brain hemisphere, that is, by hemispherectomy. Usually, this is done only in childhood because, supposedly, young brains can rewire themselves much more efficiently than older ones. Figure 3 shows the fMRI in a sample of six rare high-functioning patients after partial or complete surgical removal of one cerebral hemisphere.

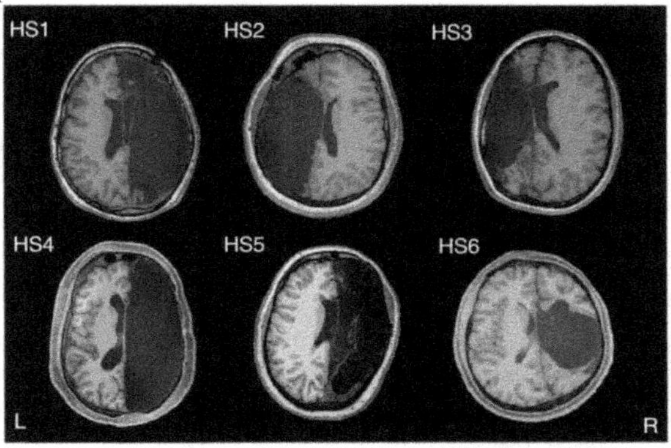

Fig. 3 Hemispherectomy Brain Anatomy - Six adults with left (HS2 and HS3) or right (HS1, HS4, HS5, and HS6) hemispherectomy. Credit: Kliemann et al. (2019).

Interestingly, Nature seems to take the left/right distinction and early plasticity hypothesis not so seriously. That the left–right brain task distribution is not an inescapable neurological dogma is testified to by people born with only one hemisphere. For example, while in healthy subjects the left visual field is represented in the right hemisphere and vice versa, someone born with only one hemisphere can develop maps of both visual fields in it (Muckli et al., 2009). Hemispherectomy on adults older than 18 years turns out to be just as safe and effective as in early childhood (McGovern et al., 2019). Even in the case of a left hemispherectomy, Broca's language area–which in normal conditions is in the left hemisphere–can be recovered in the right part of the brain (Vargha-Khadem et al., 1997). Further evidence reports of subjects in whom the frontal lobe was missing from childhood without any measurable linguistic impairments, as shown by the case of a woman who grew up without her left temporal lobe but speaks in English and Russian (Tuckute et al., 2022; Figure 4). This does not mean that persons missing a hemisphere do not suffer consequences–there is suboptimal word and face recognition (Granovetter et al., 2022) but whether it plays a role in the unity of consciousness remains to be seen.

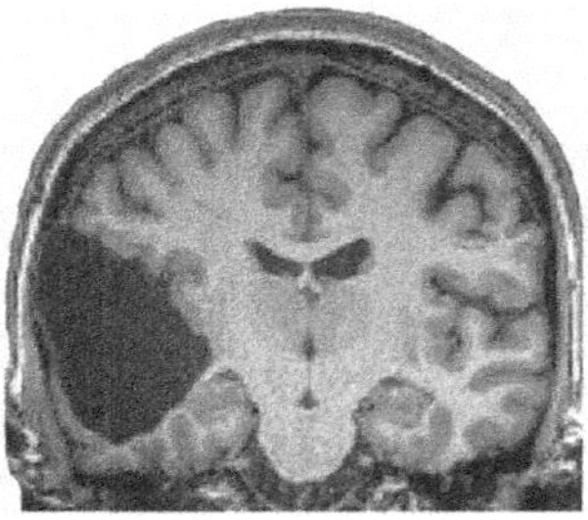

Fig. 4 Speaking without the brain's language area.
Credit: Tuckute et al. (2022).

A possible explanation is that because these patients already had severe seizures originating in one of the hemispheres, the functional rewiring on the other hemisphere began before the surgery. The findings tend to disconfirm this easy way out. Though interconnectivity inside the brain networks increases, interconnectivity between brain regions with the same function after hemispherectomy does not differ from that of two hemispheric control subjects (Kliemann et al., 2019). That plasticity alone can explain this state of affairs is far from proven (more on this later).

However, it is, most patients become seizure-free, and their cognition is relatively unchanged after surgery (some motoric and cognitive functions decrease but others improve). Overall, these patients appear to be 'normal.' Cognitive measures typically changed little between surgery and follow-up (Pulsifer et al., 2004), and in everyday life, one could not tell the difference between humans having a whole brain or only half of one. And, most notably, the subjects report no 'half-self,' 'half-awareness,' or 'half-consciousness.'

If the mind-brain identity theory is correct, and consciousness emerges as an integration of functional centers, with no particular 'seat of consciousness,' then only one brain hemisphere must be sufficient to accomplish the task.

But instances are found in which both hemispheres are severely damaged and there is not much left to integrate. Worth a reminder is how, in 1980, the British pediatrician John Lorber reported that some adults cured of childhood hydrocephaly had no more than 5% volume of brain tissue with a cerebral cortex as thin as 1 mm (Lewin, 1980). While some had cognitive and perceptual disorders and several developed epilepsy, others were surprisingly asymptomatic and even of above-average intelligence.

Then, in 2007, in Marseille, France, a 44-year-old man complaining of weakness in his left leg submitted to an MRI brain scan (Feuillet et al., 2007). As Figure 5 shows, the skull was abnormally filled with cerebrospinal fluid, leaving only a thin sheet of actual brain tissue. As an infant, he'd had a shunt inserted into his head to drain the fluid but it was removed when he was 14. Evidently, the cerebrospinal fluid build-up did not stop and ended up reducing the brain's size to 50–75% compared to its normal volume. Though he had a below-average IQ (75/100), this man had a job, a family, and a normal life.

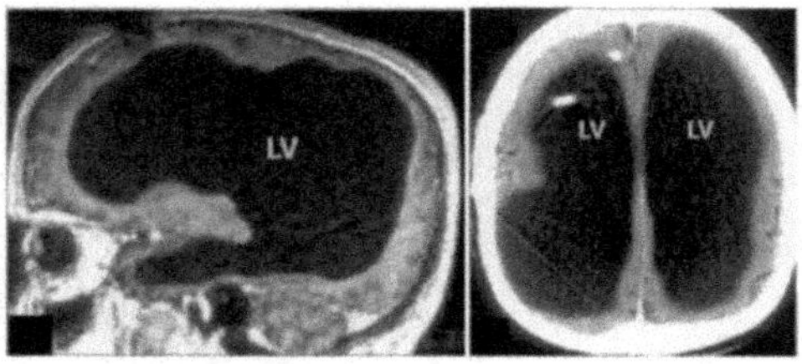

Fig. 5 MRI image of a hydrocephalus brain. Credit: Feuillet et al. (2007).

Another example that should raise doubts is the cases of children in a developmental vegetative state–that is, what the American Academy of Neurology (as declared in its guideline report in 1995 and confirmed in 2018) officially considers as being a neurovegetative state in which there is "no evidence of purposeful behavior suggesting awareness of self or environment" (Giacino et al., 2018). In other words, a universal rule reduces them to unconscious children who cannot suffer because this supposedly requires a functioning cerebral cortex.

Nevertheless, only one case showing the contrary should be sufficient to disprove a universal rule. Four such cases were brought to light in 1999 by a group led by Shewmon et al. (1999). They studied the states of awareness in congenitally decorticate children–that is, the cases of four children who were almost completely lacking cortical tissue and were neurologically certified as being in a vegetative state. Yet, the loving care of their mothers (or of someone who adopted them and bonded with them via dedicated full-time caring) could gradually 'awaken' in them a conscious awareness. From an initially unresponsive state, they showed clear signs of having developed auditory perception and visual awareness (despite the total absence of the occipital lobe that, in normal conditions, hosts the visual areas). For example, they tracked faces and toys, looked at persons they recognized, could distinguish between their mothers or caretakers, listened to music for which they manifested preferences with their facial expressions, including smiling and crying, and, at least in one case, gave clear indications of self-recognition in a mirror. Shewmoon notes: "Were they [the decorticate children] not humans studied by clinicians but rather animals studied by ethologists, no one would object to attributing to them 'consciousness' (or ability to 'experience' pain or suffering) based on their evident adaptive interaction with the environment."

These cases seem to contradict the prevailing theory, according to which the cerebral cortex generates consciousness.

One can still point out that the children were not completely decorticated, as some cortical tissue was still left. Figure 6 shows that a remnant of the frontal lobe is still present, possibly producing the conscious awareness. But that neural mechanisms of conscious function cannot be confined to the cerebral cortex alone is becoming much more plausible (Merker, 2007).

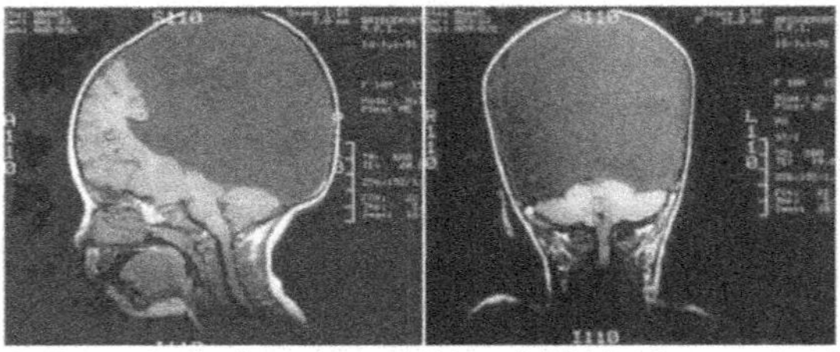

Fig. 6 Congenitally decorticate children MRI brain scan (midline sagittal and posterior coronal plane). Credit: Shewmon et al. (1999).

In fact, other speculations now retire to the last cerebral bastion for the seat of consciousness: the brainstem (Solms and Panksepp, 2012). Indeed, its stimulation can trigger intense emotions and feelings. But the question is: What property of a neural circuitry dedicated to the most physical and basal control of cardiac, respiratory, and homeostatic functions, containing mainly neurons for motor and sensory tasks, can also give rise to such an apparently immaterial and completely different and unrelated 'function' or 'property' as a conscious experience? We do not know. However, this is yet another fact telling us that we have the right, at least hypothetically, to assume that they do not and are equally allowed to study these facts in the light of a different paradigm than that of a mind-brain identity.

Overall, the cases mentioned above (except for those of the congenitally decorticate children) of people who have undergone corpus callosotomy or hemispherectomy, or people suffering from hydrocephalus, cerebellar agenesis, or several other types of brain damage, show how surprisingly intact their higher cognitive functions remain. One would expect that the first victims of such invasive neurological changes or surgical interventions would be the complex and high-demanding cognitive functions so characteristic of the mind, such as intellectual skills, abstract thinking, decision-making, reason, logically and willfully planning actions, and so on. Instead, it turns out that even if large brain masses are injured or absent, the cognitive skills of the subject remain substantially unaltered. Further empirical inquiry is needed to show if the same holds for the integrity of subjective experience and no altered states of consciousness or qualitative changes of sensory perception arise.

2.2. Further questions on the mind-brain relationship

These remarkable cases also confirm that brain size and the number of neurons in a brain do not (or, at least, do not necessarily) indicate one's intelligence. Size matters for manipulative complexity, such as the more complex hand movements in primates, which humans can develop superbly (think of the hands of an expert musician playing piano; Heldstab et al., 2020). However, a direct correlation between brain size and mental skills is not that straightforward. We like to believe that our brain size makes us human but rarely do we question what one means by 'size.' The number of neurons? The weight of the brain? Its brain-to-body ratio? Or its volume? Humans do not have the largest brain size in any of the aforementioned senses. The human brain has about 90 billion neurons, weighs ca. 1.1 to 1.4 kg, and has a volume of 1,300 cm3. However, the brain of an elephant has three times the number of neurons we have, and the weight and volume of the brain of a sperm whale are six times as much. Meanwhile, ants have a six times larger brain-to-body-mass ratio. A bit of an extreme example showing how cognitive skills and brain size are decoupled is the case of mouse lemurs, whose brains are 1/200th the size of monkeys' but that perform equally well on a primate intelligence test (Fichtel et al., 2020) Therefore, brain size alone does not make for a more developed mind, either while brain size does not scale with memory information content (Forsdyke, 2014; Forsdyke, 2015). Then what does?

It is plausible to assume that a certain degree of complexity is a mandatory factor for a brain or whatever material structure to display a form of intelligence and cognitive skills. One could think of a measure of 'brain connectivity'–that is, the number of wirings between neurons (through their axons, dendrites, and synapses) and the speed at which they transmit and receive signals–as an indicator of its complexity and see if it somehow scales with the cognitive functionality. However, MRI studies reveal that all mammals, including humans, share equal brain overall connectivity (Assaf et al., 2020). The efficiency of information transfer through the neural network in a human is comparable to that of a mouse. It is independent of the structure or size of the brain and does not vary from species to species. So, things cannot be as easy as that.

However, what the above-mentioned clinical cases have in common is the presence of the cerebral cortex. In fact, some neurologists or cognitive scientists conjecture that phenomenal consciousness resides in the cerebral cortex. This belief is not unproblematic either.

First of all, because the neocortex exists only in humans and other mammals, one must conclude that birds, fish, octopuses, amphibians, and reptiles are, per definition, all 'unconscious' and incapable of having some more or less elementary form of conscious subjective experience. There is no sentience; they do not feel pain, fear, or pleasure or have whatever feeling. They are considered Cartesian automatons or philosophical zombies.

But evidence is beginning to emerge that, for example, the neural correlate patterns of sensory perception in a corvid bird aren't substantially different from the neural correlate patterns in humans having a similar sensory conscious subjective experience (Nieder et al., 2020). Moreover, one wonders how some birds can also perform amazing cognitive feats despite their forebrains consisting of lumps of gray cells. It turns out that cortex-like circuits in avian birds exist that are reminiscent of mammalian forebrains, and the idea that advanced cognitive skills are possible only because of the evolution of the highly complex cerebral cortex in mammals is becoming less plausible (Stacho et al., 2020). Sufficiently strong evidence concludes that both cephalopods and crustaceans are sentient (Cox et al., 2021). This is unsurprising: Common sense does not really need any scientific proof to accept that ravens, crows, octopuses, or lobsters are sentient beings.

All these findings require an explanation from the physicalist viewpoint, which identifies the mind and consciousness with the brain.

Of course, one could resort to the usual conjecture that neural plasticity explains all things. Neural plasticity certainly plays a role and undoubtedly has its explanatory power. However, in most cases, it remains conjectural and is invoked to fill the gaps that save the paradigm. Some caution would be appropriate. For example, a recent study challenges the idea of adaptive circuit plasticity, according to which the brain recruits existing neurons to take over for those that are lost from stroke. Definitive evidence for functional remapping after stroke remains lacking. Undamaged neurons do not change their function after a stroke to compensate for damaged ones, as the conventional re-mapping hypothesis believed (Zeiger et al., 2021).

Moreover, it is observed that when a brain injury occurs, causing some form of amnesia, what was thought to be lost forever may reemerge into awareness, sometimes after years. Those whose loved ones suffered from dementia may have noted how memory and clarity of thought suddenly and quite surprisingly reappeared in a brief moment of lucidity, called 'paradoxical lucidity,' or even 'terminal lucidity.' Sometimes, bursts of mental clarity occur shortly before people die. Credible reports document cases in which people with dementia, advanced Alzheimer's, schizophrenia, or even severe brain damage suddenly return briefly to a normal cognitive state [for a review, (see Nahm et al., 2012); for some more recent findings, (see Batthyány and Greyson, 2021)]. It is hard to recast these brief episodes of lucidity, which last less than 1 h or even a few minutes, by resorting to brain plasticity.

One might also question if, besides the spatial distribution or localization of the neural correlates of consciousness, the intensity of its metabolic activity plays a role in generating a conscious experience. For example, it is well known how the practice of meditation or psychedelic drugs can change our brain chemistry and give rise to the dissolution of the sense of boundaries and intense subjective experiences, respectively. From the perspective of the material monist, which equates mind and brain as being one and the same thing, one assumes that the intensity of 'mind-expanding' psychedelics must be directly proportional to an increase in neural activity and connectivity. A dead brain is the cessation of any cerebral activity, in which case we assume there is no consciousness left, while an intensely subjective experience presumably involves high neural activity. One would, therefore, expect to find that the subjectively felt intensity of a hallucinogen proportionally correlates with neuronal activity.

However, the contrary turned out to be the case. A BOLD-fMRI study reported a significant decrease in brain activity–that is, decreased blood flow and venous oxygenation as being inversely proportional to the intensity of the subjective experience reported by the test subjects (Carhart-Harris et al., 2012). The authors of this research remark how this fact is reminiscent of Aldous Huxley's 'reducing valve' metaphor in the brain that acts to limit our perceptions in an ordinary state of consciousness [see also Koch's take on this (Koch, 2012)]. These findings were later confirmed by further studies with other hallucinogenic drugs such as LSD and ayahuasca (Palhano-Fontes et al., 2015; Carhart-Harris et al., 2016; Lewis, 2017). For a more detailed analysis of this rationale (see Kastrup, 2016). Kastrup also notes how several brain function impairments are accompanied by richer and more intense subjective experiences of self-transcendence (e.g., near-death-experiences associated with dramatically reduced brain function; Kastrup, 2017).

Williams and Woollacott point out how the idea of brain processes attenuating or filtering out mental acuity and broader perceptual awareness is consistent with the literature on meditation studies and Indian non-dual philosophy derived from spiritual practices: Reduced brain activity induced by reduced conceptual activity results in increased cognitive clarity, perceptual sensitivity and awareness expansion (Williams and Woollacott, 2021), suggesting that domains of awareness exist that do not depend upon brain functions.

Furthermore, a neurophenomenological study in the meditating brain showed that the reduction of beta band activity is related to a decreased 'sense-of-boundaries'–that is, to self-dissolution states giving rise to non-dual awareness (Dor-Ziderman et al., 2016). Similarly, Katyal and Goldin found that deeper meditation experiences are accompanied by increased alpha oscillations (closely linked to inhibitory processing and are often related to the suppression of distractors during attentional cognitive processing) and suppressed theta oscillations (potentially indicating reduced self-monitoring) (Katyal and Goldin, 2021).

Long-time meditators report a state of 'minimal phenomenal content', or as a 'non-dual awareness' of 'pure consciousness', and that could be posited as 'consciousness as such.' Investigations on Buddhist meditation suggest distinct correlates of nondual states exist but describe it as 'non-representational' awareness (Josipovic, 2014; Josipovic and Miskovic, 2020). Metzinger, instead, conjectures that it could be related to some neurological representational model realized in some brain region with some specific physical properties or neural signatures and correlates that have yet to be discovered Metzinger, 2020. While Katyal argues that the phenomenology of nondual meditative states suggests that a purely non-representational conscious state–that is, a 'transcendental' state beyond conscious experience–may transcend any such neural signatures altogether (Katyal, 2022).

2.3. The search for the neural correlates of memory

There remain other aspects to explain but that escape a materialistic paradigm with a strikingly similar pattern to that of consciousness and mentation: the neural correlates of memory. Also, in this case, one thing is certain: Memory is not stored in a specific brain area like it is on a digital computer. More than a century of research into the biological foundation of memory has not led to tangible results providing convincing evidence that such substratum exists. This is not a new issue. It dates back to Henri Bergson's opposition to a reductionist understanding of memory (Bergson, 1896/1912). Bergson considered memory to be of an immaterial and spiritual nature rather than being stored in the brain.

One might assume that information content should somehow scale with brain size. This is not observed, however (Forsdyke, 2014), (Forsdyke, 2016). For example, hemispherectomy in children does not lead to memory impairment (Tavares et al., 2020). How can it be that someone without half of the brain has no measurable memory impairment? We could explain this by resorting to the plasticity of the brain or the functions of residual brain tissues. Or, we could conjecture that memory is stored in both hemispheres; therefore, if one hemisphere is lost, the other remains unimpaired (a hypothesis that could also fit well with supposed evolutionary advantages). Or because it is the diseased hemisphere that is removed in all these cases, Nature might have provided a mechanism that transfers the memories to the healthy hemisphere before surgery. However, we should be aware that these are conjectures, hypotheses, and speculations, not scientifically established truths. Memory storage and retrieval in biological brains remains a largely unexplained mechanism, and no conclusive evidence exists that proves it to be of a physical nature.

Other research that might suggest how and where memories are stored in brains comes from experiments performed on freshwater flatworms called planaria. These creatures can be trained to associate an electric shock with a flash of light. Therefore, one might expect that they must have encoded the experience in their brains.

Flatworm planarians have an incredible self-regeneration ability (Ivankovic et al., 2019). If this worm is cut in half, each amputated body part regenerates as two new fully formed flatworms. Not only does the part with the head form a new tail but the remaining tail also forms a new head with a brain and eyes. In 1959, James V. McConnell showed that the newly-formed planaria with a new brain also maintained its conditioned behavior (McConnell, 1959). The newly-formed living being never received the electric shock and light flash of the training phase and yet it reacted as if it had a memory of the training it had never received.

Memories, if physical, may be stored not only in the brain but also throughout the body, in non-neuronal tissue.

McConnel's idea was that RNA molecules could transfer memory from one planarian to another as a "memory molecule." Motivated by this idea, he injected worms with RNA taken from those trained and reported that the training had been transferred. However, further research could not convincingly reproduce McConnel's experiments.

In 2013, Shomrat and Levin vindicated McConnel's first experiments by using computerized training of planarians, replacing manual procedures that caused previous test attempts to fail (Shomrat and Levin, 2013). Then, in 2018, Bédécarrats showed how the extracted RNA from a long-term trained sea slug, the aplysia, can induce sensitization in an untrained aplysia (Bédécarrats et al., 2018). This is taken as evidence for the molecular basis of memory and the hypothesis that RNA-induced epigenetic changes lead to the protein synthesis required to consolidate or inhibit memory. These local translations into synaptic proteins determining the neural structure of memory are actually the mainstream engram model.

However, the problem with this hypothesis is that the fastest protein synthesis causes cellular changes in timescales of minutes. How could it possibly be responsible for our ability to store and recall memories almost instantaneously?

Moreover, the still common idea that long-time memory is mapped as synaptic connectivity is challenged by the fact that it is possible to erase synaptic connections while maintaining the same conditioned behavior in the aplysia. Long-term memory and synaptic changes can, at least in some cases, be dissociated (Chen et al., 2014). It has also been shown that the brain tissue turns over at a rate of 3–4% per day, which implies a complete renewal of the brain tissue proteins within 4–5 weeks (Smeets et al., 2018). If the synaptic trace theory is correct, and since synapses are made of proteins, how can, in the presence of this turnover, long-time memory consolidation be achieved in synaptic strengths and neural connection patterns? Notice how the fact that proteins have short lifetimes is in line with the volatility of synaptic connections. How can considerably volatile changes in synaptic connections underlie the storage of information for long periods (even in the absence of learning; Trettenbrein, 2016; Mongillo, 2017)? If memory is physical, other physical repositories must be viable (DNA, cellular organelles, etc.), or a paradigm shift is necessary.

The search for engrams–that is, the group of neurons supposedly responsible for the physical representation of memory–resorts mostly to the correlation between the memory evaluation based on fear conditioning behavioral tasks of rodents and its presumed associated neural changes. For example, in a series of articles the group of Tonegawa claims to have discovered engram cells (Liu et al., 2012; Redondo et al., 2014; Ryan et al., 2015; Roy et al., 2016). They show how light-induced optogenetic reactivation of mice hippocampal neurons that were previously tagged during fear conditioning, induces a freezing behavior characteristic of fear memory recall. While the same activation of cells in non-fear-conditioned mice, or fear-conditioned mice in another context, did not elicit the same freezing behavior. Therefore, the activation of these context-specific neurons seems to suggest that they act like memory engrams of the specific fearful experience.

However, unclear is what really motivates the freezing behavior. The question is whether the cells' activation led to the memory retrieval of the fearful experience leading to the freezing behavior, if it activates the fear-like emotional state first before any memory retrieval, or if the mice might stop simply because they perceive an unexpected stimulus that might not be related with any fear or remembrance. Only the first case could potentially support the engram hypothesis, but lacking a first-person account, we will never know. While, on the contrary, the second case would only show that the activation of those cells triggers an emotional state that precedes the memory retrieval, and thus, the activated cells would not represent memory engrams (after all, we know that in humans also, stimulation of specific brain centers can lead to panic attacks associated with traumatic events, but these are not necessarily considered as the physical repository of the trauma memory.) While the third case questions whether mice freezing behavior correlates with fear perception in the first place. A lack of motion could be due to many things, not just fear. Moreover, besides the hippocampus, it is possible to induce freezing by activating a variety of brain areas and projections, such as the lateral, basal and central amygdala, periaqueductal gray, motor and primary sensory cortices, prefrontal projections, and retrosplenial cortex (Denny et al., 2017). It is not clear what the freezing behavior is really about.

This, again, shows how the correlation-causation fallacy based on a loss-of-function lesion rationale should be seen with a more critical eye.

Meanwhile, we are also allowed to speculate about a third complementary alternative. Memories associated to physical cues and lower cognitive processes and computational tasks for deductive, inferential, syntactic, predictive optimization problem-solving are material–that is, implemented in a synaptic and molecular basis for consolidation of learned behavior, fact learning, pattern recognition, recording and retrieval of representational content, external sensory cues and other physical information [e.g., (see Gershman, 2023), and that is also an interesting account of the puzzle of the biological basis of memory]. While other memories may be associated to higher cognitive functions involving inductive, non-algorithmic tasks and conceptualizations–that is, memory consolidation and recall of abstract thoughts, semantic categories, and non-representational forms of introspective intuitive cognition and creative expressions that may go beyond a Turing-machine-like information processing [e.g., (see Marshall, 2021), or, for alternatives such as 'extracorporeal information storage', see also (Forsdyke, 2015)].

2.4. Cognition without a brain

As a concluding note, it is worthy of mention that an increasing body of evidence shows that an at-least elementary form of cognition is already present and working in multicellular and single-celled lifeforms, without any neural substrate. Research in plant biology demonstrates how vegetal and cellular life shows elements of cognitive behavior that were not suspected or were simply considered impossible without a brain. There is extensive literature now that, especially in the last decade, has consistently shown how plants change behavior and adapt, respond predictively, possess some form of memory, resort to air and underground communication systems based on chemical, visual, and acoustic signals, have learning abilities and can evaluate their surroundings, make decisions, and have a cooperative behavior. It is not inappropriate to speak openly of a 'minimal' or 'proto-cognition' of cells, what is now called 'basal cognition'. For some reviews see Trewavas (2017), Gershman et al. (2021), and Lyon (2015).

Some climbing plants exhibit an anticipatory prehensile mechanism and able to purposefully plan its movements by an 'approach-to-grasp' behavior before having any physical contact with a support (Guerra et al., 2019). Other aspects could be mentioned, such as plants' adaptive changes that reflect developmental decisions based on 'root-perception.' Having no central nervous system or information processing centers, roots are, nonetheless, "able to integrate complex cues and signals over time and space that allow plants to perform elaborate behaviors analogous, some claim even homologous, to those of intelligent animals," as Novoplansky describes it (Novoplansky, 2019).

Several experiments with unicellular creatures have made it clear that conditioned behavior in single cells exists as well and is comparable in its complexity to that of plants.

An example could be the evidence of conditioned behavior in amoebae. It could be shown how the motility pattern of the Amoeba proteus under the influence of the two stimuli is consistent with associative conditioned behavior (De la Fuente et al., 2019).

A quite surprising 'brain-less problem-solving' was (re-)discovered in another protozoan. In 1906, the American zoologist Herbert Spencer Jennings noted how the Stentor roeselii could escalate actions to avoid an irritant stimulus by a complex hierarchy of avoidance behaviors in which the protozoan first enacts a strategy, sees if it works, and if not, resorts to another strategy in a series of attempts to solve a problem. One hundred and 13 years later, in 2019, Jennings' observations were confirmed (Dexter et al., 2019).

Another notorious example of non-brain-centered cellular cognition is that of the Physarum polycephalum, a large amoeba-like slime mold plasmodium that exhibits several skills and behavioral patterns that could be labeled as 'proto-intelligent'. For example, it can find the minimum length between two points in a labyrinth, and minimize the network path and complexity between multiple food sources (Nakagaki, 2004). Learning processes of habituation with anticipating conditioned behavior was shown as well (Saigusa et al., 2008). For an in-depth review on slime molds see also (Reid, 2023).

Finally, worth a mention is the behavior of the simplest life form, namely, bacteria. These also can sense the environment, actively move within it, target food, avoid toxic substances, and meaningfully change their swimming direction. Most evident is this behavior when they come together forming a bacterial community that shows surprising problem-solving abilities. Bacteria communicate with each other and coordinate gene expression, which determines the collective behavior of the entire community to achieve a common goal with collaborative problem-solving abilities [for a review of bacteria's behavior (see Lyon, 2015)].

If and how this basal cognition may also imply instances of phenomenal consciousness–that is, some form of more or less 'basal sentience'–is debatable but can be substantiated by arguments that aren't exclusively philosophical (Segundo-Ortin and Calvo, 2021). More recently, Parise et al. reviewed the ecological literature, suggesting the existence of an "extended cognition"–that is, a paradigm where one no longer considers the brain as the exclusive seat of cognition, but generalizes it to environmentally extended cognitive processes (Parise et al., 2023).

3. Discussion

The paper presented a series of neurological and biological observations whose implications remain controversial. This overview started by questioning the assumption of a lesion-based sufficiency criterion that identifies the causal relationship between the impairment of a specific cerebral area and the, thereby, assumed suppression of phenomenal consciousness and/or cognitive processes, as proof of a material monistic mind-brain identity interpretation. Motivated by this assumption we asked whether the idea of a specific brain area, structure, or its related activity, as being responsible for the qualitative and subjective experiences is consistent with the evidence, and pointed out the lack of conclusive evidence that the phenomenal dimension and singularity of the sense of self-hood, together with its higher cognitive functions is disrupted despite large impairments, suggesting that the hypothesis of a (local or global) brain-based 'seat of consciousness', if not inconsistent, must be too simplistic.

Some other neurological aspects of the mind/consciousness-brain relationship were investigated, such as the non-trivial scaling between cerebral size and neural complexity with intelligence, the hypothesis of the cerebral cortex as a center for subjective experience, by comparing it in humans and in other non-mammals, and we examined if and how far neural plasticity alone can be invoked to explain the recovery of cognitive functionalities. Of particular interest is the fact that, contrary to expectations, an inverse relationship between brain activity and conscious experience exists. Reduced brain activity leads to increased cognitive clarity and awareness expansion, seemingly suggesting that at least some aspects of our conscious experience do not depend upon the intensity of brain activity.

The now more than a century longstanding search for the physical basis of memory and memory engram cells was examined. While the predominant paradigm favors the engram hypothesis, here we highlighted how several findings challenge the conventional materialistic view. Observations like memory retention in hemispherectomy cases and planaria's regenerative memory, along with the limitations of protein synthesis as an explanation and volatility of synaptic connections raise doubts about synaptic trace theory.

Finally, emerging evidence in plant and cellular biology challenges the assumption that all cognition requires a neural substrate. Plant and cellular lifeforms exhibit forms of basal cognition, with abilities including adaptation, memory, communication, learning, decision-making, and problem-solving. Notable instances include the slime mold intelligent behaviors (Reid, 2023) and bacterial communities' coordinated problem-solving abilities, demonstrating that cognition is not exclusive to organisms with brains (Dinet et al., 2021).

Overall, these findings do not support the mind-brain identity ontology so straightforwardly as is commonly believed. The much too often unquestioned assumption that sees the nervous system as a sine-qua-non condition for conscious experience and cognitive behavior is challenged and we are equally allowed to consider cognition and sentience, not as emerging epiphenomena but as inherent 'pre-neuronal' aspects of life.

Of course, 'pre-neuronal' does not necessarily mean 'pre-physical.' These findings do not refute physicalism in and of themselves. Each of the cited neurobiological facts, when considered separately, may still be saved by several speculations inside the limitations dictated by material monism. The left column of the following table summarizes the findings discussed. The right column furnishes the possible interpretations that could, in principle, save a material monistic paradigm.

Apparent lack of mind-brain identity correlations	Possible interpretations that could save the mind-brain identity theory
Corpus callosotomy and hemispherectomy keep selfhood unified.	Residual subcortical structures may connect the two hemispheres preventing 'self-splitting'.
Cerebellar agenesis leads to only mild or moderate motor deficits.	Neuroplasticity: The remaining hemisphere takes over the tasks of the missing one.
Hydrocephalus can be quite extreme without necessarily leading to mental impairment.	Neuroplasticity again: Brain tissue may not be lost but only compressed maintaining its functionality.
The hypothesis of the cerebral cortex being the 'generator' of conscious experience is contradicted by research on congenitally decorticated children and non-mammalians.	What do we know about what it is like to be a bird?
Thalamus stimulation acts as a 'gate of consciousness', not as its 'generator.'	The thalamus is a hub that 'modulates' consciousness; it does not 'generate' consciousness.
Brain size (nr. of neurons, mass, volume) does not correlate with cognitive skills.	A minimal nr. of neurons is necessary, then size does not necessarily scale with intelligence.
The brain's complexity (connectivity, efficiency of information transfer) does not correlate with cognitive skills.	Complexity is more than connectivity and information transfer.
Evidence for engram cells remains debatable, and no memory loss was observed in hydrocephalus or hemispherectomy.	Progress has been made, it is only a matter of time before we will discover the physical basis for memory.
The intensity of psychedelic-altered states of consciousness inversely scale with network disruption.	Maybe psychedelic experiences are unfolding in the brain all the time in the form of unconscious processes. Psychedelics may present it to the surface awareness.

Basal cognition exists without a brain, like in plants and cells.	Will sooner or later be explained away by complicated cell signaling adaptive processes.

However, taken together the lack of these correlations, if we see things jointly in a wider context, that is, without selectively limiting our attention to the single phenomenon seen in isolation, and by taking a coherent integral view in which each phenomenon is seen collectively as the expression of a deeper causal principle underlying the entire pattern, another ontology that does not need such a plurality of physical interpretations is possible. A non-physicalist standpoint that sees mind and consciousness not as an epiphenomenon of matter but, rather, fundamental primitives that manifest through the material substrate (e.g., by what James called a 'transmissive' rather than 'generating' function) in line with a dualistic, idealistic, or other post-material worldviews. A viewpoint, that does not assume a mind-brain identity as a given apriorism but rather sees consciousness and mind as fundamental, with the brain a 'physical mind' that mediates information from and to a non-physical mind, could accommodate the above-listed lack of correlation between neurological and experiential/cognitive phenomenality inside a paradigm that does not need all these mechanistic conjectures.

Anyway, a future direction of systematic research that does not always assume the mind-brain identity as a given fact and leaves doors open to other perspectives, would be sufficient to potentially lead to powerful new insights that were previously overlooked. A possible future generalist approach, that does not necessarily impose one or another metaphysical worldview but starts with the assumption of a 'post-material psychology', could be a line of research (Beauregard et al., 2018). The mind–body problem and the hard problem of consciousness remain controversial issues more than ever, but non-physical ontologies of mind and consciousness are far from having been expunged by science. We have the right to explore these as a viable option not despite but, to the contrary, because of neuroscientific evidence that has been selectively dismissed for too long but cannot be ignored forever–if we can connect the dots.

References

Aguinaga, D. et al., 2018. Cocaine Effects on Dopaminergic Transmission Depend on a Balance between Sigma-1 and Sigma-2 Receptor Expression. Front. Mol. Neurosci., 11.

Assaf, Y. et al., 2020. Conservation of brain connectivity and wiring across the mammalian class. Nature Neuroscience, 23, pp.805-08.

Aurobindo, S., 1919. The Life Divine. Sri Aurobindo Ashram Trust.

Baars, B.J., 1988. A Cognitive Theory of Consciousness. New York: Cambridge University Press.

Batthyány, A. & Greyson, B., 2021. Spontaneous remission of dementia before death: Results from a study on paradoxical lucidity. Psychology of Consciousness, 8(1), pp.1–8.

Beauregard, M., Trent, N.L. & Schwartz, G.E., 2018. Toward a postmaterial psychology: Theroy, resaerch, and applications. New Ideas in Psychology, 50, pp.21-33.

Bédécarrats, A. et al., 2018. RNA from trained Aplysia can induce an epigenetic engram for long-term sensitization in untrained Aplysia. eNeuro, 5(3).

Bergson, H., 1896/1912. Matter and Memory. New York: McMillan.

Bogen, J.E., Fisher, E. & Vogel, P., 1965. Cerebral commissurotomy: A second report. JAMA, 194(12), pp.1328-29.

Bourget, D. & Chalmers, D., 2020. Philosophers on Philosophy: The 2020 PhilPapers Survey - (Forthcoming). Philosophers' Imprint.

Carhart-Harris., R.L. & al., e., 2012. Neural correlates of the psychedelic state as determined by fMRI studies with psilocybin. Proceedings of the National Academy of Sciences, 109(6), pp.2138-43.

Carhart-Harris, R.I. & al, 2016. Neural correlates of the LSD experience revealed by multimodal neuroimaging. PNAS, 113(17), pp.4853-58.

Chalmers, D., 1995. Facing up the problem of consciousness. Journal of Consciousness Studies, 2(3), pp.200-19.

Chen, S., Glanzman, D.L. & al., 2014. Reinstatement of long-term memory following erasure of its behavioral and synaptic expression in Aplysia. eLife, (3:e03896).

Cox, V. & al., 2021. Crustacean & Cephalopod Sentience Briefing. Conservative Animal Welfare Foundation.

Crick, F.c. & Koch, C., 2005. What is the function of the claustrum? Philos. Trans. R. Soc. Biol. Sci., 1458(360), pp.1271-79.

Cruse, D. et al., 2011. Bedside detection of awareness in the vegetative state: a cohort study. The Lancet, 378(9809), pp.2088-94.

Davis, C. et al., 2008. Speech and language functions that require a functioning Broca's area. Brain and Language, 105(1), pp.50-58.

de Haan, E.H.F. et al., 2020. Split-Brain: What We Know Now and Why This is Important for Understanding Consciousness. Neuropsychology Review, 30, pp.224–33.

de Haan, E.H.F. et al., 2021. Singularity and consciousness: A neuropsychological contribution. Journal of Neuropsychology, 15(1), pp.1-19.

De la Fuente, I.M., Bringas, C., Malaina, I. & al., 2019. Evidence of conditioned behavior in amoebae. Nature Communications, 10(3690).

Denny, C.A., Lebois, E. & Ramirez, S., 2017. From Engrams to Pathologies of the Brain. Frontiers in Neural Circuits, 11.

Dexter, J.P., Prabakaran, S. & Gunawardena, J., 2019. A Complex Hierarchy of Avoidance Behaviors in a Single-Cell Eukaryote. Current Biology, 29(24), pp.4323-29.

Dor-Ziderman, Y. et al., 2016. Self-specific processing in the meditating brain: a MEG neurophenomenology study. Neuroscience of Consciousness, Volume 2016, Issue 1, 2016, niw019, 2016(1).

Feng, Y. & al., 2015. A new case of complete primary cerebellar agenesis: clinical and imaging findings in a living patient. Brain, 138(6), p.353.

Feuillet, L. & al., 2007. Brain of a white-collar worker. The Lancet, 370(9583), p.262.

Fichtel, C., Dinter, K. & Kappeler, P.M., 2020. The lemur baseline: how lemurs compare to monkeys and apes in the Primate Cognition Test Battery. Zoological Science.

Forsdyke, D.R., 2014. Long-term memory: scaling of information to brain size. Fortiers in Human Neuorscience.

Forsdyke, D.R., 2015. Wittgenstein's Certainty is Uncertain: Brain Scans of Cured Hydrocephalics Challenge Cherished Assumptions. Biol Theory, 10, pp.336–42.

Gazzaniga, M., 1985. MRI assessment of human callosal surgery with neuropsychological correlates. Neurology, 35(12), pp.1763-6.

Gershman, S.J., 2023. The molecular memory code and synaptic plasticity: A synthesis. Biosystems, 224, p.104825.

Gershman, S.J., Balbi, P., Gallistel, C.R. & Gunawardena, J., 2021. Reconsidering the evidence for learning in single cells. eLife, 10:e61907..

Gomez-Marin, A., 2017. Causal Circuit Explanations of Behavior: Are Necessity and Sufficiency Necessary and Sufficient? In In: Çelik, A., Wernet, M. (eds) Decoding Neural Circuit Structure and Function.

Gonzalez-Castill, J. & al., 2012. Whole-brain, time-locked activation with simple tasks revealed using massive averaging and model-free analysis. PNAS, 109(14), pp.5487-92.

Granovetter, M.C., Robert, S. & Ettensohn, L..B.M., 2022. With childhood hemispherectomy, one hemisphere can support—but is suboptimal for—word and face recognition. PLOS - Psychological and Cognitive Sciences, 119(44).

Guerra, S., Castiello, U. & al., 2019. Flexible control of movement in plants. Nature Sci. Rep., 9(16570).

Heldstab, S.A. & al., 2020. When ontogeny recapitulates phylogeny: Fixed neurodevelopmental sequence of manipulative skills among primates. Science Advances, 6(30).

Huxley, A., 1954. The Doors of Perception. Chatto and Windus.

Ivankovic, M. et al., 2019. Model systems for regeneration: planarians. Development, 146(17).

James, W., 1898. Human immortality. Houghton, Mifflin and Company, The Riverside Press, Cambridge.

Jonas, E. & Kording, K.P., 2017. Could a Neuroscientist Understand a Microprocessor? PLOS Computational Biology.

Josipovic, Z., 2014. Neural correlates of nondual awareness in meditation. Ann. N.Y. Acad. Sci., 1307, pp.9-18.

Josipovic, Z. & Miskovic, V., 2020. Nondual Awareness and Minimal Phenomenal Experience. Front. Psychol., 11.

Kastrup, B., 2016. What Neuroimaging of the Psychedelic State. Journal of Cognition and Neuroethics, 4(2), pp.1–9.

Kastrup, B., 2017. Self-Transcendence Correlates with Brain Function Impairment. Journal of Cognition and Neuroethics, 4(3), pp.33–42.

Katyal, S., 2022. Reducing and deducing the structures of consciousness through meditation. Front. Psychol., 13.

Katyal, S. & Goldin, P., 2021. Alpha and theta oscillations are inversely related to progressive levels of meditation depth. Neuroscience of Consciousness, 2021(1).

Kliemann, D. & al., 2019. Intrinsic Functional Connectivity of the Brain in Adults with a Single Cerebral Hemisphere. Cell Reports, 29(8), pp.2398-407.

Koch, C., 2012. This Is Your Brain on Drugs. [Online] Available at: https://www.scientificamerican.com/article/this-is-your-brain-on-drugs/.

Krakauer, J.W. et al., 2017. Neuroscience Needs Behavior: Correcting a Reductionist Bias. Neuron, 93(3), pp.480-90.

Lewin, R., 1980. Is your brain really necessary? Science, 210(4475), pp.1232-34.

Lewis, C.R., 2017. Two dose investigation of the 5-HT-agonist psilocybin on relative and global cerebral blood flow. Neuroimage, 159, pp.70-78.

Liang, M. & al., 2013. Primary sensory cortices contain distinguishable spatial patterns of activity for each sense. Nature Communications, 4(Article number: 1979).

Liu, X. et al., 2012. Optogenetic stimulation of a hippocampal engram activates fear memory recall. Nature, 484, pp.381-85.

Lyon, P., 2015. The cognitive cell: bacterial behavior reconsidered. Front Microbiol., 6, p.264.

Lyon, P. & al., 2021. Basal cognition: conceptual tools and the view from the single cell. Philosophical Transactions of the Royal Society B, 376(1820).

Mackenzie, G., 2019. Can they Feel? The Capacity for Pain and Pleasure in Patients with Cognitive Motor Dissociation. Neuroethics, 12, pp.153–69.

Marshall, P., 2021. Biology transcends the limits of computation. Progress in Biophysics and Molecular Biology, 165, pp.88-101.

Maxwell, M.B. & al., 2022. A role for the claustrum in cognitive control. Trends in Cognitive Science, 26(12), pp.1133-52.

McConnell, J.V., 1959. Worms and things. The Worm Runner's Digest.

McGovern, R.A. & al., 2019. Hemispherectomy in adults and adolescents: Seizure and functional outcomes in 47 patients. Epilepsia, 60, pp.2416– 2427.

Merabet, L.B. & al., 2008. Rapid and Reversible Recruitment of Early Visual Cortex for Touch. Plos One.

Merker, B., 2007. Consciousness without a cerebral cortex: A challenge for neuroscience and medicine. Behavioral and Brain Sciences , 30(1), pp.63 - 81.

Metzinger, T., 2020. Minimal phenomenal experience: Meditation, tonic alertness, and the phenomenology of "pure" consciousness. Philosophy and the Mind Sciences, 1(1).

Mongillo, G.R.S.L.Y., 2017. Intrinsic volatility of synaptic connections — a challenge to the synaptic trace theory of memory. Current Opinion in Neurobiology, 46, pp.7-13.

Muckli, L. & al., 2009. Bilateral visual field maps in a patient with only one hemisphere. PNAS, 106(31), pp.13034-39.

Nagel, T., 1974. "What is it like to be a bat? Philosophical Review, 83, pp.435–456.

Nahm, M., Greyson, B., Kelly, E.W. & Haraldsson, E., 2012. Terminal lucidity: A review and a case collection. Archives of Gerontology and Geriatrics, 55(1), pp.138-42.

Nakagaki, T., 2004. Obtaining multiple separate food sources: behavioural intelligence in the Physarum plasmodium. Proc. R. Soc. Lond. B, 271, pp.2305–10.

Nieder, A., Wagener, L. & Rinnert, P., 2020. A neural correlate of sensory consciousness in a corvid bird. Science, 369(6511), pp.1626-29.

Novoplansky, A., 2019. What plant roots know? Seminars in Cell & Developmental Biology, 92, pp.126-33.

Nuerology, A.A.o., 2018. Practice Guideline Update Systematic Review Summary: Disorders of Consciousness. [Online] American Academy of Neurology Available at: https://www.aan.com/Guidelines/home/GuidelineDetail/927.

Oizumi, M., Albantakis, L. & Tononi, G., 2014. From the Phenomenology to the Mechanisms of Consciousness: Integrated Information Theory 3.0. 10(5).

Oudiette, D. et al., 2009. Dreamlike mentations during sleepwalking and sleep terrors in adults. Sleep, 32(12), pp.1621-7.

Owen, A.M. & Coleman, M.R., 2006. Detecting Awareness in the Vegetative State. Annals of the New York Academy of Sciences, 1129, pp.130-8.

Palhano-Fontes, F. & al., 2015. The Psychedelic State Induced by Ayahuasca Modulates the Activity and Connectivity of the Default Mode Network. PLoS One.

Parise, A.G., Gubert, G.F., Whalan, S. & Gagliano, M., 2023. Ariadne's thread and the extension of cognition: A common but overlooked phenomenon in nature? Front. Ecol. Evol. - Sec. Behavioral and Evolutionary Ecology, 10.

Pinto, Y. & al., 2017. Split brain: divided perception but undivided consciousness. Brain, 140(5), pp.1231–37.

Pinto, Y. & al., 2020. Unified Visual Working Memory without the Anterior Corpus Callosum. Symmetry, 12(12), p.2106.

Pulsifer, M.B. et al., 2004. The Cognitive Outcome of Hemispherectomy in 71 Children. Epilepsia, 45(3), pp.243-54.

Radek, L., 2018. Dreaming and awareness during dexmedetomidine- and propofol-induced unresponsiveness. BJA British Journal of Anaesthesia, 121(1), pp.260-69.

Redinbaugh, M. & al., 2020. Thalamus Modulates Consciousness via Layer-Specific Control of Cortex. Neuron, 106(1), pp.66-75.

Redondo, R., Kim, J., Arons, A. & al., 2014. Bidirectional switch of the valence associated with a hippocampal contextual memory engram. Nature, 513, pp.426-30.

Reid, C.R., 2023. Thoughts from the forest floor: a review of cognition in the slime mould Physarum polycephalum. Animal Cognition.

Roy, D. & al., 2016. Memory retrieval by activating engram cells in mouse models of early Alzheimer's disease. Nature, 531, pp.508-12.

Ryan, T.J. et al., 2015. Engram cells retain memory under retrograde amnesia. Science, 348(6238), pp.1007-13.

Saigusa, T. & al., 2008. Amoebae anticipate periodic events. Phys Rev Lett., 100(1)(018101).

Scheinin, A. & al., 2020. Foundations of human consciousness: Imaging the twilight zone. Journal of Neuroscience.

Segundo-Ortin, M..C.P., 2021. Consciousness and cognition in plants. WIREs Cognitive Science, 13(2).

Seth, A.K. & Bayne, T., 2022. Theories of consciousness. Nature Reviews Neuroscience, 23, pp.439–52.

Shewmon, D., Holmes, G. & Byrne, P., 1999. Consciousness in congenitally decorticate children: developmental vegetative state as self-fulfilling prophecy. Dev Med Child Neurol., 41(6), pp.364-74.

Shomrat, T. & Levin, M., 2013. An automated training paradigm reveals long-term memory in planarians and its persistence through head regeneration. Jour. of Experimental Biology, 216, pp.3799-810.

Siclari, F., Tononi, G. & al., 2018. Dreaming in NREM Sleep: A High-Density EEG Study of Slow Waves and Spindles. Journal of Neuroscience, 38(43), pp.9175-85.

Smart, J.J.C., 2022. The mind-brain identity theory - The Stanford Encyclopedia of Philosophy - Edward N. Zalta & Uri Nodelman (eds.). [Online] Available at: https://plato.stanford.edu/archives/win2022/entries/mind-identity/.

Smeets, J. & al., 2018. Brain tissue plasticity: protein synthesis rates of the human brain. Brain, 141(4), pp.1122–29.

Solms, M. & Panksepp, J., 2012. The "Id" Knows More than the "Ego" Admits: Neuropsychoanalytic and Primal Consciousness Perspectives on the Interface Between Affective and Cognitive Neuroscience. Brain Sciences, 2(2), pp.147-75.

Sperry, R.W., 1968. Hemisphere deconnection and unity in conscious awareness. American Psychologist, 23, p.723.

Sperry, R.W., 1984. Consciousness, personal identity and the divided brain. Neuropsychologia , 22(6), pp.661-73.

Stacho, M. & al., 2020. A cortex-like canonical circuit in the avian forebrain. Science, 369(6511).

Tavares, T.P., Kerr, E.N. & Smith, M.L., 2020. Memory outcomes following hemispherectomy in children. Epilepsy & Behavior, 112(107360).

Tononi, G., 2015. Integrated information theory. [Online] Available at: http://www.scholarpedia.org/article/Integrated_information_theory.

Trettenbrein, P.C., 2016. The Demise of the Synapse As the Locus of Memory: A Looming Paradigm Shift? Front. Syst. Neurosci., 10.

Trewavas, A., 2017. The foundations of plant intelligence. Interface Focus, 7(20160098).

Tuckute, G. & al., 2022. Frontal language areas do not emerge in the absence of temporal language areas: A case study of an individual born without a left temporal lobe. Neuropsychologia, 169, p.108184.

Turku, U.o., 2018. Consciousness is partly preserved during general anesthesia. [Online] Available at: https://www.sciencedaily.com/releases/2018/07/180703105631.htm.

Vargha-Khadem, F. & al., 1997. Onset of speech after left hemispherectomy in a nine-year-old boy. Brain, 120(1), pp.159-82.

Vollenweider, F.X. & Preller, K.H., 2020. Psychedelic drugs: neurobiology and potential for treatment of psychiatric disorders. Nature Reviews Neuroscience, 21, pp.611–24.

Wahbeh, H., Radin, D., Cannard, C. & Delorme, A., 2022. What if consciousness is not an emergent property of the brain? Observational and empirical challenges to materialistic models. Frontiers in Psychology 13:955594.

Wahbeh, H. et al., 2018. Exceptional experiences reported by scientists and engineers. Explore (NY), 14(5), pp.329-41.

Williams, B. & Woollacott, M.H., 2021. Conceptual Cognition and Awakening: Insights From Non-dual Saivism and Neuroscience. Journal of Transpersonal Psychology, 53(2), pp.119-39.

Zeiger, W.A., Marosi, M. & Saggi, S., 2021. Barrel cortex plasticity after photothrombotic stroke involves potentiating responses of pre-existing circuits but not functional remapping to new circuits. Nature Communications, 12(3972).

2. Vitalism and Cognition in a Conscious Universe[142]

Abstract

According to the current scientific paradigm, what we call 'life', 'mind', and 'consciousness' are considered epiphenomenal occurrences, or emergent properties or functions of matter and energy. Science does not associate these with an inherent and distinct existence beyond a materialistic/energetic conception. 'Life' is a word pointing at cellular and multicellular processes forming organisms capable of specific functions and skills. 'Mind' is a cognitive ability emerging from a matrix of complex interactions of neuronal processes, while 'consciousness' is an even more elusive concept, deemed a subjective epiphenomenon of brain activity. Historically, however, this has not always been the case, even in the scientific and academic context. Several prominent figures took vitalism seriously, while some schools of Western philosophical idealism and Eastern traditions promoted conceptions in which reality is reducible to mind or consciousness rather than matter. We will argue that current biological sciences did not falsify these alternative paradigms and that some forms of vitalism could be linked to some forms of idealism if we posit life and cognition as two distinct aspects of consciousness preeminent over matter. However, we will not argue in favor of vitalistic and idealistic conceptions. Rather, contrary to a physicalist doctrine, these were and remain coherent worldviews and cannot be ruled out by modern science

I. Introduction

From the perspective of scientific materialism, it is generally believed that metaphysical vitalism (that is, the hypothesis of an origin and life theory depending on a force or principle distinct from chemical or physical forces) or physical vitalism (a non-reductionist understanding of life which claims that an organism with its generative, developmental and living functions cannot be reduced to the sum of its parts) must both be relegated to a status as relics from the history of science (Normandin & Wolfe, 2013). Especially metaphysical vitalism was criticized beginning with the rise of positivism. Philosophers of science since the Vienna circle (Frank P. , 1941), (Schlick, 1953), and later

[142] The present article was published in *Communicative & Integrative Biology*, vol. 15, no. 1, pp. 121-136, 2021. DOI: **https://doi.org/10.1080/19420889.2022.2071102**

mid-20th century empiricists Ernest Nagel (Nagel E. , 1961), and Gustav Hempel (Hempel, 1966), have rejected neovitalist tendencies.

This exclusion is because vitalist conceptions are generally at odds with a scientific reductionist and physicalist approach while being more congruent with Western theories of philosophical idealism or Eastern philosophies.

Physicalism, also called 'material monism' or just 'materialism'[143], assumes that everything can be reduced to material constituents and that the physical world can be explained by physical laws without any reference to metaphysical concepts. Physical matter is the fundamental reality and current paradigm of scientific materialism used to define 'life', 'mind', and 'consciousness'.

On the other hand, philosophical idealism represents an alternative paradigm of reality. We will focus on a particular version of idealism, so-called 'objective' or 'ontological' idealism,[144] and call it just idealism. Idealism is a metaphysical perspective according to which reality cannot be reduced to matter but is itself fundamentally of a mental cognitive nature. It posits mind as a basic primitive of reality–that is, as something that cannot be defined in terms of previously-defined concepts. The Western conception of the word 'mind' sometimes includes 'consciousness', 'spirit', or 'will'.[145] Idealism postulates that all of reality is a form of conscious thought and cognition that exists independently of human thought and is prior to conscious life. From this idealist perspective, life, mind, and consciousness are not conceived as biological emergent phenomena or properties of matter, as physicalism posits. Rather, to the contrary, life and matter are considered to emerge from the only mental 'one-substance' (mind, consciousness, will, or spirit). Idealist philosophers, like Baruch Spinoza, Arthur Schopenhauer, George Berkely, Friedrich Hegel, and Friedrich Schelling, to mention just some, replaced material monism ("there is only matter") with a 'mental-' or 'substance-monism' ("there is only mind").

Eastern philosophies offer a different approach. In particular, the non-dual Advaita Vedanta (with many similarities to Buddhism and other Indian

[143] In the literature, the terms 'physicalism', 'material monism', and 'materialism' are frequently interchanged. The terms 'material monism' or 'materialism' were used more frequently by authors of the past centuries. However, nowadays, 'physicalism' is a more rigorous term inasmuch as modern physics describes more than just matter, including energy, force fields and physical laws.

[144] To be distinguished from 'epistemological' or 'subjective' idealism, which we will not consider here.

[145] This is a conflation we will later avoid in this paper, in section II.1. For the time being, we maintain the word 'mental' as something relating to 'cognition', 'reason', or 'intelligence', but not 'consciousness', 'sentience', 'emotions', or physical sensations.

doctrines) posits consciousness–that is, the non-dual and unconditioned and undifferentiated Brahman–as the fundamental primitive of all existence (Timalsina, 2008).[146] Contrary to the idealist, however, mind and consciousness are distinct here. Mind is a feature and a part of the cosmic illusion, 'Maya', and that should not be conflated with the only reality, the ultimate Brahman that is 'Satcitananda', namely, Consciousness-Existence-Beatitude. While in this mystical philosophy, life, mind, and matter are considered as comprising part of the cosmic manifestation in a world of ignorance, the 'veil of Avidya', they nevertheless constitute distinct 'planes' of existence. The physical plane is one of them but neither the only nor the most fundamental one. This is vividly illustrated in the Samkhya yoga philosophy that describes a cosmology in which each organism consists of a metaphysical structure made of distinct 'sheaths' or 'bodies': the material sheath ('annamaya-kosha'), the vital sheath ('pranamaya-kosha'), the mental sheath ('manomaya-kosha'), the sheet of divine knowledge or gnosis ('vijnanamaya-kosha') and, at the summit, standing above all, the 'Atman', the individuation of Satchitananda (Burley, 2007). Thus, matter, life, and mind are all derivative powers and instruments of fundamental consciousness, not separated and self-existent identities, but nevertheless to be distinguished and realized as distinct from each other. This is reminiscent of the Western Spinozian metaphysical view, in which matter and mind are just two possible 'modes' (the 'res extensa' and 'res cogitans', respectively) of the same fundamental and universal 'Substance'.

Thus, these metaphysical ontologies do not posit a priori life, mind, and consciousness as emergent properties reducible to material biological processes. They claim that a 'life-principle' and/or a 'mind-principle' exist that are inherent in organic matter, but distinct from it.

The question this paper pursues is: How far was biology effectively able to dismiss vitalistic and idealistic ontologies? Have these been disproven without appeal, or do these remain a viable option? Though we will point at some findings suggestive of this eventuality, we will not argue in favor of it. Rather, we will argue that this multidimensional conception of life and matter that one finds throughout all cultures and all times, contrary to common belief, has not been dismissed conclusively. We highlight how these conjectures about the nature of life and the mind need a more careful review, especially in light of recent findings.

In the first part, after a short review of the main historical conceptions of vitalism, we will discuss whether its core concepts have been falsified. In the same line of inquiry, in the second part, we will analyze if and how cognition,

---

[146] The ontologic framework of a fundamental non-material non-dual awareness as foundational to a self-organizing universe has also received some attention in the biological sciences, e.g., (Theise & Kafatos, 2016).

as well, might be a fundamental principle inherent but distinct from life. Conclusive remarks will follow.

II. The life in matter

1. A short history of vitalism

The history of vitalism, and the meaning attached to it, is long and variegated. We will not go into the details here (for a review, see (Wolfe & Normandin, 2013)). It may only be said that, conventionally, one distinguishes between 'physical vitalism' (or 'scientific vitalism' or 'process vitalism' or 'structural-functional vitalism') and 'metaphysical vitalism' (or 'essential vitalism').

Metaphysical vitalism is thought of as a special vital essence, the 'vis essentialis' that supposedly infuses and animates all organisms and demarcates living from non-living matter and a 'vis mediatrix' responsible for the action and coordination of bodily parts. Already, Plato and Aristotle posited that there is a force immanent to the organisms which supposedly makes life fundamentally distinct from non-life. Aristotle hypothesized that the soul (the psyche) organizes the form and structure of an organism and its purposeful activity. Later, Galen (129-210) assumed that spirit (pneuma) was the essential principle of life, an idea that continued throughout the Middle Ages. It became a matter of scientific debate when French biologist Jean-Baptiste Lamarck (1744-1829), better known for his foundational contributions to evolutionary biology, postulated the existence of an ordering 'life-power' augmented by an inner 'adaptive force' with which each organism adapts to the environment. In this view, these powers supposedly stand behind evolution and are responsible for the increasing complexity of organisms. Part of this conceptual system was also Lamarck's famous notion of soft-inheritance (contrasted with the hard-inheritance based on genetic inheritance independent from environmental factors)–that is, the notion that physical traits can be passed on to offspring even if the parent organism acquired them only in its lifetime through its interaction with the environment. The doctrine of essential life humors also found its way through medical practices. Therapeutic bloodletting (bleeding of patients) was thought of as an effective method to release the excess vital fluid considered the culprit behind every ailment and disease. This practice presumably killed more people than it helped and was finally dismissed as pseudo-science at the end of the 19th century.

Physical vitalism, instead, looked for life's mechanism, action, and system dynamics. Its main representative was Montpellier vitalism, associated with the writings of the 18th-century French faculty of medicine at the University of Montpellier, which did not show any signs of metaphysically laden 'vital forces' (Wolfe C. T., 2017). It expressed a structural-functional form of vitalism with the celebrated image of the bee-swarm: How can a swarm of bees coordinate its behavior as a unique 'super-organism'?

Physical vitalism accepts physico-chemical determinism, but differs from the physicalist's viewpoint in that it does not embrace a reductionist approach cutting down everything to physical concepts. Its most notable supporter was French physiologist Claude Bernard (1813-1878), who argued for the irreducible uniqueness of life, claiming it is impossible to see the organism as the sum of its parts, as a reductionist conception would entail. Instead, it must be seen as an integrated and harmonious whole. Bernard posited homeostasis as the foundational principle of life. It might, therefore, not come as a surprise that Bernard contributed most to the eclipse of metaphysical vitalism.

But the distinction between physical vitalism and the classic physical and mechanistic concept of life is, after all, very subtle, if it exists at all. Modern non-linear complex system dynamics theories in developmental biology ('organicism') which describe emergent properties of a system that cannot be explained in terms of the properties of its constituents (Gilbert & Sarkar, 2000) could be labeled a modern form of process-vitalism as well. Deep down, however, it is only a matter of semantics: These theoretical frameworks remain tightly anchored in the orbit of a non-reductive physicalism.

For the remainder of the article, we will focus on metaphysical vitalism and call it just vitalism, if not meant otherwise.

In 1810, Swedish chemist Jöns Jacob Berzelius postulated that organic compounds could be distinguished from inorganic ones if they require living organisms that, through a mysterious vital force, can synthesize them. However, in 1928 German chemist Friedrich Wöhler showed that urea, an organic compound ($CO(NH_2)_2$), could be synthesized in vitro from two inorganic molecules (potassium cyanate and ammonium sulfate), thus disproving Berzeliu's conjecture.

French chemist Louis Pasteur (1822–1895) kept his speculations less molecular and proposed that a 'vital action' makes life inherently different and special compared to non-living matter. Pasteur was inspired by Francesco Redi's famous experiment that disproved spontaneous generation (maggots are not generated by rotten meat, but, rather, come from fly eggs), taking this as evidence that life must always originate from life. The thesis was that there cannot be, not even in principle, the spontaneous generation of life from non-living matter.

In 1907, Henri Bergson (Bergson H. , 1907) tried to give further credence to vitalism from the philosopher's perspective. According to Bergson, the principles of Darwinian evolution are not sufficient to explain evolution's creativity. In life, there must be something more, an 'élan vital'–a 'vital impetus' or 'vital force'–responsible for the innovative complexity of nature and the morphogenesis of living beings. Bergson maintained that, also, our urges, desires, feelings, and emotions, which impel us to action, come from the same internal creative impulse. It is this self-impelled force within plants, animals, and humans that determines their effort to overcome the inertia of matter. Bergson's hypothesis was an attempt to find a compromise between a mechanical and finalistic conception, but was met with indifference or ridicule. British biologist Julian Huxley sarcastically rejected Bergson's idea, comparing the élan vital hypothesis to that of an 'élan locomotif' ('locomotive driving force') to explain the operation of a railway engine.

A more advanced neo-vitalist version came from the German embryologist Hans Driesch (1867–1941), who theorized the presence of an 'entelechy', an intensive, rather than extensive, substance or entity, determining the organic processes and reminiscent of Aristotle's entelechy–some sort of vital and self-organizing principle which aims to realize a specific design or purpose. Driesch articulated his theory with what he took to be a mindlike essence in all living things, observing the development of sea urchin embryos into whole organisms.

However, despite all the vitalistic concepts that regularly arose throughout history, in the modern scientific context, any form of vitalism has been abandoned as an obsolete working hypothesis. Meanwhile, in contemporary culture, vitalism survives in traditional healing practices, energy therapies, chiropractic therapies, biofield therapies such as Reiki, or other healing methods based on the qi or prana 'subtle energies', as postulated in Eastern cultures. It is normally assumed that the former arises as the final winner from the history of science, while the latter are only a form of popular pseudo-science or medieval superstitious healing methods based on spooks, ghosts, and spirits and where a therapeutic effect, if any, can be explained away by placebo effects.

2. The meaning of life

The dismissal of vitalism is based mainly on three pillars.

First, a non-material vitalism is metaphysical in the sense that it postulates the existence of a non-physical 'life-fluid' that, per definition, is not detectable by physical means and, thereby, positions itself beyond the boundary of conventional scientific investigation. We defer further discussion of this point to section 0, where we will question if that is really the case.

Second, vitalism is ignored by the adoption of an acritical use of principles of parsimony–namely, Occam's razor–according to which we should refrain from multiplying entities and positing any hypothesis that is not in line with the dominating paradigm. "Vitalism is an unnecessary hypothesis" is the instinctive objection. But, as we shall see, so far science has not been able to furnish any proof of the sufficiency of the opposite hypothesis either, namely, that life can be reduced to life-less mechanistic processes. While Occam's razor can offer useful methodological guidance for simplifying our conceptual frameworks and diminishing an experimental workload, it is not a criterion that makes a theory 'scientific' or a more complex working hypothesis 'false'.

Third, the exclusion of any vitalistic hypothesis is based on hope. The hope is that sooner or later everything will be recast in the current paradigm. It is a hope upheld by the belief that it is only a matter of time, funds, and research before everything will be explained inside a naturalistic mechanistic framework. However, facts have shown that the findings of the last century made vitalism even more difficult to expunge. As we will show, the organization and function of life revealed themselves to be much more complex than expected, and the goal of eliminating any vitalistic and teleological temptation appears today to be even further away than it was a century ago. The explanatory gaps did not shrink; rather, they grew.

In fact, the question is: Has vitalism been disproven?

The answer depends on what we mean by vitalism. If we assume that it is a lack of vis essentialis leading to an imbalance of Hippocratic humors (blood, phlegm, black and yellow bile) as a cause of illness and that this imbalance must be reestablished by bloodletting therapies, or that flowing through the nerves is a vis mediatrix that allegedly explains muscle contraction, then, of course, these hypotheses are no longer tenable. Or, if we posit organic vs. non-organic compounds as a demarcation line between living and non-living matter, thereby embracing Berzelius's criterion and definition for vitalism, then, obviously, with Wöhler's experiment, the case could be considered closed.[147]

But did science answer the primary and original conjectures of Aristotle and Plato first and, later on, disprove Lamarck's, Pasteur's, Bergon's, or Driesch's speculations?

Though there is no universally accepted definition, vitalism can be described in its original intent as an answer to the following intuitions.

1) Our inability to explain and define what is life is due to the fact that, in living matter, there is something fundamentally different from what is in non-living matter.

2) The inner psychological force of an organism driven by intentionality, desire, the instinct of survival, striving, will, aim, purpose, and motivations cannot be reduced to mechanistic principles alone.

3) The origin of life is not explicable by the current laws of physics and chemistry alone.

---

[147] In the present context, the word 'organic' is misleading. In chemistry, organic compounds are no longer defined as chemicals synthesized in living organisms. An organic compound is any molecule containing carbon-hydrogen bonds. Also, methane is an organic compound and no one would consider it a 'living organism'.

4) The growth and development of biological form cannot be explained by mechanistic principles alone.

5) The creative inventiveness of nature generating all the morphological diversity of species cannot be captured by mechanistic principles alone

We will argue that, contrary to common belief, these fundamental statements have not been disproven; they were simply ignored. Plato's and Aristotle's doubts have not been dispelled. Developmental biology is still far from explaining the growth and development of an organism. The Darwinian paradigm alone, especially in its 'modern synthesis' version, turned out to be insufficient to explain the variety and morphological development of species. Vitalism does not deny Darwinian evolution based on principles of natural selection, genetic drifts, random mutations, environmental factors, adaptation, etc. Rather, it finds it unconvincing that these are self-sufficient and fully explanatory principles. That these metaphysical claims were disproven was and remains an assumption unsubstantiated by scientific facts. Science does not know how to explain, in a naturalistic framework, the origin of our instincts, will-force, and cognition, let alone consciousness (the distinction between consciousness and mind or cognition will be clarified later), and is nowhere near able to generate life in a test tube.

3. A short reply to J. Huxley

First, let us consider J. Huxley's counterargument, as it is one of the most cited objections. Huxley misses Bergson's point; the question is not so much what stands behind the physical motion force of organisms, which obviously can be explained away by mere physical and chemical reactions. A simple electrical impulse can lead to a muscle contraction even in a dead body, as the Italian physicist Alessandro Volta, well known as the inventor of the electric battery, first observed by applying an electric current to the amputated legs of a frog. The question is: Where does the creative impulse of novelty and the urge to reach an ever-increasing complexity and variety of forms we observe in the evolution of life come from? Why does every living being act as if having agency, aim, and purpose if it is a bottom-up construct of aimless and purposeless molecules without agency? And what is and where does that will to survive, will to action, will to reproduce, will to grow, will to expand, will to know come from, making living matter so distinctive from non-living matter? Genetics, biophysics, chemistry, and biology explain the mechanical workings and energetic dynamics of the material substrate, but do not tell us anything about the volitional force, the coming into existence of a conscious subject, the will that determines and moves it. Vitalism does not pretend to explain the nature of the mechanical principles and forces that set an organism in motion. Rather, it posits the existence of an intentional principle, a 'desire-force' and a 'will-force' inherent to the dynamics of life. Meanwhile, there is no sign that the élan locomotif, namely, the steam's heat transformed into mechanical energy, imparts to the railway engine any creative force or will to generate new engines or leads it to a novel behavior with agency and purpose other than endlessly spinning the flywheel.

These misinterpretations come from the improper use and interpretations of phrases such as 'vital force' and 'vital energy'. Here, one means, first and foremost, a psychological property or, eventually, a subtle trans-physical 'substance' inherent in living matter, but not, or not necessarily, something to relate to the strict notion of force and energy in physics. Whether these vital 'forces' or 'energies' must be intended literally as physical properties, namely, of the ability to impel a change in motion in time or to produce physical work on material objects, is a secondary matter. The question remains: What makes living matter different from non-living matter and why does a neural network produce life instincts and a psychological 'will-force'? On the other hand, the misnomers of 'force' and 'energy', in the context of vitalism, are also not so misplaced either because, as everyone knows all too well, an 'emotion', whose etymology comes from the Latin 'emovere', meaning to stir or agitate, can be a quite powerful 'force' of life that can set us pretty much in motion.

These are details and subtleties that, nonetheless, make a whole difference, all of which Huxley ignored, commenting only on the external material dynamics of a living being, but making little or no connection to its whole internal psychological dimension.

4. Genomecentrism, morphogenesis, and synthetic biology

Until recently, biology answered to Bergson's hypothesis of a 'creative morphogenesis' and to Driesch's 'vital self-organizing principle', more elegantly than Huxley, with a naturalistic counter-hypothesis: There is no life force because every morphological organization and its development can be explained in terms of genetic expressions, and their mutation in time is determined by their interaction with the environment.

This was a hypothesis that could, in principle, explain the appearances. The premise was that all our morphological traits are encoded in the DNA molecule, like a 'book of life'. This was held for a long time as the most plausible working hypothesis. But the data that emerged from the Human Genome Project shortly after the turn of the millennium showed unequivocally how this idea, which was dominant throughout the 20th century, required substantial revision. One of the main findings of the genome mapping and analysis of the past two decades was that the complexity of an organism does not scale with the number of genes in the genome (the 'C-value enigma'). The DNA's functionality is different than previously imagined and much more flexible than thought; it is no longer felt by many to be a control kernel of life. It serves mostly as a template for amino acids that make up polypeptide chains transcribed into RNA, then translated into proteins. (In fact, only a small fraction of the genome is actually translated; the rest is 'non-coding DNA'–the infamous and inappropriately named 'junk DNA'.) It is not the genotype that determines the phenotype–that is, the DNA does not specify how proteins will have to be assembled to create the anatomical architecture of a fully developed organism. DNA is not a blueprint and does not function like a 'computer program' that codes for the morphology of the body (for reviews, see (Jose, A framework for parsing heritable information, 2020), (Jose, Heritable Epigenetic Changes Alter Transgenerational Waveforms Maintained by Cycling Stores of Information, 2020 b), (Cohen, Atlan, & Efroni, 2016), or a summary of these (Cutlip, 2020), (Pistoi, 2020).)

Meanwhile, the instructions to form an organism arise from an extremely complex network of interactions and relations between components, such as a myriad of cell signaling factors, enzymes, other proteins, amino acids, vitamins, minerals, etc. Genes are a passive database to build proteins, but they neither design nor control the shape, form, and structure of a living being. We now know that the DNA codes for the ingredients not for the recipe. From epigenetics, we now know that heritable phenotype changes are possible without altering the DNA sequence, involving instead the gene activity and expression which, among other things, may result from how the organism responds to external or environmental factors. There are complicated non-genetic factors that cause the organism's genes to behave differently.

Of course, physical traits like hair, skin, and eye color, facial features, and so on depend on the inherited gene pool. But none of these traits originates from a single gene; rather, they are the result of several streams of chemical synthesis, controlled by regulatory networks whose dynamics govern the genetic transcription. Even this results, in most cases, in a statistical outcome, not a certain and predetermined fact. More evolved functions depend upon even vaster regulatory networks and thousands of genes. Genes are only one of the players in an incredibly complex process inside the whole cell. Moreover, genetically identical individuals do not grow identically, even if subjected to the same environment, while large genetic variations can lead to the same phenotypical trait via multiple alternative pathways (Wagner & Wright, Alternative routes and mutational robustness in complex regulatory networks, 2007). Cells and organisms can also alter their own DNA, rewriting their genome throughout life (Shapiro J. A., How life changes itself: The Read–Write (RW) genome, 2013). By reading the DNA, one cannot determine the shape and size of an organ of a creature. DNA does not even tell us if there is a particular organ at all because it is not the DNA molecule that determines this.

As genomic studies have shown, we humans share 98.9% of our DNA with chimpanzees (Suntsova & Buzdin, 2020). The remaining 1.1% codes for olfactory receptors, some having to do with the size of the pelvic arch, which allows us to walk upright, for fur, and differences in the immune system. It is not clear if the cognitive abilities between humans and chimps are a matter of genetic difference. What is known is that the genome determines a cognitive difference because of genes coding for a higher number of rounds of cell division during fetal brain development, leading to thrice as many neurons in the human brain as in that of a chimp brain. The genes encode only for a quantity which, much later, enables a quality, but there is nothing known in the genome that determines that quality. The latter must somehow be acquired with experience by our human neocortex, which embodies the main cerebral differences between our brain and those of other animals. So far, nothing has been found that codes for these skills other than saying "multiply chimp neurons by three".[148]

Therefore, the popularly conceived genomecentric notion of genes as a blueprint, encoded in the DNA as a program determining the organism's structure, its development, its variations up to every physical trait, and even our psychological inclinations, does not exist. We are not nearly as determined by our genes as once thought.

Another aspect that turned out to be hard to expunge is the (apparent or real) teleological dimension of life. An intriguing example of this comes from morphogenesis–that is, how, from a single fertilized cell, a highly complex self-assembling pattern emerges, developing the organism appropriate to its species. What are the mechanisms and underlying biophysical and chemical phenomena that preside over complex and apparently goal-driven pattern formations, like the self-assembly of structures such as an eye, a limb, and the entire 'bodyplan' during embryogenesis? How do organs regenerate after injury? How can large numbers of cells coordinate their individual activity to assemble themselves into organs and achieve geometric and functional goals that are defined at a macroscopic scale of the whole, but that cannot be found anywhere at the cellular level? Nowadays, we know that there are also correction mechanisms with adaptive decision-making capabilities within living tissues able to correct and adjust embryonic development despite forceful induced defects (Vandenberg, Adams, & Levin, 2012). An inherent 'goal anatomy' that does not stop at the formation of an adult organism, as well-known from the salamander's amazing ability to regenerate an amputated limb.

Contrary to popular belief, we are far from having any coherent theory capable of explaining how all this works. What we know is that it looks like the self-monitoring and repair of complex multi-tissue organ systems and their

---

[148] But does size really matter? Not really, as the brain of an elephant has three times the number of neurons that humans have, and the weight and volume of a sperm whale brain measures six times as much. The neuroanatomical uniqueness of humans is in the number of neurons in the cerebral cortex: about 16.3 billion neurons vs. the 9 billion within a gorilla and the 6 billion within chimpanzees.

pattern formation involves a bioelectric code driving and changing the cell's transmembrane electric potentials. The large-scale anatomical pattern formation is regulated by very complicated networks of bioelectric signaling among cells which determine the differentiation and regulation of embryonic development and the regeneration of injured tissue and even avoid tumor formation (Tseng & Levin, 2013).

Science is now trying to reduce these processes to ordinary electro-chemistry by speaking of hugely complex 'long-range signals', 'planar polarity of proteins on cell surfaces', 'standing waves of gene expression', 'trans-membrane voltage potential and tensile forces', and 'chemical morphogen gradients carrying information about both the existing and the future pattern of the organism' (Levin M. , Morphogenetic fields in embryogenesis, regeneration, and cancer: Non-local control of complex patterning, 2012). Some speak of a 'teleophobia' that *"has gone too far, putting biologists into a straitjacket that prevents them from exploring the most promising hypotheses"* (Levin & Dennet, Cognition all the way down - Aeon, 2020).

Other researchers have placed biological development and morphogenesis in terms of an information architecture. The multicellular organism is considered beyond the material aggregation of material units: It is a whole complexity of a communicative assembly of cells seen as an integrated cellular information field. Individual cellular information fields aggregate into an architectural matrix that enables organism-wide information management (Miller, Torday, & Baluska, The N-Space Episenome Unifies Cellular Information Space-Time within Cognition-Based Evolution, 2019), (Miller, Baluska, & Torday, Cellular senomic measurements in Cognition-Based Evolution, 2020), (Baluškaa & Miller, 2018). From this perspective, multicellularity is collaborative cellular management directed towards the optimization of information quality through its collective (internal and external) measured assessments. The biological organization represents a dual heritable system: It constitutes a biological materiality and also a conjoining information architectural matrix, with morphogenesis deriving from the reciprocations between these two inter-related facets and thereby yielding coordinated multicellular growth and development. Hence, an information architecture could serve as a morphogenetic template. This could be deemed a natural bridge between vitalistic concepts and physical/energetic realities. The logic is that in the cognitive frame, information is physical, not ethereal.

Thus, while previously all hope was laid into a gene-centric approach, and that sooner or later would have abolished any vitalist temptation, now developmental biology resorts to bioelectric molecular signal functions at long range or information-theoretic approaches. To date, however, morphogenesis, despite a century of studies, is to a large degree still shrouded in mystery. The original Aristotelian vitalist claim that the creative generation of the form of living organisms needs a vital force and/or a goal-directed process, and that there must be a 'mind-like essence', an indwelling 'idea-plan' of assembly, without which life cannot develop, is far from having been expunged.

This also raises the question of whether the generation of life by artificial means would dispel any form of vitalist conceptions. The recent developments in synthetic biology come to mind. Synthetic biology is a rapidly growing field of science that aims to redesign organisms for medical and agricultural applications, employing biotechnology, genetic engineering, molecular biology, and other methods of chemical, biological, or computer engineering. A 2007 Nature editorial declared synthetic biology as providing "a welcome antidote to chronic vitalism" (Nature, 2007). The claim was that once science became able to synthesize life, such as building protocells from macromolecular constituents, that would finally discard any form of vitalism. This was questioned more recently, showing that a synthetic approach would nonetheless be consistent with a physical–that is, a functional or organizational–form of vitalism (Bognon-Küss, Chen, & Wolfe, 2018). Meanwhile, the question of whether it would dismiss metaphysical vitalism is not that straightforward either. It may only be said that the ability to engineer, design, recreate, or reproduce something does not necessarily imply that one can explain it and understand which forces are, or are not, at work. Also, relatively simple natural phenomena can be reproduced artificially without the ability to explain the mechanisms that make them come into being and which are the underlying forces and efficient causes.[149]

However, for the time being, this would be a purely speculative discussion. While, to a certain extent, synthetic biology made impressive progress, the creation of primitive cells does not go further than the design and synthesis of a minimal bacterial genome (Hutchinson, Chuang, & Noskow, 2016), protometabolic self-replicating molecules (Ottelé, Hussain, Mayer, & S., 2020), (Santiago, Liu, Browne, & Otto, 2020), or primitive membranes, such as lipid vesicles (liposomes), encapsulating macromolecules (D'Aguanno, et al., 2015). But creating the genetic material to enact the molecular machinery for protein synthesis and the metabolic pathways' complex network biochemical reactions, with all the associated functions performed by the cell's organelles and substructures (nucleus, ribosomes, mitochondria, cytoskeleton, etc.), and, finally, obtaining a structurally stable protocell able to grow and reproduce, pass on heritable traits to offspring, replace worn parts by repair mechanisms, and respond adequately to environmental changes remains a very distant achievement, if possible at all. Therefore, we feel it premature to

[149] For example, one can produce electric currents without even suspecting the existence of electron flows and electric fields, or one can reproduce chemical reactions without knowing anything about the laws of atomic physics and quantum mechanics determining it–let alone understand how, deep down, these necessitate the existence of nuclear forces as well.

speculate on the question of whether synthetic biology repudiates metaphysical vitalism, and leave it to posterity.

a. What is life?

The other claim of the vitalist is that there is a fundamental difference between living and non-living matter. However, also, the attempt to abolish this distinction by reducing the former to the latter remains an elusive undertaking.

Famous is Erwin Schrödinger's question "What is Life?" (Schrödinger E., What is life? The Physical Aspect of the Living. Aspect of the Living Cell, 1943), which precisely tried to address this issue, without any convincing final answer. Biology tried hard to define what should be considered a living organism in naturalistic terms. However, there is no consensus regarding a universal definition of life. Nowadays, biology identifies the 'living', loosely speaking, as anything capable of metabolizing, eating, excreting, maintaining homeostasis, growing, adapting and responding to the environment, reproducing, and evolving. NASA adopted Carl Sagan's original proposal to define life as a "self-sustaining chemical system capable of Darwinian evolution" (Benner, 2010). But one could present several counterexamples of non-living entities which nevertheless satisfy several of the above criteria.

Viruses have no metabolism and remain inactive without reproducing as long as they do not encounter a host cell. And yet, they are capable of Darwinian evolution. Whether or not a virus is to be considered a life form is a matter of debate.

In a sense, cars 'eat', 'metabolize', and 'excrete', but no one would recognize them as resembling life. One might object that cars do not reproduce. But while reproduction is a common trait among all living beings, it cannot be the decisive aspect individuating life because the opposite does not hold: Living beings can also be infertile or sterile. Moreover, computer programs can simulate 'artificial organisms'–digital cellular automatons–fitting the above definition, though only at a much lower level of complexity than that of a real living cell. Whether such 'virtual lifeforms' could be designated as 'living organisms' is highly doubtful.

Thus, any definition of life based on such exclusively material functionalism always fails to capture something of what it is trying to define. For some reason, there is something undefinable, ineffable, and intangible in life, escaping a rigorous and universal scientific definition inside a purely reductionist and physicalist paradigm. The demarcation line between what is living and non-living remains unclear. Why is it so hard to define something whose existence is undeniable and whose distinctiveness so evident?

Because of these philosophical issues, others, such as molecular biologist Andrew Ellington, declared: *"There is a more obvious conclusion to be drawn from our failure to define life: there is no such thing as life. Life is a term for poets, not scientists. There are only replicators with different degrees of complexity. PS: Many of you are closet vitalists"* (Christine, 2010). Indeed, if something cannot be explained inside a particular normative view, one might legitimately conclude that it is a misplaced abstraction and deny its objective existence in the first place, thereby avoiding the burden of defining or explaining it.[150]

However, one may submit another conclusion to be drawn from the failure to define life: Vitalism is always lurking behind the scenes because, contrary to the accepted common narrative, there has never been any serious refutation of it. In life, we see consciousness, will, desire, motivation, cognition, mind, and goal-driven behaviors. Naturalism's aim to reduce these psychological traits to functional descriptions of material processes remains unsuccessful.

b. The mystery of sleep

The author would like to add a short personal conjecture: To the best of my knowledge, no one has connected vitalism to the function of sleep. Let me outline the rationale behind this.

Science is not driven solely by a purely objective, third-person study of reality, but has always been conditioned by the way we perceive the world from a subjective first-person perspective. Consciousness studies are a paradigmatic example of this. From an exclusive third-person physicalist perspective, consciousness should be considered 'non-existent', as we do with vitalism. It is only because of our first-person subjective phenomenal experience that we admit something like consciousness, mind, emotions, feeling, and, more generally, sentience as being a subject worthy of scientific and empiric study. Otherwise, we would not allow these ineffable and undefinable quiddities to be part of scientific analysis.

On the other hand, we do not perceive 'vital energies' as being part of our inner first-person experiential realm and, thereby, the analogy may not hold. The question, however, is: Do we really not perceive it? Are our inner emotional sensations, feelings, instincts, desires, and all those drives we would label as 'vital' of the same nature as the outer physical sensations of touch, sight, sound, smell, and taste? Does the fact that both manifest inside a bodily boundary prove them to be material?

For example, does that feeling of 'freshness', that sensation of having 'recharged the batteries' we know from our personal experience after a good sleep, come only from the physical? Conversely, does that perception of a 'lack of vitality' we know from sleep deprivation indicate only an organic lack of

[150] This approach is reminiscent of the doctrine of eliminative materialism in the philosophy of mind. Because phenomenal consciousness refuses to be cast into a strict naturalist account, the eliminativist contends that our subjective experiences, such as pleasure and pain, joy and grief, and the existence of our mental states, are just inexistent 'illusions'.

energy? If so, why can that metabolic deficiency not be replaced by something metabolic, say, by nutritional means, and instead requires an apparently passive function such as sleep? When we do not sleep enough, we increasingly feel a sense of a lack of 'vitality', 'energy-deprivation', and 'inertia' that no substance, food, or metabolic process in a waking state can regenerate; only the sleep cycles can restore this balance.

One of the unresolved issues in modern biology and psychology remains the nature and function of sleep and dreams. Findings suggest that dreams have the functional role of processing our emotional waking-life experiences to avoid an informational and experiential overload that we could otherwise barely handle. Scientists agree that sleep has the purpose of repairing and reorganizing neural pathways, consolidating memories acquired during the waking state, filtering out redundant information, restoring the body and mind, and serving other metabolic, physiological, and psychological functions. As is well known, sleep deprivation leads to severe psychophysiological disorders and, in extreme cases, leads to death. However, there is now a growing consensus that this cannot be the whole story and that the real function and purpose of sleep remain elusive (for a review, see (Freiberg, 2020), (Frank & Heller, 2018), (Cao & al., 2020)).

In fact, we do not know why all the functions of sleep mentioned above could not be performed in a waking state as well. Also puzzling is the evolutionary origin of sleep. From a naturalistic evolutionary perspective, sleep looks like an outsider. According to Neo-Darwinism, everything–including sleep–must have evolved from natural selection and random mutations to allow for the best survival and reproduction chances. But sleep is a risky habit in a prey-predator environment, especially for those that are at the bottom of the food chain. Yet, there is no living organism, from cyanobacteria all the way to humans, that is not subjected to a circadian clock that, in brains, expresses itself as what we call 'sleep'.

The assumption that identifies sleep uniquely as a cerebral cycle is questioned by findings that show how the waking-sleep activity is not something proper only to organisms with a brain; those without a brain also show waking-sleep cycles. A sleep-like state has been observed in the cnidarian Hydra vulgaris, a small freshwater polyp that has only a primitive nervous organization (Kanaya & al., 2020). It is now known that sleep is also present in animals, such as the jellyfish Cassiopea, which possess neurons organized into a non-centralized nerve system, but still have no brain. Their pulsing behavior, alternated by periods of quiescence at night, is consistent with waking-sleep cycles. When deprived of these quiescence periods, their activity and responsiveness decrease, indicative of a sleep-like state and supporting the hypothesis that sleep arose prior to the emergence of a centralized nervous system (Nath R. D., 2017). This suggests that sleep serves a more fundamental function than neuronal reorganization, primarily metabolic functions (Anafi & al., Exploring phylogeny to find the function of sleep, 2019).

The question remains, however: Why must this metabolic function occur with an altered mind, body, and consciousness state? The vitalistic hypothesis that sleep may serve as 'recharging' the 'life-sheet' or 'vital body' with its 'vital energy' is not falsified.

5. The mind in life

a. Consciousness and mind

Vitalism cannot be abstracted from the so-called 'hard problem of consciousness' (Chalmers D. , 1995), and that also asks for a satisfying closure of the explanatory gap between neural correlates and the subjective experiential and mental dimension of phenomenal consciousness and our inner subjectivity. Life stands in front of us with its cognitive and conscious dimensions as well. Every form of life, including the most primitive unicellular organism, as we will see, displays different degrees of cognitive behavior.

We previously distinguished between life and mind, with the former being characterized by qualities such as desire, emotions, instinct, strife, or will, while the latter relates to cognition, reason, or intelligence. Let us complement this distinction with a few clarifying remarks on the terminological and qualitative distinction between consciousness and mind as well.

In modern consciousness studies, consciousness is usually regarded as the ability to have any subjective experience, sentience, or perception–what the American philosopher Thomas Nagel called "something it is like to be" (Nagel T. , What is it like to be a bat?, 1974). If there is something, someone, or anything it is like to have an experience in and of itself, then one is said to be conscious. Loosely speaking, consciousness is what allows us to be a sentient, an experiencing subject witnessing qualitative phenomenal contents–so-called qualia–such as perceptions, feelings, pain and pleasure, sensory data, the perceptions of qualities like the redness of a rose or the sweetness of sugar, emotions, and mental contents.

But consciousness is also conscious *of* something: the fleeting thoughts, the impression of emotions, the sensory perceptions, the film of life, etc. In particular, it is conscious *of* mental events.

At a higher level, mind is the rational intellect, the thinking reason, that organizes and structures into representations, ideas, concepts, and meanings, that analyzes, reflects, makes more or less abstract logical inferences by analytic and deductive thought processes, etc. Cognition is a process of mind. It is the mental act or process of knowing, understanding, and perceiving through thought, experience, and the senses. As we are going to see in the next section, at a lower level, one speaks of 'basal cognition', meaning rudimentary learning and decision-making processes, beginning from the dynamic problem-solving capacities of cellular and subcellular forms (Lyon P. , The biogenic approach to cognition, 2006), (Ginsburg, Jablonka, & Zeligowski, 2019).

Therefore, mind must be kept distinct from consciousness insofar that mind is an instrumental entity of consciousness with the cognitive function of thinking and knowing. Consciousness is distinct in the sense that it apprehends (but is not) mental states such as concepts, ideas, images, imaginations, etc. The distinction is necessary not only for semantic clarity, but also because, in hindsight, it is a common first-person experience: While mental contents such as thoughts and ideas come and go, the conscious individual subject witnessing those impermanent thoughts and ideas does not change its identity. To put it metaphorically: Consciousness is the canvas, mental phenomenality is the painting.

According to scientific materialism, mentality, in its low or high-level cognitive function, must be explained in the frame of a reductionist and physicalist formulation. All those cognitive tasks we characterize as 'mental' are considered epiphenomena and emergent properties, resulting from biophysical interactions of networked elementary units, such as molecules or neurons, for adaptation purposes of the organism. Meanwhile, the mind-body problem of the philosophy of mind remains, more than ever, an actual debate failing to find a commonly accepted resolution. Despite the enormous advances in neuroscience and consciousness studies within the last decades, consciousness and mind remain one of the most elusive aspects of reality.

This led to the recent revival of metaphysical ontologies, such as panpsychism or idealistic and panentheistic conceptions of a universal mind. Most notably, panpsychism has been reconsidered in its different forms by modern leading philosophers in the field, like Thomas Nagel (Nagel T. , Panpsychism, 1979), Galen Strawson (Strawson G. , 2006), David Chalmers (Chalmers D. , Panpsychism and Panprotopsychism, 2015), and Philip Goff (Goff, 2017), and on the other side of the spectrum, theories of cosmic consciousness, such as Ithai Shani's cosmopsychism (Shani I. , Cosmopsychism: A Holistic Approach to the Metaphyscis of Experience, 2015) or Bernardo Kastrup's analytic idealism (Kastrup B. , The Universe in Consciousness, 2018), to mention only the most noted. These posit not matter, but, rather, consciousness or mind, as fundamental.

b. Cellular cognition and consciousness and plant basal cognition revisited

There is now sufficient empiric evidence suggesting how at least forms of 'basal cognition' necessitate neither a brain nor a nervous system. Mounting evidence suggests how at least cognition might be a fundamental aspect of life, before the emergence of neural activity, while until recently it was taken for granted that any form of cognition could emerge solely from a brain or that it requires at least a neural substrate.

However, biology is discovering that cells have some primitive ability to learn and associate, resembling conditioned behaviors or changing them by anticipatory skills. Within the turn of the millennium, a renewed interest in this field gained momentum, especially due to the new findings that are transforming our understanding of how mentality emerges in living organisms and even questions the very notion of 'intelligence' and 'mind' itself.

The most notorious example of cellular intelligence is that of the 'Physarum polycephalum', a large amoeba-like slime mold 'plasmodium'–a fungal cytoplasm containing several nuclei, but enclosed in a single membrane – that can be considered a single giant cell. It changes its shape as it crawls in search of food, as a yellow network of tube-like structures that grow a few centimeters per hour, a movement that can be captured via time-lapse recordings. This slime mold has several skills and behavioral patterns that could be labeled 'proto-intelligent' and that one would hardly associate with such a primitive creature. For example, it can find the minimum length between two points in a labyrinth (Nakagaki, Yamada, & Tóth, 2000). Further research showed that P. polycephalum can minimize the network path and complexity between multiple food sources–it can solve the 'traveling-salesman problem', commonly known in both biology and computer architecture (Nakagaki T. , 2004).

Conditioned behavior was shown as well. When it is exposed to life-threatening electric pulses at constant time intervals, it reduces its speed of growth or stops entirely for a while before starting to grow again. Once conditioned, it learns to anticipate the arrival of the shock if one administers a series of pulses and leaves out the last one (Saigusa & al., 2008). P. polycephalum is also able to adjust to unfavorable circumstances. If one forces it to cross an agar bridge with caffeine or quinine at toxic, but not killing, concentrations, it first slows down but, after repeated attempts, nevertheless crosses the bridge at the same speed in the absence of the irritant substances (Boisseau, Vogel, & Dussutour, 2016). It was believed that this was something only neural networks could do: learning to ignore repeated negative stimuli. That is, a learning process of habituation took place.

Another quite surprising behavior was (re)discovered recently in another protozoan. Already in 1906 (for a short historical account, see (MS, 2019)), the American zoologist, Herbert Spencer Jennings, noted how the ciliate 'Stentor roeselii' is capable of escalating actions to avoid an irritant stimulus. One hundred and thirteen years later, Jennings' observations were confirmed (Dexter, Prabakaran, & Gunawardena, 2019). Indeed, this unicellular organism can change its behavior about how to respond to the environment in an escalation of actions that, to date, represent the most complex behavior known for a single cell.

Empiric evidence suggests that conditioned behavior in other unicellular organisms exists, such as in amoebae. A Spanish group analyzed the motility pattern of the 'Amoeba proteus' under the influence of stimuli, consistent with associative conditioned behavior (De la Fuente, Bringas, Malaina, & al., 2019).

Similar abilities have been observed in bacteria. Bacteria are considered the most elementary form of life because they are prokaryotic cells. Nevertheless, it has been shown that they can sense the environment, actively move within it, target food, avoid toxic substances, and meaningfully change their swimming direction. Most interesting is their behavior when they come together and form a bacterial community, which shows surprising problem-solving abilities. Bacteria communicate with each other and coordinate gene expression, which determines the collective behavior of the entire community by a 'quorum sensing'. They can cooperatively self-organize into complex colonies through a communication-based intercellular interplay (Ben-Jacob & Levine, 2006).

This leads to a collaboration that allows the community to achieve a common goal. These communities elaborate functional structures to determine if other microorganisms nearby are threatening their survival and eventually secrete antibiotic compounds toxic to other species except their own, anchor in a place or stay motile, divide for the growth of the community, release spores, etc. This allows them to work together to survive in stressful environments. Analogous to P. polycephalum, bacteria's collective intelligence becomes evident when they are confronted with relatively complex task-solving problems, such as route-finding in mazes and fractals (Phan & al., 2020). Cells were observed in sensing a shortcut using self-generated gradients and selecting a new minimal route (see (Tweedy & al., Seeing around corners: Cells solve mazes and respond at a distance using attractant breakdown, 2020) or for a review of bacteria's basal cognitive skills, see also (Lyon P. , The cognitive cell: bacterial behavior reconsidered., 2015)).

Whether unicellular forms of life can be considered conscious in the sense of having some form of a sentient experience is something hard to establish given the inherently subjective nature of consciousness. Nevertheless, this evidence of cellular intelligence prompted hypotheses regarding the cellular origin of consciousness as well. Noteworthy is the Cellular Basis of Consciousness (CBC) model, which challenges the assumption that subjective awareness emerged late in evolution (Reber & Baluška, 2021). Conscious experience may already appear in cellular form, co-terminus with life. Processes taking place in excitable cellular membranes may lead to a basal form of proto-consciousness with intentional and cognitive capacities (Baluška, Miller, & Reber, 2021). In this theoretical framework, the plasma membrane, together with the nuclear envelope/centrosome/microtubules complex, are seen as the two different 'nanobrains' generating a 'cellular self' with a purposive agency.[151] Individual cells maintain their identity by measuring information to sustain a self-referential homeostatic equipoise in reaction to the external environment (Miller Jr, Baluška, & Torday, 2020). A self-referential awareness that *is* self-organization, and in which every cell embodies an elemental cognition. From this perspective, one could conceive of a Cognition-Based-Evolution (CBE), complementing Darwinian natural selection processes, in which receptive intelligent cells measure the environment–that is, assess information–and consequently learn and react to uphold self-identity. In a sense, life could be seen as a 'continuous measurement of information'. For a thought-provoking summary of the CBC theory, see also (Reber A. , 2019).

What about multicellular organisms? In this regard, it might be worth noting how there is evidence that non-neuronal multicellular organisms also show forms of intelligence. At the turn of the millennium, terms like 'plant neurobiology'[152] emerged, drawing parallels between the complex information processing and signaling system in plants and the animal's neuronal activity (Brenner & al., 2006), (Calvo, 2016).

To provide some examples, it was shown that garden pea seedlings (Pisum sativum) change their foraging behavior–their direction of growth–if trained to associate a running fan with a light source, shining an hour after the fan's operation (Gagliano & al., 2016). Another example that raises important questions, not only about the predictive abilities of plants, but also about how they perceive the environment, was an experiment analyzing the goal-directed movements of the same pea plants, which showed that the climbing plant searching for a support to attach to exhibited an anticipatory prehensile mechanism, which gave it the ability to plan its movements before having any physical contact with the support (Guerra, Castiello, & al., 2019). There is now an extended literature that, especially in the last decade, has consistently shown how plants change behavior and adapt, respond predictively, possess some form of memory, resort to an air and underground communication system based on chemicals, have visual and acoustic signals, have learning abilities and can evaluate their surroundings, make decisions, and even have a social life cooperating with one another (for a not-too-long review, see (Trewavas, 2017).)

The overall picture that arises from these findings is that, among living beings without a brain, an intimation of collective intelligence arises when

---

[151] Also others, independently from the CBC model, point out how mitochondria already exhibit social behavior by communicating with each other and with the cell nucleus, exhibiting group formation and interdependence, synchronizing their behaviors, and engaging in functional specialization (Picard & Sandi, 2021).

[152] Speaking of 'plant neurobiology' for an organism without neurons is an obvious misnomer. This, however, has become a common nomenclature.

several units or 'nodes'–unicellular organisms–connect and form an informational network. Once several of these nodes signal to each other in a complex connectome, a new order and level of functional skills arise that the single cell does not have. The single cell already seems to have some elementary cognition (and, perhaps, a proto-consciousness), but something new and qualitatively different emerges once these little entities connect in a complicated communication web.

c. What is cognition?

These recent discoveries on monocellular or multicellular behaviors raise questions. Is a tiny unicellular creature, which swims through a fluid, hunts for its prey, avoids obstacles, has a memory, and can even predict events in advance, just a piece of complex biophysical machinery? Or should we ascribe to it, at least, some form of 'basal cognition' and eventually even some elementary form of sentience? Is a climbing plant that nervously flutters its tendrils throughout space, analyzes the environment, grows towards a support it apparently 'sees', and begins to grab for it before even touching it only a machine driven by a chemical reaction, or something to which we could ascribe a form of conscious cognition?

As with the notion of life, the unambiguous definition of what is commonly called 'cognition' remains a matter of considerable debate (for a couple of recent works on this matter, see (Lyon & al., Reframing cognition: getting down to biological basics, 2021) and (Regolin & Vallortigara, 2021).) Broadly speaking, we can maintain an understanding of cognition as the action or faculty of learning, decision making, sensing and responding, communicating, information processing, and having memory, agency, and associative skills, including high-level forms of cognition that result in analytic thinking and reason, known as the 'mind'.

A first standpoint, that of a physicalist approach, is that which does not consider the above described cellular behavioral phenomena as proof of what is commonly meant by 'cognition', 'intelligence', or even less a 'mind' or 'consciousness'. One might regard such a standpoint as an attempt to anthropomorphize non-human behavior and, instead, express cognitive functions in organisms without a brain in terms of system theory, reducing all complex systems dynamics to mere elementary processes: as an emergent property instantiated as an adaptive behavior of a complex nonlinear dynamical system, which rearranges its internal state in response to the stimuli of an external environment. For example, the system theory of autopoiesis, by Chilean biologists Humberto Maturana and Francisco Varela (Maturana & Varela, 1980), refers to the process of self-creation and self-preservation of living systems, where cognition is seen as a self-referencing mechanism determined by a structural coupling to the environment. It works along the line of Gilles Deleuze, who considered the tendency of life to move towards a greater self-organizing complexity that maximizes difference (the genesis and process of ongoing biological differentiation), with its dynamic potentiality to develop beyond its actuality (life's tendency to build new organic structures and properties that allow it to perform new functions), and renamed it 'vitality', a form of physical vitalism ((Deleuze, 1966), (Grosz, 2004)).

The second standpoint, that of a metaphysical approach, is the idea of matter having in itself, a priori, mental or experiential properties; this was considered only by some philosophers, like Gottfried Wilhelm Leibniz, Baruch Spinoza, Alfred North Whitehead, and William James, who expressed a panpsychist view, namely, that everything is fundamentally a form of consciousness or mind. But these speculations did not have, until recently, much influence on the overall established naturalistic paradigm. The science of biosemiotics, a branch of biology that interprets living processes as the production and interpretation of signs, codes, information, meanings, habit formation, and their communication in the biological realm (for an introductory essay, see (Else, 2010)), came somewhat closer to the idea of cognition as a fundamental aspect of reality, but refrains from making the decisive step towards the reversion of the naturalistic paradigm.

The (more or less implicit) premise of present science is the physicalist and reductionist view in which we must recast everything into a causational chain determined by a system of processes that lead to cognition as an emergent secondary epiphenomenon–that is, something that works by a unidirectional bottom-up process, such as: aggregation of microscopic fundamental interacting units into large-scale network processes → information/code/semantics → cognition. The plausibility of the hypothesis that goes the other way around is rarely addressed (for an interesting exception, see (Marshall P. , 2021)), namely, that cognition is fundamental, and even more fundamental than matter, and that the causational chain works as: cognition → semantics/code/information → aggregation of microscopic fundamental interacting units into large-scale network processes.

The observational data that suggest cognitive properties in non-neuronal tissue allow us to posit that cognition might not be just an emergent behavior determined by a complex network of processing units, but, rather, that it is an inherent and basic feature of living matter itself. There might exist not only a life-principle, but also a mind-principle, both expressing a vital-mental polarity.

6. Conclusion

We presented some arguments and scientific evidence aimed at critically reviewing the widely held belief that vitalism is dead, with mentality and consciousness having no place in Nature other than being emergent epiphenomenon. This belief holds fast to the prophecy of Francis Crick: *"And so to those of you who may be vitalists I would make this prophecy: what everyone believed yesterday, and you believe today, only cranks will believe tomorrow"* (Crick F. , Of Molecules and Men, 1966).

Upon closer inspection, however, one realizes that there is no end in sight for the search for a mechanistic explanation that reduces life to non-life and dismisses Bergson's 'creative evolution' or Driesch's entelechy as self-organizing principles, furnishing a naturalistic foundation of the volitional and cognitive behavior in living organisms. For the same reason, modern findings on the relatively complex cognitive behavior of simple life forms without a nervous system also shift us away from a naturalistic reduction of mental phenomena. Life, mind, and consciousness are far from having been naturalized, even from a philosophical perspective. Whatever the truth, as of the date of this writing, their naturalization cannot be considered a closed case. Metaphysical vitalism and idealism remain viable options. Nothing in science prevents us from seeing will, cognition, and sentience as powers of a subconscious Nature trying to express itself in different biological forms. A theoretical framework is possible that sees life and mind as the two aspects of a conscious universe and as two fundamental inherent properties that emerge in and through matter rather than by matter alone. Life and mind, as two distinct aspects of reality, both having consciousness as their origin and matter as their supporting basis.

Moreover, the trouble with expunging teleological temptations might well be due to humans' tendency to anthropomorphize Nature, as Ellington implied, but nothing prevents us from recognizing life with its impulsive dimension (will, desire, emotions, motivations, etc.), mind with its cognitive dimension (learning, problem-solving and decision-making processes, information and thought processes, reason, intelligence, knowing), and consciousness with its sentient dimension (experience, sensations, subjectivity) as fundamental principles that are already inherent in Nature and that impel the organization and function of living matter itself, rather than being mechanistic emergent appearances alone. Denying these features of the natural world as being exclusive to humans could be seen as a form of anthropocentrism as well.

The dismissal of these paradigms might have been too premature, and they may soon find their way back into our world-constructs, at least in matters of philosophical interest in biology, and, like idealism, panpsychism, and various conceptions of universal consciousness, are already doing in the philosophy of mind. The phenomenon of life emerging from non-living matter, and that of mind from non-mental stuff, is no less perplexing than that of consciousness emerging from non-conscious matter. There is not only a hard problem of consciousness, but also a hard problem of life, and a hard problem of mind.

Seeing life, mind, and consciousness as appearance arising out of a mechanistic clash of material forces and particles, or, on the contrary, as fundamental properties already inherent in matter itself, is not just a philosophical musing, but something that may influence our way of seeing biology and determine the very practical aspects of science and its future progress or stagnation.

Acknowledgement
The author is grateful to the anonymous referees for their careful reading of the paper and for valuable suggestions and comments.

References

Anafi, R. C. & al., e., 2019. Exploring phylogeny to find the function of sleep. *Nature Reviews Neuroscience,* Volume 20, pp. 109-116.

Baluškaa, F. & Miller, W., 2018. Senomic view of the cell: Senome versus Genome. *Commun Integr Biol.,* 11(3).

Baluška, F., Miller, W. B. & Reber, A. S., 2021. Biomolecular Basis of Cellular Consciousness via Subcellular Nanobrains. *Int. J. Mol. Sci.,* 22(5), p. 2545.

Ben-Jacob, E. & Levine, H., 2006. Self-engineering capabilities of bacteria. J. Royal Soc. Interface, 3(6), 197-214. *J R Soc Interface.,* 3(6), p. 197–214.

Benner, S. A., 2010. Defining Life. *Astrobiology,* 10(10), pp. 1021-1030.

Bergson, H., 1907. *Creative Evolution.* 1st ed. s.l.:s.n.

Bognon-Küss, C., Chen, B. & Wolfe, C., 2018. Metaphysics, function and the engineering of life: the problem of vitalism. KAIROS-JOURNAL OF PHILOSOPHY & SCIENCE, 20(1), pp.113-140.. *Kairos-Journal of Philosophy & Science,* 20(1), pp. 113-140.

Boisseau, R. P., Vogel, D. & Dussutour, A., 2016. Habituation in non-neural organisms: evidence from slime moulds. *Proc. Roy. Soc. B,* 283(1829).

Brenner, E. D. & al., 2006. Plant neurobiology: an integrated view of plant signaling. *Trends in Plant Science,* 11(8), pp. 413-419.

Burley, M., 2007. *Classical Samkhya and Yoga: An Indian Metaphysics of Experience.* s.l.:Routledge.

Calvo, P., 2016. The philosophy of plant neurobiology: a manifesto. *Synthese,* Volume 193, p. 1323–1343.

Cao, J. & al., e., 2020. Unraveling why we sleep: Quantitative analysis reveals abrupt transition from neural reorganization to repair in early development. *Science Advances,* 6(38).

Chalmers, D., 1995. Facing up to the problem of consciousness. *Journal of Consciousness Studies,* 2(3), p. 200–219.

Chalmers, D., 2015. Panpsychism and Panprotopsychism. In: *Consciousness in the Physical World: Perspectives on Russellian Monism.* s.l.:New York: Oxford University Press., pp. 246-276.

Christine, A., 2010. *Harnessing non-modernity: a case study in artificial life - Dissertation Thesis,* s.l.: s.n.

Cohen, I. R., Atlan, H. & Efroni, S., 2016. *Genetics as Explanation: Limits to the Human Genome Project.* [Online].

Crick, F., 1966. *Of Molecules and Men.* s.l.:Seattle: University of Washington Press.

Cutlip, K., 2020. *DNA may not be life's instruction book—just a jumbled list of ingredients.* [Online]
Available at: https://phys.org/news/2020-04-dna-life-bookjust-jumbled-ingredients.html

D'Aguanno, E. et al., 2015. Physical Routes to Primitive Cells: An Experimental Model Based on the Spontaneous Entrapment of Enzymes inside Micrometer-Sized Liposomes. *Life (Basel),* 5(1), p. 969–996.

De la Fuente, I., Bringas, C., Malaina, I. & al., 2019. Evidence of conditioned behavior in amoebae. *Nature Communications,* 10(3690).

Deleuze, G., 1966. *Bergsonism.* s.l.:Boston, MA: Zone Books..

Dexter, J. P., Prabakaran, S. & Gunawardena, J., 2019. A Complex Hierarchy of Avoidance Behaviors in a Single-Cell Eukaryote. *Current Biology,* 29(24), pp. 4323-4329.

Else, L., 2010. A meadowful of meaning. *New Scientist,* 207(2774).

Frank, M. & Heller, H., 2018. The Function(s) of Sleep. In: *Sleep-Wake Neurobiology and Pharmacology. Handbook of Experimental Pharmacology - Landolt HP., Dijk DJ. (eds)* . s.l.:Springer, Cham..

Frank, P., 1941. The Law of Causality and Experience (1908). In: *Between Physics and Philosophy.* s.l.:Cambridge, MA: Harvard University Press., pp. 17-28.

Freiberg, A. S., 2020. Why We Sleep: A Hypothesis for an Ultimate or Evolutionary Origin for Sleep and Other Physiological Rhythms. *J Circadian Rhythms,* 18(12).

Gagliano, M. & al., 2016. Learning by Association in Plants. *Sci. Rep.,* Volume 6:38427.

Gilbert, S. F. & Sarkar, S., 2000. Embracing complexity: Organicism for the 21st century. *Developmental Dynamics.*

Ginsburg, S., Jablonka, E. & Zeligowski, A., 2019. *The evolution of the sensitive soul: Learning and the origins of consciousness.* s.l.:Cambridge, Massachusetts: The MIT Press..

Goff, P., 2017. *Consciousness and Fundamental Reality.* s.l.:New York: Oxford University Press.

Grosz, E., 2004. *The Nick of Time: Evolution and the Untimely.* s.l.: Durham NC: Duke University Press..

Guerra, S., Castiello, U. & al., 2019. Flexible control of movement in plants. *Nature Sci. Rep.,* 9(16570).

Hempel, C., 1966. *Philosophy of Natural Science.* s.l.:Englewood Cliffs (NJ): Prentice Hall..

Hutchinson, C. A., Chuang, R. & Noskow, V. V. C., 2016. Design and synthesis of a minimal bacterial genome. *Science,* 351(6280).

Jose, A. M., 2020 b. Heritable Epigenetic Changes Alter Transgenerational Waveforms Maintained by Cycling Stores of Information. *Bioessay,* 42(7).

Jose, A. M., 2020. A framework for parsing heritable information. *Journal of the Royal Society,* 17(165).

Kanaya, H. J. & al., e., 2020. A sleep-like state in Hydra unravels conserved sleep mechanisms during the evolutionary development of the central nervous system. *Science Advances,* 6(41).

Kastrup, B., 2018. The Universe in Consciousness. *Journal of Consciousness Studies,* 25(5–6), p. 125–55.

Levin, M., 2012. Morphogenetic fields in embryogenesis, regeneration, and cancer: Non-local control of complex patterning. *Biosystems,* 109(3), pp. 243-261.

Levin, M. & Dennet, D., 2020. *Cognition all the way down - Aeon.* [Online] Available at: https://aeon.co/essays/how-to-understand-cells-tissues-and-organisms-as-agents-with-agendas

Lyon, P., 2006. The biogenic approach to cognition. *Cogn. Process.,* Volume 7,, p. 11–29..

Lyon, P., 2015. The cognitive cell: bacterial behavior reconsidered.. *Front Microbiol.,* Volume 6, p. 264.

Lyon, P. & al., e., 2021. Reframing cognition: getting down to biological basics. *Phil. Trans. Royal Soc. B,* 376(1820).

Marshall, P., 2021. Biology transcends the limits of computation. *Progress in Biophysics and Molecular Biology.*

Maturana, H. & Varela, F., 1980. *Autopoiesis and cognition: The realization of the living.*. s.l.:Boston: Reidel.

Miller Jr, W. B., Baluška, F. & Torday, J. S., 2020. Cellular senomic measurements in cognition-based evolution. *Progr. Biophys. Mol. Biol.,* Volume 156, pp. 20-33.

Miller, W., Baluska, F. & Torday, J., 2020. Cellular senomic measurements in Cognition-Based Evolution. *Progress in Biophysics and Molecular Biology,* 156(3).

Miller, W., Torday, J. & Baluska, F., 2019. The N-Space Episenome Unifies Cellular Information Space-Time within Cognition-Based Evolution. *Progress in Biophysics and Molecular Biology,* 150(2).

MS, H., 2019. *New study hints at complex decision making in a single-cell organism.* [Online]
Available at: www.sciencedaily.com/releases/2019/12/191205113129.htm

Nagel, E., 1961. *The structure of science. Problems in the Logic of Scientific Explanation.*. s.l.: London, Routledge-Kegan Paul.

Nagel, T., 1974. What is it like to be a bat?. *The Philosophical Review,* 83(4), p. 435–450.

Nagel, T., 1979. Panpsychism. In: *Mortal Questions.* s.l.:Cambridge University Press, p. 181–195.

Nakagaki, T., 2004. Obtaining multiple separate food sources: behavioural intelligence in the Physarum plasmodium. *Proc. R. Soc. Lond. B,* Volume 271, p. 2305–2310.

Nakagaki, T., Yamada, H. & Tóth, Á., 2000. Maze-solving by an amoeboid organism. *Nature,* 407(470).

Nath, R. D. e. a., 2017. The Jellyfish Cassiopea Exhibits a Sleep-like State. *Current Biology,* 27(19), pp. 2984-2990.

Nature, 2007. Meanings of 'life'. *Nature (editorial),* Volume 447, p. 1031–1032.

Normandin, S. & Wolfe, C., 2013. Introduction. In: *Vitalism and the Scientific Image in Post-Enlightenment Life Science, 1800-2010.* s.l.:Springer.

Ottelé, J., Hussain, A. S., Mayer, C. & S., O., 2020. Chance emergence of catalytic activity and promiscuity in a self-replicator. *Nature Catalysis,* Volume 3, p. 547–553.

Phan, T. V. & al., e., 2020. Bacterial Route Finding and Collective Escape in Mazes and Fractals. *Phys. Rev.,* X(10), p. 031017.

Picard, M. & Sandi, C., 2021. The social nature of mitochondria: Implications for human health. *Neuroscience & Biobehavioral Reviews,* Volume 120, pp. 595-610.

Pistoi, S., 2020. *DNA Is Not a Blueprint.* [Online]
Available at: https://blogs.scientificamerican.com/observations/dna-is-not-a-blueprint/

Reber, A., 2019. *The First Minds: Caterpillars, Karyotes, and Consciousness.* s.l.: New York: Oxford University Press.

Reber, A. S. & Baluška, F., 2021. Cognition in some surprising places. *Biochem. Biophys.l Res. Commun.,* Volume 564, pp. 150-157.

Regolin, L. & Vallortigara, G., 2021. Rethinking cognition: From animal to minimal. *Biochemical and Biophysical Research Communications.*

Saigusa, T. & al., 2008. Amoebae anticipate periodic events. *Phys Rev Lett.,* 100(1)(018101).

Santiago, G. M., Liu, L., Browne, W. & Otto, S., 2020. Emergence of light-driven protometabolism on recruitment of a photocatalytic cofactor by a self-replicator. *Nature Chemistry,* Volume 12, p. 603–607.

Schlick, M., 1953. Philosophy of Organic Life. In: Appleton-Century-Crofts, a cura di *Readings in the Philosophy of Science.* New York: H. Feigl and M. Brodbeck (eds.), p. 523–536.

Schrödinger, E., 1943. *What is life? The Physical Aspect of the Living. Aspect of the Living Cell.* s.l.:s.n.

Shani, I., 2015. Cosmopsychism: A Holistic Approach to the Metaphyscis of Experience. *Philosophical Papers,* 44(3), pp. 389-437.

Shapiro, J. A., 2013. How life changes itself: The Read–Write (RW) genome. *Physics of Life Reviews,* 10(13), pp. 287-323.

Strawson, G., 2006. Realistic Materialism: Why Physicalism Entails Panpsychism. *Journal of Consciousness Studies,* 13(10-11), p. 3–31..

Suntsova, M. V. & Buzdin, A. A., 2020. Differences between human and chimpanzee genomes and their implications in gene expression, protein functions and biochemical properties of the two species. *BMC Genomics,* 21(535).

Theise, N. D. & Kafatos, M. C., 2016. Fundamental awareness: A framework for integrating science, philosophy. *Commun Integr Biol.,* 9(3).

Timalsina, S., 2008. *Consciousness in Indian Philosophy: The Advaita Doctrine of 'Awareness Only'.* s.l.:Routledge.

Trewavas, A., 2017. The foundations of plant intelligence.. *Interface Focus,* 7(20160098).

Tseng, A. & Levin, M., 2013. Cracking the bioelectric code. *Commun Integr Biol.,* 6(1).

Tweedy, L. & al., e., 2020. Seeing around corners: Cells solve mazes and respond at a distance using attractant breakdown. *Science,* 369(6507).

Vandenberg, L. N., Adams, D. S. & Levin, M., 2012. Normalized shape and location of perturbed craniofacial structures in the Xenopus tadpole reveal an innate ability to achieve correct morphology. *Developmental Dynamics,* 241(5), pp. 863-878.

Wagner, A. & Wright, J., 2007. Alternative routes and mutational robustness in complex regulatory networks. *Biosystems.,* 88(1-2), pp. 163-172.

Wolfe, C. & Normandin, S., 2013. Vitalism and the scientific image: an introduction. In: *Vitalism and the scientific image in post-Enlightenment life science,.* s.l.:Springer.

Wolfe, C. T., 2017. Models of organic organization in Montpellier vitalism. *Early Scienece and Medicine,* Volume 22(2-3), pp. 229-252.

3. Quantum Indeterminism, Free Will, and Self-Causation[153]

Abstract

A view that emancipates free will by means of quantum indeterminism is frequently rejected based on arguments pointing out its incompatibility with what we know about quantum physics. However, if one carefully examines what classical physical causal determinism and quantum indeterminism are according to physics, it becomes clear what they really imply–and, especially, what they do not imply–for agent-causation theories. Here, we will make necessary conceptual clarifications on some aspects of physical determinism and indeterminism, review some of the major objections against libertarian conjectures, and show that there is no conceptual incompatibility preventing us from taking a 'quantum-libertarian' approach to the problem of free will. In particular, we will illustrate the possible role of self-causation (causa sui) as a potential solution to otherwise apparently incompatible or even paradoxical statements concerning free will and quantum indeterminism.

I. Introduction

While quantum mechanics (QM) became one of the most successful scientific theories of the 20th century, and the predictive power of its mathematical foundation is unparalleled and unquestioned, its conceptual foundation and philosophical implications are notoriously less unambiguous. Classical realism based on concepts of local causality, determinism, non-contextuality, and what Einstein, Podolsky, and Rosen (EPR) famously called "*elements of physical reality*" (Einstein, Podolosky, & Rosen, 1935), has been replaced by a 'quantum realism' based on notions of non-locality, indeterminism, and contextuality. The experimental demonstration that Bell's inequalities (Bell, 1964) are violated by QM implies that either reality is non-local, or reality is irreducibly indeterministic, or both.[154] Though it is still possible to frame various interpretations of QM that save a deterministic form of realism (e.g., the de Broglie-Bohm theory), quantum indeterminism is

[153] The present article was published in the *Journal of Consciousness Studies*, vol. 30, no. 5-6, pp. 32–56, 2023. DOI: **https://doi.org/10.53765/20512201.30.5.032**

[154] Therefrom Bell's theorem: "No physical theory of local hidden variables can ever reproduce all of the predictions of Quantum Mechanics."

considered by most physicists to be an inherent and irreducible aspect of reality. This is also the reason why, in an experimental context requiring highly reliable random number generation (e.g., in cryptography or for Bell-tests), quantum random generators are preferred over classical pseudo-random ones.

Less consensus meets the conjectures about the possible role that quantum effects could play in brain activity, the emergence of consciousness, and its implications for free will. While quantum biology is an emerging field of investigation supported by some empiric evidence (for an update and perspective, see (Youngchan, Bertagna, D'Souza, & Heyes, 2021), or for a review (McFadden & Al-Khalili, 2018)), the hypothesis that there could be a relationship between quantum mechanical phenomena in cellular and subcellular structures, and our mental phenomenality emerging from neuronal activity, remains a controversial matter. One of the main concerns is related to the question of how quantum effects in the brain could withstand environmental thermal decoherence processes. Another objection, which received less attention but is frequently invoked as a knockdown argument, opposes theoretical frameworks that link quantum indeterminism to free will. Compatibilists argue that determinism is compatible with free will and that there is no need to invoke classical or quantum indeterminism (e.g., see (Strawson G. , 2000)), whereas libertarians (or 'incompatibilists') hold that free will and determinism are mutually exclusive and, consequently, it must be rooted in some form of indeterministic phenomena (e.g., see (Kane, 2016)). There is no consensus on whether the non-deterministic aspect of QM is or is not compatible with a libertarian conception of free will.

The 'quantum free will' is a much more longstanding debate than normally assumed, not rarely colored by ideological standpoints, and dates back to the 1920s, with the first speculations of Arthur Eddington, Arthur Compton, and Pascual Jordan (for an overall historical account, see (Kožnjak B. , 2020)). Nonetheless, since the times of E. Schrödinger (Schrödinger E. , 1951, pp. 164-166), several physicists tend to object to any eventuality that quantum indeterminism could have anything to do with volition (see also, Sean Carroll (Carroll, 2016, p. 381), Sabine Hossenfelder (Hossenfelder, 2020), B. Greene (Greene, 2020, p. 180)). The general argument is, roughly speaking, that randomness can't be a source for a controlled agency. If quantum random effects were to play a role, they would only cause us to act randomly–that is 'lawlessly'–and lead to the violation of the presently known laws of physics. Therefore, quantum indeterminism can't be compatible with free will.

The aim of this paper isn't so much to convince the reader of the opposite, but rather to highlight how these arguments rest on a priori assumptions. In the first part, we show how these are based on unwarranted assumptions and logical extrapolations resting on a popular misunderstanding of what quantum indeterminism implies for causation and free will. We first offer a short conceptual clarification of what determinism and indeterminism really mean from the perspective of classical and quantum physics. This is because misunderstandings frequently stand in the way of developing clear comprehension of the real nature of quantum indeterminism, its implications, and especially, what it does not imply.

In the second part, we will reveal how this also shows that there is no logical incoherence in rejecting the randomness argument, the second argument of the standard arguments against free will, and that there is no conflict between QM and free will supposedly violating the laws of physics. Another aspect that has been posited almost axiomatically in the literature is self-causation as an impossibility. Nietzsche notoriously stated, "*The causa sui is the best self-contradiction that has been conceived so far*" (Nietzsche, 1886), while G. Strawson posited this impossibility as a 'basic argument' (Strawson G. , 1994) from which all other arguments on moral responsibility are then extrapolated. Nonetheless, we show how in contrast to event-causation (an event causes an event), agent-causation interpretations (a 'being' who is not an event causing events) are not only compatible with QM in the frame of a self-causative framework, but also offer a natural interpretation of quantum indeterminism itself. A few conclusive comments will follow.

II. Classical and quantum randomness: what they are, and what they are not

Before proceeding to our main arguments, we deem it necessary to make some conceptual clarifications that are frequently lost sight of, thereby leading to unaware conflations and logical fallacies, including among well-trained scientists and philosophers. In fact, the arguments against the hypothesis that quantum indeterminism may play a role in free will are usually based on metaphysical assumptions and conceptual unwarranted extrapolations that rely on the improper use of statistical notions.

1. Classical indeterminism: A few conceptual clarifications

Within physics, there is no universal definition of determinism or randomness. The simplest idea of what we usually mean by 'random events' is a set of outcomes to which equal probability is assigned, also called 'white noise'. However, the general notion of randomness involves probability distributions that are a measure of unpredictability, with 'unpredictable' meaning everything that can't be predicted with 100% certainty.

Roughly speaking, one considers as 'deterministic' all those processes which dynamical evolution can, at least in principle, be predicted and described by natural laws in the form of a set of differential equations having a unique solution once the initial and boundary conditions are given.

These seemingly sufficient characterizations have, nevertheless, a caveat. They are as valid in classical physics as they are in quantum physics. In the former case, one uses the differential equations whose solution leads to non-probabilistic functions with definite-valued states (e.g., the equations of motion of point particles), whereas in the latter case, these are replaced by probability distributions or probability densities obtained by applying the Born rule, after solving Schrödinger's differential equations[155], which, however, can have a unique solution as well.

Therefore, so defined, the distinction between classical determinism and quantum indeterminism remains vague. Because, in classical physics[156], determinism always presupposes value-definiteness–that is, an isomorphic identity between the physical state of an object and the measurement outcome of some of its properties in terms of physical quantities[157].

Thereby, the classical physical causal determinism implies that at any moment in time, the present state of a system (a point-particle, an ensemble of point-particles, but also a complex system, like a brain), with its (eventually temporally dependent) constraining boundary conditions, uniquely determines its future state and its related measurement outcomes, as a time-parametrized injective function. The future state and our measurements are already pre-determined by the present, with no other possible alternatives. Each state of a dynamical system is uniquely determined by its past state and condition, while any measurement yields only one possible outcome. In short: A single value-definite cause determines only one possible value-definite effect, with the present physical state and its related measurement outcomes being a function of the past.

One can express this in a somewhat more vivid manner, highlighting two aspects.

First is the initial state at time t_0 and the final state at time t_f of the system, each corresponding to only one unique point in a $\mathbb{R}^{2n}$-dimensional phase-space. Because, contrary to QM, for each classical state, when a physical quantity is measured, only one possible definite measurement outcome is possible. Second, the dynamical evolution of the system is determined by a set

---

[155] Or fields in Dirac's equations or in Lagrangians in the case of relativistic quantum field theories.

[156] By 'classical physics' here we generally mean 'non-quantum'–that is, relativity is seen as part of non-Newtonian but still classical physics.

[157] Because measurement inaccuracy is not the cause of quantum mechanical indeterminism, throughout this paper, we will assume only ideal measurements, (meaning that, according to the projection postulate, an ideal measurement projects a quantum state onto one of its eigenstates of the measured observable, leaving the system in a pure state that is exactly correlated with the measurement outcome.)

of differential equations which analytic solutions are unique and, consequently, represent a unique one-dimensional phase-space trajectory between these two zero-dimensional points.

Thus, in no way are these notions related to propositions having any teleological significance or implication. Neither in physics nor in mathematics is the indeterministic character of a phenomenon a measure of its supposed lack of 'volition', 'purpose', or 'aim'. It is this unaware conflation between a popular and a rigorous scientific understanding of these notions that we must be aware of.

The popularly accepted meanings of terms like 'randomness', 'pure chance', and 'coincidence' can be found in dictionary definitions. According to the online Cambridge dictionary, 'random' is, among other things, something "*happening, done or chosen by chance rather than according to a plan*"–that is, "*without choosing intentionally.*" The Oxford English Dictionary instead defines 'random' as anything "*having no definite aim or purpose; not sent or guided in a particular direction; made, done, occurring, etc., without method or conscious choice.*"

These conventional notions always suggest (more or less implicitly), by a non-sequitur fallacy, a lack of will, agency, directedness, aim, moral responsibility, reason, plan, and purpose, equating it with capricious, uncontrolled, irrational behaviors. They present themselves as a test guaranteeing the absence of any teleological force standing behind any phenomenality. This is something that one is not allowed to do in a statistical physical context, where there is no connection between randomness with notions of agency in the sense of a common-sense everyday understanding.

The concept of probability dates back to the 17th century with Blaise Pascal (1623–1662), Pierre de Fermat (1601–1665), and Christiaan Huygens (1629–1695). The latter introduced, in 1657, the concept of 'expectation value', the average value of a random variable over a large number of experiments (for a historical account, see (Hacking, 2006) and (David, 1998)). What we normally have in mind is the 'frequentist interpretation of probability' of Richard Von Mises (1883–1953), who, in 1936, defined the probability of an event as the limit of its relative frequency in many trials (Von Mises, 1936). Nowadays, however, in exact sciences, such as mathematics and physics, the von Mises frequentist interpretation has given way to a purely axiomatic approach introduced by A. Kolmogorov (1903–1987).

Arguments for or against speculations that directly or indirectly involve a conscious deliberation have no place in these mathematical formalizations.

In this sense, probability is only an indicator of an epistemic limitation measured by a statistical prediction of specific occurrences. It is an epistemic or Bayesian probability, not an inherent physical property of the system. The closely related notion of randomness is, therefore, also an epistemic, not a naturalizable inherent property of things.

Perhaps the most notorious and rigorous definition of randomness is that based on the works of Gregory Chaitin. Chaitin's 'algorithmic information theory' established a link between the concepts of randomness and complexity (Chaitin G. , 1975), (Chaitin G. J., 2007), (Downey & Hirschfeldt, 2010). To make a long story short: Random is everything that is represented by an incompressible string and, thereby, has maximum complexity. While this seemed to be a powerful and rigorous way of defining randomness, it soon turned out to be a Pyrrhic victory. This is because Chaitin himself was also able to prove that there is no way, not even in principle, to determine whether a program has this shortest length, as it is forbidden by the famous 'incompleteness theorem' of Kurt Gödel (Gödel, 1931). However, setting aside the absence of this conceptual closure, again, also in this theoretical framework, there is no logical connection between a supposed lack of agency and randomness.

Moreover a measure of (un-)predictability is a relative concept: What appears to be totally unpredictable in some context might well turn out to be predictable in another one. From there comes the expression of 'deterministic randomness' which, actually, is an oxymoron. For example, we label the outcome of tossing a dice as 'random' because it is unpredictable. However, at least in principle, if one had all the physical parameters, the initial and boundary conditions of the dice, as well as a powerful numerical tool that calculated the dynamics of the tossing event, one could ideally predict the outcome. Would we then still label this process as 'random'? In classical physics, we don't perceive a contradiction between determinism and randomness.

Still, there is no direction in a random outcome of coin or dice tosses and which are non-volitional. This simple fact, however, is not an argument that allows us to imply the opposite, namely that a random sequence can't be the expression of goal-directed agency. An easy counterexample that shows how an apparently random phenomenon is fully volitional, can be found in encryption technology. Mathematical encryption models have been created just for a purpose: to transform sensible information into a stream of data that a complicated informational protocol optimizes to appear by all means patternless–that is, random noise.

illustrates a self-explanatory example.

What in the encryption is seen as just noise without any regularity and meaningful structure–that is, a stream of apparently 'purely random' data without meaning and purposeful agency–suddenly makes sense once decoded with a key or password and reveals the presence of something meaningful.

From this follows the notion that deterministic randomness is in the eye of the beholder according to what one knows, not something inherent in a process or the 'world out there'. There are things that could be labeled 'random'–

meaning unpredictable or having a certain degree of unpredictability–but that nevertheless are fully directed by a volitional agent.

Fig. 1 A meaningful symbol and its encrypted image.

Classical statistical mechanics is conceptually and historically rooted in a Laplacian worldview of a deterministic and, at least in principle, completely predictable universe by the almighty Laplace demon *(Laplace, 1814)*.

Thus, in science, and even less in physics, equating any statistical notion of randomness with any form of intentionality or directedness is an unwarranted conceptual conflation. The difference has no ontological correspondence and appears only in our minds.

2. Quantum indeterminism: A few conceptual clarifications

So, what about quantum physics? Here, one says that quantum indeterminism is 'truly random' because its unpredictability is not due to our lack of knowledge but, rather, is 'intrinsic.' We contend, however, that this is, again, a bit too popular of an understanding of what the scientific theory really tells us.

Saying that a phenomenon is quantum indeterministic, or just 'truly random', means that the theory is a complete description of reality and has no hidden variables. The wave function completely describes the physical system, and the measurement outcomes for an observable can only be predicted by a probability distribution.

The questions, then, are: What is a 'hidden variable'? What kind of 'reality' are we talking about, what does all this allow us to conclude, and, more importantly, what kind of inferences we are *not* allowed to make?

Let us gain conceptual clarity first.

A physical theory is a mathematical model based on physical quantities–that is, measurable entities (in QM, the observables) quantifiable with a numerical value (in QM, an eigenvalue), amenable to algebraic manipulation–and that fits the observed measurements predicting the dynamical evolution of the physical system.

A classical theory is a complete description of the physical system when it contains all the physical quantities necessary to describe at all times the measurement outcomes in terms of definitive-valued classical observables (real-valued functions on an appropriate phase space), and once the initial and boundary conditions of the system are given. Whereas a quantum theory is a complete description of the physical system when it contains all the physical quantities necessary to describe the measurement outcomes in terms of expected values of quantum observables (bounded self-adjoint operators acting on a Hilbert space) which are obtained by applying the Born rule, and once the initial and boundary conditions of the system are given. Notice that at the level of the Schrödinger equation the dynamics is deterministic; the state vector evolves from an input state to a unique output state. In QM it is the application of the Born rule that 'breaks' the deterministic dynamics: It's only when one considers a particular observable and projects the state vector onto a particular eigenvector that one gets an amplitude.

While, a theory that is not complete is a theoretical model that does not contain all the necessary information for a complete description. There exist unknown physical quantities–commonly called 'hidden variables'–that determine the observed measurement outcomes and that, therefore, the model can predict only with limited accuracy. The variables are 'hidden', meaning the current theoretical framework does not incorporate them because of a lack of knowledge, but we can visualize them, say, with a microscopic model of reality having some invisible properties. They exist and, at least in principle, are still knowable.

If this were the case for QM, it would mean that the wave function emerges as a statistical quantity from an underlying hidden variable reality[158] (such as in thermodynamics) and that a theory exists that can supersede QM. However, as explained above, we have good reasons to believe that this is not the case, and more sensitive becomes the question of how to interpret a theory that, is complete but without definite-valued physical quantities–that is, preserving a statistical formal structure capable of making only probabilistic predictions (unlike thermodynamics).

The conceptual difficulty involved in ascribing an ontology to it arises because we have nothing to visualize since there is no missing information and no invisible properties to talk about in the first place. We cannot resort to a form of realism describing quantum phenomena in terms of a classical chain of causes and effects. No underlying ontology can be framed in our naïve human physical-mind realism. In QM, this means that a quantum mechanical system is completely specified by the state vector (or wave function) that evolves satisfying the time-dependent Schrödinger equation and that, despite its completeness, makes (by applying the Born rule) only probabilistic predictions via expectation values. Quantum indeterminism (such as Heisenberg's uncertainty principle, the random measurement of spin states, the collapse of the wave function, etc.) is not due to ignorance. The probability

[158] In Bell's original paper, a hidden variable appears as a continuous parameter, λ, in a probability density $\rho(\lambda)$ (Bell, 1964).

distributions involved in the theory are no longer epistemic; rather, they are inherent in the ontology–that is, are 'ontological' or 'ontic' probabilities.

As a corollary, it might also be useful to recall that according to the Kochen–Specker theorem (Kochen & Specker, 1967), assigning definitive values at all times to observables, and independently from the measurement context, is in conflict with the quantum mechanical Hilbert space formalism (there could be no decomposition of the projection operators on the system's Hilbert space). We can't have it both ways: Either we must give up value-definitiveness or we must abandon contextuality. While Bell's theorem states the mutual exclusivity between local realism and hidden variables theories, the Kochen-Specker theorem adds an exclusion rule between the latter and a non-contextual ontology. In other words, the idea that the pointer of a measuring device reveals to us an element of physical reality, or what we call in our everyday parlance a 'property' of a physical object, that is independent of the measurement act and its context–that is, how the measurement is performed–is a form of non-contextual causal realism that is incompatible with QM.

Nevertheless, in QM, the dynamical evolution of each state vector in a Hilbert space representation is uniquely determined by a time parameter: The Hamiltonian operator generates the time evolution of a quantum state with the unitary time evolution operator.[159] Or, equivalently, one calculates the propagator with the path integral formulation that replaces the unique classical trajectory with a sum over all the quantum-mechanically possible trajectories. However, this yields probability distributions for the observables–that is, of possible measurement outcomes: A single cause determines more than one possible effect. For example, in the double slit experiment, preparing a particle in some initial state (its position, its momentum–that is, its de Broglie wavelength, etc.) and setting the boundary conditions (the width of the slits, their distance, etc.) will uniquely determine the interference pattern (the probability distribution) on a detecting screen. However, at which interference fringe the single particle will be detected is an entirely probabilistic process.

Therefore, while the probability distribution is determined by the past, the potential measurement outcomes are not. Eigenstates are statistically constrained but are not functions of the past. There are aspects of the present reality with no causal dependence between its past and future. In other words, there could be observable effects that have no antecedent causes. A quantum theory without hidden variables describes an 'acausal' ontology in which

[159] Notice that, from a formal point of view, the evolution operator which acts on the wave function, generates a unitary time evolution that is uniquely determined by its Hamiltonian, and is reversible: It propagates a unique quantum state to another unique one. Therefore, the dynamics is deterministic. The indeterministic character of QM always emerges in the measurement process which, formally, is represented by the application of the Born rule.

events just happen and must not necessarily be what they are because of preceding causal relations with past events.

It is this temporal independence from the past, together with value-indefiniteness, that makes quantum indeterminism so peculiarly different from classical forms of indeterminism. This is what sparked the conventional expression of 'true' or 'inherent' quantum randomness, and QM being an 'inherently' or 'truly' indeterministic theory–not any ontologies supposedly telling us anything about agent-causality.

Statistically, there is nothing wrong in regarding quantum indeterminism as being 'truly' or 'inherently' random, but philosophical inferences ruling out agent-causation are unwarranted conceptual extrapolations that are based on neither the theory nor the phenomenology of QM. Unpredictability is not a signature, a proof, or a litmus test that allows us to jump to conclusions regarding a supposed lack of agency. It just means what it means: unpredictable. That's what famously puzzled Einstein, who stated: "*God, doesn't play dice.*" Bohr's answer became equally famous: "*Stop telling God what to do with his dice.*"

III. Quantum indeterminism and agent-causation: Objections answered

1. The (inconclusive) standard argument against free will

Keeping in mind this distinction, which avoids any straightforward conflation between scientific statistical concepts and the lack or presence of agent-causation interpretations, we are now in a position to briefly review the standard arguments against free will.

British-Australian philosopher J. J. C. Smart framed the two 'standard arguments against free will' in 1961 (Smart, Free-will, praise and blame, 1961).

The first objection against free will is the '*determinism objection*': If all events are caused deterministically, then all our choices must be predetermined, and there can be no free will.

The second is the 'randomness objection': We can be free only if our choices are free, and these are free only if they are neither deterministically caused nor nomically necessitated by antecedent events. However, a free choice that is not causally determined by antecedent events is a random occurrence and, therefore, cannot be regarded as a free action. Indeterminism can't be the cause of our freedom because any random factor induces a loss of control rather than empowering it–as signal processing scientists know all too well when they must deal with noisy sources. It is just 'noise' that has no meaning and, thereby, can't be considered a source of volition or creative power. Moreover, if any right choice we make were determined by some underlying random processes, it would, thereby, be just a matter of luck (this is the so-called 'luck objection'; for a recent review, see (Moore D. , 2022)).

In short, it doesn't matter if our actions are caused deterministically or non-deterministically; either way, we can't have any control over it.

Though the main subject matter of this paper deals with the second objection, let us briefly address the first one. The first objection remains a matter of debate. Notoriously, compatibilism that sees determinism and free will as compatible has a great number of supporters (for a review, see (McKenna & Coates, 2019)). The author, however, can't find a sensible reason for how this could be reconciled with a notion of classical physical causal determinism. We have seen that once the initial conditions and the constraining boundaries of the physical system are given[160], a non-quantum Newtonian form of determinism uniquely fixes between two zero-dimensional points a definite one-dimensional trajectory in phase-space. Therefore, there could be no 'choice' intervening in-between, making things happen differently. Otherwise, it is not a physical casual determinism. There can be no compatibility of determinism, or at least with this kind of classical physical causal determinism, with any notion of free will, because any volitional aspect has been expunged from the outset.

We fully agree with the determinism objection.

Let us now focus on the random objection, rephrasing it in the context of QM.

If everything is ruled by indeterministic microscopic processes, where everything seems to be ruled by blind chance, by pure coincidences, just random events, no form of will–that is, a consciousness applying itself to a goal-directed work and a result–then no control over one's choices could conceivably emerge from it. If QM were a complete theory without hidden variables where all physical laws are ultimately rooted in quantum processes, any volitional act emerging from quantum indeterminism would lead to erratic and uncontrollable behaviors. This supposedly implies that postulating any connection between quantum mechanics and causal autonomy is an untenable logical fallacy.

This line of reasoning assumes (subconsciously) the naïve conception of randomness and chance as being synonymous with an undirected phenomenon, an indicator of the lack of any conscious agency. However, as we have seen, this remains an unwarranted extrapolation. There is nothing in the formalism and the conceptual foundations of QM that hooks up quantum indeterminism and quantum 'true' randomness–that is, an inherent unpredictability arising

---

[160] One might object that there should be no constraints, since an action arising due to external forces is, by definition, no longer free to do otherwise. However, we are making a concession to the compatibilist, not imposing a limitation. Because any physical system that is not subjected to external forces is even more predictable and deterministic.

due to the absence of hidden variables–to any conceptions of an (un-)guided or (un-)directed agency, no more and no less than in classical physics. The action, communication, or any phenomena stemming from conscious agent-actions can appear random–that is, unpredictable–to the best statistical analysis. Randomness means only indeterminism and is not an intentional notion. Unpredictability is no sign of lack of agency or lack of control, and there is no compelling reason to believe that things are different in QM.

Thus, the standard argument is, if not flawed, at least inconclusive.

2. Quantum indeterminism and free will: No violation of physical laws

Nevertheless, one possible objection is the following (for example, see (sider, 2007)): While we might concede quantum randomness making room for agent-causation, we shouldn't forget that QM assigns probability distribution to each of these possible 'free choices' constrained by the strict laws of QM: The solution of differential equations, such as the Schrödinger equation, and the Born rule. Though the so-obtained probability distribution does not say which event will become actual, the overall behavior of the system is strictly constrained by these quantum laws. For example, the photon diffracted at the double slit will have to hit one of the interference fringes; it can't displace itself, 'at will', on a dark band. This would induce statistical divergences in the probability distribution or density function. Therefore, agent causation could not peacefully coexist with quantum mechanics because it makes it a slave to quantum-mechanical probabilities or leads to the violation of physical laws. QM not only fails in helping the agent-causation theorists, but it also makes them 'anti-scientific'.

However, careful analysis of this statistical objection reveals the fallacy.

This is something vividly illustrated in a famous essay by E. Anscombe in 1971, in which she cast doubts about whether there is a contradiction between classical micro-indeterminism and freedom of action (Anscombe, 1971). Anscombe imagined a glass box full of tiny colored particles randomly shaken and that, by statistical fluctuations, could form patches of uniform color. What if a meaningful word appears? It isn't obvious that this implies a violation of physical or statistical laws. More recently, her claim was substantiated by computer simulations: It is possible to implement an external control that leads to the creation of meaningful pattern formation without violating the micro-statistical laws (Müller, 2022). There is no incompatibility between stochastics and causal agency control.

In the quantum mechanical context, R. Clarke (Clarke R. , 2010) showed that "probabilistic laws of Nature do not require, for any finite number of trials, a precise distribution of outcomes." Several highly unlikely outcomes may occur, without violating the overall statistical law. A conscious agent may choose a prescribed distribution law of outcomes but there is no prescribed order to do that. An agent can freely choose how to approximate a target distribution in each case.

A distribution is neither determined nor undetermined for any finite number of repetitions. It is only constrained to match the distribution law after a large number of 'choices'. Free will is compatible with quantum indetermination as long as it does not cause any observable deviations from quantum laws.

Moreover, R. Kastner strengthened the case studying the particular case of the possible violation of the Born rule by agent-causation (Kastner R. E., 2016). Choices of complex macro-creatures are not accurately modeled by unique quantum observables on quantum states, as Sider presupposes. Human choices can't be tested by Born probabilities and Sider's claim that we are 'slaves to the probabilities' is off track.

To illustrate how there are still many 'degrees of freedom' (here, the physical and psychological connotation coincide) and why there isn't any conflict between an agent-causation theory and statistical laws, we can use an intuitive analogy, originally proposed by F. Faggin (Faggin, The primordial quantum language, 2021), applying it to quantum field theories. It is something we know well from our everyday experience and that shows how, whatever probability distribution is given, it can be taken neither as a signature for lack of agency nor as a violation of natural laws: the probabilistic distribution of the alphabet's letter in a written text.

Statistical analysis shows that, given a sufficiently long text in a specific language, regardless of who writes and what one writes, the probability distribution describing the letter frequency will always be the same. For example, in the English language, no matter who is writing and no matter what he/she is writing, in the large number limit, the most probable occurrence will always be the letter "e", appearing with a probability of about 12.7% (see *Fig.* top). Without knowledge of the language, one would see only the occurrence of an unpredictable sequence of symbols according to a probability distribution and may conclude that no conscious volition and willful semantic agent stands behind the occurrence of these 'truly randomly' appearing letters. There is a probability distribution with the occurrence of the single letter remaining completely unpredictable and that can be characterized only by an expectation value.

From the statistical point of view, this is no different than the notion of a probability distribution and expectation value that describes quantum indeterminism. To illustrate this, consider Fig. 2 bottom. It shows the emission line spectra in the visible light of an atom (Neon). It is a snapshot resulting from an exposure that collects the flux of photons in a time interval. The horizontal position of the line represents the wavelength of the photons, while its intensity (the width of the line) is proportional to the number of photons collected per unit time interval.

This also dictates the probability that one will observe the single photon hit the specific line every time the atom emits a photon. Which photon will hit which line next is an exclusively quantum process; one can speak only of the probability with which the next photon will appear on one or the other spectral line, but no matter how much we know about the quantum system, the single event remains intrinsically unpredictable.

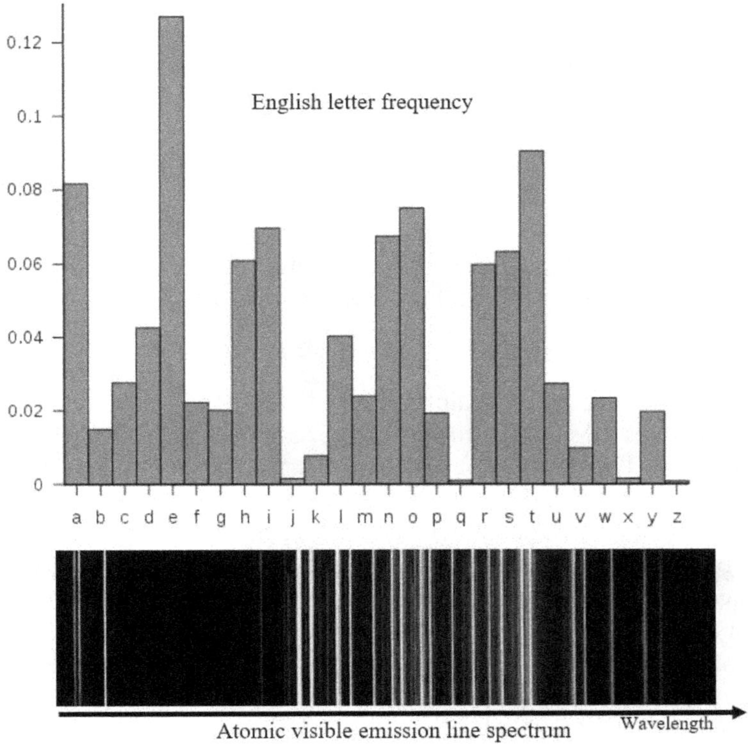

Fig. 2 Top: Probability distribution of the English alphabet. Bottom: Atomic spectral emission lines. Credit: Wiki-Commons.

Obviously, this is not to say that language has anything to do with quantum physics. However, this analogy highlights how there is no conceptual difference between the two statistical tests: the statistics of the occurrence of the alphabet's letters in a text and the occurrence of that of the photon's wavelength by an atomic emission. If the former is not a 'litmus test' that allows for any extrapolations against teleological speculations, the latter isn't either. Arguments based on quantum indeterminism or randomness supposedly violating physical laws and/or demonstrating a lack of agency, directionality, will, or final causes are flawed logical inferences.

This also suggests a resolution of the interactionist or energy conservation objection and the luck objection at once.

A theory positing free will in quantum indeterministic phenomena cannot be dismissed on the grounds of arguments resorting to energy conservation violation, no more and no less than random quantum events do. For example, in quantum field theories, one frequently gets to know about so-called 'quantum fluctuations' and 'virtual particles', which are often misrepresented as popping in and out of existence by a temporary violation of the energy conservation principle for a fleeting time interval allowed by the time-energy uncertainty relation. However, these are not 'fluctuations' or 'particles' at all, but only superpositions of eigenstates and computational devices in Feynman diagrams, respectively. What fluctuates in time are the measurement outcomes, not the state of the measured quantum object.

For the same reason, since, from this standpoint, a choice is not determined by a quantum indeterministic process, but rather *is* the very same quantum process, there is no luck problem left. Every choice is no more and no less a matter of 'luck' than the choice to use the letter "e" in the English language with a 12.7% probability. It is not a matter of luck. It is a matter of will.

3. The self-causation hypothesis

The standard argument against free will hides an unaware assumption. Whoever invokes these standard objections assumes them to be exhaustive of all possibilities. There can be only determinism or indeterminism. What else?

By maintaining this background assumption, several approaches have investigated if and how quantum indeterminism may allow for freedom of action (for a couple of reviews, see, e.g., (Acin, Gisin, Brassard, & Raymond-Robichaud, 2012), (Hodgson, 2012)). What they have in common is that quantum unpredictability is taken as an external factor to the agent, something which we could consciously recruit to build alternatives, such as the mind manipulating quantum indeterminism to produce new ideas and to generate unexpected and unpredictable behavior, or a source to exploit–for example, by a conscious agent that makes decisions using quantum effects, or controlling the quantum uncertainty and modifying the probabilities of events. Most notoriously, R. Penrose and S. Hameroff developed their 'orchestrated objective reduction' theory (Orch OR), which identifies quantum state reduction as quantum computations in microtubules inside neurons with discrete conscious moments, thereby supposedly rescuing free will (Hameroff S. , 2012). Quantum randomness is conceived of as something that instantiates a conscious moment or that the 'being' can access in order to freely choose, but it is not considered an inherent quality in the being itself. In all these models, free will is seen, in one way or another, as assisted by randomness, and where the agent is influenced by undetermined events. The idea that this indetermination is the manifestation of conscious causal agency–that is, will itself–is therefore neglected.

Could this latter point of view be a viable option? Let us unpack it in more detail.

From a purely mechanistic and physicalist point of view, a complete theory without hidden variables, random events just happen and are seemingly 'causeless'. Our deterministic and mechanistic understanding of the world where everything that happens must be determined by a chain of causes and effects preceding it in time, and where there must always be a reason for whatever happens, no longer holds. Leibniz's principle of sufficient reason, which states that everything must have a reason, cause, or ground, is questionable in this domain. In QM, an atomic nucleus that decays at time t_1 rather than time t_2 doesn't need 'reasons' to do so. There are only potentialities inherent in the system actualizing or not actualizing something, and which are not reducible to some mechanistic causal relations yet to be discovered. In QM, every physical event has a 'potentiality', 'propensity', 'aptitude', or 'disposition' to manifest, but must not. This is a 'cause-less' and 'reason-less' phenomenality that inflicts violence on our binary deterministic vs. indeterministic mindset and that makes the standard arguments against free will appear inescapable.

However, if we relax our physicalist causality, we may admit to a third possibility that great minds such as Plotinus, Spinoza, Descartes, St. Augustine, Schelling, Hegel, Schopenhauer, Whitehead, Husserl, and the seers of the Upanishads contemplated: the 'causa sui'–that is, something generated within itself, is self-caused, or the cause of itself and independent of any other ground, yet containing within itself a sufficient explanation of its own being, and the source of every self-modification as a self-creative process. It is neither determined nor undetermined, but self-determined.

Spinoza begins his magnum opus, "Ethics", stating: "*By cause of itself (causa sui) I understand that whose essence involves existence; or, that whose nature cannot be conceived except as existing*" (Spinoza, Ethics: Demonstrated in Geometrical Order, 1677) I, Def. I. For Spinoza, the causa sui was the very notion of Substance, or God, as the first cause of all things, and also the cause of Itself. This Substance, however, was not meant by Spinoza as a resting being, but as an unconditional and temporal power of action, a creative and dynamic force. "*That thing is called free, which exists solely by the necessity of its own nature, and of which the action is determined by itself alone*" (Spinoza, Ethics: Demonstrated in Geometrical Order, 1677) I, Def. VII.

Whitehead, who took a reductionist and atomistic perspective, claimed that the actual entities, the '*atomic occasions of experiences*', are self-creative or causa sui: "*Self realization is the ultimate fact of facts. An actuality is self-realizing, and whatever is self-realizing is an actuality*" (Whitehead A. N., Process and Reality: An Essay in Cosmology, 1978). For Whitehead, each actual entity creates its own identity and strives for self-actualization by self-transformation. It achieves its subjective aim by becoming actual: Every process is active in and of itself and reveals a drive to realize its potentialities.

According to the Upanishads there is a principle of will of our own becoming, the *svabhava* (self-expression, self-being, real-nature, truth of being) and that works out of latency by the law of action the *svadharma* (self-law, real -action, self-shaping).

The question, then, is: Could we, at least in principle, interpret quantum indeterminism in the frame of a 'self-causal' ontology? Something along the lines of Leibnitz's panpsychist monadology where each Monad is an actual entity not only having a passive proto-conscious atomic occasion of experience but also being an active proto-volitional self-determining will? Quantum fields not just having but being minutest 'causa-sui-propensities'?

If one posits consciousness with an inherent free-agency quality as an ontological fundamental primitive, no physical law and no logical reason prevents us from conjecturing that quantum randomness might be a local expression of a self-determining and selecting will, some sort of primitively volitional capacity, intrinsic in matter and space-time. This would, at least in part, be in line with a Spinozian universal 'substance', a Natura naturans at work by a consciousness-force and consciousness-will, or with Schopenhauer's vision of the 'World as Will'. Alternatively, it could make sense from a panpsychist perspective or, conversely, in a cosmopsychist setting, or some other metaphysical construct. We are not going to prefer any of these but, rather, point out that these alternatives are fully compatible with what modern science knows.

Self-causation has mostly been looked upon with suspicion, if not outright rejection, as the 'best self-contradiction so far conceived.'[161]

However, if instead of asking how free will relates to causation, or a lack of causation, but, rather, posit causation itself as an aspect of a volitional power expressing in what we see as quantum potentialities, its paradoxical implications with regards to an apparent incompatibility with both determinism and indeterminism dissolve.

It is a view that contrasts with and even replaces the event-causation of the mechanistic Laplacian clockwork universe–with the agent-causation–where 'being' is the agent that causes an event. In a quantum indeterministic context, the single event is a potentiality out of many (or infinitely many) possible ones with no physical selecting mechanism other than will *being* the selecting 'mechanism'. Several metaphysical approaches can be taken in accord with such a theoretical framework. Such as a Spinozian cosmology where the 'agent' is Nature itself manifesting in an infinite number of 'Substance-modes'

[161] We guess this to be the same reason why Richard Feynman considered the idea of the electron's self-interaction, which is nowadays an accepted aspect of quantum field theories, as a "silly and unnecessary idea." (Feynman, 1966)

actualized in microscopic primary acts of volition. For example, an atom 'choosing' when and which energy transition occurs next and, thereby, determining when a photon is detected at which spectral line (such as those in *Fig.* .) Something that, before the actual event, to an observer can appear only as stochastic and unpredictable potentialities.

Heisenberg came close to this idea with his suggestion that the statistical nature of quantum theory can be interpreted as a form of Aristotle's 'potentia'– that is, not actualities but, rather, dynamic possibilities of nature "*standing in the middle between the idea of an event and the actual event, a strange kind of physical reality just in the middle between possibility and reality*" (Heisenberg W. , 1958, p. 41) (for a more in-depth historical account and recent debates, see also (Kastner, Kauffman, & Epperson, Taking Heisenberg's potentia seriously, 2018), (Jaeger G. , 2017), (Kožnjak B. , 2020)). There is, however, no teleological dimension in Heisenberg's interpretation.

Yet, here we go a step further and posit that the 'potentia' *is* self-caused will, or a 'proto-will', expressing some primitive form of volition, aim, and telos. A system in quantum superposition collapses into one outcome by no other 'hidden variable', 'reason', or 'motive' than for a self-determined act of self-actualization. This fits the definition of free will as the 'ability to do otherwise', and much more. There is nothing that contradicts the laws of QM. Quite the opposite; the most natural interpretation of what a statistical theory that is complete and without hidden variables implies *is* a theory of self-causation.

A view in which quantum indeterminism is seen from the perspective of self-causation invalidates the standard argument against free will. Because will isn't a hidden variable, it cannot be just a hidden parameter in a mathematical expression, or some unknown gear mechanism determining a probability distribution, and it doesn't need to appear as such in a wave function or density matrix. A probability density without hidden variables is a measure of unpredictability; it doesn't tell us anything more than that. From this standpoint, quantum indeterminism reflects the existence of volition. Will doesn't require any causality. Here, we align with Schopenhauer: causality as an emergent property of will. Quantum indeterminism differs from classical indeterminism inasmuch as it is not only unpredictable for an outside observer, but is also self-determined from the inside, as there is nothing other than itself as a possible causal source or agent and that is not dependent on prior events. Will is an inherent aspect of consciousness and is free precisely and only because it is the cause of itself and doesn't need anything beyond itself or a reason or motivation to will. Or, as Schopenhauer used to say: "*Motivation is causality seen from within*" (Schopenhauer, 1847). And, in the context of a more recent Eastern metaphysical vision, the Indian mystic and philosopher Aurobindo Ghose, describes *"the finite as a frontal aspect and a self-*

determination of the Infinite", where *"e*ach self-determination of the self-being must have its own awareness." (Aurobindo G. , 1919)

After all, it is self-determination that we consider in our everyday parlance as the true meaning of what we call 'freedom.' The strive for freedom *is* the strive for a self-actualization that enacts a self-realization.

4. Conclusion

We argued against the habitual tendency to connect stochastic processes to statements against teleological ontologies. There is no principle of mutual exclusivity. A scientific and philosophical approach that starts from the more or less abstract and loose notion of randomness, connecting it to unguided processes or a lack of purpose, meaning, or any kind of volition, is misplaced. Using unpredictability as a measure of a lack of intentionality leads to unaware but unwarranted and inaccurate statements. Statistical concepts such as randomness, unpredictability, noise, and chance do not exclude free will and conscious creativity and are not at odds with the physical laws we know so far. Thus, the arguments by randomness against free will hypotheses in any scientific context are a common fallacy lacking inferential value or explanatory power.

Quantum micro-indeterministic process that appears locally as unguided randomness can direct the formation of a non-random and ordered macro-phenomenality. What looks like microscopic noise can turn out to be a macro-structure forming process. Microscopic randomness can be a bottom-up creative power that appears random only because of the complexity of its action, which is beyond our discernable comprehension. The randomness objection based on arguments of quantum indeterminism does not hold under careful scrutiny.

Moreover, conceptual difficulties arise when we exclude a priori the possibility of going beyond a deterministic vs. indeterministic dichotomy and conception of reality. Self-causation is not a new metaphysical category but is worth being rediscovered with its potential explanatory power to be studied further in the context of quantum-consciousness studies. What we call 'unpredictability, 'noise', 'randomness', and 'chance' might well be the backdoor for free agency.

Of course, quantum indeterminism is not evidence of free will in Nature, either. QM does not force us to such a conclusion, but nothing disallows for teleological and metaphysical speculations. Ideally, science should refrain from jumping to conclusions based on one's own metaphysical belief system and, in this regard, maintain an agnostic attitude, as long as these are not presented as a scientific fact that excludes the opposite point of view.

Acknowledgments

I would like to thank one of the anonymous referees whose very helpful and clarifying suggestions enhanced the quality of this paper.

References

Acín, A., Gisin, N., Brassard, G. & Raymond-Robichaud, P. (2013) Quantum physics and free will, in Suarez, A. & Adams, P. (eds.) *Is Science Compatible with Free Will? Exploring Free Will and Consciousness in the Light of Quantum Physics and Neuroscience*, pp. 7-61, Berlin: Springer. doi: 10.1007/978-1-4614-5212-6

Anscombe, G.E. (1971) Causality and determination, in *The Collected Philosophical Papers of G.E.M. Anscombe*, **2**, pp. 133–147, Oxford: Oxford University Press.

Aurobindo, G. (1919) Brahman, Purusha, Ishwara — Maya, Prakriti, Shakti, in Aurobindo, G. *The Life Divine*, pp. 336-379, Pondicherry, India: Sri Aurobindo Ashram Press.

Bell, J.S. (1964) On the Einstein Podolsky Rosen paradox, *Physics Physique Phyzika,* **1**, pp. 195-200. doi:10.1103/PhysicsPhysiqueFizika.1.195

Carroll, S. (2016) *The Big Picture: On the Origins of Life, Meaning, and the Universe Itself*, New York: Dutton.

Chaitin, G.J. (1975) A theory of program size formally identical to information theory, *Journal of the Association for Computing Machinery*, **22** (3), pp. 329–340. doi:10.1145/321892.321894

Chaitin, G.J. (2007) *Algorithmic information theory: Some recollections*. doi:10.48550/arXiv.math/0701164

Clarke, R. (2010) Are we free to obey the laws?, *American Philosophical Quarterly*, **47** (4), pp. 389-401.

David, F.N. (1998) *Games, Gods & Gambling: A History of Probability and Statistical Ideas*, New York: Dover Publications.

Downey, R.G. & Hirschfeldt, D.R. (2010) *Algorithmic Randomness and Complexity*, Berlin: Springer Science & Business Media. doi:10.1007/978-0-387-68441-3

Einstein, A., Podolsky, B. & Rosen, N. (1935) Can quantum-mechanical description of physical reality be considered complete?, *Physical Review*, **47** (10), pp. 777–780. doi:10.1103/PhysRev.47.777

Faggin, F. (2021) *The primordial quantum language*, [Online], https://www.essentiafoundation.org/the-primordial-quantum-language/reading/ [23 May 2021].

Feynman, R.P. (1966) The development of the space-time view of quantum electrodynamics, *Science*, **153** (3737). doi: 10.1126/science.153.3737.699

Gödel, K. (1931) Über formal unentscheidbare Sätze der Principia Mathematica und verwandter Systeme, I, *Monatshefte für Mathematik und Physik*, **38** (1), pp. 173–198. doi:10.1007/BF01700692

Greene, B. (2020) *Until the End of Time: Mind, Matter, and Our Search for Meaning in an Evolving Universe*, New York: Alfred A. Knopf.

Hacking, I. (2006) *The Emergence of Probability: A Philosophical Study of Early Ideas about Probability, Induction and Statistical Inference*, 2nd ed., Cambridge: Cambridge University Press. doi:10.1017/CBO9780511817557

Hameroff, S. (2012) How quantum brain biology can rescue conscious free will, *Frontiers in Integrative Neuroscience*, **6** (93). doi:10.3389/fnint.2012.00093

Heisenberg, W. (1958) *Physics and Philosophy*, New York: Harper Row.

Hodgson, D. (2012) Quantum physics, consciousness, and free will, in Kane, R. (ed.) *The Oxford Handbook of Free Will*, 2nd ed., pp. 56-83, New York: Oxford University Press. doi:10.1093/OXFORDHB/9780195399691.003.0003

Hossenfelder, S. (2020) *You don't have free will, but don't worry*, [Online], http://backreaction.blogspot.com/2020/10/you-dont-have-free-will-but-dont-worry.html [10 Oct 2020].

Jaeger, G. (2017) Quantum potentiality revisited, *Philosophical Transactions of the Royal Society A*, **375** (2106), pp. 1-14. doi:10.1098/rsta.2016.0390

Kane, R. (2016) On the role of indeterminism in libertarian free will, *Philosophical Explorations*, **19** (1), pp. 2-16. doi:doi.org/10.1080/13869795.2016.1085594

Kastner, R.E. (2016) The Born rule and free will: Why libertarian agent-causal free will is not "antiscientific", in Aerts, D., de Ronde, C., Freytes H. & Giuntini R. (eds.) *Probing the Meaning of Quantum Mechanics*, pp. 231-243, Singapore: World Scientific. doi:doi.org/10.1142/9789813146280_0009

Kastner, R.E., Kauffman, S. & Epperson, M. (2018) Taking Heisenberg's potentia seriously, *International Journal of Quantum Foundations*, **4** (2), pp. 158–172. doi:10.1142/9781786346421_0011

Kim Y., Bertagna F., D'Souza E.M., Heyes D.J., Johannissen L.O., Nery E.T., Pantelias A., Sanchez-Pedreño Jimenez A., Slocombe L., Spencer M.G., Al-Khalili J., Engel G.S., Hay S., Hingley-Wilson S.M., Jeevaratnam K., Jones A.R., Kattnig D.R., Lewis R., Sacchi M., Scrutton N.S., Silva S.R.P. &

McFadden J. (2021) Quantum biology: An update and perspective, *Quantum Reports*, **3** (1), pp. 80-126. doi:10.3390/QUANTUM3010006

Kochen, S. & Specker, E.P. (1967). The problem of hidden variables in quantum mechanics, in Hooker, C.A. (eds.) The Logico-Algebraic Approach to Quantum Mechanics. The University of Western Ontario Series in Philosophy of Science, vol 5a. Berlin: Springer. doi:10.1007/978-94-010-1795-4_17

Katnjak, B. (2020) Aristotle and quantum mechanics: Potentiality and actuality, spontaneous events and final causes, *Journal for General Philosophy of Science*, **51** (3), pp. 459-480, Berlin: Springer Nature. doi:10.1007/s10838-020-09500-y

Katnjak, B. (2020) The earliest missionaries of 'quantum free will': A socio-historical analysis, in Grgić, F. & Pećnjak, D. (eds.) *Free Will & Action. Historical-Analytical Studies on Nature, Mind and Action*, pp. 131-154, Berlin: Springer. doi:10.1007/978-3-319-99295-2_10

Laplace, P.S. (1814) *A philosophical essay on probabilities,* New Jersey: John Wiley & Son, and London: Chapman & Hall.

McFadden, J. & Al-Khalili, J. (2018) The origins of quantum biology, *Proceedings of the Royal Society A*, **474** (2220). doi:10.1098/rspa.2018.0674

McKenna, M. & Coates, D.J. (2019) Compatibilism, *Stanford Encyclopedia of Philosophy*, [Online], https://plato.stanford.edu/entries/compatibilism/ [16 Nov 2019].

Moore, D. (2022) Libertarian free will and the physical indeterminism luck objection, *Philosophia*, **50**, pp. 159-182. doi:10.1007/S11406-021-00370-2

Müller, T. (2022) "Let's build an Anscombe box": Assessing Anscombe's rebuttal of the statistics objection against indeterminism-based free agency, *Synthese*, **200** (74). doi:10.1007/s11229-022-03565-6

Nietzsche, F. (1886) *Beyond Good and Evil Prelude to a Philosophy of the Future*, New York: Warbler Press, reprint ed. 2015.

Schopenhauer, A. (1847) *On the Fourfold Root of the Principle of Sufficient Reason,* Cambridge: Cambridge University Press, reprint ed. 2015.

Schrödinger, E. (1951) *Nature and the Greeks and Science and Humanism*, Cambridge: Cambridge University Press, reprint ed. 2014.

Sider, T. (2007) Free will and determinism, in Conee, E. & Sider, T., *Riddles of Existence*, New York: Oxford University Press.

Smart, J.J.C. (1961) Free-will, praise and blame, *Mind*, **70** (279), pp. 291-306. doi:10.1093/MIND/LXX.279.291

Spinoza, B. (1677) *Ethics: Demonstrated in Geometrical Order*, London: Penguin Classics, reprint ed. 2005.

Strawson, G. (1994) The impossibility of moral responsibility *Philosophical Studies*, **75**, pp. 5-24. doi:10.1007/BF00989879

Strawson, G. (2000) The unhelpfulness of indeterminism, *Philosophy and Phenomenological Research*, **60** (1), pp. 149-155. doi:10.2307/2653435

Von Mises, R. (1936) *Probability, Statistics and Truth*, New York: Dover Publications, reprint ed. 1981.

Whitehead, A.N. (1978) *Process and Reality: An Essay in Cosmology*, New Jersey: Prentice Hall.

4. Quantum Indeterminacy and Libertarian Panpsychism[162]

Abstract

The consequence argument", together with the "luck objection", which are summed up by the "standard argument against free will", state that if our volition were dependent on physical causally indeterministic processes, our actions would lack control and, thereby, result in random behavior that would be a mere matter of luck and chance. In particular, quantum indeterminacy is supposed to be of no use in support of libertarian agent-causation theories because any volitional act interfering with the probability distributions defining quantum laws would lead to its violation. Building upon recent conjectural work questioning this assumption (Clarke 2010, Kastner 2016, Masi 2023), it is shown, with a concrete example involving quantum indeterminacy, how a hypothetical agent with access to the temporal ordering of events can pre-determine the result of a process taking place in time without modifying the probability laws defining it. This conclusion is then taken as a basis for a libertarian panpsychist interpretative model.

I. Introduction

The question of whether a relationship exists between free will and quantum mechanics is longstanding. One of its motivations is to present an alternative to compatibilist accounts which see determinism as reconcilable with free will, while an in compatibilist or libertarian view rejects such compatibility. Compatibilists and libertarians both believe in free will, but they diverge on whether physical causal determinism is compatible with notions of free agency, or whether (classical or quantum) indeterministic processes are necessary as a "reservoir" allowing an agent to make free choices unconstrained from other external factors. The latter in compatibilist position differentiates event-causation from agent causation libertarianism.

Event-causation libertarians posit that actions are caused solely by events. Actions are conceived of as agent-involving events caused solely by prior events, and an agent's freedom of choice is reducible to events causally involving him in making the choice. However, if so, such libertarianism is arguably inconsistent because the extent of the agents influence on his decision is limited to the causal impact of states and events directly involving him.

[162] The present article was published in *Mind & Matter*, vol. 22, Nr. 1, pp. 31-50(20). DOI: **https://doi.org/10.53765/mm2024.31**

Meanwhile, agent-causation libertarians do not start from an event causing another event but, rather, from a "being" who can settle the matter of which options for action should occur. Agent-causation suggests that an agent has a unique causal power to choose without being causally predetermined to do so. However, the issue with agent-causation is that any undetermined agent-caused action seems to imply physical changes that would lead to divergences from the known physical laws (Pereboom 2001, p. 81; 2014, pp. 65–69).

The present paper aims to demonstrate, with a specific example from quantum physics, how libertarian agent-causation that doesn't imply violations of physical laws is naturally recovered if we do not abstract from the temporal dimension of the processes ruled by probabilistic laws and how it lends itself to a panpsychist metaphysical framework.

However, before getting there, it might be instructive to briefly review the main objections that stand in the way of this worldview.

The canonical version of the incompatibilist argument is the "consequence argument" presented by van Inwagen (1983, p. 3) as follows:

> *"If determinism is true, then our acts are the consequence of laws of Nature and events in the remote past. But it's not up to us what went on before we were born, and neither is it up to us what the laws of Nature are. Therefore, the consequences of these things (including our present acts) are not up to us."*

On the other hand, the challenge to libertarianism is the "luck objection", according to which there cannot be any exercise of free will based on causally undetermined actions, as these would be based on mere luck or chance, for which no one is morally responsible (Shabo 2020).

I would like to present this issue in a more physical context of microphysical causation. To this purpose, the so-called "standard argument against free will" neatly summarizes the two perspectives. Roughly speaking, it goes as follows (for an in-depth analysis see Doyle 2011). The libertarian challenges the compatibilist with the "determinism objection": If the universe is ruled by a physical causal determinism, all its dynamics, from the micro- to the macro-cosmos, are ruled by a strict physical bottom-up causality. Then no room is left for free agency, not even in principle. This is because if everything is determined from the atomic scale upwards to the brain, then all our actions must be determined as well. Therefore, if we have free will, it must be rooted in non-deterministic processes such as those physics describes at a quantum level.

This latter hypothesis, however, isn't unproblematic either and is challenged by the "randomness objection": If our brain states are ultimately dependent on some random processes, then our thoughts, desires, and will to act would result in a random and uncontrolled series of behaviors. Random events are no more within our control than deterministic ones. Thus, either way, there is no freedom of action, neither in determinism nor in indeterminism. Free self-determination seems to be incompatible with modern physics.

There have been different attempts to overcome one or the other objection (for an overview, see Kane 2012b). However, they mostly rely on notions of "indeterminism" that are rooted in pseudo-random processes. Also, in a deterministic universe dominated by the physical causal determinism of classical physics, the "indeterministic noise" that the agent "harnesses" on a macroscopic higher-level, isn't ultimately indeterministic; rather, it is grounded in a lower-level micro-physical determinism and, thereby, remains inside a deterministic paradigm. The switch from a classical to a quantum indeterministic ontology is, in my view, an unavoidable step one must still take. A causal classical physical deterministic low-level ontology, even if based on pseudo-random events, on the unpredictability of a non-linear complex system and deterministic chaos, or whatever kinds of classical stochastic processes, does not become consistent with any higher-level free-will theoretical framework. Especially from the event-causation libertarian perspective, the agent's causal power over this stochasticity remains unclear. Therefore, here, I would like to investigate further the possible logical relationship between quantum indeterminacy and agent-causation.

Yet, there are conceptual hurdles to overcome. Already, giants of science such as Schrodinger (1951, pp. 162) commented on this, asking: "Would physical indeterminacy give free will a chance?" He concluded that it is an impossible solution because "the direct stepping in of free will to fill the gap of quantum indeterminacy does amount to an interference with the laws of Nature".

While Searle (1986, pp. 86) once stated that

> *"even if there is an element of indeterminacy in the behavior of physical particles – even if they are only statistically predictable – still, that by itself gives no scope for human freedom of the will; because it doesn't follow from the fact that particles are only statistically determined that the human mind can force the statistically determined particles to swerve from their paths. Indeterminism is no evidence that there is or could be some mental energy of human freedom that can move molecules in directions that they were not otherwise going to move. So it really does look as if everything we know about physics forces us to some form of denial of human freedom."*

This view is reiterated by other prominent scientists and philosophers and remains, to date, a widely accepted opinion, and that Muller called the "statistics objection": Any interference by a willful agent with (classical or quantum) indeterministic events would inevitably lead to the violation of the statistical laws. Such an interference would lead to the modification of the prole of the probability distribution (PD) describing it. Thereby, this would render any libertarian view of free will based on principles of indeterminism in conflict with the laws of physics. In other words, we must conclude that we are "slaves to the probabilities".

Clark, however, realized that things aren't as simple as that, pointing out that an agent, even if subjected to strict probabilistic laws, is, nevertheless, not instructed to follow any particular sequence to produce a predetermined distribution, as
"probabilistic laws of Nature also do not require, for any finite number of trials, a precise distribution of outcomes" (Clarke 2010, p. 390). We are free to choose the order in which things happen provided that they approximate a prescribed PD in the
limit of large numbers. Kastner further showed that the violation by agent-causation of the Born rule, which is the formal basis for every PD for measurement outcomes in quantum mechanics, can't be taken as a litmus test for or against free will. Choices of complex macro-creatures are not accurately modeled by unique quantum observables on quantum states and, thereby, can't be tested by Born probabilities (Kastner 2016). Masi reviewed these aspects by adding some thoughts, clarifying first the distinction in physics between classical and quantum indeterminism and reviewing the old metaphysical idea of self-causation (causa sui) (Masi 2023).

Here I will complement the above insights, especially expanding on Clark's claim. While it is true that we are not forced to obey probabilistic laws following any particular predetermined series of outcomes, it isn't entirely clear how much freedom of choice this leaves. To investigate this aspect, the present work has been inspired by an original idea of British analytic philosopher Elizabeth Anscombe who wondered about this in the form of a classical thought experiment, nowadays known as the "Anscombe box" (Anscombe 1971, for a review of the Anscombean perspective see also Mulder et al. 2022). She imagined a box filled with colored particles whose dynamics are ruled by statistical laws. Suppose it displays on its sides a particular pattern with slightly varying shapes and sizes but, nevertheless, has certain recognizable regularities. Her thesis was that "it is not at all clear that those statistical laws [...] would have to be supposed violated by the operation of a cause for this phenomenon" (Anscombe 1971, p. 146, as cited by Mulder et al. 2022).

Anscombe did not further develop her thoughts on this. Recently, however, Muller made a computer simulation of such a box and showed that, indeed, it is possible to intervene in the dynamics by flipping the position or colors of some particles making appear repeatedly similar pattern without changing the statistics of the process (Muller 2022). As Mulder et al. (2022, p. 9) summed it up:

"...agency might be a higher-level, emergent power that realizes possibilities left open by the lower level laws that govern, say, agents' bodies."

Anscombe was more concerned with defining the principle of causality and its relation to determinism rather than applying it to philosophical questions regarding free will. Van Miltenburg further worked on her legacy, discussing an alternative to determinism in terms of probabilistic causation and different kinds of causality (van Miltenburg 2022), and Mulder pointed out how life as such requires physical indeterminism (Mulder 2021). While the present approach could be useful in clarifying aspects of the randomness objection. I present a similar but more general argument and simulation that aim at a deeper understanding of the relationship between indeterminacy and agency.

Clark, however, realized that things aren't as simple as that, pointing out that an agent, even if subjected to strict probabilistic laws, is, nevertheless, not instructed to follow any particular sequence to produce a predetermined distribution, as "probabilistic laws of Nature also do not require, for any finite number of trials, a precise distribution of outcomes" (Clarke 2010, p. 390). We are free to choose the order in which things happen provided that they approximate a prescribed PD in the limit of large numbers. Kastner further showed that the violation by agent-causation of the Born rule, which is the formal basis for every PD for measurement outcomes in quantum mechanics, can't be taken as a litmus test for or against free will. Choices of complex macro-creatures are not accurately modeled by unique quantum observables on quantum states and, thereby, can't be tested by Born probabilities (Kastner 2016). Masi reviewed these aspects by adding some thoughts, clarifying first the distinction in physics between classical and quantum indeterminism and reviewing the old metaphysical idea of self-causation (causa sui) (Masi 2023).

The present work is subdivided as follows. Section 2 provides a more general and formal description, which elucidates how there is virtually an infinite number of possible ways to approximate a probability law without modifying its representation, that is, leaving it invariant. This aspect frequently remains unnoticed because we tend to think of probability functions independently of their temporal dimension. Section 3 seeks to clarify this further with the specific example of Young's famous double-slit experiment. With a simple computer simulation, it is possible to show how an agent can choose the appropriate sequence of outcomes subjected to a probability law inducing a random walk of a particle, which, nevertheless, can be directed towards a chosen target without violating the laws determining its random dynamics. Section 4 will provide a brief tentative suggestion of how the obtained results could be implemented inside a libertarian panpsychist framework. If quantum mechanics could, at least in principle, open a door to micro-physical agent-causation, this could naturally lend itself not only to some form of micropsychism but also to a cosmopsychist metaphysical paradigm. Concluding remarks then follow.

II. Randomness, Probability Distributions, and Time: A Brief Formal Reminder

Despite their pervasive use in our common language, the words "randomness" and "indeterminacy" have no unique, rigorous, and universally accepted definition. Perhaps the most widely accepted version of randomness is that which is based on the works of Chaitin's algorithmic information theory and which links what he called Kolmogorov randomness to the notion of complexity: Random is everything that is represented by an incompressible string and, thereby, that has maximum complexity (Chaitin 1975 a,b). However, this definition is of little practical use because it is limited by Godel's theorem of incompleteness, which forbids even in principle the knowing of whether a program's length is the shortest that describes the data in the first place. Therefore, let's take a simpler but still widely accepted understanding.

In our common understanding, a process is random when all its possible outcomes have equal probability of being realized, that is, what is called white noise. This, however, is a much too restrictive definition. In fact, in statistical mathematics, one speaks of random variables linking it to a PD. It is of more practical use to label a process as having a degree of randomness or indeterminacy when its outcomes are not predictable with certainty, but that can, nevertheless, be described statistically by a PD (or continuous probability density functions).[163]

Let us then consider a discrete random variable $X : \Omega \rightarrow \mathbb{R}$ that is, a measurable function X from a countable sample space Ω (i.e., a finite or countable infinite number of distinct values) to the real numbers R of the set of possible outcomes. Let P(X) be its discrete PD[164] defined as the set of

[163] Notice how even this is not completely unproblematic because, for example, a sequence of equally probable binary digits of length N, such as 10101010101010101010..., could hardly be called random, with its two digits appearing with equal relative frequency. However, this sequence can still be called random, provided that, on average, it doesn't appear more than once over 2N times. With a general PD, things would be formally more complicated but the underlying principle does not change. These kinds of issues inspired Kolmogorov and Chaitin to search for a rigorous definition of randomness. However, these formal aspects are not of concern here as long as this provision is observed.

[164] Here discrete PDs are assumed, if not stated otherwise.

nonzero probabilities that the discrete random variable X assigns to a countable number of distinct values $x_i(i = 1, \ldots, c)$ as[165]

$$P(X = x_i) = p_i = \frac{n_i}{N} \qquad (1)$$

with X = x_i a particular realization of X, n_i the number of favorable outcomes for the ith possible outcome, N the total number of events or measurements, and c the number of all possible outcomes, such that, in the limit of large numbers
N $\rightarrow \infty$, for a normalized PD, $\Sigma_i\, p_i = 1$.

Typical examples are the two possible outcomes of a coin toss, or, in quantum mechanics,[166] the measurements of the spin of a particle ($p_i = 1/2$; $i = 1,2$), or a dice toss ($p_i = 1/6$; $i = 1,\ldots, 6$), etc. A more general case of a discrete PD is the Poisson distribution, which expresses the probability of observing a certain number of occurrences once the expected mean rate of occurrences is known (e.g., the number of photons emitted by a light source in some time interval). Equally, a PD could represent the discretization of continuous probability density functions. Several examples of this kind exist in classical and quantum statistics (we will see a concrete example in the next section). The normal (Gaussian) symmetric
and bell-shaped distribution, for example, could describe the state probability of a particle whose measured values have a variance centered on their mean value. The exponential distribution finds its most eminent application in describing transition probabilities such as the decay rate of atomic nuclei, once their specific mean lifetime of radioactive decay is known. The power-law distribution describes the distribution of energy levels of a quantum system and does not always possess a well-defined variance.

These are just a few examples of distributions that arise in classical and quantum random processes. However, the precise mathematical function and which statistical properties it possesses are, for our purposes, not so essential. This is because the argument we are putting forth here is of a general character that applies to any PD representing the predictability (or uncertainty) of specific events after the collection of a certain number of measurements in time has been performed. PDs are, so to speak, the convenient "reduction to a single frozen graphical time snapshot" of the sequence of multiple processes unfolding in time.

---

165 A probability space with well-defined variables need not necessarily appeal to a frequentist interpretation. But in quantum physics, this is always assumed.

166 Of course, in quantum mechanics, probabilities are given by the Born rule and complex probability amplitudes, defining self-adjoint operators on a Hilbert space. However, in the present context, we do not need to dwell on these formal aspects beyond ordinary discrete PDs.

In fact, it is frequently left in the background that in a spatio-temporal world, all these events are time series. The number of favorable outcomes n_i in Eq. (1) is a count of the measured events corresponding to the ith possible outcome manifesting at n_i instants in a given time interval T corresponding to the duration of the experiment. For most applications, knowing this time series isn't of much interest. A measurement device accumulates the measurements for each outcome and waits until the end of the experiment, then reads out the n_i values defining the PD, without any need to know when each event has taken place singularly. This is also how all probabilistic laws in quantum mechanics are described: by probability functions describing what to expect from a series of N measurements in an interval of time T, but without any prescription as to the order of the outcomes and its actualizations in time. No physical law is defined by such a temporal order. Nevertheless, for our purposes, taking into account this temporal dimension makes the difference.[167]

Let us investigate this first with a simple example. Consider tossing a coin with its two possible outcomes, namely, heads (H) and tails (T). Suppose you toss the coin 10 times, at subsequent times t = {t_1; t_2, ..., t_{10}}, and obtain a sequence

$$S = \{H, H, T, H, T, T, H, T, H, T\}.$$

More precisely, we could write

$$S = \{t_1^H, t_2^H, t_3^T, t_4^H, t_5^T, t_6^T, t_7^H, t_8^T, t_9^H, t_{10}^T\},$$

with t_j^{ij} meaning that we had outcome $i = H, T$ at time $t_j (j = 1, \dots, 10)$. Eq. (1) asks for the number of trials (i.e., instants in time) returning heads or tails, $n_H = n_T = 5$ on a total of N = 10 trials, obtaining equal probability for both events, $p_H = p_T = 0{:}5$. However, whatever kind of permutation one applies to sequence S, it will not change the relative frequencies.

If the trials had a different order of outcomes, say,

$$S' = \{t_1^H, t_2^H, t_3^T, t_4^H, t_5^T, t_6^T, t_7^H, t_8^T, t_9^H, t_{10}^T\},$$

nothing would change as to the probability law. Notice that this is true for any PD in general. Whatever permutation of the set of outcomes (i.e., their order in time of appearance) leaves the number of heads and tails invariant. Eq. (1) needs only the subsets

$$S_H = \{t_1^H, t_2^H, t_4^H, t_7^H, t_9^H\},$$

and

[167] The solution of the time-dependent Schrodinger equation may furnish a time dependent PD, describing, for example, a transient phenomenon, but nothing in it determines the event-time of the outcomes.

$$S_T = \{t_3{}^H,\ t_5{}^H,\ t_6{}^H,\ t_8{}^H,\ t_{10}{}^H\}.$$

It is the cardinality of S_H and S_T that determines a PD, not any order. Nevertheless, as we will see, this "freedom to permute" can determine a freedom of agency with a causal power that remains "invisible" to any statistical law of nature rooted in PDs.

Let us now investigate this in a more general setting and formal language. In the real world, the frequency of each outcome and the temporal ordering of which outcome actualizes at which time is determined by its likelihood, defined by Eq. (1). Denote with i_j the ith outcome (measurement "channel" with i = (1, …, c)) at the jth time of measurement as t_j^{ij}. Represent the temporally ordered time series S of all the N measurements in the time interval T as:

$$S = \{t_1^{i1}, \ldots, t_N^{iN}\} \qquad (2)$$

This is a set that contains more information about a stochastic process than necessary for a PD because it counts not only the outcomes, but also all the instants at which they occur. A mathematically equivalent representation of S is to order the events, not according to their time of actualization, but rather by the order of the outcomes and the series of instants at which they occurred. We can collect the measurement times for a specific outcome n_i as

$$S = \{t_1^{i1}, \ldots, t_N^{iN}\}\ ;\ S = \bigcup_i S_i \qquad (i = 1, \ldots, c). \qquad (3)$$

S_i is an ordered time series, that is, $t^i_1 < \ldots < t^i_{ni} \leq T$, with an average time between events $\bar{t}_i = t^i_{ni} / n_i$, but the temporal separation between each couple of events is scattered stochastically. This confers the uncertainty of the time of actualization and measurement that makes each event unpredictable.

However, in which order we count those events is irrelevant, as we will always obtain the same PD. Were the events to take place with a permuted temporal ordering such as

$$S'_i = \{t_1^{\prime i}, \ldots, t_{ni}^{\prime i}\}, \qquad (4)$$

this wouldn't change the probabilities of occurrence of the different possible outcomes because, for whatever event-time series, their cardinality n_i remains the same:

$$n_i = card(S_i) = card(S'_i) \qquad (5)$$

and the PD of Eq. (1) remains invariant.

Note that this is an exact invariance, provided condition (5) holds. Thus, if the temporal ordering of the outcomes changes, this doesn't perturb the PD, not even negligibly. It simply doesn't modify it at all and, thereby, doesn't constitute a violation of any physical law defined by a probabilistic law.

III. Interfacing the Double Slit to a Particle Random Walk

Let us illustrate the above reasoning with a concrete example relevant to the quantum context. One such example that helps with intuitive visualization could be the temporal unfolding of the photons' detection at the interference fringes from a double-slit experiment. Young's double-slit experiment can also be described by means of classical electromagnetism and wave interference. However, which photon hits which fringe at what time is a purely quantum stochastic process. It is only in the limit of large numbers (that is, when $N \to \infty$ or, equivalently, $T \to \infty$) that the ordinary wave interference pattern becomes visible on a detection screen. This is a macroscopic manifestation of quantum processes. The preference for quantum indeterminism over classical indeterminism is dictated by the fact that the former is believed to be ruled by ontic probability laws, that is, by causally independent events in which the indeterminacy is not due to the lack of knowledge about hidden variables.[168]

Let us generalize this to a discrete *PD* first in the conventional way. In this specific example, the random variable X defined on a countable sample space over c possible outcomes is the subdivision of the detection screen in a discrete horizontal coordinate system x_i; ($i = 1, \ldots, c$) (for example, a line of pixels of a CCD camera). Let's call them detection channels.

The normalized PD is defined by the probability of detecting a number of *ni* photons at those detection channels x_i as $p_i = \frac{n_i}{N}$, with $N = \Sigma_i\, n_i$ the overall number of photons measured over all the possible outcomes, in some time interval T, and such that $\Sigma_i\, p_i = 1$.

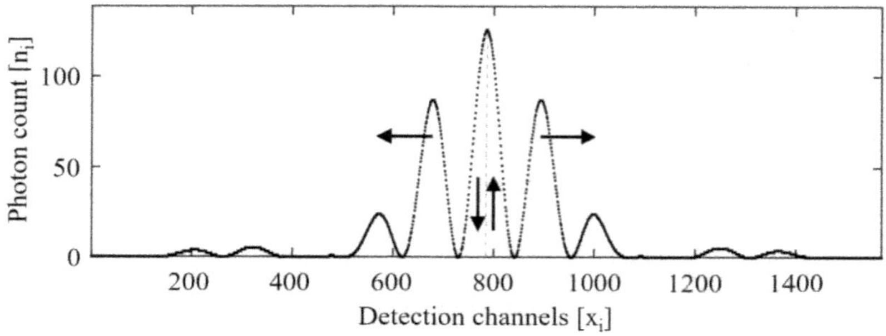

Figure 1: Photon counts for the double-slit experiment. Arrows: Direction of movement.

---

[168] Contrary to pervasive belief, there are still minority voices claiming that Bell's theorem does not conclusively rule out interpretations of quantum mechanics based on local and/or deterministic ontologies (e.g., see Hance and Hossenfelder 2022.) Here, however, we assume quantum mechanics is a theory without hidden variables.

Fig. 1 shows a particular example. It is the graph of the interference pattern, that is, a PD in terms of an intensity profile represented by $N = 20.000$ photons for two slits with aperture size a = 3λ and distance between the slits of d = 9λ, with an λ arbitrary wavelength. The sample space over $[-\frac{\pi}{4}, \frac{\pi}{4}]$ radians is projected on the x-axis in steps of 10-3 radians, resulting in c = 1571 channels. This is the conventional statistical representation of Young's double-slit experiment after a sufficiently large number of measurements for a sampling time *T*.

However, an equivalent representation in temporal ordering is also possible. One can represent in an event-time space the overall experiment denoting the ordered time series

$$S= \{t_1^i, ..., t^i_N\}, \qquad i \in [1, ..., c = 1571],$$

which are the N time-steps (the N measurements) corresponding to every photon detection at whatever position, that is, whenever the detector clicks. One obtains Fig. 2 with the vertical axis representing the same process in time

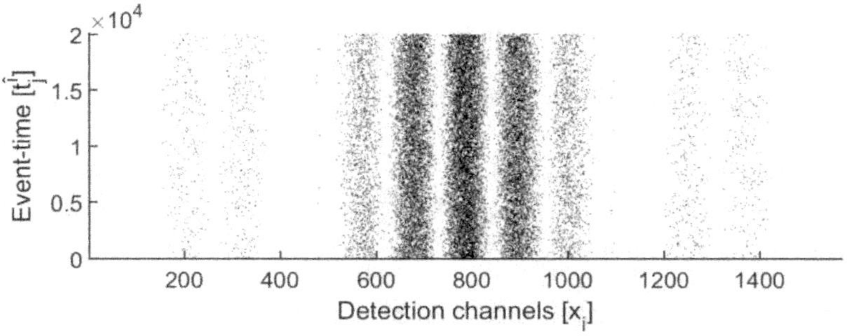

Figure 2: Temporal representation of interference fringes for the double-slit experiment.

This representation contains more information than conventional PDs or the pictures that represent spectral lines. The latter (not to be confused with Fig. 2) are only integrations over time represented by intensity fringes while, in the present case, the time of arrival of each photon is shown. These instants remain unpredictable due to quantum uncertainty. In these processes, not only is the outcome subjected to quantum uncertainty, but so is its temporal dislocation, that is, the times of arrival of the photons during the experiment (even though the average time is a calculable constant of the theory). And, again, the temporal sequence is irrelevant for determining the PD. Permuting or flipping along the vertical axis of Fig. 2 the dots of the interference fringes would have no impact on the pattern of the PD. Their count at each x_i will always be n_i.

Thus, while nature is constrained by the PD, it is free to choose this order without any need to change the PD itself. The question, however, is: How large is that freedom and can it be used to originate anything useful other than yet another random result?

To answer this question, let us imagine a simple toy-model double-slit experimental setup in which each photon detection at a specific interference fringe triggers a dot movement on a screen in a 2D space. For example, consider the PD of Fig. 1 and say that whenever a photon is detected at the secondary fringe left (channels 619–730), this determines a movement leftwards, whenever one is detected at the first half of the central fringe (channels 731–786), this determines an equal shift downwards, whenever one is detected at the second half of the central fringe (channels 787–841), this determines a shift upwards, and whenever one is detected at the secondary fringe left (channels 842–953), this determines a shift rightwards.

For the sake of simplicity and graphical representation, only constant two-unit lengths displacements are chosen. The probability of a turn left or right is about 24% each, while the

probability of a move upwards or downwards is about 17% each. This means that in about 82% of cases, a photon hit produces a movement, while in the remaining 18% of cases, the dot remains immobile. This connects physically random quantum events, that is, the order of the position outcomes of photons and its temporal sequence in a double-slit experiment, to the motion of a digital object.

From a statistical standpoint, this process can be investigated with a simple computer simulation. To do so, one must include the temporal dimension in the PD of Fig. 1 with the time series given by Eq. (2), that is, assigning for each photon detected at the ith channel also its jth time of detection, as represented in Fig. 2. This time series can be constructed as the union of time series at each ith channel. The latter is generated by a random

generator which distributes n_i events temporally at each ith channel between 0 and T, for every $i = 1, \ldots, c$ (or, equivalently, distributes randomly the n_i pixels of Fig. 2 along each vertical column). This enables the creation of the sets S_i and their union S of Eq. (3), which then can be temporally ordered.

Once this array of N ordered events is known, the sequence is played out on the graphical interface: A vector is traced along one of the four directions in the temporal ordering of S, according to each photons positions in the above-described fringe intervals. The result is shown in Fig. 3 left.

The starred dot in the origin is the starting point, while the squared dot is the point of arrival after N = 20.000 time steps. The image doesn't suggest any ordering. It is a typical random walk through a 2D surface, determined by the random hit of the photon on one or the other fringe.[169]

[169] The analogy with the trajectory of a particle subjected to Brownian motion isn't entirely coincidental. Both are random walks. Of course, the underlying physical principles are entirely different.

Directing the path towards a target point, say, the cross at coordinate (50, 50), would be a quite difficult task dependent on extreme luck if we have no control over the temporal sequence of the photons hitting the ith channel at the jth time.

However, suppose a sort of "quantum Maxwell's demon" has control over which photon hits which fringe at what time by permuting the temporal series on each channel as described by the temporal permutation leading from Eq. (3) to Eq. (4), with the only provision that it doesn't alter the number of n_i photons randomly distributed in the time interval between 0 and T.

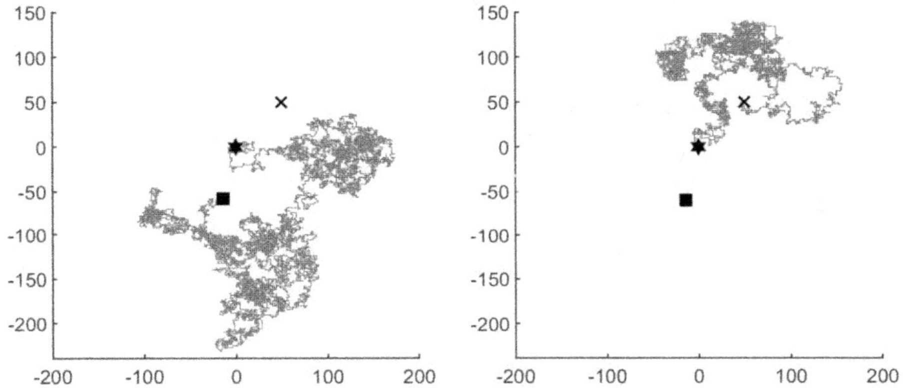

Figure 3: Random walk starts from the starred dot at (0,0) and always ends after 20.000 steps at the squared dot. Left: Target destination x failed, the particle moves until the end of the simulation. Right: Target destination reached, stop at time-step 10091.

Then it also has control over the random walks. Again, this does not alter the PD; it only changes the sequence in time and the channels with which it builds up the interference pattern and, therefore, does not violate any statistical law.

In fact, note that in the context of this experimental toy model the starting and ending points (the random walk from the starred to the squared dot) are always the same. This is because, with whatever random permutation the temporal ordering of the events is realized, the sum of the series of the associated four directional vectors always ends up at the same end point, the square in Fig. 3.[170] Nevertheless, each permutation corresponds to a different path connecting these two points. Also, there exists a virtually infinite number of possible permutations and, consequently, a virtually infinite number of paths (N!, to be

[170] Mathematically speaking, this is a consequence of the fact that vector addition in a flat space is a commutative algebra.

precise, which for N = 20.000 is an outrageously huge number) and with which one can scan the 2D space inside an average radius.[171]
Therefore, while one will never end at the target point, nevertheless, following a Monte Carlo method simulation, after not too many trials it is easy to find a path passing through the desired target point, such as that in Fig. 3 right. This different random walk encounters the target point at time step 10091. (The plots from time steps 10091 to N = 20.000 have been omitted for clarity.) This intervention with control over the temporal ordering of the events permutes the vertical photon distribution of Fig. 2 along the time-step ordinate axis, but does not change the profile of Fig. 1.
Thus, a quantum probability law is a constraint on the possible future outcomes and, in this sense, limits the potential future happenings. Nevertheless, if quantum mechanics is, indeed, a theory without hidden variables, that is, the microscopic laws of nature are ontologically indeterministic, then probability laws merely limit the future to a hypothetical self-determining agent-causation, but don't disallow the agent intention to act freely inside these constraints, eventually also by "moving molecules in directions they were not otherwise going to move".
Moreover, notice how this could potentially offer a different perspective on another long-standing debate in the philosophy of mind regarding the question of the violation of energy conservation in interactive dualism.[172] There is no need to violate conservation laws to actualize a specific history over many potential paths once the demon has access to the temporal unfolding of the quantum processes that determine it. This kind of mental causation doesn't lead to the interaction problem.

IV. A Hypothesis for a Libertarian Panpsychist Framework

The numerical simulation in Sec. 3 further underpins Muller's model of an Anscombe box implemented in a computer program. Intentional action at the micro-physical level does not necessarily violate the constraints of micro-statistical laws. However, the difference here is twofold. Firstly, Müller worked with a model that did not venture into an ontology based on a quantum theory without hidden variables. Secondly, and most importantly for our

[171] This is somehow reminiscent of Feynman's path integral formulation, where classical particle trajectories are replaced by the sum over an infinite number of quantum paths connecting an initial and final point in spacetime.

[172] For example, in the time-independent Schrödinger equation $H|\Psi\rangle = E|\Psi\langle$, once the Hamiltonian and the state vector (which, ultimately, determines the probability via the Born rule) are determined, the energy eigenvalue is fixed.

considerations, at this point, the question is: What or who is supposed to be the "demon"?

Muller assumed an agent-causal libertarianism in terms of a rational will acting as an external cause on the physical processes – an agent external to the micro-statistical process that, nevertheless, if operating carefully enough from the outside, could lead to the appearance of regular structures on the box filled with colored particles without violating the statistical laws. This is something he admits is "suggestive perhaps of certain variants of dualism" (Müller 2022). On the other hand, one can also interpret the room left to free agency at the micro-phenomenal level of quantum laws as something suggestive of certain variants of panpsychism, such as micropsychism, with micro-level entities having agential causal powers. This metaphysical perspective is in line with some form of panpsychist conception positing will, consciousness, and a volitional agency as a fundamental primitive finding its way through a reality constrained by the laws of physics.

Panpsychism is the view that consciousness is fundamental and ubiquitous (for recent reviews see Seager 2019, Skrbina 2017, or for a short introduction see Goff 2019). Constitutive panpsychism postulates a fundamental form of consciousness that is ubiquitous throughout nature, not only proper to living beings. It is usually seen from the bottom-up perspective of micropsychism, namely, the view that all macroscopic facts, such as consciousness in biological organisms, are grounded in micro-phenomena involving consciousness. Micropsychists postulate micro-level entities, such as elementary particles, instantiating very basic forms of consciousness. This is contrasted with the top-down perspective of cosmopsychism, a holistic form of panpsychism seeing the universe as a ubiquitous field of consciousness (e.g., see Shani 2015). Varieties of cosmopsychism can be found in the Indian tradition such as the ancient Veda, the Upanishads, and Advaita Vedanta philosophies (for contemporary work on cosmopsychism connecting to the Indian tradition see Ganeri and Shani 2022).

However, micro- and cosmopsychist perspectives need not necessarily be seen as mutually exclusive. Rather, they can be complementary. Even panentheism can be reconciled with panpsychism within an idea where the universe itself is causa sui (Bruntrup et al. 2020). The universe as a whole is considered a conscious subject, with all conscious entities and properties an aspect of it. Every conscious mind is an individuation of a cosmic mind, as every particle displays some degree of proto-conscious agency. This perspective is reflected in, for example, the Indian mystic Sri Aurobindo, according to whom the universal consciousness individuates into various distinct consciousnesses by limiting itself through a process of "exclusive concentration" (Medhananda 2022).

In a sense, this micro- and macro-scale duality is reflected in quantum physics. Quantum mechanics describes phenomena at the micro-physical level of elementary particles, atomic nuclei, atoms, and molecules. Meanwhile, quantum entanglement is a phenomenon in which two or more particles share the same quantum state over large distances but, once measured, result in locally distinguishable objects with correlated states. Quantum processes were relevant from the first instant of the Big Bang to the formation of large-scale structures of the universe. QFT conceives of universal quantum matter and force fields that are subjected to quantum rules (e.g., symmetry principles and conservation laws) and that also display a stochastic character in the form of random quantum fluctuations that locally determine the quantum vacuum state (sometimes also called the zero-point field). These are responsible, for example, for the spontaneous emission of photons from excited atomic states, for radioactive decay, or for quantum tunneling processes – that is, a universal quantum field determines in space and time the outcome of events at the micro-physical atomic level.

We usually do not see anything volitional in this, precisely because of an implicit randomness objection that prevents us from taking this logical step. Yet we have seen that it is possible to circumvent this objection and conceive of a micro-phenomenal libertarian agent causation, such as a particle finding its way toward a target generating a proper quantum random sequence that does not violate the imposed probabilistic laws and is in line with the known laws of nature. This is in line with Heisenberg's idea that the stochastic aspect of quantum mechanics could be seen as the dynamic possibilities of nature, a form of Aristotelian potentiality that brings a potentiality, or an idea, into actuality (Heisenberg 1958, p. 41).

However, we are also allowed to speculate about libertarian agent causation at the macrolevel over the time series of these events without interfering with the quantum probabilistic laws by conceiving every random quantum fluctuation as the local expression of a pervading cosmic agent. It is not an electron that chooses its spin but the universal consciousness operating through it. Here, we embrace a sort of micro-physical quantum agent-causation hypothesis in which the microscopic quantum field is the expression of a self-determining intentional universal agency. Randomness is not the determinant process, with volition its determination; rather, quantum indeterminacy is the local expression of a global agent-causation determining will. In this sense, a libertarian panpsychist theoretical framework is a viable hypothesis.

V. Conclusion

We have seen how no fundamental logical reason or physical law prevents us from speculating about theories connecting agent-causation and indeterminacy. Upon closer scrutiny, it becomes clear that ample room for free agency exists and this does not alter the profile of a naturally imposed PD. Obeying the quantum laws does not prevent freedom of action.

The understanding of this fact requires essentially a subtle but decisive change in perspective. The randomness objection turns out to be a fallacy based on a lack of awareness of how statistical processes develop over time. We must take into account the temporal dimension of the sequence of outcomes.

We can't limit our outlook to the final PD independent of its dynamic development and the ordering of the events in time. In this context, the statistical mathematical aspect of the temporal developments of physical events cannot be ignored.

Inspired by this result, a possible link between quantum indeterminacy and panpsychism has been shortly addressed. In this paper I argue that there is no inconsistency in entertaining a hypothesis of libertarian panpsychism – that is, a theoretical framework conceiving of an agent causation at the micro-level and macro-level, where micropsychism and cosmopsychism appear as complementary aspects rather than opposites.

Quantum effects are usually thought to be relevant only at the microscopic scale due to thermal quantum decoherence. The brain is thought to be a too-warm environment and too large an object to allow quantum phenomena to play any appreciable role. Nevertheless, some have taken seriously the hypothesis that quantum effects may also play a role in neural processes (for some aspects of this see Adams and Petruccione 2020). For instance, some contemporary libertarians suggest that human decision-making may be traced back to an indeterminism at the neuronal level that might be sensitive to quantum indeterminacy (e.g., Kane 2012a). Meanwhile, there is extensive literature in the field of quantum biology (for a review see Youngchan et al. 2021).

While all these approaches seek quantum effects in matter, the present approach considered quantum indeterminacy itself as the expression of an agent-causation determining will. If true, this would also imply that what appears to be pure randomness – that is, just noise – could hide a creative power. Inside the domain of classical physics, this is what biologists begin to suspect. Stochastic fluctuations during development might be a deciding factor for non-genetic sources of variation

and diversity in anatomical and behavioral traits (see, e.g., Eldar and Elowitz 2010, Bierbach et al. 2017, Linneweber et al. 2020). Noble and Noble (2018) even conjecture that organisms may "harness stochasticity" to have agency and make novel choices. Contrary to common wisdom, noise and randomness in living systems can play a functional role in biological processes (Roy and Majumdar 2022).

Whether microscopic quantum stochastic phenomena might be amplified up to a macroscopic scale, remains a matter of speculation to date. Nevertheless, it was a conjecture that some of the founding fathers of quantum physics were actively pondering. It dates back to the times of Arthur Eddington, Arthur Compton, and Pascual Jordan. (For an overall historical account see Koznjak 2020). More recent theoretical investigations tend to confirm this to be the case. Boekholt et al. (2020) showed that the dynamics of some chaotic three-body systems of stellar size (!) are, on longer time scales, numerically time-irreversible up to the Planck length. This means that even these must be sensitive to quantum uncertainties, leading to the breakdown of the classical Laplacian determinism at all scales. Bandak et al. (2024) have shown how spontaneous stochasticity at the molecular level amplified es molecular noise even to the largest scales of turbulence. Since the molecular scale is the interface between quantum and classical phenomena, quantum spontaneous stochasticity might be amplified at the macro-scale as well.[173]

Thus, while quantum coherence needs low thermal noise at low temperatures to survive and scale up to mesoscopic systems, the effects of quantum indeterminacy are not limited by these requirements and can take advantage of the strong sensitivity to small perturbations of nonlinear systems. Thereby, on the base of what has been outlined here, it can link micro-phenomenal randomness with notions of agency in the frame of a libertarian free will ontology. This might not be palpable to everyone's philosophical inclinations and might put into question the significance and nature of the sourcehood of free will. However, the fact remains that, to date, nothing in the presently known laws of nature prohibits us from entertaining such a metaphysical standpoint.

References

Adams B. and Petruccione F. (2020): Quantum effects in the brain: A review. AVS Quantum Science 2, 022901.

Anscombe G.E.M. (1971): Causality and determinism. In The Collected Philosophical Papers of G.E.M. Anscombe, Vol. 2, Oxford University Press, Oxford, pp. 133–147.

Bandak D., Mailybaev A., Eyink, G. and Goldenfeld N. (2024): Spontaneous stochasticity amplifies even thermal noise to the largest scales of turbulence in a few eddy turnover times. Physical Review Letters 132, 104002.

Bierbach D., Laskowski K. and Wolf M. (2017): Behavioural individuality in clonal fish arises despite near-identical rearing conditions. Nature Communications 8, 15361.

[173] Indeed, in a preprint, Eyink and Drivas (2015) argue, starting from the same principle of spontaneous stochasticity, that quantum indeterminism persists at the classical limit.

Boekholt T., Portegies S. and Valtonen M. (2020): Gargantuan chaotic gravitational three body systems and their irreversibility to the Planck length. Monthly Notices of the Royal Astronomical Society 493(3), 3932–3937.

Bruntrup G., Gocke B.P. and Jaskolla L., eds. (2020): Panentheism and Panpsychism, Mentis, Paderborn.

Chaitin G.J. (1975a): A theory of program size formally identical to information theory. Journal of the Association for Computing Machinery 22(3), 329–340.

Chaitin G.J. (1975b): A theory of program size formally identical to information theory. Scientific American 232(5), 47-53.

Clarke R. (2010): Are we free to obey the laws? American Philosophical Quarterly 47(4), 389-401.

Doyle B. (2011): The standard argument against free will. In Free Will: The Scandal in Philosophy, I-Phi Press, Cambridge, Chapter 4.

Eldar A. and Elowitz M.B. (2010): Functional roles for noise in genetic circuits. Nature 467: 7312.

Eyink G. and Drivas T. (2015): Quantum spontaneous stochasticity. Accessible at arXiv:1509.04941.

Ganeri J. and Shani I. (2022): What is cosmopsychism? The Monist 105(1), 1-5.

Golf P. (2019): Cosmopsychism, Micropsychism, and the Grounding Relation. In The Routledge Handbook of Panpsychism, ed. by W. Seager, Routledge, London, pp. 144–156.

Hance J. and Hossenfelder, S. (2022): Bell's theorem allows local theories of quantum mechanics. Nature Physics 18, 1383.

Heisenberg W. (1958): Physics and Philosophy, Harper Row, New York.

Kane R. (2012a): Can a traditional libertarian or incompatibilist free will be reconciled with modern science? Steps toward a positive answer. In Is Science Compatible with Free Will? – Exploring Free Will and Consciousness in the Light of Quantum Physics and Neuroscience, ed. by A. Suarez and P. Adams, Springer, Berlin, pp. 255{272.

Kane R. (2012b): The Oxford Handbook of Free Will, Oxford University Press, New York.

Kastner R. (2016): The Born rule and free will: Why libertarian agent-causal free will is not “antiscientific”. In Probing the Meaning of Quantum

Mechanics, ed. by in D. Aerts, C. de Ronde, H. Freytes and R. Giuntini, World Scientific, Singapore, pp. 203-212.

Koznjak B. (2020): The earliest missionaries of `quantum free will': A sociohistorical analysis. In Free Will & Action. Historical-Analytical Studies on Nature, Mind and Action, ed. by F. Grgic and D. Pecnjak, Springer, Berlin, pp. 131–154.

Linneweber G.A., Andriatsilavo M., Dutta S.B., Bengochea M., Hellbruegge L., Liu G., Ejsmont R.K., Straw A.D., Wernet M., Hiesinger P.R., and Hassan B.A. (2020): A neurodevelopmental origin of behavioral individuality in the drosophila visual system. Science 367(6482), 1112–1119.

Masi M. (2023): Quantum indeterminism, free will, and self-causation. Journal of Consciousness Studies **30**(5-6), 32–56.

Medhananda S. (2022): The playful self-involution of divine consciousness: Sri Aurobindo's evolutionary cosmopsychism and his response to the individuation problem. The Monist 105(1), 92–109.

Mulder J. (2021): Animals run about the world in all sorts of paths: varieties of indeterminism. Synthese 199, 1945–11961.

Mulder J., Muller T., Ometto D., and Miltenburg N. (2022): Causality and determinism, powers and agency: Anscombean perspectives. Synthese 200: 452.

Muller T. (2022): "Let's build an Anscombe box": Assessing Anscombe's rebuttal of the statistics objection against indeterminism-based free agency. Synthese 200: 74.

Noble R. and Noble D. (2018): Harnessing stochasticity: How do organisms make choices? Chaos 28(10), 106309.

Pereboom D. (2001): Living without Free Will, Cambridge University Press, Cambridge.

Pereboom D. (2014): Free Will, Agency, and Meaning in Life, Cambridge University Press, Cambridge.

Roy S. and Majumdar S. (2022): Noise and Randomness in Living System, Springer, Berlin.

Schrodinger E. (1951): Nature and the Greeks and Science and Humanism, Cambridge University Press, Cambridge.

Seager W., ed. (2019): Routledge Handbook of Panpsychism, Routledge, London.

Searle J.R. (1986): Minds, Brains and Science, Harvard University Press, Cambridge.

Shabo S. (2020): The two-stage luck objection. Nous 54(1), 3–23.

Shani I. (2015): Cosmopsychism: a holistic approach to the metaphysics of experience. Philosophical Papers 44, 389–437.

Skrbina D. (2017): Panpsychism in the West, MIT Press, Cambridge.

van Inwagen P. (1983): An Essay on Free Will, Oxford University Press, New York.

van Miltenburg N. (2022): Causality, determination and free will: Towards an Anscombean account of free action. Synthese 200: 279.

Youngchan K., Bertagna F., D'Souza E., Heyes D.J., Johannissen L., Nery E., Pantelias A., Jimenez A., Slocombe L., Spenceret M., et al. (2021): Quantum biology: An update and perspective. Quantum Reports 3(1), 80–126.

5. The Interaction Problem and Mental Causation: A Quantum Field Theoretical Perspective[174]

Abstract

A fundamental issue in the philosophy of mind, particularly within dualist theories, is the distinction between the mind and body as separate 'substances.' The challenge lies in elucidating how an immaterial or unphysical mind can causally interact with a physical body or its brain states, considering that they seemingly belong to fundamentally different ontological categories. Questions regarding mental causation often operate under assumptions rooted in a macroscopic worldview based on our everyday intuitions and classical physics, which we believe have clear and defined meanings. However, upon closer inspection within an extended quantum field theoretical context, these concepts lose significance. When viewed through the lens of modern physics, the conceptual categories defining this debate acquire more complex nuances than we might initially assume. In light of these clarifications, a new understanding of mental causation is presented that reconciles the dualist perspective with the probabilistic nature of quantum mechanics.

Introduction

Although contemporary philosophy has largely moved away from the original Cartesian substance dualism, non-physicalist theories remain prominent. Examples include dual-aspect monism, panpsychism, cosmopsychism, idealism, substance monism, and various theistic or spiritualist accounts. They contend that the mind is neither reducible to nor entirely dependent on the physical brain or body. Mental properties are understood as non-physical properties, with the body and mind regarded as distinct and separable entities or, at the very least, not reducible to exclusively physical properties.

Despite their differences and nuances, these theories within the philosophy of mind share a common rejection by physicalist mind-body identity theories, which posit that physical processes or properties can fully explain mental

[174] The present article was published in *Qeios*. DOI: **https://doi.org/10.32388/78F7FZ.4**

phenomena. This is because asserting a fundamental distinction between the mental and the physical implicitly endorses some form of 'interactionism' or 'interactive dualism'—the notion that the body and mind causally interact with one another despite being fundamentally different in nature. The same question applies not only to the mind but also to any entity or purportedly metaphysical substance, such as a soul, spirit, or 'immaterial' or 'unphysical' consciousness. This leads to one of the most enduring philosophical challenges, known as the 'interaction problem' or the 'problem of mental causation,' framed in 1643 by Princess Elisabeth of Bohemia, who presented a pointed challenge to Descartes regarding his *Meditations on First Philosophy*: *"Given that the soul of a human being is only a thinking substance, how can it affect the bodily spirits in order to bring about voluntary actions?"* (Shapiro, 2007).

In more modern terms, the issue can be expressed by asking how putatively non-physical mental states—such as beliefs, desires, and intentions—can exert any causal influence in a physical world governed by physical laws. The dualist causal heterogeneity problem arises: If mental properties, states, or substances are radically heterogeneous from material objects and physical forces, they lack the commonality necessary for interaction. If the mind is immaterial, it remains unclear how it can causally interact with a material body without violating fundamental principles of physics and maintaining metaphysical coherence. For an overview, see (Robb, 2023) and (Cucu, 2022), the classic work by Popper and Eccles (Popper, 1977), or (Lindahl, 1994), (Libet, 1994), and (Libet, 2006).

Another common objection to the concept of mental causation, dating back to Leibniz[175], is that such interactions would violate the principle of energy conservation. If physics encompasses all interactions among particles within the brain, then an immaterial mind influencing the brain's biochemistry would introduce a novel form of interaction at the microscale, ultimately violating the principle of energy conservation. For this long-standing dispute, see (Averill, 1981), (Larmer, 1986), (Kim, 1993a), (Wilson, 1999), (Montero, 2006), (White, 2016), (Cucu, 2019), or (Seager, 2022).

Given that, despite centuries of debate, these fundamental issues remain unresolved and no consensus has been reached, the interaction problem is frequently invoked as a knockdown argument against any form of dualism and as a compelling rationale in support of physicalist mind-brain identity theories.

This paper seeks to highlight various semantic ambiguities related to the categories involved in the interaction problem, thus providing conceptual clarity in light of modern physics. It will emphasize the distinction between what we consider 'material' versus 'immaterial' and 'physical' versus 'unphysical,' clarify what an 'interaction' is according to the standard model of particle physics, and question what an extended versus non-extended

[175] See the introduction of (Pitts, 2020) and references therein for a historical overview.

'substance' could possibly mean in the context of contemporary physics, particularly quantum field theory (QFT). Finally, we will shortly reexamine a principle of mental causation grounded in stochastic phenomena associated with quantum effects.

Whether physics can tell us anything at all about these philosophical quests regarding the mind, the nature and origin of consciousness, and related metaphysical speculations is debatable. However, if we build theoretical frameworks about the interaction problem of mental causation arguing in terms of the physical sciences, we cannot be confined to the classical physics of the 19th century or the early 20th century, as many philosophers of mind tend to do, but must instead incorporate contemporary understandings of the world.

I. When Words Matter: Pursuing Conceptual Clarity in Light of Contemporary Physics

1. What Does 'Immaterial' Mean?

From a physics perspective, the interaction problem is frequently articulated using ambiguous and potentially misleading terminology. For example, the inquiry regarding how an 'immaterial' mind can causally interact with a 'material' body implies a clear and well-defined distinction between the two, or at least suggests that this distinction is self-evident and requires no further elucidation. Eventually, these opposing categories become conflated with the related notions of 'physical' versus 'unphysical.' However, at least from the perspective of the physical sciences, this conceptual ambiguity prevents a deeper understanding of the subject.

In physics, the term 'material' encompasses all entities composed of matter—specifically, those that possess mass. Mass serves as a measure of inertia, which is the property of a particle or body that enables it to resist changes in its state of motion. A massive particle can always be accelerated or decelerated, moving at speeds less than the speed of light. In contrast, light has no rest mass—i.e., it lacks inertia (it does not undergo acceleration or deceleration)—and always travels at about 300,000 km/s in any frame of reference. In the framework of special relativity, the concept of rest mass takes on a more abstract role as a relativistic scalar invariant. However, one does not need to pin down rigorous definitions to recognize that physics is replete with entities that are 'immaterial' yet nonetheless real, concrete, and not 'unphysical.'

For example, light is a massless—that is, 'immaterial'—electromagnetic field that propagates throughout space and interacts with material particles or bodies. From the perspective of quantum electrodynamics (QED), electromagnetic waves are mediated by light particles—the massless gauge bosons, commonly known as photons—which are immaterial entities that, as we know from everyday experience, have considerable causal power over material ones.[176]

The same can be said about the action of a gravitational field. Gravity is a massless force field that interacts with material objects. The causal influence of this 'immaterial' field on massive bodies, such as planets and stars, is well established. In general relativity, gravity is represented as a curved 4D differentiable spacetime Riemannian manifold whose curvature is determined by mass or energy and along which everything follows the shortest geodesic path. Taylor and Wheeler summed it up concisely: *"Spacetime tells mass how to move, mass tells spacetime how to curve"* (Taylor and Wheeler, 1992). The status of the gravitational force relative to the three other fundamental forces (the electromagnetic and the strong and weak nuclear forces) is a matter of some debate; however, the relevant fact is that a material object can interact with another material object without direct contact by warping its surrounding vacuum.

The question of how causal influence can occur between material bodies without direct contact through empty space, solely through an immaterial field or an even more ethereal notion of spacetime curvature, is not a novel philosophical issue. When Newton was asked about the nature of gravity's 'action at a distance,' his official response in the Principia was: *"Hypotheses non fingo"("I do not frame hypotheses").* However, his private correspondence reveals that framing hypotheses was likely one of his major occupations: *"That gravity should be innate, inherent, and essential to matter, so that one body may act upon another at a distance through a vacuum, without the mediation of anything else, by and through which their action and force may be conveyed from one to another, is to me so great an absurdity that I believe no man who has in philosophical matters a competent faculty of thinking can ever fall into it. Gravity must be caused by an agent acting constantly according to certain laws; but whether this agent be material or immaterial, I have left open to the consideration of my readers"* (Cohen, 1978).

Unfortunately for Newton, for centuries, nearly all scientists with a "competent faculty of thinking" have accepted this "great absurdity" without further questioning it.

The same phenomenon presents itself at the microscopic scale in the realm of nuclear forces. The stability of matter relies on the strong nuclear force,

---

[176] One could argue that, according to the mass-energy equivalence principle, matter and energy can be converted into one another (e.g., during annihilation processes between particles and their antiparticles); their ability to interact might stem from a shared commonality at a fundamental level. However, adopting this perspective could provide more reason to embrace a panpsychist viewpoint, positing that mind and matter share a similar equivalence. This interpretation renders the interaction problem no more problematic than any interaction within physical theories.

which binds quarks through asymptotic confinement to form protons and neutrons in atomic nuclei. This interaction is mediated by massless gluons, the gauge bosons responsible for the strong nuclear force.

The fact that massless fields can interact with material particles rarely raises concerns because modern physics takes a pragmatic approach, accepting this state of affairs as a given. Instead, it focuses on the mathematical formalization that allows for empirically verifiable predictions, largely overlooking the philosophical issues regarding how immaterial entities can causally affect material objects. No explanation regarding the principle of causal heterogeneity is deemed necessary.

Yet, for some reason, when a similar conceptual issue arises in the philosophy of mind—specifically regarding how an immaterial mind could interact with a material brain or body—it is presented as a seemingly insurmountable objection against any form of dualism.

Of course, one might object that the central issue in the debate over mental causation should not be framed as a conflict between 'material' and 'immaterial' physical entities. Instead, it involves the question of how an 'unphysical substance' can influence a 'physical substance.' Indeed, that is what we will see next.

However, this terminological clarification was necessary. It underscores that when we deal with more foundational philosophical questions, category conflation can lead to conceptual confusion. It also highlights that a broader interaction problem has existed throughout the history of science yet has not been considered particularly worrisome.

2. What Does 'Unphysical' Mean?

The same question arises regarding the distinction between what is meant by something being 'physical' versus 'unphysical' or 'metaphysical.' When discussing whether a non-physical mind can influence a physical brain, we cannot exempt ourselves from clarifying the distinction between what 'physical' and 'unphysical' mean. It is precisely the nature of this distinction, along with the more or less unaware premises underlying it, that determines the character of our reasoning, conjectures, and conclusions. If such categories are based solely on a vague commonsense understanding of everyday experience, then speculations about the mutual interaction of physical and unphysical entities, or about causal closures, could become misleading or devolve into mere wordplay.

For example, an unaware but not uncommon fallacy lurking behind the debates on mental causation is that of positing a priori that anything that cannot interact with the physical world is considered 'unphysical' and then asking how an unphysical mind could interact with a physical body. This results in circular reasoning such that the conclusion is implicitly assumed in the premise. While coming from the via negativa, defining the physical as anything that is not mental or experiential merely reinforces the notion of the mind as non-physical and does not clarify its causal relationship.

On the other hand, there is no universally accepted definition of what constitutes the 'physical,' nor is there a clear distinction between physics and metaphysics or an agreement on the foundational doctrines of physicalism. The term 'physical' varies in interpretation depending on the context and discipline, such as science, metaphysics, philosophy of mind, and philosophy of science. This definitional challenge is not new; its history is encapsulated in Hempel's dilemma (e.g., see (Stoljar, 2010), (Ney, 2014), (Elpidorou, 2018), (Judish, 2007)). Hempel's dilemma poses a challenge to physicalism and, more broadly, to naturalism, due to the ambiguity surrounding the definitions of 'physical phenomenon' and 'natural.' If we define 'physical' according to current physics, physicalism is likely false, as no existing physical theory incorporates the qualitative mental dimension of consciousness within its framework. Conversely, future theories may alter our current understanding. Indeed, a future comprehensive physics could potentially include entities that we typically do not consider physical. As a result, the dilemma compels physicalists to choose between a definition that renders their thesis false or one that renders it vacuous.

Nonetheless, while one could agree that physics alone cannot define the 'physical' (Crook and Gillett, 2001), one could also say that it helps us gain conceptual clarity.

The aim of this section is not to provide a definitive answer, as, even in physics, there is no unique and rigorous characterization of the physical appropriate for mind-matter conjectures. Instead, the aim is to clarify the quantum field theoretical foundations that physics uses to describe the physical world while acknowledging its limitations. Investigating how far physics can lead us might allow us to expand (or restrict) or further define the boundaries and foundations of other metaphysical non-physicalist theoretical frameworks, an example of which I will present in the second part.

Conventional physics textbooks do not define physics in contrast to metaphysics, except perhaps through a historical note that the term 'physics'—and, thus, 'physical'—originates from Aristotle, who coined it to mean "the study of Nature." Nonetheless, a characterization in line with contemporary science could be to describe as 'physical' anything that can be empirically verified, existing in space and time, and governed by the laws of physics. Spacetime, matter (or energy, according to the mass-energy equivalence), and the laws of physics (or causality) are the ingredients of the typical understanding of classical mechanics, which is often assumed in dualist debates.

Meanwhile, in physics, the term 'unphysical' does not carry a metaphysical implication; it usually refers to meaningless solutions of differential equations, such as negative or infinite energy states, complex-valued lengths or time intervals, probabilities exceeding one, the violation of conservation or symmetry principles, etc.[177] An 'unphysical solution' is a mathematical limitation of the formalism describing impossible physical processes or nonexistent entities and does not hint at anything metaphysical.

Therefore, if we accept this definition and consider the mind to be an unphysical substance, we either refer to something that does not exist—a self-negating paradox—or imply something that transcends spacetime, matter, energy, and the laws of physics. Once again, this leads us to assume an interactive impossibility from the outset, resulting in the same circular reasoning mentioned earlier.

However, if we are looking for the most fundamental primitives of physical reality from the perspective of modern physics, this is an outdated view.

Firstly, as we have seen in the previous section, physics is full of immaterial things that are not made of matter. Matter is not fundamental and must be replaced by something else. The contemporary theory that describes reality at the most fundamental physical level, which is the standard model of particle physics based on QFT, describes matter, energy, forces, and all interactions with quantum fields.

What is a quantum field in QFT?[178]

Technically speaking, quantum fields are operator-valued distributions defined in spacetime that operate on a Hilbert space (or a Fock space in the context of multiparticle systems). Specifically, quantum field operators are defined at a given point in space and time and can be expressed in terms of creation and annihilation operators, which correspond to raising or lowering a particle's state. These fields can be scalar fields (such as the Higgs boson) or vector fields with integer spin (bosons like photons and gluons) or half-integer spin (fermions, such as quarks and leptons). Within this theoretical framework, quantum fields are modeled as the sum of quantized harmonic oscillators (field modes), with particles representing wave-like quantized excitations of the ground state (the vacuum in its unexcited condition), with a distinct quantum field assigned to each type of particle (e.g., the photon corresponds to the electromagnetic field, while the electron is associated with the Dirac field, etc.).

---

[177] However, it is standard practice to utilize negative energy states for the description of potentials. Moreover, history shows that, eventually, such solutions might turn out to be physical if adequately interpreted. The paradigmatic example is the negative energy states predicted by Dirac's equation, which are now interpreted as representing antiparticles.

[178] Whether the ultimate ontology of QFT is based on particles, fields, quanta, 'bundles of properties,' or whatever other kind of noumenal entities is debatable, e.g., see (Kuhlmann, 2010). However, despite the existence of various interpretations, the mathematical formalism of QFT is unambiguously articulated in the language of fields. Given the wider support for this perspective, I will adopt the field as the conceptual foundation of reality in QFT.

However, we need not engage in technicalities; for our purposes, we can consider quantum fields to be the foundational fabric of what we refer to as 'physical' reality. In this framework, particles can be intuitively visualized as traveling-wave perturbations of the respective particle field parameterized over a spacetime manifold (sort of localized excitation or fluctuation, just imagine vibrating 'bumps'), with interactions occurring when these excited states couple with one another. (Section 0.0 contains more on the nature of interactions).

Secondly, while in philosophy the concepts of 'laws of Nature' and 'causality' have also given rise to longstanding debates, in QFT the laws of physics are determined by symmetry conditions, and causality refers to the interaction of fields constrained by a relativistic microcausality condition. (No influence can travel superluminally, more in this later.)

All the laws of physics emerge from the requirement of symmetry principles—that is, the principle of covariance, global and local gauge invariance, and spontaneously broken symmetries. Ultimately, what we call the 'laws of Nature' and 'causality' boil down to symmetry principles involving quantum field operators constrained by algebraic rules (canonical commutation relations) imposed to preserve the wave-like aspect of the fields and particles.

Thus, according to QFT, spacetime and (wave-like) quantum fields subjected to symmetry principles and relativistic causality are what contemporary physics considers 'physical' at the most fundamental level. In contrast, whatever transcends spatial and temporal relationships and cannot be modeled by a quantum field subjected to these symmetry and causality principles may be classified as 'unphysical.'

That said, this viewpoint is also not without ambiguities.

Interestingly, QM and QFT inherently possess a Platonic flavor, with unobservable abstract entities that are not directly observable or verifiable nevertheless being part of the physical theory. Examples include the ambiguous epistemic vs. ontological status of the wave function, the ghostly appearance and disappearance of virtual particles in the quantum vacuum, and the non-unitary evolution resulting from the measurement process (the measurement problem and the infamous 'collapse' of the wave function). Opinions vary widely on whether these entities or processes should be regarded as 'physical' or 'real' beyond useful mathematical abstractions.

A clear example of this situation is the well-known wave-particle duality, in which particles like electrons and photons display both wave-like behaviors (such as interference and diffraction) and particle-like behaviors (such as localized interactions on detectors). In QM, particles' states are described by the wave function, while in QFT, this duality is reconciled by viewing particles as discrete excitations of underlying continuous wave-like quantum fields. However, the ontology of these concepts remains ambiguous. Although in QM and QFT wave functions and quantum fields are taken as the fundamental entities that describe reality, our observations are limited to the detection of localized interactions impacting detectors that we reify as point-like particles or as particle tracks rather than waves or fields. The question of whether wave functions or operator-valued fields should be regarded as 'real' and 'physical' objects or merely as abstract mathematical tools devoid of ontology is ultimately a matter of personal philosophical preference.

Then, while the microcausality condition is preserved, long-distance correlations can arise between spacelike separated entangled particles. This means that quantum theory is inherently non-local, which complicates the classical understanding of spatial relationships. More on this later.

A further ambiguity arises when we consider that QFT is embedded in a special relativistic framework where space and time are not absolutes but relative. If we consider anything that cannot be described in spatiotemporal terms as unphysical, should we then view the interior of a black hole beyond the event horizon—where our understanding of spacetime breaks down—as ‘unphysical’? QFT is not a general relativistic theory, so we cannot definitively answer this question. However, it is unlikely that general relativistic—let alone classical—notions of spacetime survive in such extreme conditions. This makes the characterization of physicality tied to spatiotemporal concepts even more questionable.

Meanwhile, some contemporary efforts to unify quantum physics with general relativity into a theory of quantum gravity are raising questions about the nature of space and time themselves. Rather than being fundamental structures of reality, they might be emergent properties arising from a deeper layer of existence that transcends spatial and temporal dimensions. If a theory of quantum gravity were to be successfully confirmed—one that views spacetime not as fundamental but as an emergent property—would the criteria used to differentiate between 'physical' and 'unphysical' entities still hold? Especially the principle of causal closure, which states–to quote Jaegwon Kim (Kim, 1993b), who includes the temporal dimension–"Any physical event that has a cause at time t has a physical cause at t," becomes less clear-cut than commonly assumed.

Taken together, these aspects of QM and QFT should prompt us to take some precautions regarding the conceptual categories we employ when we speculate about unphysical minds interacting with physical bodies. The categories with which we discriminate between what is physical and unphysical are ideas tied to our subjective perceptions of reality and are mental abstractions that are useful for navigating our everyday lives; however, they have little to do with the more fundamental structure of reality. Nonetheless, despite its ambiguities, QFT provides a stronger foundation for constructing models of interactive dualism that are more consistent with current scientific knowledge.

3. What Are 'Interactions' and 'Substances'?

Given that the problem of mental causation contemplates the interaction between minds and bodies or brains, it might be helpful to clarify what 'interaction' or 'causation' means in contemporary physics. Particularly in QFT, this concept differs significantly from our ordinary intuition.

The notion of causation refers to the propagation of 'influences'—i.e., interactions and correlations—governed by specific laws. It involves an ordered sequence of events within a spacetime framework, where one event can affect another that falls within its future light cone—i.e., the region of spacetime containing all points that can be reached from that event by signals or particles moving at or below the speed of light. It pertains to physical interactions that travel along timelike or lightlike paths—i.e., trajectories through spacetime along which the separation between events is such that massive (or massless) particles, moving less than (or equal to) the speed of light, could travel. As anticipated previously, in QFT, the relativistic microcausality condition is defined by the commutation relations of operators associated with measurements or events occurring at spacelike-separated points. In simple terms, physical causality is about no faster-than-light cause-and-effect relationships between events.

While interactions in QFT are represented as the coupling of fields in scattering processes. They are defined by the higher-order (non-quadratic) terms in the Lagrangian density (a mathematical function that summarizes the dynamics of a physical system), representing nonlinear couplings between fields—that is, the exchange of energy and momentum. These are useful for describing scattering processes in which particles deflect each other, eventually producing new particles in creation or annihilation processes. Meanwhile, weak force interactions determine decays, with unstable particles transforming into lighter ones.

The interaction Lagrangian determines the scattering matrix, or S-matrix, which is a mathematical object that encodes the probabilities for all possible initial particle states to scatter into final particle states, as a weighted sum of all possible 'histories.' Computing its matrix elements provides a way to calculate scattering amplitudes—that is, the transition probabilities from an initial state to a final state of each of these possible 'histories'—for which perturbative expansion terms are graphically represented as Feynman diagrams. These are pictorial representations of the interaction by an exchange of virtual particles. Symmetries constrain the possible forms of the interaction Lagrangian and, thereby, determine the structure of these interactions.

The strength of the interaction is quantified by a coupling constant. For instance, in the interaction between an electron field and a photon field, the fundamental electric charge of the electron, denoted as e, serves as the coupling constant in quantum electrodynamics. Meanwhile, in a Feynman diagram, the electromagnetic force is represented as a virtual photon exchange between charged particles. Fig. 83 illustrates a scattering process, with time flowing along the vertical axis, between two electrons (depicted as straight lines) through the exchange of a virtual photon (shown as a wavy line). Virtual particles can be considered temporary disturbances of a field that exist for only extremely short time intervals during the interaction. They are called 'virtual' because they are a mathematical tool used in perturbation theory, not measurable entities corresponding to stable, propagating, physical objects.

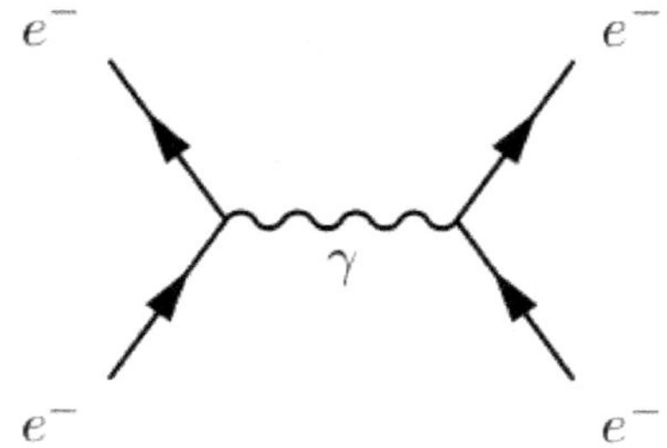

Fig. 83 Scattering between two electrons via the exchange of a virtual photon.

Fig. 84 presents a simplified version of a more general interaction process as conceived of in QFT. Two incoming particles (represented by straight lines p_1 and p_2) scatter in a spacetime region (the gray circle), where they exchange energy and momentum through a complex process, resulting in two outgoing—potentially different—particles (the continuous lines p_3 and p_4). The energy-momentum transfer, mediated by the exchange of virtual particles (which are not necessarily virtual photons), can occur in various ways. The first possible history (first-order scattering amplitude) is that of Fig. 83, but an infinite number of other possible interaction schemes are possible.

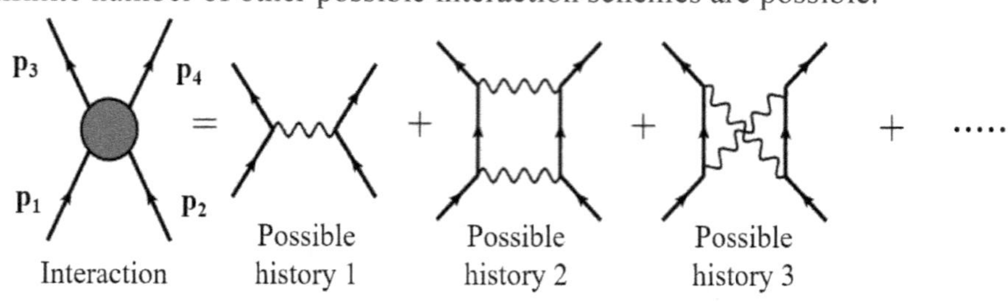

Fig. 84 In QFT, an interaction is represented as the sum over all possible scattering processes.

This counterintuitive conception of what an interaction is in QFT, while perfectly coherent from a theoretical and mathematical standpoint and at the basis of one of the most successful theories of all time (namely, the standard model of particle physics), highlights that we should not interpret these graphical representations as having any ontology other than fictions that serve our naïve intuition. However, we must recognize that our everyday experience of 'interactions' in terms of substances 'colliding' or 'contacting' each other is merely a mental abstraction based on subjective sensations. This conception, which more or less implicitly often still survives in the interactive dualism debates, lacks significance at a more fundamental level; it arises on a macroscale from a collective phenomenon of countless microscopic quantum scattering processes. There is no solid, point-like particle that bounces off another point-like or extended rigid object, nor is there a classical understanding of an extended continuous substance occupying a definite volume of space. There is only a complex interaction of fields that manifests through a localized exchange of forces in space and time.

From this viewpoint, even the classical notion of 'matter' as a substance becomes problematic. Matter itself represents an expression of a force field, always reducible to an exchange of forces—specifically, the mutual interactions of these fields. Positing matter in a definition of what must be considered 'physical' is misleading; force fields are more fundamental.

Notice that Nature doesn't provide a solution to the heterogeneity problem; it simply ignores it. In QFT, quantum fields do not need to have anything in common to interact with each other. For example, an electron, which is a localized excitation of the electron field, interacts with a photon, the excitation of the electromagnetic field. In contrast, gluons (the carriers of the strong force) and the W^+, W^-, and Z^0 gauge bosons (the carriers of the electroweak force) couple with quarks but not with electrons and all the other leptons. On the other hand, photons, despite their commonality, do not couple with one another.[179] What determines these interactions involves a complex series of factors, such as particle charge, the gauge symmetries governing the scattering process, higher-order correction terms in the Lagrangian, etc. However, aspects like 'commonality,' 'sameness,' and 'diversity' of the fields do not play a role. At the most fundamental level, physical reality does not align with our classical understanding of how 'substances' interact.[180]

And what qualifies as 'substance'?

In this framework, the notion of 'substance' also tends to become vague or even redundant. The problem of mental causation investigates how two fundamentally different substances can exert causal influence on one another. However, it often obscures the lack of philosophical consensus regarding the definition of 'substance.' This debate is as ancient as philosophy itself, tracing

---

[179] However, they do weakly interact in very high-energy scattering processes.

[180] That said, it should be noted that, while the current standard model of particle physics does not conceive of a singular quantum field, as it assigns a separate quantum field to each type of particle, one can expect that a future theory of quantum gravity might unify these fields as specific states of a single universal quantum field. This would reintroduce a sense of commonality that aligns with our intuitive understanding of causal influences. On the other hand, it would also suggest viewing reality through the lens of substance monism instead of substance dualism. An investigation of the interaction problem from the perspective of a Spinozist worldview within modern QFT would be worthy of more attention.

its origins from Aristotle to post-Roman Neoplatonism, encompassing the contributions of Arabic and Eastern traditions, and continuing through thinkers such as Locke, Hume, Kant, Spinoza, and Descartes. For a brief introduction, see (Hoffman and Rosenkrantz, 2009). Contemporary discourse has witnessed a resurgence of neo-Aristotelian hylomorphic interpretations (e.g., (Simpson, 2024)).

Broadly speaking, the ontology of substance can be conceptualized as an underlying entity, substratum, a haecceity, or noumenal Kantian 'thing in itself,' distinguished from its attributes and properties. It signifies something invariant that nonetheless maintains properties subject to change.

The curious aspect of QFT is that the 'substances' involved (massless or massive particles) and their interactions are ultimately both manifestations of quantum fields. In QFT, all interactions are understood as being the exchange of force-mediating particles between other particles, both of which are described as excitations of fields. No clear distinction exists between substance and interaction; both are made of the same object.

Thus, if we overlook the fundamental aspects of physical reality that modern science highlights and entertain ourselves with no better-defined 'interactions' between physical and metaphysical 'substances' while adhering to conventional everyday life interpretations of these terms or having in mind 17th- to 18th-century physics, we will likely encounter persistent conceptual challenges and paradoxes.

QFT is certainly not the final theory; a theory of quantum gravity is likely to offer deeper insights into the nature of reality. Nevertheless, it is the best model we currently have, and its conceptual foundations can't be ignored in metaphysical speculations regarding interactions between 'immaterial,' 'unphysical,' and 'non-extended' minds and 'material,' 'physical,' and 'extended' substances.

4. What Is a 'Non-Spatial' Substance?

The debate framed within Cartesian dualism becomes equally vague when we address how a non-extended mind—meaning a mind that is 'non-spatial'—can interact with entities existing in a spatial realm. What might initially appear to be a well-defined question loses clarity when examined through the lens of the non-local nature of QM.

Let's explore the most distinctive quantum effect in this regard: quantum entanglement. Quantum theory states that when entangled particles undergo a measurement process, their previously non-separable pure state is instantaneously projected into a mixed state of distinguishable particles—i.e., the particles acquire definite properties. This results in an instant correlation (or anti-correlation) between the particles, even if they are light years apart—as if there were no extension separating them. This non-local aspect of QM famously prompted Einstein to complain about "spooky actions at a distance."

Thus, whether these non-local quantum phenomena can be promoted to 'non-spatial' phenomena is debatable. Quantum entanglement correlates measurements across space, yet it describes a physical process that cannot be framed within the classical or relativistic concepts of spatiotemporal causality. The holistic character of long-range correlations between coherent states doesn't fit into our conventional notions of spacetime locality.[181]

Consequently, in the context of modern physics, this becomes one more reason for caution. The question is: Can a non-local mind qualify as 'non-spatial' or 'non-extended'? This is a mind that, while not situated in space, would still have the causal power to determine long-range correlations between physical systems.

Several objections to this hypothesis can be raised.

Mind-matter causality often relies on the principle of transference, which posits that for A to be causally effective on B, A must transfer something to B (Hoffman and Rosenkrantz, 1991). Causation appears to require something like a flow of energy or a force interaction between objects to produce physical effects. However, physical forces, along with energy, linear or angular momentum, spin, and other dynamical variables, are meaningful only within a context involving spatial relations parameterized over time. This is a context that would not only reintroduce the ontology that the hypothesis seeks to transcend but also violate conservation laws, particularly principles of energy and momentum conservation. This is why, since the time of Schrödinger, critics, including popular physicists like Sean Carroll, have argued that incorporating mental causation would necessitate a modification of the natural laws of physics (Schrödinger, 1951), (Carroll, 2021). Meanwhile, some authors argue that the principles of energy and momentum conservation might not be the inviolable axioms they are commonly thought to be (Cucu and Pitts, 2019), (Pitts, 2020), (Pitts, 2021), (Pitts, 2022).[182] Energy conservation assumes that a system is closed. However, it is unclear how this applies to mind-body interactions. Moreover, in general relativity, energy conservation is no longer universally applicable. According to Noether's theorem, energy conservation stems from time-translation invariance, which does not hold in the dynamic spacetime of general relativity. In this framework, energy and momentum depend on how spacetime behaves in which they are embedded. In particular, in the cosmological context of an expanding universe, total energy

[181] One of the interpretations of QM is the Transactional Interpretation of QM, which suggests the existence of a quantum substratum of a pre-spatiotemporal and pre-empirical realm of possibilities rather than that of a mechanistic causal determinism mappable onto a Cartesian coordinate grid (Kastner and Kauffman, 2018), (Kastner, 2022).

[182] Carroll acknowledges this point (Carroll, 2010) but contends that consciousness cannot exist beyond the laws of physics (Carroll, 2020). In response, Pitts argues that this position is implicitly based on an overly simplistic view of Cartesian dualism (Pitts, 2022).

is not conserved; for example, photons lose energy as they are redshifted. In a sense, one could say that energy conservation is not an absolute truth, it is conditional. Under certain conditions, it fails. But general relativity doesn't violate the law of energy conservation, the law simply doesn't apply (Koperski, 2022). Thus, the issue of energy conservation is more complex than commonly believed, and there is no widespread agreement on it.

Additionally, one should always keep in mind that non-local quantum phenomena, which correlate the observable properties of particles, cannot be used to exchange information. Although a measurement or interaction can cause a state projection that correlates the states of distant particles, the specific state into which each particle will be projected is an exclusively stochastic process. As a result, no useful signal can be sent from one particle to another.

However, elsewhere (Masi, 2023; 2024) I have shown how these objections, while technically correct, can be circumvented by the inherent stochastic aspect of QM and QFT. There is no need to conceive of mental causation that violates energy conservation or to extend it to conservation principles in curved spacetime, to which our bodies are not subjected anyway. A notion of causality that goes beyond the principle of transference and reexamines the entire subject, including the temporal dimension, can provide a simpler answer. For completeness and to make this paper self-contained, I will only briefly sketch its rationale here.

II. Causality Without Interactions?

When analyzing principles of mind-matter interaction within a quantum theoretical framework, we cannot abstract from the fact that the concept of physical causality in quantum physics is significantly different from its classical and relativistic counterparts due to its stochastic nature (except for the aforementioned microcausality precept). The differential equations that govern the unitary time evolution of a quantum state are 'deterministic' only in a statistical sense; they uniquely specify the wave function (or state vector) and, consequently, the probabilities derived from the Born rule, not which event will be observed next. While the timing and nature of single events—that is, the instant of the state vector projection (or 'collapse' of the wave function)—are constrained probabilistically, they are not uniquely predetermined by the past conditions. Physicists often refer to a quantum theory as a theory in which stochasticity lacks hidden variables, without delving into the philosophical implications.[183] Quantum uncertainty is framed in terms of potentiality rather than actual events (Kastner and Kauffman, 2018), (Kožnjak, 2020b).

For example, consider the radioactive decay of the element Bismuth-212. A single Bi-212 atomic nucleus has a lifetime of about an hour, which means that after an hour, there is a 50% chance it has decayed and a 50% chance it has not. The time of decay is described by a probability function (e.g., there is an approximately 16% chance of decay within the first 15 minutes and a roughly 3% chance that the nucleus is still intact after five hours, and so on.) Moreover, there are two competing decay modes: a 64% chance of decaying via a β^--decay into Polonium-212 and a 36% chance of decaying via α-decay into Thallium-208.

What determines which of the possible decays will take place and when? According to standard quantum theory, the answer is that, apart from the constraints described by probability functions, there is no other underlying cause or local hidden variable determining the time and kind of decay; it is an ontologically random occurrence. If we assume quantum theory is a non-local theory without hidden variables there is an inherent non-spatial and 'acausal' or 'self-causal' kernel within quantum phenomena.

The essential point relevant to the interactionist debate is that the diverse and seemingly insurmountable conceptual and metaphysical difficulties arising in the classical physical context can be resolved by recognizing that a system's state depends not only on the interactions that determine its evolution but also on the temporal order of these interactions. The history of the world is shaped not only by the events that have influenced it, but also by the sequence in which these events have unfolded. The kinematic order of events is as causally significant as the type and magnitude of the events themselves.

858585868600To illustrate this, consider the previously mentioned decay of Bi-212 in the following thought-experiment.

For N atoms, there will be N decays in a time series (T_1, T_2, … , T_N). Suppose that the α-decays trigger a maneuverable apparatus, traveling at constant speed, to turn left (L), while the β^--decays cause it to turn right (R). The movement of such a device, resulting from sequential changes in direction and distance, depends solely on when it receives the command to steer left or right. It could follow a path described by the series, such as (T_1,L), (T_2,R),

[183] Some might favor interpretations of QM, such as the Many-Worlds Interpretation, de Broglie-Bohm mechanics, superdeterminism, or other speculative frameworks with local hidden variables that, at least partially, recover our everyday classical spatiotemporal realism. Analyzing the issues raised in this paper from the perspective of all these interpretations would exceed the paper's scope and space limitations. Instead, I will align with those interpretations that endorse a non-local, ontic indeterministic realism without hidden variables.

(T_3,L), (T_4,R),… (T_N,L), or a different path determined by an entirely different time series and turn, such as (T'_1,R), (T'_2,R), (T'_3,L), (T'_4,L),… (T'_N,R).

Every sequence—of which there are infinitely many—leads to different paths traveled. The vastly different trajectories, i.e. 'histories' that emerge do not depend on varying types of interactions or forces; instead, they are determined exclusively by the combinatorial order of which event actualizes at which time.

If we identify quantum uncertainty itself with a form of volition or a quasi-Schopenhauerian "will of Nature" that can permute the sequence of possible quantum events, the interaction problem does not arise because this is a causal principle that does not need 'interactions' or 'transference' of 'substances' in the first place.[184] In such a model, the agent does not choose among actualized outcomes or "exploit," "harness," or "piggyback" quantum stochasticity; the agent chooses which potentiality should become actuality. [185]

This shifts the causal principle of a mental agent, whether spatial or non-spatial, from interactions based on physical forces—i.e., from a principle of transference—to potentialities based on the temporal ordering of quantum microphenomenal events. It constitutes a form of 'combinatorial causality' that selects and actualizes an agential history from a set of many possible agential histories, despite being limited by the laws of quantum physics that determine the probability function. Most importantly, it does not alter the frequency of outcomes and, therefore, does not affect the wave function or its associated probability function. As a result, no laws of physics must be modified, and there are no issues with conservation principles.

It is a hypothesis or conjecture that still does not provide a comprehensive theoretical framework for mental causation. Nonetheless, it could serve as a starting point, allowing us to explore where it leads.

For more details refer to (Masi, 2023), (Masi, 2024).

III. Conclusion

This paper called for a reevaluation of traditional frameworks addressing the longstanding philosophical challenge known as the interaction problem exploring its semantic ambiguities, particularly through the lens of QFT.

The distinction between a 'material' and 'immaterial' mind is discouraged because of its misleading significance within a scientific context where

[184] Technical considerations support the idea that quantum indeterminacy persists across all scales and, consequently, may influence neural activity. It is possible to demonstrate that even mesoscopic objects, such as a chaotic pendulum, are sensitive to noise at the Planck scale (Masi, 2025). For an assessment from the neuroscientific perspective, see (Jedlicka, 2017).

[185] Uncertainty, randomness, or what is commonly referred to as "chance," are not necessarily indicative of a lack of volition, aim, or purpose. Consider the white noise that an encrypted message represents.

massive versus massless particles are an entirely normal state of affairs and do not raise any conceptual difficulties typically associated with the problem of mental causation. More challenging is the terminological ambiguity concerning the distinction between what is meant by 'physical' versus 'unphysical,' which is necessary to avoid conceptual confusion in philosophical debates. The definitions based on classical physics, characterizing physical entities as those that occupy space and are governed by causal laws (suggesting that anything beyond these may be considered unphysical), are no longer adequate from the perspective of modern physics. The nature of causality and spacetime in quantum physics differs from classical notions.

Then, the concepts tied to quantum field dynamics in QFT, such as the coupling of fields depicted in Feynman diagrams as virtual particle exchanges and the notion of 'contact' or 'collision' between objects framed in terms of quantum processes, along with the nature of matter as a substance expressed through force fields, render classical notions of interaction and causality inadequate. Interactions are independent of the heterogeneity of quantum fields. Moreover, the same ontological status of substance and interaction that one finds in QFT makes them speak a very different language than that to which we were accustomed in debating interactionist ontologies. The vagueness of 'non-spatial' substances, once we consider quantum non-locality and entanglement in the context of Cartesian dualism, further alters the conceptual frame.

By exploring the concept of causality in a quantum theory with ontic indeterminism and emphasizing how the order of events is crucial in determining a system's evolution, we could propose a 'combinatorial causality' concept based on an interaction-less form of causality, rooted in the probabilistic nature of QM, in line with physical laws, and not giving rise to issues involving violations of conservation principles.

Taken together, these facts not only challenge traditional conceptions of space and substance but, if not acknowledged, also hinder a deeper understanding of the interaction problem.

There is a long tradition that has used quantum mechanical effects to try to overcome the challenges of interactive dualism (e.g., (Bass, 1975), (Mattuck, 1982), (Stapp, 1993), (Stapp 2017), (Burns, 2002), (Schwartz, 2005)). Theories suggesting potential quantum effects in neurons and other biological cells—such as wave function collapses in microtubules, ion channels, or other molecular complexes, as well as the influence of quantum uncertainty on the macroscopic behavior of large structures like brains—are not new[186]. I believe

[186] The most notable example of the former is Penrose and Hameroff's Orch OR theory (Hameroff and Penrose, 2014). For an interesting historical overview of the latter, see (Kožnjak, 2020a).

that these theoretical frameworks could also benefit from greater conceptual clarity, and, eventually, consider whether and how the principle of 'combinatorial causality,' as outlined here, could integrate them.

Advancements in physics, particularly regarding new discoveries and the development of a future theory of quantum gravity, might reshape several points in this article. The primary aim here was to present the interaction problem from a more contemporary viewpoint. While concepts of mental causation might seem more complex, especially in light of current quantum theories that demand greater effort to understand, they also open up new and unexpected paths for exploring theories of mental causation. I hope that this exploration will inspire further research in this direction.

References

Averill, E., Keating, B.F. (1981). "Does interactionism violate a law of classical physics?" Mind, **XC**: p. 102-107.

Bass, L. (1975). "A quantum mechanical mind-body interaction." Foundations of Physics, **5**(1): p. 159-172.

Burns, J.E. (2002). "Quantum fluctuations and the action of the mind." Noetic Journal, **3**(4): p. 312-17.

Carroll, S. (2010). "Energy Is Not Conserved." Preposterous Universe https://www.preposterousuniverse.com/blog/2010/02/22/energy-is-not-conserved/

Carroll, S. (2021). Consciousness and the Laws of Physics. Journal of Consciousness Studies, Vol. 28, 9-10, pp. 16-31(16).

Cucu, A.; Pitts, J.B. (2019). "How Dualists Should (Not) Respond to the Objection from Energy Conservation." Mind & Matter, 17(1), pp. 95-121.

Cucu, A. (2022). "Interacting Minds in the Physical World." PhD thesis. doi: 10.13140/RG.2.2.23398.29767

Cohen, I., ed. (1978). "Isaac Newton's Papers & Letters on Natural Philosophy and Related Documents." Reprint 2014 ed. edition. Harvard University Press.

Crook, S.; Gillett, C. (2001). "Why Physics Alone Cannot Define the 'Physical': Materialism, Metaphysics, and the Formulation of Physicalism." Canadian Journal of Philosophy, **31**(3):333-359.

Elpidorou, A. (2018). "Introduction: The Character of Physicalism." Topoi **37**(435–455) Doi: https://doi.org/10.1007/s11245-017-9488-2

Hoffman, J.; Rosenkrantz, G.S. (1991). "Are Souls Unintelligible?" Philosophical Perspectives, Vol. 5, Philosophy of Religion, Ridgeview Publishing Company, pp. 183-212.

Hoffman, J.; Rosenkrantz, G.S. (2009) "Substance." A Companion to Metaphysics, 2nd ed. Wiley-Blackwell, p. 83-94.

Jedlicka, P. (2017). "Revisiting the Quantum Brain Hypothesis: Toward Quantum (Neuro)biology?" Front. Mol. Neurosci., Sec. Molecular Signalling and Pathways, Vol. 10.

Judisch, N. (2007). "Why 'non-mental' won't work: on Hempel's dilemma and the characterization of the 'physical'." Philosophical Studies **140**(299–318). Doi:

Kastner, R.E.; Kauffman, S. (2018). "Taking Heisenberg's potential seriously." International Journal of Quantum Foundations 4:2, p. 158-172.

Kastner, R.E. (2022). "The Transactional Interpretation of Quantum Mechanics", Cambridge.

Kim, J. (1993a), "The Non-Reductivist's Troubles with Mental Causation." Mental Causation (Oxford, 1993; online edn, Oxford Academic, 31 Oct. 2023)

Kim, J. (1993b). Supervenience and mind: Selected philosophical essays. Cambridge: Cambridge University Press.

Koperski, J. M. (2022) "Does Physics Forbid Libertarian Freedom?" Filozoficzne Aspekty Genezy vol. 19, no. 1 pp. 111-132.

Kožnjak, B. (2020a). "The earliest missionaries of `quantum free will: A sociohistorical analysis." In: Grgić, F., Pećnjak, D. (eds) Free Will & Action. Historical-Analytical Studies on Nature, Mind and Action, vol 6. Springer, Cham. Doi: https://doi.org/10.1007/978-3-319-99295-2_10

Kožnjak, B. (2020b) "Aristotle and Quantum Mechanics: Potentiality and Actuality, Spontaneous Events and Final Causes." J. Gen. Philos. Sci. 51, 459–480. Doi: https://doi.org/10.1007/s10838-020-09500-y

Kuhlmann, M. (2010). "The Ultimate Constituents of the Material World: In Search of an Ontology for Fundamental Physics." Berlin, Boston: De Gruyter, 2010.

Larmer, R. (1986), "Mind-body interaction and the conservation of energy." International Philosophical Quarterly.

Libet, B., (1994). "A testable field theory of mind-brain interaction." Journal of Consciousness Studies, **1**: p. 119-126.

Libet, B., (2006). "Reflections on the interaction of the mind and brain." Progress in Neurobiology, **78**(3-5): p. 322-6.

Lindahl, B.I.; Arhem, P. (1994). "Mind as a force field: comments on a new interactionistic hypothesis." Journal of Theoretical Biology. **171**(1): p. 111-22.

Masi, M. (2023). "Quantum Indeterminism, Free Will, and Self-Causation," Journal of Consciousness Studies, vol. 30, no. 5-6, pp. 32–56, 2023.

Masi, M. (2024). "Quantum Indeterminacy and Libertarian Panpsychism," Mind & Matter, vol. 22, Nr. 1, pp. 31-50(20).

Masi, M. (2025). "Quantum Indeterminacy Persists Across Scales." International Journal of Quantum Foundations, Vol. 11, Issue 4 pg. 492-514.

Mattuck, R. (1982). "A crude model of the mind-matter interaction using Bohm-Bub hidden variables." Journal of the Society for Psychical Research, **51**(790): p. 238-245.

Montero, B., (2006). "What does the conservation of energy have to do with physicalism?" Dialectica, **60**(4): p. 383-96.

Ney, A. (2014). "Metaphysics: An Introduction." New York, NY: Routledge.

Hameroff, S.; Penrose, R. (2014). "Consciousness in the universe: A review of the 'Orch OR' theory," Physics of Life Reviews, vol. 11, no. 1, pp. 39-78.

Pitts, J.B. (2020). "Conservation Laws and the Philosophy of Mind: Opening the Black Box, Finding a Mirror." Philosophia, 48:673–707.

Pitts, J.B. (2021). "Conservation of Energy: Missing Features in Its Nature and Justification and Why They Matter." Found Sci 26, 559–584.

Pitts, J.B. (2022). "General Relativity, Mental Causation, and Energy Conservation." Erkenntnis **87**, p. 1931–1973.

Popper, K.; Eccles, J.C. (1977). "The Self and its Brain." Springer International.

Robb, D.; Heil, J.; Gibb, S. (2023). "Mental Causation." The Stanford Encyclopedia of Philosophy (Spring 2023 Edition), Edward N. Zalta & Uri Nodelman (eds.), URL = **https://plato.stanford.edu/archives/spr2023/entries/mental-causation/**

Schrödinger E. (1951): "Nature and the Greeks and Science and Humanism." p. 162, Cambridge University Press.

Seager, W. (2022). "Panpsychism and Energy Conservation." Mind and Matter, 20(1): p. 17-34.

Simpson, W. (2024). "Cosmopsychism and the Laws of Physics: A Hylomorphic Perspective." Journal of Consciousness Studies, **31**(9-10), pp. 132-157(26).

Shapiro, L. (2007). "The Correspondence between Princess Elisabeth of Bohemia and René Descartes." University of Chicago Press; New ed.

Schwartz, J.M.; Stapp, H.P.; Beauregard, M. (2005). "Quantum physics in neuroscience and psychology: a neurophysical model of mind-brain interaction." Philosophical Transactions Royal Society London, B Biol Sci, **360**(1458): p. 1309-27.

Stapp, H.P. (1993). "Mind, matter, and quantum mechanics." New York: Springer-Verlag.

Stapp, H.P. (2017). Quantum Theory and Free Will: How Mental Intentions Translate into Bodily Actions. Springer.

Stoljar, D. (2010). "Physicalism." Routledge.

Taylor, E.F.; Wheeler, J.A. (1992). "Spacetime Physics." Second Edition, p. 275, New York, W. H. Freeman and Co.

White, B. (2016). "Conservation Laws and Interactionist Dualism." The Philosophical Quarterly, **67**(267): p. 387-405.

Wilson, D.L. (1999). "Mind-brain interaction and violation of physical laws." Journal of Consciousness Studies, **6**(8-9): p. 185-200.

6. Quantum Indeterminacy Persists Across Scales[187]

Abstract

The paper examines the amplification of quantum fluctuations in a mesoscopic nonlinear chaotic system, showing that they are not averaged out by classical noise. Minimal fluctuations—from the Planck scale to the size of cells—can significantly affect system dynamics within one minute to a few seconds. The study uses a driven damped pendulum (DDP) as a model, highlighting the impact of continuous noise that is 27 orders of magnitude smaller than thermodynamic stochasticity. This underscores the cumulative nature of stochastic events that scale from quantum domains, thus suggesting implications for understanding noise in complex nonlinear systems in physics and biology. Additionally, these results challenge the traditional notion of determinism, suggesting that it emerges from mesoscopic and macroscopic realities rather than being intrinsic to Nature. Classical deterministic evolution appears to be an idealization rather than a fundamental aspect of reality.

1. Introduction

Complex nonlinear dynamical systems in chaotic regimes—that is, characterized by high sensitivity to initial conditions—exhibit long-term unpredictable behavior. This phenomenon is observed in various natural and artificial contexts, including weather patterns, turbulent fluids, ecosystems, and even economic markets. A well-known extreme dynamical sensitivity to slight modifications of nonlinear chaotic systems was first evidenced by Lorenz with his homonymous Lorenz attractor in the 1960s. The background idea is famously illustrated by the proverbial 'butterfly effect,' in which even an almost infinitesimal change in the initial conditions can, in the long run, lead to very different dynamical outcomes.

The question is whether, and to what extent, thermal stochasticity and even smaller quantum fluctuations could influence the dynamical evolution and function of bodies that are generally considered purely classical.

For example, conjectures emerged over the past three decades that have garnered attention regarding quantum effects in the brain (for a comprehensive

[187] The present article was published in the *International Journal of Quantum Foundations*, Vol. 11 (4), pg. 492-514.

review of these theoretical frameworks, see (Adams & Petruccione, 2020)). However, these ideas face considerable skepticism. It is widely believed that the brain's thermalized noisy environment quickly disrupts quantum coherence. Although, emerging evidence from the field of quantum biology indicates that quantum mechanics might subtly influence biological processes in ways that are not yet fully understood (Kim, et al., 2021), the ongoing conjectures remain largely speculative.

Note, however, that not all of these models require physical conditions for long-range and/or long-lasting quantum coherence but rely on classical and/or quantum indeterminacy. Moreover, a noisy environment does not affect a system only in its initial conditions; rather, it subjects the system to continuous change at every instant during its temporal evolution in phase space. It is a cumulative process of random events. A stochastic amplification effect does not necessarily require the persistence of quantum coherence; rather, it is based solely on statistical phenomena combined with nonlinear systems operating in chaotic regimes.

The idea that an amplification of quantum noise could have a functional role linked to the unpredictability of life is not new. In fact, it was one of the earliest conjectures originating with Pascual Jordan in the 1930s. Jordan proposed that the unpredictability of life, our behavior, and the mind's apparent indeterminacy might stem from a macroscopic amplification of the chance and unpredictability inherent in the quantum world, scaled up within living organisms. He referred to this concept as the "amplifier theory" (Jordan P. , 1941). A. Eddington and A. Compton entertained similar hypotheses. (For a historical review of the earliest missionaries of this idea, see (Kožnjak B. , The Earliest Missionaries of 'Quantum Free Will': A Socio-Historical Analysis, 2018).

More recently, research findings suggest also that noise, regardless of whether it stems from quantum or classical deterministic randomness, can be a source of innovation. Modern developmental biology is increasingly recognizing that, in addition to genetic and environmental factors, fluctuations in biological processes driven by Brownian motion and other stochastic phenomena significantly contribute to diversity and variation in a species' phenotypic traits. Contrary to common belief, noise and randomness play a functional role in biological processes (for a review, see (Roy & Majumdar, 2022)).

One might also wonder whether, beyond living systems, all macroscopic physical phenomena and processes are influenced by microscopic classical or quantum stochasticity, which might play a role in shaping the world's destiny.

Numerical investigations have shown that the dynamics of certain chaotic three-body systems of stellar size(!) are numerically time-irreversible over extended time scales, down to the Planck length (Boekholt, Zwart, & Valtonen, 2020). This suggests that these macroscopic objects are sensitive to quantum uncertainties, resulting in a breakdown of classical Laplacian determinism across all scales. Furthermore, Bandak et al. have demonstrated that spontaneous stochasticity at the molecular level amplifies molecular noise, affecting even the largest scales of turbulence (Bandak, Mailybaev, Eyink, & Goldenfeld, 2024). Because the molecular scale acts as the interface between quantum and classical phenomena, quantum spontaneous stochasticity might also be amplified, allowing quantum indeterminacy to persist at the classical limit (Eyink & Drivas, 2015).

Taken together, these findings suggest that, contrary to common belief, even the tiniest fluctuations play a role in the temporal evolution of all events in the universe. Rather than being deterministic, as classical physics suggests, all physical events—at both meso- and macroscopic scales—follow an 'open-ended history' that is rooted not only in classical stochasticity, but is fundamentally linked to a quantum unpredictability.

A common objection is that these small stochastic processes cannot influence biological or any macroscopic dynamical systems. Like thermal noise, quantum fluctuations are viewed as self-averaging. The much noisier and warmer environment tends to 'drown' quantum noise and, ultimately, cancel it out. Therefore, given that it lacks causal power, it can be considered completely negligible (Clarke P. G., 2014). However, more recently, Jedlicka re-examined how *nonlinear chaotic dynamics can amplify lowest-level quantum fluctuations upward, modulating even larger scale mesoscopic and, maybe, macroscopic neural activity*, and highlights that this could be especially pertinent for systems operating near criticality or during phase transitions on the brink of chaos and instability, i.e., at the boundary between order and disorder, with heightened sensitivity to minor fluctuations that could initiate cascading phenomena (Jedlicka, 2017).

The influence of microscopic stochasticity on macroscopic systems is not just a matter of interest for the life sciences; it pertains to our understanding of the evolution of all dynamical systems and the conceptual foundations of physics. In a more general and philosophical context but, as we will see later, also relevant to the present discussion, Potter et al. explore how our understanding of physical causal determinism colors philosophical debates about free will and propose an alternative view of indeterminism, termed "pervasive indefiniteness." They argue that indeterminism shouldn't be pictured by positing a fundamentally deterministic framework, to which we merely add randomness as an "*intrusion that disrupts the otherwise deterministic goings-on*" (Potter, Ellis, & Mitchell, 2025).

The aim of this paper is to further explore this aspect through numerical evaluations. After a short recap of how the quantum-to-classical correspondence emerges, it is shown how even extremely weak quantum noise, originating from the Planck scale, can impact the dynamics of a mesoscopic nonlinear chaotic system within a minute, challenging the traditional view that quantum indeterminacy can be disregarded in macroscopic systems. This sensitivity will be examined not only through changes in initial conditions but also by applying continuous noise from a stochastic source. The paper will conclude with a brief discussion of the possible conceptual implications.

2. On the Quantum-to-Classical Correspondence

There are some common misconceptions about the quantum-to-classical correspondence that need to be addressed[188]. One misconception is that decoherence cancels or reduces microscopic quantum indeterminacy, rendering it irrelevant at larger scales. Another is that, quantum nonlinear chaotic effects on macroscopic objects are impossible because for a statistical ensemble, any stochastic fluctuations at the microscopic scale are averaged out in the limit of large numbers and cannot influence macroscopic behavior. The next three sections will aim to bring some conceptual clarity regarding these statements.

2.1 Decoherence Doesn't Suppress Quantum Indeterminacy

Informally, the quantum-to-classical transition can be described as a quantum decoherence process that occurs when a system becomes entangled with its environment, leading to a loss of interference. The destruction of the system's phase coherence is a process of information transfer; it effectively 'washes out' the phase information due to ensemble dephasing caused by the system's entanglement with a large environment. The environment can be viewed as a measuring or monitoring apparatus that 'observes' or 'performs measurements' on the state of the system, thereby destroying coherent superpositions.

Let's sketch this briefly in more formal terms (for a more detailed analysis, see (Schlosshauer, 2019)).

Denote the quantum state of the joint non-separable entangled state-environment system as

$$\Psi = \sum_i |s_i\rangle|e_1\rangle,$$

with $|s_i\rangle$ and $|e_i\rangle$ being the orthonormal bases of the system S and environment E defined in the Hilbert spaces H_S and H_E, respectively. Then,

[188]The conventional understanding of a quantum-to-classical transition is Bohr's correspondence principle asserting that the predictions of quantum mechanics should align with classical physics when quantum numbers are large or when examining large scales. However, the principle isn't accepted universally. Here, the term "correspondence" means that quantum behavior approximates classical results as systems become macroscopic.

the composite system-environment is initially in a pure state described by the time-dependent density matrix:

$$\rho_{SE}(t) = |\Psi(t)\rangle\langle\Psi(t)|.$$

The reduced density matrix ρ_S which partially traces out from ρ_{SE} the degrees of freedom of the environment and contains all the necessary information about the local statistics of the system, is:

$$\rho_S(t) = \mathrm{Tr}_E\big(\rho_{SE}(t)\big).$$

Let's denote with τ the decoherence time.[189] On the onset of the environmentally induced quantum decoherence process, i.e., for a time $t < \tau$, the system is entangled with the surroundings, and its corresponding density matrix will contain diagonal and non-zero off-diagonal terms as:

$$\rho_S(t) = \underbrace{\textstyle\sum_i \, |c_i(t)|^2 |s_i\rangle\langle s_i|}_{\text{Diagonal terms}} + \underbrace{\textstyle\sum_{i\neq j} \, c_i(t)c_j(t)^* \langle e_j \mid e_i\rangle |s_i\rangle\langle s_j|}_{\text{Off-diagonal terms}}.$$

For a sufficiently long time, i.e., for $t \gg \tau$, due to the system-environment interaction, the progressive suppression of quantum interferences is reflected in the exponential decay of the off-diagonal terms representing the superposition states. These are time-averaged over the combined system-environment state vector, initially in a product state, and become negligible, i.e., the system states become macroscopically distinguishable. Since the overlap $\langle e_j \mid e_i\rangle \to 0$, the pure states that begin as a superposition quickly decay into a classical mixture represented by a time-independent density matrix:

$$\rho_S = \textstyle\sum_i \, p_i |s_i\rangle\langle s_i|, \qquad \text{XIII.1}$$

with, as usual, $p_i = |c_i|^2 (\sum_i p_i = 1)$ the probability to find the system in eigenstate $|s_i\rangle$.

This suppression of quantum interferences leads to an environment-induced superselected stable outcome eigenbasis ('einselection') that 'survives' the monitoring by the environment, causing the system's quantum state to exponentially localize basis states that are least entangled with the surroundings, i.e., the ones whose information spreads out more efficiently across the system and environment.

At this point, the system is separable, but no measurement has occurred yet, meaning that equation XIII.1 still describes a mixture of possible outcomes that, when a measurement device interacts with the system, are referred to as

---

[189] Usually extremely short time spans. In most physical conditions, the destruction of a system's states phase coherence takes place on time scales that are much shorter than those of the relaxation times to thermal equilibrium.

the 'pointer basis.' These states represent coherent states in which open quantum systems can be observed, continuing to exist despite environmental monitoring (Paz & Zurek, 1999).

This is to say that it is necessary to keep in mind that the diagonal terms are *not* affected by decoherence (in the absence of dissipation). During a measuring process, the joint system-environment Hamiltonian can be separated into three distinct components: the self-Hamiltonian H_s of the system, of the environment H_E, and an interaction Hamiltonian H_{SE}. Both the self-Hamiltonian and interaction Hamiltonian contain the kinetic energy operator $\frac{\hat{p}^2}{2m}$ and the potential energy $V(r)$ operator, which do not commute. Consequently, the position and momentum observables are never an exact pointer basis, not in the sense that a pointer observable is meant to be an exact observable of the system. They remain affected by quantum indeterminacy that persists despite decoherence, with the system's observables ensemble width determined by Heisenberg's uncertainty principle or, more generally, by the generalized uncertainty principle. Say, the state vector of the system (in the limit $t \gg \tau$) is $|\Phi\rangle = \sum_i c_i |s_i\rangle$, with $|s_i\rangle$ the pointer basis of the measuring device, then the variance of an observable O will be

$$(\triangle O)^2 = \langle\Phi|(O - \langle O\rangle)^2|\Phi\rangle,$$

where $\langle O\rangle = \langle\Phi|O|\Phi\rangle$ is the expectation value, as described in every standard textbook.

In the case of continuous spectra, such as in measurements involving position and momentum operators, or functions of them (Hamiltonians, angular momenta, etc.), which is for practical purposes always the case, their determination always leads to quantum uncertainties due to the natural fact that they are non-commuting observables[190].

Moreover, it might be worth recalling that the measurement, say of some observable O_k, that projects the system's state vector to an eigenbasis with a measurement outcome with eigenvalue λ_k as: $O_k|s_k\rangle = \lambda_k|s_k\rangle$, is the last step of a stochastic process 'collapsing' a quantum state to a single outcome, and which is *not* accounted for by any quantum decoherence theory. Decoherence alone isn't sufficient to explain why only one of the outcomes is measured, resulting in a statistical distribution rather than a definite value. It tells us why the off-diagonal elements of the density matrix are suppressed, not why, once the measurement eigenbasis is chosen, its diagonal element is 'zeroed out' in the measurement process, apparently violating unitary evolution. An issue that is famously known as the 'measurement problem' and that highlights how decoherence cannot be the whole story that seeks to reconcile the laws of

---

[190] Something nicely visualized in the literature with the Wigner representation; not shown here (e.g., see Fig. 2, Fig. 3 in (Schlosshauer, 2019)).

quantum theory with the classical description for macroscopic objects and how a measurement comes into being (for a review, see, (Tomaz, Mattos, & Barbatti, 2025).

Thus, decoherence accounts for the suppression of state superposition of macroscopic objects, but it should not be misinterpreted as a 'collapse in uncertainties.' Instead, every observable should be treated as a statistical state ensemble, maintaining a statistical distribution of possible outcomes. Decoherence does not cancel, average, or 'wash out' this inherent indeterminacy, nor does it clarify the distinction between quantum indefiniteness and classical definiteness that characterizes the universe.

The question is whether these uncertainties affect the meso- and macroscopic scales, and if they do, to what extent.

2.2 Decoherence Prevents Chaos from Disrupting the Quantum-to-Classical Correspondence

Classical determinism is captured by Liouville's theorem, which states that the phase-space distribution function $\rho(p,q,t)$—the density of statistical ensemble points in phase space and time—remains constant along the trajectories of an isolated Hamiltonian system H. That is:

$$\frac{\partial \rho(t)}{\partial t} + \{\rho(t), H\} = 0,$$

with $\{\cdot,\cdot\}$ Poisson's brackets. It is a continuity equation that emphasizes the incompressible nature of classical Hamiltonian phase space flow. It directly results from the uniqueness of phase space trajectories, which never intersect. If trajectories were to merge, it would imply that two or more distinct initial states evolve into the same state at a later time, causing an increase in point density at the intersection. This would violate Liouville's theorem and undermine the classic notion of physical causal determinism, which holds that when a system is constrained by initial and boundary conditions at the present moment, its future dynamical evolution follows a fixed and unique trajectory in phase space.

In quantum mechanics, the analog of Liouville's equation for a closed system, is the Liouvillevon Neumann equation:

$$\frac{\partial \rho(t)}{\partial t} = -\frac{i}{\hbar}[H, \rho(t)], \qquad \text{XIII.2}$$

with $[\cdot,\cdot]$ denoting the commutator brackets.

For open systems, eq. XIII.2 must be extended to a Lindblad form $\partial_t \rho = (-i/\hbar)[H,\rho] + D$, with D standing for decoherence and dissipative non-unitary terms that include the interaction with the environment. This is where the whole analysis on decoherence starts, and we can't exhaust it here, not even superficially. It may only be said that the dynamics is approximately described by Markovian master equations, where the environment is treated as 'memoryless,' meaning it does not retain information about its interaction with the system. As a second approximation (the Born approximation), it is assumed that there is weak coupling between the system and a sufficiently large environment. (For an in-depth discussion, see (Schlosshauer, 2019) and references therein.)

Nonetheless, to understand how nonlinear chaotic behavior disrupts the quantum-classical correspondence, it is sufficient to examine the Liouville-von Neumann equation for closed systems. Equation XIII.2 is closely to Ehrenfest's theorem, which states that the time evolution of expectation values behaves similarly to classical variables. Specifically, for a sufficiently narrow wave packet representing the canonical coordinates of a particle, the mean position and momentum in the quantum state will follow a classical Newtonian trajectory as:

$$\frac{d\langle x\rangle}{dt} = -\frac{i}{\hbar}\langle[H,x]\rangle = \frac{\langle p\rangle}{m}.$$

This might look appealing and provide reassurance regarding the emergence of classicality from the quantum world. However, this correspondence is only approximate and breaks down when the probability density support in phase space is exponentially stretched, which is precisely the case for chaotic systems. In this case a small uncertainty in position δx_0 leads to an exponential divergence in time evolution as

$$\delta x = \delta x_0 e^{\lambda t},$$

with λ the Lyapunov exponent and, to preserve the phase space volume, i.e. $\delta x \delta p \sim \hbar$, simultaneously leads to an exponential narrowing in the momentum as $\delta p = \delta p_0 e^{-\lambda t}$. Then, the wave packet spreads over a spatial region of

$$\delta x \sim \frac{\hbar}{\delta p_0} e^{\lambda t}, \qquad \text{XIII.3}$$

unless (or until) it becomes confined by a potential. If the spatial extension χ of a potential $V(x)$ is much larger and doesn't change significantly in δx, the correspondence approximately still holds because the relatively small wavepacket can still be considered a localized 'spike' in phase space subjected to an extended force field that acts like a gradient $\partial_x V$. If, instead, $\delta x \gtrsim \chi$, the wavefunction is spread over a relatively large region where the force field is varying, i.e., where the classical notion of a force expressed as the local gradient vector field acting on a point particle no longer holds, then the quantum to classical correspondence breaks down. From XIII.3, the time when this breakdown is going to happen (also called Ehrenfest time since until that point trajectories could be considered approximately Newtonian) is

$$t_B = \frac{1}{\lambda} ln \left(\frac{\delta p_0 \chi}{\hbar}\right). \quad \text{XIII.4}$$

This is the timescale at which the wave function of a chaotic system becomes 'smeared' throughout the available phase space, ultimately folding back onto itself.

This phenomenon is not limited to microscopic objects. Adopting a simplified approach[191], Zurek considered the case of Saturn's moon Hyperion—
a body with a prolate shape subjected to tidal forces arising from its interaction with Saturn's gravitational field, resulting in chaotic dynamics of its orientation — and obtained a breakdown time of about 20 years (Zurek & Paz, 1999); (Zurek, 2003). After this time, Hyperion would be in a coherent superposition of orientations distributed along an angle of 2π radians. However, even in deep space, objects are open systems that constantly interact with the environment, and only weak thermal radiation, such as the radiation from the cosmic background, or particles from cosmic rays, or a micrometeorite colliding with Hyperion's surface, quickly decoheres their wavefunction.

In other words, in principle, we should expect wavepackets to spread exponentially over time due to chaos, leading to quantum delocalization becoming a reality even for macroscopic objects. The reason we do not observe this in our everyday lives is that large-scale bodies function as open systems that are constantly monitored by their environment. This interaction causes decoherence, which transforms a pure quantum state into a mixed state almost instantly. Quantum decoherence is so effective because tiny environmental interactions cause quantum information to spread irreversibly into an enormous space of environmental degrees of freedom, suppressing interference and producing classical-like behavior on extremely short timescales. Chaotic folding and twisting of the wavepacket's phase space support is second only to quantum decoherence and would, otherwise, dominate and determine the reality of complex systems.

2.3 Statistical Averaging Doesn't Lead to Unique Histories

In statistical thermodynamics, we work with statistical ensembles that are macroscopically indistinguishable. These ensembles are defined by observable state variables, such as pressure, temperature, energy, and volume, which, in the thermodynamic limit, emerge as expectation values derived from a vast

---

[191] It was further shown that 2.4 has an upper bound determined by $t_r = \frac{1}{\lambda}\ln\left(\frac{LP}{\hbar}\right) = \frac{1}{\lambda}\ln\left(\frac{I}{\hbar}\right)$, where L and P represent the range of values of the coordinates and momenta in phase space of the system, and I is the action (Berman & Zaslavsky, 1978); (Berry, 1979). This simplifies the estimate of XIII.4.

number of microscopic configurations.[192] This understanding allows us to view quantum micro-indeterminacies as largely disappearing, since they are self-averaging fluctuations. For example, white noise consists of randomly fluctuating positive and negative values that occur with equal likelihood. According to the law of large numbers, for N particles, relative fluctuations scale as $1/\sqrt{N}$, and as N approaches infinity, the variance converges to zero. The central limit theorem applies, indicating that microscopic randomness is 'washed out' by the large number of degrees of freedom. The resulting random deviations become so minuscule that they can be considered negligible, leading to a form of 'statistical determinism' at the macroscopic level.[193]

It is in this sense that we understand microscopic classical as quantum events being 'damped out' due to self-cancellation, resulting in deterministic dynamics at the macroscopic level. Consequently, in a fully decohered object, localized microscopic events cannot 'percolate up' to affect the macroscopic pointer states, and the future state of a macroscopic system is solely determined by its classical macrostate variables, which cannot be 'disrupted from below.' We can confidently state that the past and present macrostate uniquely determines the future, leading to a single outcome rather than a probability function, once the initial and boundary conditions are established. Collisional decoherence continually localizes the quantum wavepacket of every molecule, atom, and particle, the system's 'history' is effectively reduced to a one-dimensional trajectory in phase space.

However, it is worth recalling how statistical mechanics depicts the microscopic behavior of objects undergoing stochastic fluctuations in a more nuanced way. The paradigmatic example is the statistical behavior of a particle immersed in a thermal bath, which undergoes random jittering, as described by classical Brownian motion. This phenomenon is represented as a continuous-time stochastic Wiener process, first formalized by Einstein through the diffusion equation:

$$\frac{\partial \rho(r,t)}{\partial t} = D\,\nabla^2 \rho(r,t),$$

with $\rho(r,t)$ the probability distribution to find the particle in r at time t, and D the diffusion coefficient. Diffusion means that the particle doesn't stay there. There is no such thing as an 'averaging out' that keeps the particle at rest in time. If the particle is at rest at time $t = 0$, so that $\rho(r,0) = \delta(r)$, the probability to find the particle at an Euclidean distance $|r|$ in a 3D space from

[192]Notoriously, an outrageous large number of microstates. From Boltzmann's entropy formula: $W = e^{S/k_b}$, where k_b is Boltzmann's constant and S is the entropy.

[193] For instance, for a mole of air at room temperature in thermodynamic equilibrium, the expected macroscopic fluctuations in temperature are approximately $\triangle T \approx 10^{-21}$ K (Mishin, 2015).

the origin is determined by a (width time-dependent) Gaussian probability density function

$$\rho(r,t)=\frac{1}{(4\pi Dt)^{3/2}}e^{-\frac{|r|^2}{4Dt}};\ D=\frac{k_BT}{6\pi\eta a},$$

with η the solvent's viscosity, a the radius of a spherical particle, k_B Boltzmann's constant, and T the temperature in Kelvin. The mean square displacement in time is: $\langle\triangle r(t)^2\rangle=6Dt$, which accounts for the particle's random walk.

This idea has been part of conventional wisdom since Einstein's studies on Brownian motion, but a lesser-known aspect is the correspondence between classical and quantum mechanics. For instance, in the context of thermal collisions, we can model a particle's motion as being weakly coupled to a thermal bath composed of non-interacting harmonic oscillators. In this scenario, decoherence and dissipation arise from the scattering of environmental particles, such as gas molecules, off a central particle. This scattering process carries away the which-path information and, as a result, the ensemble width in position space scales with time according to the relation $\Delta r(t)\approx\sqrt{t}$. This model exhibits the same scaling behavior as the random walk diffusion found in classical Brownian motion, which is why it is referred to as 'quantum Brownian motion'. Similarly, the classical Maxwell-Boltzmann distribution can be derived as a limiting case of a quantum linear Boltzmann equation (Gamble & Lindner, 2009); (Schlosshauer, 2019); (Vacchini & Hornberger, 2009). This indicates that quantum effects are an inherent part of classical processes, and which determine the random walks. Each particle's random walk (or those of the same particle over different time intervals) represents a unique path that never repeats—essentially, each particle has its own distinct 'history'.

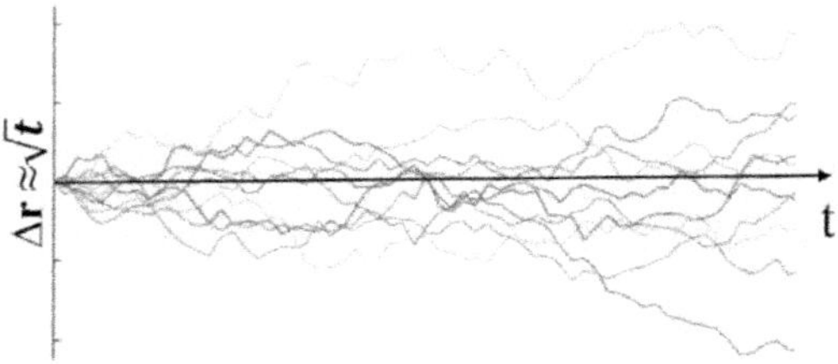

Fig. 1: Different diffusive paths ('histories') of a (classical or quantum) Brownian motion in form of a Wiener process random walk.

While this is what is going on smaller scales, on larger scales, these stochastic micro-fluctuations do not result in multiple possible dynamical histories; instead, they are 'damped out' into a single trajectory in phase space. This certainly holds provided that the system's equations of motion are linear or only weakly nonlinear and near-integrable. It is this linearity that makes the system insensitive to such deviations. The thermodynamic limit depends on this linearity to guarantee effective statistical averaging.

However, while decoherence prevents the nonlinear chaotic 'contortions' of the wavepackets' phase space support of microstates from manifesting as coherent states at the macroscopic level, as emphasized in the previous two subsections, it does not influence the effects of microscopic classical or quantum indeterminacies. Moreover, nonlinear chaotic systems are not only sensitive to minute changes in initial conditions; they are also continuously sensitive to the underlying micro-statistical noise present throughout the state evolution. What we are going to show is that this noise can 'steer' the system's fate, leading to random walk behavior on both mesoscopic and macroscopic scales. The interplay between classical and quantum micro-indeterminacies is amplified, causing the system's evolution to diverge along trajectories shaped by both classical pseudo-indeterminism and quantum indeterminism.

The following section elaborates on this using a specific dynamical toy model.

3. Sensitivity of Mesoscopic Nonlinear Chaotic Systems to Extremely Weak Noise

A comprehensive and rigorous analysis quantifying the propagation of quantum fluctuations in a complex system would require an assessment that cannot be fully explored here. However, one can begin with a proof of concept, demonstrating that even the smallest fluctuations influence the large scale dynamics of relatively simple systems. We examine the behavior of a specific nonlinear chaotic mesoscopic dynamical system as a representation of a general phenomenon in all nonlinear chaotic systems of similar size and Lyapunov exponents. High-precision numerical integrations can show that the macroscopic size of a material body is far less effective at 'averaging out' or 'canceling out' microscopic fluctuations than is conventionally assumed. Furthermore, simplifying a system by minimizing its degrees of freedom—such as by reducing the number of particles or bodies involved, and consequently the number of differential equations governing its motion does not guarantee the long-term suppression of quantum indeterminacy at larger scales in the sense of Bohr's correspondence principle. The only condition sufficient for the emergence of quantum effects beyond the microscale is the presence of nonlinearity and a high sensitivity to changes in initial conditions.

This can be demonstrated using a well-studied nonlinear chaotic oscillator: the driven damped pendulum (DDP). The DDP is a paradigmatic textbook example of a classical nonlinear mechanical system that, under particular conditions, exhibits complex chaotic behavior. (For a detailed description, see, for example, (Taylor J. R., 2005)).

It consists of a pendulum of length L influenced by three primary forces. First is the gravitational force, which acts as the natural restoring force (the weight mg, with m being its mass and g the gravitational acceleration), which pulls it back toward its equilibrium position. Second is the damping force, such as friction or air resistance, which opposes the pendulum's motion and causes its oscillations to decay over time (the resistive force bv, with b being a damping force parameter and v the instantaneous velocity of the tip of the pendulum). Finally, a driving periodic force continuously adds energy to the system (the driving force $F(t) = F_0\cos(\omega t)$, with F_0 being the drive amplitude and ω the drive frequency).

The interaction of these forces can result in various behaviors. With small driving forces, the pendulum typically settles into a regular, periodic oscillation. However, increasing the driving force or adjusting other parameters can induce chaotic behavior, causing the pendulum's motion to become highly sensitive to initial conditions.

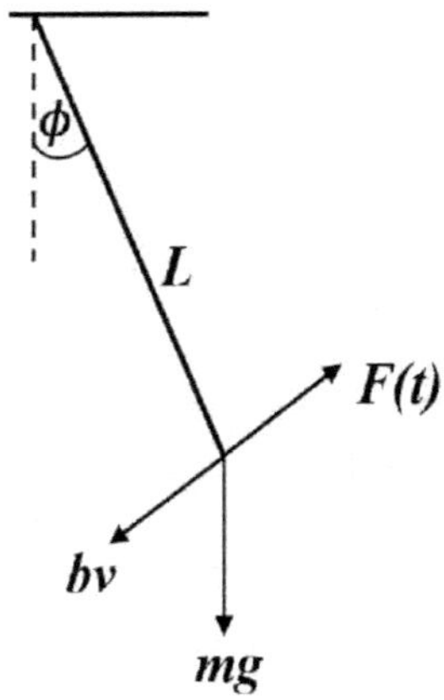

Fig. 2: Forces acting on the DDP.

It can be shown that the equations of motion of the DDP are governed by the following differential equation:

$$\ddot{\phi} + 2\beta\dot{\phi} + \omega_0^2 sin\ \phi = \gamma\omega_0^2 cos\ (\omega t), \qquad \text{XIII.5}$$

with ϕ the angular displacement, $\beta = \frac{b}{2m}$ the damping constant, $\omega_0 = \sqrt{\frac{g}{L}}$ the simple pendulum natural frequency, and $\gamma = \frac{F_0}{mg}$ a dimensionless measure of the strength of the driving force. If, as in our case, $\gamma > 1$ it exceeds the pendulum's weight and can eventually produce larger scale motions; i.e., the pendulum is pushed all the way over the top at $\phi = \pi$. One possible choice of parameters that highlights the chaotic behavior with strong sensitivity to initial conditions is, for example: $\omega = 2\pi; \omega_0 = \frac{3}{2}\omega = 3\pi; \beta = \frac{\omega_0}{4} = \frac{3}{4}\pi; \gamma = 1.16$. These describe a pendulum of mass $m = 50g$, length $L \approx 11cm$, subjected to a drive amplitude $F_0 \approx 0.57N$, with resisitve coefficient $b \approx 0.34\frac{1}{s}$. Thus, the physical magnitudes involved are those of a fully classical mesoscopic non-quantum system.

High numerical precision and accuracy are essential for analyzing nonlinear chaotic systems due to their high sensitivity to truncations resulting from the limited decimal representation of real numbers in a computer. To prevent error propagation from exceeding the accuracy of interest, the numerical precision was set to 100 decimal significant digits[194]. It is possible to obtain the "true" solution $\phi(t)$ for a time interval of 100 seconds with a precision of about 10^{-2}

[194] The numerics were performed with the Julia programming language (Bezanson, Edelman, Karpinski, & Shah, 2017), which allows for arbitrary-precision integers and floating point numbers, limited only by hardware specifications.

rad using a standard desktop PC[195]. The numerical solver employed was Feagin's 14th-order explicit Runge-Kutta method, which is slow but particularly suitable for problems requiring highly accurate solutions within a reasonable time frame (Feagin, An explicit Runge–Kutta method of order fourteen, 2009); (Feagin, 2012). The solution was fitted using a fifth-order Hermite interpolation, ensuring that the values and their first and second derivatives matched exactly at the data points.

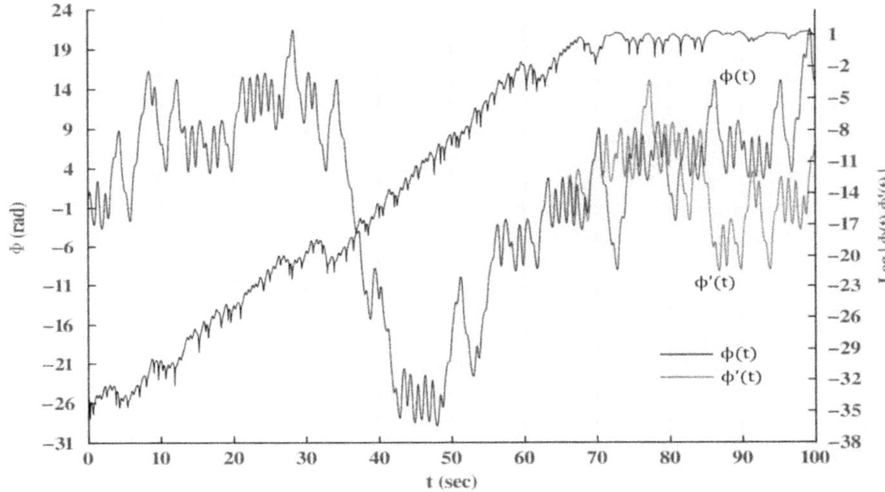

Fig. 3: Blue curve $\phi(t)$: unperturbed DDP; Red curve $\phi'(t)$: the DDP with angular initial condition differing by a Planck arch length; Black curve: the difference between the two curves in logarithmic scale.

The blue curve $\phi(t)$ in the linear graph of Fig. 3 (unperturbed DDP) shows the result starting from initial conditions $\phi_0 = 0°, \dot{\phi}_0 = 0° \frac{1}{s}$, for the first 100 seconds

As expected, the system appears to be highly unpredictable and chaotic. The key question is how sensitive it is to minor changes in its initial conditions. In theory, there is no limit to the smallness of the 'perturbation'; classical physics is based on idealizations involving quantities that can be expressed with an infinite number of digits. However, we can consider the Planck scale as the smallest layer of reality beyond which current quantum mechanics breaks down.

By shifting the DDP's initial angular position by an extremely small Planck arch length[196] (i.e., $\phi(t_0) = \phi_0 \to \phi'(t_0) = \phi_0 + \delta_0$ with $\delta_0 = 1.474 \times$

[195] The conventional method for verifying this is by observing the convergence of the solutions as the number of digits increases, and the tolerance and time-step size decrease.

[196] The Planck length is 1.616×10^{-35} m; for a pendulum of 11 cm , the corresponding angular change is 1.474×10^{-34}rad. The same unit transformation will apply later to Table 1.

10^{-34}rad), we can run the same integration process. The results are illustrated by the red curve $\phi'(t)$ in the linear graph of Fig. 3 (perturbed DDP). Around 65 seconds into the simulation, the difference in angular displacement between the perturbed and unperturbed DDP becomes noticeable. The black 'bumpy' curve in the same graph, displayed on a logarithmic scale, highlights the divergence from the start of the simulation.

This shows how even the smallest fluctuations at the Planck scale become significant beyond a certain time horizon. It illustrates that quantum uncertainties are amplified to meso- and macroscopic scales within minutes, or even less.

From the slope of the logarithmic graph, one can roughly estimate a Lyapunov exponent of about $\lambda \approx 1.17$, i.e., a Lyapunov time of $\approx 0.85s$[197]. In general, once the initial separation δ_0 between the initial conditions of the unperturbed and perturbed systems is known, the Lyapunov exponent λ quantifies the trajectory divergence $\delta(t)$ after a time t as:

$$\delta(t) = \delta_0 e^{\lambda t}.$$

Therefrom, the time t_d necessary to reach a separation of, say, $\delta(t_d) = d = 0.1\text{rad}$, is:

$$t_d = \frac{1}{\lambda} \ln \left(\frac{d}{\delta_0}\right).$$

From this, one can infer the time of divergence between the perturbed and unperturbed DDP for different micro-physical initial conditions. Table 1 demonstrates the DDP's sensitivity to even the slightest changes in initial conditions within a time horizon t_d. The system exhibits this sensitivity to molecular and atomic-scale structures within seconds.

However, as previously noted, real dynamical systems are not only affected by perturbations in their initial conditions but also embedded in a noisy environment that subjects them to a thermal bath of continuous stochastic fluctuations throughout their evolution. Additionally, classical noise, typically of a thermodynamic nature (e.g., Brownian motion), overlaps with quantum random fluctuations. At room temperature, the classical noise presents a much stronger stochastic signal than the quantum fluctuations.

The question at hand is how the evolution of a nonlinear chaotic system is influenced by two noisy signals whose amplitudes differ by several orders of magnitude, while both remain relatively weak compared to the magnitude of the system's canonical coordinates.

---

197 For comparison, the Lyapunov exponent of the chaotic Lorentz attractor is $\lambda \approx 0.9$ (Strogatz, 2018).

Table 1: Time horizon t_d when the divergence between the perturbed and unperturbed DDP becomes discernable for different initial conditions δ_0 (d=0.1rad; λ=1.17).

Description	Length	Arch length δ_0	t_d
Planck length	1.616×10^{-35} m	1.474×10^{-34}rad	$\approx 66s$
Proton radius	0.84×10^{-15} m	7.64×10^{-15}rad	$\approx 26s$
Bohr radius	0.39×10^{-9} m	4.81×10^{-10}rad	$\approx 16s$
Average cell size	50×10^{-6} m	4.54×10^{-4}rad	$\approx 5s$

In principle, the DDP enables the investigation of this by modifying equation XIII.5 into a stochastic differential equation. However, stochastic differential equation numerical solvers are not very efficient at handling chaotic systems and will exhibit $O(1)$ divergence after a few Lyapunov time periods. For our situation, a more convenient approach is to approximate the effects of noise through piecewise integration. This involves dividing the total integration time T in N time-intervals $\Delta t_i (i = 1,..,N)$, and solve eq. XIII.5 by adding a tiny Gaussian random displacement to the i-th position $\phi_i = \phi(t_i)$ and i-th initial momentum $\dot{\phi}_i = \dot{\phi}(t_i)$. Ideally $N \to \infty$; here we chose $N =$ 10.000.

One approach is to analyze the system in broad terms, without quantifying a specific physical scenario, and to examine how the DDP behaves in the presence of two types of stochastic processes: one with relatively large noise (but still small compared to the angular displacements) and the other with significantly smaller noise. The first represents classical variability (Wiener process or thermal noise, biological activity, environmental stochastic factors, etc.), while the second signal is many orders of magnitude smaller, representing quantum fluctuations that are typically considered negligible. For instance, consider two normally distributed with mean zero random series ϕ_1^{rnd}, and ϕ_2^{rnd}, with minimum and maximum standard deviation of σ_1 and σ_2, respectively. These are summed over the DDP's angle $\phi(t)$ resulting in a noise-affected function $\phi^N(t) = \phi(t) + \phi_1^{rnd} + \phi_2^{rnd}$. In this study $\sigma_1 =$ 10^{-3}rad and $\sigma_2 = 10^{-30}$rad were chosen (see the appendix for details.)

The solution is found piecewise for $\phi_i^N = \phi^N([t_i, t_{i+1}])$. At t_{i+1}, the final displacement $\phi^N(t_{i+1})$ and momentum $\dot{\phi}^N(t_{i+1})$ serve as the new initial conditions for obtaining $\phi_{i+1}^N = \phi^N([t_{i+1}, t_{i+2}])$, to which two noisy signals are added again. This process continues until the integration concludes at $t_N =$ T. The results are illustrated in *Fig. 4*.

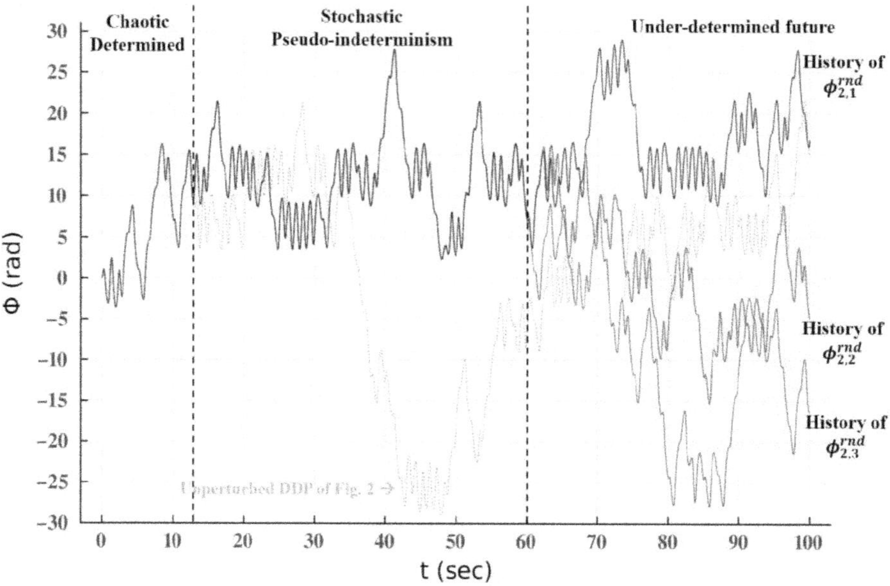

Fig. 4: The chaotic dynamics of the DDP affected by a strong and a weak noise source.

For approximately 13 seconds (within the graphical resolution; here, ≈ 0.1rad), there is no discernible difference between the evolution of the unperturbed system and that subjected to the perturbing white noise.[198] Up to this point, the curve overlaps with the chaotic solution in Fig. 3 , i.e., $\phi^N \approx \phi$, indicating dynamics characteristic of a chaotic yet determined system. Around the time of 12-13 seconds, as indicated by the dashed vertical line, the divergence between the two systems becomes noticeable in the graph. The more intense stochastic source, ϕ_1^{rnd} with $\sigma_1 = 10^{-3}\text{rad}$, causes the first divergence from the 'true' solution. (The blue curve in Fig. 3 is represented as a thin gray dotted line in the background of *Fig. 4* for comparison). This can be viewed as the role of a classical stochastic process that is chaotic yet pseudo-indeterministic (e.g., classical Brownian motion.) It is considered pseudo-indeterministic because, ultimately, all underlying causal mechanisms are modeled by a physical causal determinism ruled by the laws of classical physics. The system's behavior appears random and unpredictable due to complex chaotic dynamics and incomplete information, representing epistemic indeterminacy. Once the initial conditions are established, the future trajectory of the dynamical evolution in phase space is, nonetheless, determined, leading to only one possible 'history.'

However, the system is also influenced by the smaller stochastic and indeterministic process that embodies quantum indeterminacy. Three

[198] A log-scale graph would be necessary to detect any differences (not shown; qualitatively similar to the curve in Fig. 3).

numerical integrations are then performed by generating first a normal distributed random sequence (Wiener or Gaussian process) $\phi_{2,1}^{rnd}$ with $\sigma_1 = 10^{-30}$rad. Then, another two random sequences are obtained, not by new random number generations, but rather by two random permutations of the same times series $\phi_{2,1}^{\mathrm{rnd}}$, i.e., $\phi_{2,2}^{rnd} = \pi_1(\phi_{2,1}^{rnd})$, and $\phi_{2,3}^{rnd} = \pi_2(\phi_{2,1}^{rnd})$, with π_1 and π_2 two random permutations. Permuting a random time series, rather than generating a new one, preserves the shape of the probability density function, creating three statistically identical and indistinguishable noise sources that, nonetheless, lead to three different trajectories in phase space. This is a numerical method to simulate an ontological indeterminacy by actualizing multiple possible histories that constantly under-determine the future, i.e., leaving the future 'open.'

This approach enables us to identify the second point of divergence in time, with the curve splitting along three different potential histories caused by the three random sources $\phi_{2,i}^{rnd}$(i = 1,2,3). In the current simulation, this branching occurs at around 60 seconds, marked by the second dashed vertical line in *Fig. 4* where the three colored curves diverge.

Therefore, despite the two noise signals being separated by 27 orders of magnitude, the smaller ones do lead to different actualizations beyond some time interval. The amplification of continuous quantum fluctuations in a nonlinear chaotic system at mesoscopic scales over short timescales is not only plausible but also inevitable. Even if the signal-to-noise ratio (SNR) between thermal and quantum effects is huge, rendering each quantum fluctuation statistically insignificant and effectively 'averaging it out,' this does not mean that the system will settle into a state describable by a statistical ensemble of macroscopically indistinguishable microstates. Instead, each state within the ensemble represents a potentiality that actualizes into a single discernible macro physical state, beyond a certain time interval.

4. Discussion and Conclusion

The fact that even the tiniest microscopic uncertainties can quickly magnify into domains relatable to our everyday experience raises questions about the conceptual foundations of determinism and its philosophical implications. For instance, if classical and quantum stochasticity ultimately determine the history of all complex processes, from where does the notion of physical causal determinism originate? One straightforward answer is to note that, historically, the mathematical foundations of physics have relied on ideal systems, where the equations of motion are governed by linear differential equations. Once the initial conditions are set, the evolution of the system is uniquely determined in phase space. However, the real world is governed by nonlinear differential equations that shift between regular and chaotic regimes. In this sense, determinism is considered an idealization that arises when the human mind simplifies, approximates, and reduces a system's degrees of freedom to facilitate formal statistical analysis.

Moreover, as is well known, there exists a multitude of interpretations of quantum mechanics, some of which attempt to preserve determinism and/or locality (e.g., superdeterminism, the de Broglie-Bohm pilot wave theory, the Many-Worlds Interpretation, etc.). Nonetheless, if one accepts that quantum mechanics is a theory without hidden variables governed by an ontological indeterminism, we must conclude that, on the basis of what we have just seen, even at macroscopic scales, the history of the world is not merely unknown or uncertain in an epistemic sense but may be ontologically under-determined. If the probabilities (or probability densities) derived from the wave function through the Born rule do not represent an underlying unknown stochastic process, but instead reflect an inherent 'potentiality,' 'propensity,' 'disposition,' or 'latency' of Nature[199], then all physical events are not merely (pseudo-)indeterministic but are instead open-ended histories[200]. Even if Laplace's demon were omniscient about the state of the universe, it could not predict an open and under-determined future.

Thus, Jordan's intuition that the unpredictability of life and behavior relates to a macroscopic amplification of the unpredictability of quantum events might not be so far off.

Anyway, regardless of our philosophical commitments, we must acknowledge that noise influences the state of a system across all scales. The sensitivity to initial conditions is a wellknown characteristic of nonlinear chaotic systems. It is not surprising that stochastic processes at the microscopic levels can influence the behavior of macroscopic bodies. However, the extent of this sensitivity is rarely appreciated or is only implicitly assumed. This study quantified this aspect using a toy model, specifically the classical DDP, and examined how a steady weak to extremely weak noise signal can affect its evolutionary trajectories over time. The results demonstrate that even the tiniest uncertainties, down to the Planck scale, persist across all scales and influence the dynamics of much larger bodies within time horizons of a minute or less, challenging traditional views of determinism in mesoscopic and macroscopic systems.

A general state of affairs has been explored through a numerical analysis that provides at least an order-of-magnitude estimate of how quantum fluctuations can impact the dynamics of large objects. If displacements at the Planck scale and noisy effects spanning 27 orders of magnitude can produce a macroscopically discernible divergence in the trajectories of a classical pendulum, it is reasonable to expect that this effect would be even more pronounced in complex systems, such as cellular structure, in which

---

[199] For an interesting analysis of this philosophical aspect, see (Suárez, 2007); (Karakostas, 2007); (Dorato, 2011); (Jaeger G. , 2017); (Kastner & Epperson, 2018); (Kožnjak B. , 2020).

[200] For an in-depth analysis of these aspects, see (Potter, Ellis, & Mitchell, 2025).

randomness is not a flaw but rather a functional aspect of biological processes. The notion that stronger stochastic processes, like thermal effects, 'cancel out' quantum randomness is not entirely accurate. Beyond a certain time horizon, regardless of the weakness of the background perturbation, it will inevitably lead to divergent dynamical histories. This amplification of quantum uncertainties suggests that, contrary to quantum coherence, quantum indeterminacy persists across scales.

This paper aimed to demonstrate that the prevalent belief that quantum fluctuations are self-averaging and negligible because they are 'drowned' in a much noisier environment is only a half-truth. The situation is more complicated and nuanced than that, and it is also much more interesting, with potential implications that have not yet been fully explored.

Appendix: Evaluation of σ_1 and σ_2

To assess how macroscopic nonlinear chaotic systems respond to extremely weak noise signals, particularly constant stochastic perturbations like thermal and quantum noise, it is crucial to first evaluate the relevant orders of magnitude.

We already considered quantum fluctuations on the Planck scale as a lower bound of $\sim 10^{-34}$ rad. Let us take a more conservative value for ϕ_1^{rnd} with $\sigma_1 = 10^{-30}$ rad.[201]

To understand the extremely weak thermal stochastic effects on the DDP subjected to a heat bath, one could first assess the impact of a single photon reflecting off the tip of the pendulum. For instance, let us consider a photon in the visible light spectrum, say of wavelength $\lambda = 600$ nm . An easy calculation shows that this produces a change in angular velocity of the order of $d\omega \approx 4 \times 10^{-25}$rad/s.[202] In our model this would describe the random displacements caused by 100 photons per second ($N = 10.000$ on 100 seconds numerically adds a random displacement every 0.01 s), and would add up to $\sigma = 4 \times 10^{-23}$rad in 100 s, which is still an extremely small thermal effect. While, the internal Brownian motion of the molecules in a rigid body is isotropic and does not affect the overall dynamics of the center of mass.

However, the DDP aims to serve as a mathematical model for understanding sensitivity to stochastic effects in systems governed by nonlinear chaotic dynamics. It can be viewed as an anharmonic nonlinear

[201] The choice is motivated by numerical reasons. Beyond an integration time of 100 s , numerical accuracy becomes less reliable, and to show the evolution of the trajectories between $60 - 100$ s, a noise function four orders of magnitude larger was chosen.

[202] For the simple pendulum of mass m and length l, with moment of inertia $I = ml^2$, its change in angular momentum $dL = I \times d\omega = ml^2 d\omega$. The photon's impulse $p = h/\lambda$, and its reflection causes $dL = 2\frac{h}{\lambda}l$. Therefrom $d\omega = \frac{2h}{\lambda ml}$.

oscillator, and studying it could provide insights into how noise affects other physical systems - both macro and microscopic - that are governed by nonlinear differential equations, such as equation XIII.5. Driven and damped oscillatory forces are ubiquitous (e.g., tidal dissipation and locking dynamics, biological oscillators and neuron models, fluid dynamics, nonlinear chemical models, seismological models, etc.) A larger upper bound, indicating a more robust noisy signal, is likely to influence real-world systems, particularly in the field of biology.

Therefore, let us roughly estimate the magnitude of Brownian noise's influence on a cell's electrical physiology. The membrane resistance of a neuron is determined by its surface area and the density of ion channels. It varies between $R = 10 \div 100M\Omega$ for typical neurons (Spruston & Johnston, 1992). A neuron spike is typically a 1 ms pulse with a sharp rise and fall in membrane potential, from $V_m \approx -70mV$ to $+50mV$ (Alberts, et al., 2002). Because a spike is a brief, transient voltage pulse, its Fourier spectrum spans a broad range of frequencies. The dominant energy spectrum lies in the range of $\Delta f = 100 - 1000$ Hz (Orcioni, Paffi, Apollonio, & Liberti, 2020). The RMS-electric thermal noise (Johnson-Nyquist noise) is given by $V_{RMS} = \sqrt{4k_BTR \triangle f}$. For $R = 100M\Omega, \triangle F = 1kHz$, and V_m = 50mV, the resulting ratio between the neurons' electric thermal noise and the working membrane's action potential is about $\sigma = V_{RMS}/V_m \approx 0.8 \times 10^{-3}$. This inspired as a rough analogy the choice of $\sigma_2 = 10^{-3}$rad in the context of the DDP.

References

Adams B., Petruccione, F. (2020). Quantum effects in the brain: A review. AVS Quantum Sci.1 2 (2): 022901. DOI: **https://doi.org/10.1116/1.5135170**

Alberts B., Johnson A., Lewis J., et al. (2002). Molecular Biology of the Cell. 4th edition. New York: Garland Science; 2002.

Bandak D., Mailybaev A., Eyink, G., Goldenfeld N. (2024): Spontaneous stochasticity amplifies even thermal noise to the largest scales of turbulence in a few eddy turnover times. Physical Review Letters 132, 104002. DOI: **https://doi.org/10.1103/PhysRevLett.132.104002**

Berry M.V. (1979). Evolution of semiclassical quantum states in phase space. Phys. A: Math. Gen. 12 625. DOI: **https://doi.org/10.1088/0305-4470/12/5/012**

Berman G.P., Zaslavsky G.M. (1978). Condition of stochasticity in quantum nonlinear systems. Physica A: Statistical Mechanics and its Applications, Vol 91(3–4), pg. 450-460. DOI: **https://doi.org/10.1016/0378-4371(78)90190-5**

Bezanson J., Edelman A., Karpinski S., Viral B. S. (2017). Julia: A Fresh Approach to Numerical Computing. SIAM Review, 59: 65–98. DOI: **https://doi.org/10.1137/141000671**

Boekholt T., Portegies S., Valtonen M. (2020): Gargantuan chaotic gravitational three bodysystems and their irreversibility to the Planck length. Monthly Notices of the Royal Astronomical Society 493(3), 3932-3937. DOI: **https://doi.org/10.1093/mnras/staa452**

Clarke P.G. (2014). Neuroscience, quantum indeterminism and the Cartesian soul. Brain Cogn. 84(1):109-17. DOI: **https://doi.org/10.1016/j.bandc.2013.11.008**. - Epub 2013 Dec 18. Erratum in: Brain Cogn. 2014 Jul;88:91. PMID: 24355546.

Dorato, M. (2011). Do Dispositions and Propensities Have a Role in the Ontology of Quantum Mechanics? Some Critical Remarks. In: Su´arez, M. (eds) Probabilities, Causes and Propensities in Physics. Synthese Library, vol 347. Springer, Dordrecht. DOI: **https://doi.org/10.1007/978-1-4020-9904-5_9**

Eyink G., Drivas T. (2015) Quantum spontaneous stochasticity. DOI: **https://doi.org/10.48550/arXiv.1509.04941**

Feagin T. (2009). An Explicit Runge-Kutta Method of Order Fourteen. Numerical Algorithms. Feagin T. (2012). High-order Explicit Runge-Kutta Methods Using M-Symmetry. Neural, Parallel & Scientific Computations, Vol. 20, No. 4, pp. 437-458.17

Gamble J.K., Lindner J.F (2009). Demystifying decoherence and the master equation of quantum Brownian motion. Am. J. Phys., 77 (3): 244–252. DOI: **https://doi.org/10.1119/1.3043847**

Jaeger G. (2017). Quantum potentiality revisited. Phil. Trans. R. Soc. A. 375:20160390. DOI: **http://doi.org/10.1098/rsta.2016.0390**

Jedlicka P. (2017). Revisiting the Quantum Brain Hypothesis: Toward Quantum (Neuro)biology? Front. Mol. Neurosci., Sec. Molecular Signalling and Pathways. DOI: **https://doi.org/10.3389/fnmol.2017.00366**

Jordan P. (1941). Die Physik und das Geheimnis des organischen Lebens, Braunschweig, Germany: Friedrich Vieweg & Sohn.

Karakostas V. (2007). Nonseparability, Potentiality, and the Context-Dependence of Quantum Objects. J Gen Philos Sci 38, p. 279–297. DOI: **https://doi.org/10.1007/s10838-007-9050-9**

Kastner R., Kauffman S., Epperson M. (2018). Taking Heisenberg's Potentia Seriously International Journal of Quantum Foundations 4 (2):158-172. DOI: **https://doi.org/10.1142/9781786346421_0011**

Koznjak, B. (2018). The Earliest Missionaries of 'Quantum Free Will': A Socio-Historical Analysis. In: Grgic, F., Pecnjak, D. (eds) Free Will & Action. Historical-Analytical Studies on Nature, Mind and Action, vol 6. Springer, Cham. DOI: **https://doi.org/10.1007/978-3-319-99295-2_10**

Koznjak, B. (2020). Aristotle and Quantum Mechanics: Potentiality and Actuality, Spontaneous Events and Final Causes. Journal of General Philosophy of Science, Vol. 51, p. 459–480. DOI: **https://doi.org/10.1007/s10838-020-09500-y**

Mishin Y. (2015). Thermodynamic theory of equilibrium fluctuations. Annals of Physics, vol. 363, pg. 48-97. DOI: **https://doi.org/10.1016/j.aop.2015.09.015**

Orcioni S., Paffi A., Apollonio F., Liberti M. (2020). Revealing Spectrum Features of Stochastic Neuron Spike Trains. Mathematics, 8(6), 1011. **https://doi.org/10.3390/math8061011**

Paz J.P., Zurek W.H. (1999). Quantum Limit of Decoherence: Environment Induced Superselection of Energy Eigenstates. Physical Review Letters, 82(26): 5181–5185. DOI: **https://doi.org/10.1103/physrevlett.82.5181**

Potter H.D., Ellis G.F.R, Mitchell K.J. (2025). Reframing the Free Will Debate: The Universe is Not Deterministic. DOI: **https://doi.org/10.48550/arXiv.2503.19672**

Roy S., Majumdar S. (2022). Noise and Randomness in Living System, Springer, Berlin.

Schlosshauer M. (2019). Quantum decoherence, Physics Reports, Volume 831, pp. 1-57. DOI: **https://doi.org/10.1016/j.physrep.2019.10.001**

Spruston N., Johnston D. (1992). Perforated patch-clamp analysis of the passive membrane properties of three classes of hippocampal neurons. Journal of Neurophysiology, 508-529, vol. 67(3). DOI: **https://doi.org/10.1152/jn.1992.67.3.508**

Su´arez, M. (2007). Quantum propensities. Studies in History and Philosophy of Science Part B: Studies in History and Philosophy of Modern Physics, 38 (2). pp. 418-438. DOI: **https://doi.org/10.1016/j.shpsb.2006.12.00318**

Strogatz, S.H. (2018). Nonlinear Dynamics and Chaos. p. 328, CRC Press. Taylor & Francis Group Taylor J.R. (2005) Classical Mechanics, University Science Books.

Tomaz A.A, Mattos R.S., Barbatti M. (2025). The Quantum Measurement Problem: A Review of Recent Trends. DOI: **https://doi.org/10.48550/arXiv.2502.19278**

Vacchini B., Hornberger K. (2009). Quantum linear Boltzmann equation. Physics Reports, Vol. 478, Issues 4–6, Pg. 71-120. doi: **https://doi.org/10.1016/j.physrep.2009.06.001**

Youngchan K., Bertagna F., D'Souza E. et al. (2021): Quantum biology: An update and perspective. Quantum Reports 3(1), 80-126. DOI: **https://doi.org/10.3390/quantum3010006**

Zurek W.H., Paz J.P. (1999). Why We Don't Need Quantum Planetary Dynamics: Decoherence and the Correspondence Principle for Chaotic Systems. In: Greenberger, D., Reiter, W.L.,

Zeilinger, A. (eds) Epistemological and Experimental Perspectives on Quantum Physics.

Vienna Circle Institute Yearbook [1999], vol 7. Springer, Dordrecht. DOI: **https://doi.org/10.1007/978-94-017-1454-9_13**

Zurek W.H., Paz J.P. (2003).Decoherence, einselection, and the quantum origins of the classical. Rev. Mod. Phys., vol. 75(3). DOI: **https://link.aps.org/doi/10.1103/RevModPhys.75.71519**

7. The Integral Cosmology of Sri Aurobindo: An Introduction from the Perspective of Consciousness Studies[203]

Abstract

In the contemporary philosophy of mind and consciousness studies, views such as panpsychism or theories of universal consciousness, have enjoyed a recent renaissance of metaphysical speculations in Western philosophy. Its similarities with Eastern philosophical traditions went not unnoticed. However, the potential contribution that the evolutionary cosmology of the Indian poet, mystic and philosopher Sri Aurobindo can offer to these ontologies, remains largely unknown or unexplored. Here, consciousness, mind, life, matter and evolution are interpreted in an extended metaphysical framework, uniting Western and Eastern traditions inside a modern philosophical system where materialism, panpsychism, cosmopsychism and philosophical idealism are not seen as irreconcilable exclusives, but different points of view that can be reconciled in a synthesis of knowledge, as envisioned from higher states of consciousness. This integral view of reality offers a conceptual platform that can complement and incorporate current theoretical approaches in modern philosophy, where their present internal inconsistencies do not emerge, and can be suggestive for scientists as well, especially in the fields of biology and physical sciences.

I. Introduction

Much has been written about Sri Aurobindo's whole-person-centered spiritual teachings, which are first and foremost a basis for spiritual practice. However, his integral mystics, philosophy, poetry, and vision have also become a pole of attraction for the Western academic world. Aurobindo is particularly known among psychologists (for a review of 'integral psychology,' see, e.g., Banerji, 2020 or Salmon, 2007). He is also receiving increasing attention from pedagogues in terms of an 'integral education' (e.g., Masi, 2018; Shirazi, 2011; Zulaski, 2017). Some scholars also pointed out his actuality in the field of international relations and the social development of humankind (e.g., Aurose, 2021; Shah & Jacoby, 2013). However, despite its relevance, his cosmology received much less attention in the philosophy of science until recently (with few exceptions, e.g., Banerji, 2015; Lester, 2011)

[203] The present article was published in the *Integral Review*, Vol. 18, No.1.

The main motivation leading the author to write this paper was to compensate for this shortcoming and furnish an introduction to the metaphysical framework of Sri Aurobindo for philosophers and scientists, especially those working in the field of consciousness studies and evolutionary biology. These could profit from Aurobindo's vision as a source of inspiration and as an original scheme that could potentially resolve several unanswered questions.

In particular, in contemporary consciousness studies, we see a multiplication of theories and speculations trying to relate that ineffable and undefinable experiential entity we call 'consciousness' to brain mechanisms and their neural correlates. However, despite its impressive progress in mapping brain activity, after decades of research, neuroscience made no tangible progress in closing the explanatory gap dividing the third-person neuroscientific perspective and our first-person subjective experiential dimension. From a naturalistic standpoint, the question of how consciousness, sentience, subjective experience, and the perception of qualities ('qualia') come into being, supposedly because of some yet undiscovered neural machinery, remains unanswered more than ever. This also led David Chalmers to write his seminal paper on the famous 'hard problem of consciousness' (Chalmers D. , Facing up to the problem of consciousness, 1995): how and why does the seemingly immaterial phenomenal experience arise out of material brain activity? Tackling this problem, the modern philosophy of mind adopted mainly a scientific and philosophical approach founded on the Western tradition. Though it drew inspiration from the old school, such as Descartes' dualism, Berkeley's ontological idealism, Kant's transcendental idealism, and up to modernity with Husserl's phenomenology, Whitehead process philosophy, or Gebser's 'integral structure of consciousness,' to mention just a few, it only rarely admits for post-material frameworks. It did, and to a large degree still does, rely almost exclusively on a neuroscientific approach, eventually augmented by a Darwinian evolutionary paradigm, usually only inside a strictly reductionist and physicalist perspective. This resulted in a plethora of theories, conjectures, and the most diverse speculations. For example, highly rated theories of consciousness are Baars' 'Global Workspace Theory' (Baars, 1988), Tononi's 'Integrated Information Theory' (Oizumi, Albantakis, & Tononi, 2014), Penrose and Hameroff's 'Orchestrated Objective Reduction' (Penrose, 1994; Hameroff & Penrose, 2014), and Hoffman's 'User Interface Theory' (Hoffman, 2008), to mention just a very few of them (for a review, see also (Seth & Bayne, Theories of consciousness, 2022) or (Wahbeh, Radin, Cannard, & Delorme, 2022). However, nothing indicates that we are getting anywhere closer to a converging understanding of what a consistent theory of consciousness might look like.

Given this lack of progress, in more recent times Western schools of thought expanded their horizons towards metaphysical speculations positing consciousness, not matter, as the fundamental primitive. Most notably, panpsychism has been resurrected by some modern leading philosophers, like Nagel, Strawson, Chalmers, and Goff (for a modern review, see Seager, 2020; Skrbina, 2017) – something reminiscent of Leibniz's monadology but recast into modern reductionist bottom-up perspective, and in the light of modern scientific findings. On the other hand, top-down holistic approaches have also been considered, inspired by Eastern philosophies such as Velmans' post-Kantian 'reflexive monism' that conceives the subject as a differentiated part of the one universe (Velmans, Reflexive Monism: psychophysical relations among mind, matter and consciousness, 2012), Shani's holistic-monistic variant of panpsychism, 'cosmopsychism' (Shani I., 2015; Shani & Keppler, 2018), Kastrup's modern version of Schopenhauer's idealism, analytic idealism (Kastrup, 2018), Taylor's idea of a pervasive spirit in all things, 'panspiritism' (Taylor S. , 2020), and other lines of research inspired by the Asian tradition, such as Albahari's mystical solution to the mind-body problem (Albahari, 2019) or investigations on the parallel between panpsychism or cosmopsychism and Advaita Vedanta (e.g., Gasparri, 2019; Vaidya, 2020).

While a cross-cultural opening to Eastern philosophy and mysticism among Western scholars is certainly to be welcomed, one can't fail to notice how its conceptual structures are still limited to a classic Vedantic non-dual outlook. They lack an evolutionary (and 'involutionary') perspective or, at any rate, they do not expand on the evolution of consciousness to any degree similar to that of Aurobindo. On the other hand, although Western science and philosophy placed evolution, especially in its modern neo-Darwinian incarnation, at the center of their attention, the emergence of consciousness in the evolutionary drama is an accepted datum and their substantial and temporal relation remains scarcely addressed. The most notable exceptions are Teilhard de Chardin's cosmic theology (Chardin, 1955), or Gebser's historical view of the emergence of the 'structures of consciousness' unfolding towards an 'integral consciousness' (Gebser, 1985). And noteworthy is that the seed-idea was already present, especially among the German idealists, like Fichte, Schelling, or Hegel (for an interesting historical account of what Murphy called an 'evolutionary panentheism', see (Murphy, 2015). However, these metaphysical paradigms remain lower dimensional coarse-grained models– that is, much too often one sees the tendency to consider mind and consciousness as synonymous, while the role of higher cognitive states is ignored if admitted at all. Lastly, what Aurobindo called the 'planes of consciousness and parts of being', while well known in several spiritual and mystic traditions, remains, especially in Western philosophy, an unknown or ignored ontological multidimensionality. The old dualistic school or, at best, a dual-substance ontology that doesn't go beyond the dichotomy between mind and matter (or consciousness, or spirit, and that, again, are taken as synonymous) still prevails.

Only very recently attempts have been made to connect Aurobindo's spiritual emergentism to the Western philosophy of mind, such as Medhananda's evolutionary cosmopsychism (Medhananada, 2021, 2022), (Medhananada, 2024)–Shani's search for a connection between cosmopsychism and non-dualism (Shani, 2022).

This paper aims at furnishing a first general introduction to Aurobindo's cosmology (for a more detailed analysis see also the author's work Masi, 2022a), and that may help to build a bridge making this connection. Aurobindo's vision and deep insights into the meanderings of our conscious, subconscious, and subliminal life, and that consciousness studies try so hard to untangle, remain largely unknown or ignored despite the fact that it appears to be the most comprehensive synthesis of knowledge. It allows for an integration between East and West, encompassing contemporary philosophical thought from an evolutionary perspective and in line with current scientific findings. What I call 'integral cosmology'–the theoretical framework emerging from Aurobindo's contemplative practice, which he called 'integral yoga'[204] (also called 'purna yoga')–has potentially interesting applications that could extend the current lines of Western metaphysics, such as panpsychism, pantheism (God is Nature and Nature is God) and panentheism (Nature is in God), or complement the above-mentioned forms of philosophical idealism based on universal consciousness.

Also biologists, physicists, and other scientists may find, in this review, a source for contemplation as well. The evolutionary perspective of Aurobindo, inside a paradigm of spiritual emergentism, and a metaphysical vision where matter and the laws of physics are only a small subset of a much wider reality, may not appeal to the orthodox materialist mindset. Nevertheless, the scientist who feels inclined to explore post-material views of evolution and reality may find in him interesting suggestions.

The paper is organized as follows.

After a short biographical note, in the second part, follows an introduction to Aurobindo's spiritual ontology, his notion of evolutionary progress, from the Absolute to the universal, or cosmic consciousness and the planes and parts of being–the different levels of consciousness in its individuated and cosmic dimension. I will emphasize how, in this integral vision, our identity is not confined neither to a one-dimensional materialistic monism, nor to a two-dimensional dual-aspect or neutral monism, rather it is a multidimensional structure that transcends beyond a simple mind, body, and soul ontology. A 'concentric system' of inner, outer and inmost identities will be examined. The centrality of the 'psychic being'–the notion of an evolutionary soul–will be analyzed with particular attention because of its vast implications leading to an

[204] The terms 'integral cosmology' and 'integral Vedanta' are of the author, while 'integral yoga' is Aurobindo's terminology.

extended concept of evolution. An account of the diverse levels of cognition beyond mind, such as overmind, cosmic mind and supermind will be discussed.

Reminiscent of Teilhard de Chardin's evolutionary philosophy, central is also Aurobindo's claim that the human represents only a transitional being, not the apex of evolution, and that will be surpassed by a coming future species he called the 'gnostic being'. An evolutionary vision towards higher states of consciousness, which includes a material transmutation of matter and our bodily existence by a 'supramental transformation'. A vision that can be grasped in its completeness only once the distinction between evolution and involution is made on all planes of existence.

In the third part, we will explore in-depth central concepts of Aurobindo's mystic insights. The multiplicity in unity, the coexistence of the personality and impersonality of the Divine, the transcendent and immanent, the supramental vision of space-time and the twofold identity of Nature.

This will clarify consciousness and the emergence of life from a perspective that bridges Aurobindo's cosmology with modern theories of cosmic idealism and modern evolutionary theories in a post-materialistic ontology.

An overview will follow showing how, taking the integral perspective, the monistic reductive materialism, philosophical idealism and the first-person mystical approaches are the three complementary forms of cognition that can converge in a synthesis of knowledge, both unifying and finally surpassing them all.

Few concluding remarks will complete this introductory paper.

II. Short Bio of Sri Aurobindo

The Indian mystic, yogi, poet and nationalist Aurobindo Ghose (1872-1950), more commonly known as Sri Aurobindo, was born in Kolkata (former Calcutta). At the age of seven, he was sent to Cambridge, England, to study English literature and to protect him from, what his father considered, a backward mystic Indian culture. Ironically, Aurobindo would become one of the most influential yogis–a practitioner of yoga – of the 20th century. After having perfectly absorbed the British language and culture and returning to India 14 years later, he worked as a teacher and, in about 1906, became directly involved in nationalist politics by joining the Indian revolutionary movement against British rule.

However, it soon became evident that a matter much more interesting and urgent would determine the rest of his life. In 1908, Aurobindo was arrested for his political engagement and spent about a year in jail. During that time, he went through powerful spiritual experiences, which became the basis of a vision of human progress and spiritual evolution. In 1910, Aurobindo left politics and moved to Puducherry, in the south Indian state of Tamil Nadu, a place he would never leave again. Here, he developed the principles of an integral yoga, a form of ascesis that incorporated the former Indian spiritual practices and, most importantly, expanded them to a discipline aimed also at physical transformation. He founded an ashram (a spiritual hermitage for yoga practitioners), nowadays called the 'Sri Aurobindo Ashram'.

In 1926, Mirra Alfassa (1878-1973), who became his collaborator and spiritual partner, also called by their disciples "The Mother," would take over the management of the ashram. This allowed Aurobindo to dedicate himself more intensely to his sadhana (the yogic spiritual discipline) and to what would later be called the 'cell yoga' – an attempt at spiritual transmutation of the physical body–determining his seclusion for the rest of his life. He maintained, however, an intense written correspondence with his disciples and the world.

Aurobindo was an extremely prolific writer and became the author of an enormous number of texts, mainly about spiritual topics, which highlighted his spiritual vision that sees the human race ascending to increasingly higher states of consciousness within an evolutionary process that would unite spirit and matter. His three main magnum opuses are "The Synthesis of Yoga" (Aurobindo, 1914-21) – a practical guidance to integral yoga, "The Life Divine" (Aurobindo, 1914-19) – the description of his spiritual vision based on his inner experiences framed in a philosophical language – and "Savitri" (Aurobindo, 1950) – a spiritual epic poem of about 24,000 verses. Aurobindo might be described as a 'mystic empiricist'. He noted his inner spiritual experiments in a diary, the 'Record of Yoga' (Aurobindo, 2002; Banerji, 2012). The core of the vision and teachings contained in these writings were published around 1915-1920 but were also amended and revised thereafter, until his departure. An important aspect to keep in mind is that despite the philosophical relevance of his teachings, he never considered himself a philosopher. Several of his writings have been written with a philosophical style but relied exclusively on a first-person spiritual and mystic experience. It would be impossible to fully comprehend his message relaying exclusively on an analytic study.

III. The Integral Structure of Being

1. The Notion of 'Progress' in the Evolutionary Integral View of Sri Aurobindo

What differentiates Aurobindo's evolutionary spiritual vision from other paths is the concept of the existence of a spiritual evolution and progress in Nature. Not only a liberation ('moksha' or 'nirvana') from the bonds of matter of the classical Asian paths, but also an integral transformation of our mental, emotional and physical nature, culminating in the transmutation of the material body itself into the image of the Spirit by means of a 'cellular yoga'. An 'integral Vedanta' that distances itself from the traditional Indian conception of life and the universe as the cosmic 'illusion of Maya'. Aurobindo rejects the idea common to many mystics of his time that matter is an illusion and, to the contrary, elevates it to the container of an ultimate physical and spiritual transformation.

According to Aurobindo, evolution is a form of 'yoga of Nature' and insists on a world-affirming view or, as he puts it, 'all life is yoga'. Humans are transitional beings undergoing an evolutionary process that must ultimately lead us to the next evolutionary step, which will transcend reason and the analytic physical-mind, to a form of (non-religious) gnosticism. A spiritual and material progress that should not be a privilege for just a few but that will be generalized to humanity as a whole. It is a synthesis of progressive inner and outer evolution aimed at a long-term spiritual and physical transformation, where matter and spirit are not exclusives but complementary destined to merge. An ascesis towards higher states of consciousness and forms of cognition, that distinguishes the spiritual practice of integral yoga from other spiritual paths, which, however, are not denied, but rather embraced and extended.

Furthermore, this evolutionary vision does not reject the scientific concept of Darwinian evolution but rather extends and includes it. Darwinian evolution could be integrated within Aurobindo's evolutionary cosmology. The main difference, however, is that while evolutionary biology does not accept a teleological dimension, conceiving evolution only as a change in the physical heritable characteristics and genetic mutation in time, in this integral perspective it also represents the unfoldment and the progress of a trans-physical indwelling spirit, driven by a concealed omniscience having an aim and working towards a goal.

2. The Integral Yoga of Sri Aurobindo in a Nutshell

A thorough introduction to the practical aspects of the integral yoga would exceed the scope of this paper and is not necessary to grasp the metaphysics of the integral cosmology that would be outlined here. It may only be said that this spiritual practice has a double aspect. On one side, integral yoga accepts and, to a large degree, even follows the paths of other contemplative traditions, especially those of the Eastern Hindu and Vedantic schools and yoga philosophical approaches as the Advaita Vedanta and Samkhya philosophy. Techniques of concentration and meditation, like in the classical raja yoga or jnana yoga, aimed at calming the mind is a basic requisite here too. The purpose is to reach an inner spiritual realization that transcends the ordinary states of consciousness emerging into the ultimate reality, which is the Brahman, the 'One without a second' of the Indian Vedic tradition, the non-dual teachings or the Buddhist nirvana.

However, Aurobindo claims that, while this is a necessary step along the path of self-realization, Nature also aims to achieve a much higher level of self-perfection that does not exclude the outer being. By a more integral practice of karma yoga, the 'yoga of divine works' of the Bhagavad Gita, the principles of bhakti yoga and the spiritual path of devotion, Aurobindo delineated, especially in his 'Synthesis of Yoga', that there is a larger and more encompassing form of spiritual practice aimed at an ascension and also a descension of a higher Consciousness that works on our inner and outer being. Central to this discipline is the act of surrender and self-giving to the Divine. By a constant single-mindedness of our whole being onto the Divine and by transforming our emotional and still limited mental consciousness, a divine Shakti, i.e., the universal Conscious-Force, the 'Chit-Tapas' or 'Chit-Shakti' of the Shakti-Advaita, can transform and transmute our untransformed nature. The yoga is 'integral' inasmuch that its purpose is the transmutation of our whole psychological, and also material, aspect of existence, aiming at the generalization of yoga to the whole of humanity. It is from the attained higher state of consciousness that Aurobindo was able to gain a vision which he outlined in his writings.

3. Planes of Consciousness and Parts of Being, the Psychic and Spiritual Transformation

If one wants to comprehend the philosophical background of Aurobindo's integral yoga and its cosmology and cosmogony, it is necessary first to get acquainted with its complex 'topography of consciousness'. But before outlining the differences from orthodox traditions, it is especially important to consider which aspects of his vision align with these traditions.

The traditional Indian yoga philosophy of the Upanishads conceives of the being made of 'vehicles', 'bodies', or 'sheets of consciousness', called 'koshas' covering our soul like the rings of an onion. Matter is not the only substance that exists. The coarse-grained metaphysical dualistic model that reduces life to just a soul and a body is expanded. Once the consciousness of the mystic ascends beyond the material realm, more subtle forms of matter, substances and forces of different grossness, which exist beyond the physical sense, become part of the experiential reality. Our physical body appears as the outer layer of other, more subtle, 'bodies'. In descending order (according to the Upanishadic classification with the correspondent 'aurobindonian' plane, see later for more details):

Anandamaya kosha,	Satcitananda
Vijñānamaya kosha,	Supermind
Manomaya kosha,	Mental
Pranamaya kosha,	Vital
Annamaya kosha,	Physical

In integral yoga, the five kosha-structure of classical yoga was embraced and further refined, distinguishing between the 'higher hemisphere' and the 'lower hemisphere'. By heightening one's awareness ascending to higher states of consciousness through a meditative practice, Aurobindo claims that one meets planes, parts, domains and layers of existence that we are not conscious of in our ordinary cognitive state. *"Below us there are lower ranges to which we are insensible and these we call subconscience or inconscience. Above us are higher ranges which are to our inferior nature an unseizable superconscience"* (Aurobindo, 1910-1950, pp. 247, pt.II, sec.I).

One might draw an analogy in the Western tradition with Spinoza's 'modes of substance.' According to Spinoza, there is only one Substance–that is, God or Nature – which, however, can acquire infinitely many modes and attributes that express the divine essence. All is a mode of that one Substance. Similarly, here also the lower and higher 'planes' are the different expressions of one universal Consciousness on different ranges of self-awareness.

The first two universal superconscient planes, namely Satcitananda and the Supermind, make part of what Aurobindo called the 'higher hemisphere'. We will look at these in more detail later. Let us first consider the lower hemisphere.

a. Planes and Parts of the Lower Hemisphere

The physical plane: The annamaya kosha, the 'food-sheet', is our familiar physical body. The universal plane in which it resides is that of the physical universe made of matter, the layer of existence that science investigates.

The vital plane: The pranamaya kosha, the 'life-force sheet', what Aurobindo termed as the 'vital', is the sheet and plane responsible for our emotional consciousness. It is not made of matter, rather, its substance is of the nature of a more subtle 'vital energy' also commonly known as 'prana' or as the Chinese 'ki', which exists independently from the permanence of the physical body. Note that life is not understood as derivative from matter, contrary to what is commonly assumed in a Darwinian evolutionary paradigm. The universal vital plane–the life-plane–emerges in and through matter by an evolutionary process but is distinct from it. Life is always indwelling in all matter and will always manifest, once the conditions for its emergence are realized. In fact, throughout his writings, Aurobindo uses such terms as 'vital plane,' 'Life-Force,' and 'Life-Energy.' The vital is used by the mind as an instrument to establish a contact with the physical world; for it is that by which the mind receives the sensory information through the organs of sight, hearing, touch, smell, and taste. This seems to fly in the face of modern biology which explicitly distances itself from any form of vitalism. However, it is a common misconception that modern science has disproved vitalism or, at least, the form of vitalism this integral cosmology posits. Contrary to a physicalist doctrine, these were and remain coherent worldviews and cannot be ruled out by modern science, as I have outlined more rigorously here (Masi, 2022b).

The mental plane: The manomaya kosha, the 'mental sheet' is the plane of 'manas', more commonly named 'mind'. It is the sheet of reason and of analytic thinking that is responsible for our thoughts.

This metaphysics is obviously at odds with any physicalist conception of reality. All these planes, except the physical one, are non-physical. Mind and brain are part of the mental and physical plane, respectively, and are two parts to be distinguished.

It is not a Cartesian dualism that distinguishes only between mind and body, rather a multi-layered reality reminiscent, again, of the Spinozian God, where mind (res-cogitans) and matter (res-extensa) are only two of its infinite 'modes of Substance'. Moreover, in Eastern traditions, consciousness is the very nature of Satcitananda and not to be conflated with mind, as it is frequently in neuroscience, psychology, consciousness studies, and the philosophy of mind. Here mind is considered a mutable form of cognition and not even the highest one, it is just a 'sheet' that connects the vital with higher forms of cognition. Whereas consciousness is the transcendent and atemporal 'witness', the immutable observer.

Aurobindo enlarges the traditional Samkhya classification to subconscient and inconscient planes even below matter and the bodily consciousness. These stand as antipodal to the superconscient 'planes of Knowledge' and constitute the unenlightened 'planes of Ignorance'.

The subconscient: With similarities, but also differences, to Western psychology the subconscient is the submerged part of our being dominated by incoherent thoughts, irrational fears, impressions of traumatic experiences, habitual movements, and disorganized emotions that we are mostly unaware of in our surface waking state. The subconscient records and stores experiences, impressions, and emotions throughout our life, especially those of particular intensity or traumatic character. Aurobindo maintains that the influence of our subconscient goes down to cellular processes, viewing it in a wider evolutionary perspective than Western psychologies.

That part of us which we can strictly call subconscient because it is below the level of mind and conscious life, inferior and obscure, covers the purely physical and vital elements of our constitution of bodily being, unmentalised, unobserved by the mind, uncontrolled by it in their action. It can be held to include the dumb occult consciousness, dynamic but not sensed by us, which operates in the cells and nerves and all the corporeal stuff and adjusts their life process and automatic responses. It covers also those lowest functionings of submerged sense-mind which are more operative in the animal and in plant life; in our evolution we have overpassed the need of any large organized action of this element, but it remains submerged and obscurely at work below our conscious nature. (Aurobindo, 1914-19, pp. 763, pt. II, ch. XIX)

The inconscient: The inconscient is a plane that is positioned even deeper than the subconscient, insofar that its movements become even more mechanical and reactive, almost impossible to change.

The nescient: At the lowest rung of the ladder of consciousness, Aurobindo sometimes mentions what he called the 'nescient'. It is the most inert, passive, obscure, mechanical and obnubilate involved consciousness, the bottom of creation, the root of matter's inertia itself. It is a substance almost completely forgetful of its original source, the characteristic trait of matter.

b. The Concentric System of Being, Soul and Nature

By proceeding with the psycho-physiological ascesis, on each plane, one begins to discover further inner realms of being that are concealed from the ordinary surface awareness. For example, there is further metacognitive surface mind and a whole mental domain we are not directly aware of. The same is applicable to our 'emotional body', i.e. the vital. Our surface emotions are largely driven from behind by a variety of impulses we are unaware of. The spiritual practitioner, however, can come into closer contact with what Aurobindo called the 'inner being', distinguishing it from the 'outer being'. The distinction goes as follows.

The outer being: Our outer being – our surface personality – is dominated by half-controlled mental and emotional impulses, desires, fears, and more or less reactive mental patterns, restless thoughts, the physical sensory reasoning and all those psychological traits we call our 'personality' or ego or self (lower case 's').

The *'outer mental'* is what we know well as our *'sense-mind'*–the mind that organizes and conceptualizes our sensory perceptions–closely related to the *'physical mind'*, which is the mind that believes only in the physical and cannot even comprehend anything beyond this. This sense-mind is responsible for mentalizing and organizing all the perceptions it receives in forms and shapes, presenting it to the physical mind that is unable to conceive anything other than what the sense-mind presents – it takes the physical world and its mechanical workings for all that there is.

There is also the emotional mind, which Aurobindo termed the *'vital mind'*. It is dominated by the life-impulses; it is this irrational part of the mind that thinks instinctively and imagines emotionally, with restless thoughts in an unreflective manner. The *'outer vital'* or *'lower vital'* is the source of what we term, in our everyday parlance, 'negative emotions', such as anger, fear, selfishness and greed. However, the lower vital can eventually express somewhat nobler emotions of human love, albeit tainted by possessive tendencies or philanthropic actions driven by subconscious egoistic motives.

The *'outer physical'* is just the gross physical body.

The inner being: This inner spiritual consciousness, which Aurobindo also termed as the 'subliminal being' or simply the 'subliminal', has a different significance than in Western psychology. This 'subliminal being' is a much vaster and luminous *"place of deep peace, light, happiness, love, closeness to the Divine or the presence of the Divine."* (Aurobindo, 1927-50, pp. 307, vol. I)

It is important to note that in integral yoga, the subliminal does *not* stand for the sub- or unconscious but, on the contrary, it is a veiled and more intuitive inner domain of consciousness compared to our metacognitive access of consciousness. It is a wider 'intelligence' behind the surface mind and opens up to higher degrees of intuition and insight. The subliminal possesses all and perhaps more than high-level functions, such as thought, problem-solving, decision-making and creativity. It is a non-metacognitive higher-level function which we fail to report.[205]

The subliminal self stands behind and supports the whole superficial man; it has in it a larger and more efficient mind behind the surface mind, a larger and more powerful vital behind the surface vital, a subtler and freer physical consciousness behind the surface bodily existence. And above them it opens to higher superconscient as well as below them to lower subconscient ranges. (Aurobindo, pp. 612, vol. IV)

In contrast to the outer mental, the *'inner mental'* is a silent, calm, unbound, unattached mind free from the influences and errors of the ignorant thought and will of the outer mind, directly open to the knowledge and guidance from above. It is the analytic mind, the cherished reason of the Enlightenment, but in its detached state from any emotional impulses of the vital mind or other conditionings of the outer mind.

The *'inner vital'* or *'higher vital'* is that emotional part in us that once it comes to the front, expresses itself with finer emotions of love, beauty, peace and calm. Meanwhile, the *'inner physical'* or *'subtle physical'* is a body made by a more subtle substance that is much more plastic, having the freedom and power that the physical gross body does not possess. It is a physical existence parallel to the material one that is accessible only by developing subtle senses and rising to the highest levels of consciousness.

In the vision outlined by Aurobindo, the inner being is open to cosmic influences. The subliminal is in contact with the mental, vital and physical environment surrounding it. Every individual receives subliminally subtle forces, thoughts, and emotions via this *'environmental consciousness'* or *'circumconscience'*. Aurobindo maintains that this environmental consciousness receives influences beyond the physical body and can catch

[205] A concept that modern science is now referring to as the 'no-report paradigm' (Tsuchiya, Wilke, & al., 2015).

emotions, thoughts, passions, suggestions, or forces of illness from others, which can affect our subconscient and physical beings and eventually cross to the outer being, into our outer conscious awareness.[206]

Inmost being: Our 'inmost being' is the true being and individuality in us, not to be confused with the inner mind, vital or body. Also, the noblest thoughts and emotions of the inner being are the reflection of the qualities of our inmost being, our true evolutionary soul, which Aurobindo called the *'psychic being'* (the 'chaitya purusha', as the rishis, the ancient seers of the Indian Upanishad, termed it). The psychic being requires special attention; therefore, it will be described in depth within the next section.

But each individual is a being embedded in, and subjected to, the dynamics of a cosmic environment. As in the Samkhya philosophy, Aurobindo's integral cosmology also distinguishes between Soul and Nature.

Prakriti: What distinguishes the outer and inner being from the inmost being is that the former take part in the dynamical universe and are active in the temporal manifestation, belonging to what the Samkhya yoga system called *'Prakriti'*, the cosmic universal Nature, or just Nature. However, this conception of Nature is not limited to the Western naturalistic idea of the physical natural world. As we will see in more detail later, it is not a transcendent but also not a purely material phenomenal world. Prakriti contains all the universal dynamical manifestation, which is not solely material but vital and mental as well as beyond it, into the higher and lower planes of being.

Purusha: One's true inmost being, the psychic being, supports all actions of Nature, but is beyond it and transcends the temporal manifestation, it is what the Samkhya philosophy termed as the *'Purusha'*. Purusha is the Soul or Spirit-side, as opposed to the Prakriti or Nature-side. The psychic being is the Purusha in the heart center, but besides it, the Purusha can be mental or vital, standing behind these planes as an unaffected 'soul-witness' to the activities on these planes of being. By stepping back and going inwards, one learns to distinguish the phenomenal universe (Prakriti, Nature) on its physical, vital or mental plane with a mental, vital and physical 'witness awareness' that observes, but is unaffected by any clash of forces it experiences. *"The Yogi arrives first at a sort of division in his being in which the inner Purusha fixed and calm looks at the perturbations of the outer man as one looks at the passions of an unreasonable child..."* (Aurobindo, 1927-50, pp. 658, pt. IV).

[206] Something along the lines of the "psi-field" theory of psychologist Jim Carpenter (Carpenter, 2015).

This triple system of the outer, inner, and inmost beings is called the *'concentric system of being'* and is summarized in Table 1.[207] It represents the spiritual structure of the individuated being inside a larger cosmic context.

Table 1. The concentric system of being according to Sri Aurobindo.

Prakriti		***Purusha***
Outer being	**Inner (subliminal) being**	**Inmost Being**
Outer mental (sense - and physical mind + emotional or vital mind)	Inner mental (reason, intellect)	Psychic being (Chaitya purusha)
Outer vital (lower vital)	Inner vital (higher vital)	
Outer physical (gross body)	Inner physical (subtle physical)	

Therefore, according to this hierarchy of metaphysical structures of integral yoga, the psychological existence of the ordinary human is a mixture, an interdependent cross-interaction of all these beings. We are many different intermixed personalities at different levels, but we are only aware of those elements that arise to our outer surface nature. Behind our waking consciousness, exists an inner subliminal consciousness, with the subconscient and inconscient below and the superconscient above. But we misinterpret our surface emotions, such as love and hate, joy and sadness, euphoria, and depression, for the true emotions which define us, as that ego or little self which makes up our 'personality', and that we take for ourselves, believing it to be our true essence and nature. *"This chaos we call our soul"*, was Aurobindo's biting conclusion (Aurobindo, 1914-21, pp. Pt. II, Ch. VIII, p. 351). But our true essence is the psychic being, which should not be confused with what we call our 'character', 'personality' or 'ego-sense'. While its distinctive trait is of a divine emotional nature, our surface emotions are determined mostly by our outer vital impulses, mechanical habits, physical and emotional preferences, mental tendencies, unaware prejudices, and convictions. In Aurobindo's integral cosmology, these are only aspects of our surface untransformed personality.

---

[207] The table is inspired by (Teklinski, 2016).

This is an aspect that the modern philosophy of mind still struggles to consider and discriminate, as is vividly exemplified when it comes to the question of whether we have free will. Tomes have been written on this topic but rarely do we question who is supposed to be free? Free from what? And what is will? Is the entity supposed to be free (or not), the inner or the outer being? Is it its vital, mental, or physical part? Or are we speaking about the freedom of a no-better-defined 'soul'? Is the freedom of the lower vital a real 'freedom'? This lack of distinction and discrimination have led, again, to endless debates marked by confusion and inconsistencies which affect Western philosophy on the issue of free will. It might, instead, find a source of inspiration in Aurobindo's less coarse-grained psychology, potentially leading to more clarity and deeper insights.

As a side consideration, it should be briefly mentioned that integral yoga naturally integrates the 'seven chakra system', the spiritual centers of the ancient yoga meditation practices, but diverges in the practice of awakening the dormant 'Kundalini' – the spiritual cosmic energy, the universal 'Shakti' – which resides at the base of the spine and once awakened by the yogic practice, runs along the spine opening the other centers above, allowing us to reach the state of 'samadhi' – merging our subjective identity in the Absolute by coalescing it with Satcitananda. In integral yoga, the awakening of Kundalini is neither a necessary nor an encouraged practice because it relies on the purification and transformation of the being, performed by a spiritual Shakti acting from above downwards.

c. The Psychic Being and Psychic Transformation

Integral yoga aims to transform the outer surface consciousness to the image of the inner divinity by obtaining self-perfection and gaining higher states of consciousness that can lead us beyond our present life-mind nature. Aurobindo called his own process of ascesis, a 'triple transformation': the psychic, the spiritual and the supramental transformation.

The first transformation, 'the psychic transformation', brings the soul, our inmost spiritual individuality, our 'mystical heart' – the psychic being–to the front. It will then take the lead and determine the transformation of all other parts. The qualities of the psychic being are primarily of a higher emotional nature, such as love, beauty and a perception of truth that does not need mental inferences. It has a sense of good, harmony, emotional equanimity, sweatiness, peace, and an intimate relationship with the Divine. *"The pure psychic being is of the essence of Ananda* [Bliss], *it comes from the delight-soul in the universe."* (Aurobindo, 1914-21, p. 737)

In our intimate nature, these qualities are always present but hidden behind the veil of our untransformed mental, emotional, and physical activities. Only once this inmost true individuality has sufficiently developed itself can it come to the surface and impose itself over our outer superficial nature. *"It is one of the decisive moments of the integral Yoga when this psychic being, liberated, brought out from the veil to the front, can pour the full flood of its divinations, seeings and impulsions on the mind, life and body of man and begins to prepare the upbuilding of divinity in the earthly nature"* (Aurobindo, 1914-21, p. 150). At this point, it can begin to govern and transform our whole being, according to higher directives, without the lower impulses acting as tyrant external rulers, working from a direct action from the inside-out, which aims to turn every plane and part of being towards the Divine. This is the psychic transformation that transforms our state of ignorance into a state of knowledge. But, as long as the psychic being remains hidden behind the veil of activities of the material existence made of bodily sensations and appetites, the gross emotional impulses and the mental activities, it is subjected mainly to a passive evolution. Therefore, all that we are (or believe to be), depends on the clash between the outer mental, vital and physical forces and the outer events that escape our control in a domain of ignorance made of alternating pains and pleasures, rest and stress, ups and downs, which are determined by the mechanical aspect of Nature.

In fact, the psychic being is behind the physical heart, residing in the heart chakra. This heart center is also described in different forms throughout many traditions, particularly in Sufism, which speaks of the 'five levels of the heart' and is not alien to the Western mystical teachings that encourage the practice of the 'Prayer of the Heart'.

Aurobindo's psychic being must not be confused with the traditional Indian notion of the '*Self*' of the eternal and immutable Brahman, and which is said to identify with the '*Atman*', or '*Jivatman*', also just '*Jiva*', and which is the individuated reflex of the Self. Non-dual mystics refrain from speaking of a 'soul' and if they do, they do not mean the personality identifying itself with an "I", having a feeling of selfhood. Here, and in Western metaphysics and psychology as well[208], the concept of the soul, if accepted at all, is usually considered a static, passive, and immutable witness, representing an immaterial and spiritual individuality that preserves its existence after physical death. According to this Eastern philosophical tradition, it then continues its journey in a cycle of reincarnations but is never conceived of as a spiritual individuality able to change and evolve through this process of transmigration. The notion of an evolutionary soul is absent.

Whereas Aurobindo contends otherwise. His cosmology integrates this classical perspective, inasmuch that it also recognizes the existence of an evolutionary soul, the psychic being, a pure 'inner flame' that is not just a static

[208] Also, among those forms of Western psychology open to Eastern psychology, for example, the transpersonal psychology of Ken Wilber or the psychosynthesis of Roberto Assagioli, the tendency is to attribute little or no importance to the soul, let alone ascribe to it an evolutionary nature and purpose. For an interesting review of this aspect see the research of Teklinski (Teklinski, 2016; 2019).

and eternal entity; rather, it is the very evolutionary essence of each individual that evolves life after life in the process of transmigration. While both the Jiva – the 'central being' – and the psychic being are a 'spark of the Divine' and are two aspects of the same individuality, the former is non-evolutionary; it is above evolution and projects, or 'individuates' an evolutionary '*psychic entity*' into Nature by what Aurobindo called an 'exclusive concentration' or 'self-limitation' of the Self. Or, taking a bottom-up perspective, it is the Jiva that merges itself in identity with the Divine and knows itself as a center of the multiple Divine, while the psychic being is the evolutionary outgrowth of the psychic entity, our true inmost individuality, that links the impersonal Absolute with a personal identity in a material, vital and physical individuality.

It is this psychic entity which, during its life experiences and its evolutionary journey of transmigrations, forms the psychic being. It grows and gains *"strength every time there is a higher movement in us, and, finally, by the accumulation of these deeper and higher movements there is developed a psychic individuality – that which we call usually the psychic being"* (Aurobindo, 1927-50, p. 103, vol. I).

This evolutionary soul progressively develops out of the nescience of the mineral, first in the plant, then in the animal, now in the human, and finally in the future 'gnostic being'. It is in this sense that the integrality of Aurobindo's metaphysics does not abolish or contradict the past spiritual knowledge, rather, it extends, complements and integrates it.

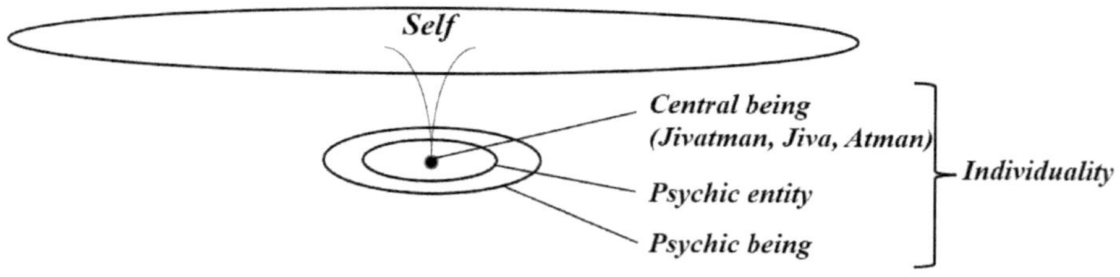

Figure 1. The evolutionary soul as a 'spark of the Divine'.

In anticipation to what we will examine in more detail later, note how here, Aurobindo's mystic cosmology begins to suggest a solution to the 'decombination' or 'individuation problem' that the cosmopsychist encounters. The soul-individuation is not a separation from its Source; rather, it emerges due to an individuation of the cosmic consciousness into itself by a process of 'exclusive concentration' leading to internal differentiations.

d. The Spiritual Transformation

While the psychic transformation is more of a divine emotional nature, the second transformation, which Aurobindo called the 'spiritual transformation', has as its main objective, an ascension towards higher states of cognition beyond the ordinary mental one.[209]

In the Aurobindonian integral scheme, several gradations exist between unconsciousness and supra-consciousness. Aurobindo described the gradations of the 'higher hemisphere', as those surpassing the mental consciousness as a series of sublimations. There is not just mind and a trans-rational mind, but several steps on the ladder of consciousness. He described it in the Life Divine, and named it, "The Ascend towards Supermind" (Aurobindo, 1914-19, p. ch. XXVI), according to an ascending scheme that, at the bottom, places the mind, the conventional reason and analytic mind, still in the lower hemisphere. Only by going beyond the mind we enter into the trans-rational higher hemisphere where one first encounters the 'higher-mind', but also can continue beyond it, towards even higher cognitive states, namely an 'illumined mind', an 'intuitional mind', an 'overmind' and, at the summit, the 'Supermind' or 'divine Gnosis'.

Let us unpack this in more detail.

The higher mind: The first layer above mind is the 'higher-mind'. The peculiar aspects of the higher-mind (which must not be confused with intuition, which we will deal with later) is a large "*clarity of the spirit*" capable of a spontaneous inherent knowledge by a *"luminous thought-mind,"* a mind of *"spirit-born conceptual knowledge."* This kind of cognition is the first that meets us when we rise from a conceptive and rational mind and is considered the first step out of ignorance. It is a knowledge-power that tips into the first layers of a spiritual realm. It no longer proceeds by a stepwise logical process of inference and deduction, but rather, by a greater thought which does not need a seeking ratiocination because *"this limping action of our reason is a movement of ignorance searching for knowledge."* Instead, it proceeds by a *"harmony of significances put into thought-form."* However, this higher-mind is not just a passive tool of cognition. It also possesses an inherent power of action and creation that expresses itself in feelings, in life and even in the body, through the power and force of the idea it presents.

The illumined mind: The higher-mind is only one rung of the ascending ladder towards a full gnostic knowledge. The 'illumined mind' surpasses the vision of the higher-mind, transcending from the higher thought into a cognition guided by a *'spiritual light'*. It is characterized by a "*clarity of the spiritual intelligence [...] and illumination of the spirit: a play of lightnings of spiritual truth*". It is a spiritual-cognitive calm and wide enlightenment that operates not by thought, but by vision. In this spiritual light, there are no verbal

[209] This is not to say that the psychic transformation isn't 'spiritual' in the ordinary sense of the term. Aurobindo distinguishes the two types of transformations as 'psychic' and 'spiritual' because of their distinct source and action: one coming from the heart center, the other from the centers above the head.

representations; it is a consciousness that proceeds by an inner and more direct sight, way beyond any intellectual thought-conception. It is a *"spiritual sense that seizes something of the substance of Truth."* As the higher-mind, the illumined mind is not just a passive container, but brings with its spiritual vision, a light and energy into the heart. It infuses itself in the whole being and transmutes even the physical mind – enlarging it by *"replacing its narrow thought-power and its doubts by sight and pours luminosity and consciousness into the very cells of the body."*

Intuition: At a higher summit, we find the 'intuitional being'. It is *not* about an instinct or the proverbial 'gut feeling'. Intuition, in the terminology of integral yoga, refers to the step that comes close to an original knowledge by identity. The *"consciousness of the subject meets with the consciousness in the object, penetrates it and sees, feels or vibrates with the truth of what it contacts"*. In other words, intuition is the first step from a phenomenal to a noumenal awareness, a phase towards a direct contact and 'seeing' of Kant's 'thing-in-itself'. This intuition, this 'lightning-flash' that emerges from the contact between the subject's consciousness and the object it contacts, is also a cognition of the hidden forces behind the appearances it meets, a first spark of intimate 'truth-perception', a penetrating and revealing touch beyond sight and conception. It is a light of truth, a 'truth-remembrance', so-called because it does not know by learning but, rather, creates a contact with something we already knew to be inherently true. It is an automatic certitude that does not proceed from any reasoned conclusion or verified conjecture. It is a supra-rational source of pure and native intuition, in which the lower intelligence of reason can be only an observer or registrar. Aurobindo describes it as having four powers: a power of 'revelatory truth seeing', a power of 'inspiration' or 'truth-hearing', a power of 'immediate seizing of significance' and a power of 'true and automatic discrimination'. Therefore, it can perform all the action of reason and much more than that. Also, intuition is not limited to a passive cognition of things but takes up and transforms the other parts of the being, namely the mind, the vital and the physical senses, imparting to it a deeper perception, leading us towards a divination of the body and a greater integrality and perfection. Thus, it can change and recast the whole consciousness, from a state of ignorance into the stuff of intuition transmuting our will, feelings, emotions, actions, and the very workings of the body consciousness.

The overmind: However, intuition is still not the true Knowledge by identity. Our subconscient and inconscient basis is too vast, deep and solid to be altogether penetrated and transformed. The next step in the ascent brings us to the overmind, a power of cosmic consciousness. When the overmind descends, the centralizing ego-sense is finally abolished and replaced by a *"wide cosmic perception and feeling of a boundless universal self."* – *"Many motions that were formerly ego-centric may still continue, but they occur as* *currents or ripples in the cosmic wideness"*. Thoughts are no longer perceived as an individual act of cognition of a separate self but realized as 'cosmic Mind-waves'. *"Feelings, emotions, sensations are similarly felt as waves from the same cosmic immensity"* with the body being a mere point of action for a vast cosmic instrumentation. The sense of individuality, even that of our body and mind individuality, entirely disappears and reveals itself as the delightful play of cosmic forces alone.

The overmind consciousness is an unlimited consciousness of unity with the sense of the universe in oneself, or as oneself as a cosmic being. The nature of the being with all its thoughts, feelings, and bodily activities becomes more universal, all-understanding, all-embracing, cosmic, and infinite. This overmind is governed by the directions of the cosmic Self, where the body is a physical instrument recognized as *"something instrumental to the action of a Transcendent and Universal Being"*, a cosmic center of the action of the Infinite. It is experienced as a consciousness of Light and Truth with a sensation of beauty and delight.

Here also, *"all essential experiences belonging to the mind, life, body are taken up and spiritualized, transmuted and felt as forms of the consciousness, delight, power of the infinite existence."* The other forms of cognition and active powers below, those of Intuition, illumined sight and thought, are themselves enlarged by the action of the overmind.

Of relevance for the individuation problem with which the Western philosophy of mind tackles with, is Aurobindo's observation that the overmind is not only a unitary cosmic consciousness but, conversely and complementarily, also the first principle of separate dynamism. It is by the action of the overmind that all dualities emerge from the 'One without a second' by an exclusive concentration. The overmind is a source of creative power in the manifestation of multitudes it has created, and that *"though luminous itself, keeps from us the full indivisible supramental Light, depends on it indeed, but in receiving it, divides, distributes, breaks it up into separated aspects, powers, multiplicities of all kinds"* (Aurobindo, 1927-50, pp. 138, vol. I).

This overmental transformation completes the spiritual transformation.

e. The Supermind

Therefore, the overmind still contains the seeds of a separative and divisive character. Despite its basis in a cosmic unity, its action is one of division and interaction in the cosmic play of the multiplicity. Whereas, at the highest step of the ascending ladder of consciousness, Aurobindo posits the Supermind.

It is the Vijñānamaya kosha of the Upanishads. Aurobindo perceived it as a divine Gnosis or 'Truth-Consciousness', the *"divine Nature in which there can be no place for the principle of division and ignorance; it is always a full light and knowledge superior to all mental substance or mental movement."*

(Aurobindo, 1927-50, pp. 146, Vol. I) The state of consciousness of the future post-human species, the gnostic being.

The Supermind is *"the supreme power of the principle of unity taking all diversities into itself and controlling them as parts of the unity, which must be the law of the new evolutionary consciousness"* (Aurobindo, 1914-19, p. 989, pt. II, ch. XXVI). It is only the Supermind, the *"supreme self-determining truth-action and the direct power of manifestation of that Transcendence"*, that can accomplish the final transmutation – the supramental transformation. While the overmind allows the individual consciousness to be universalized and, to some degree, transform the lower parts of the being, a basis of untransformed nescience in matter remains, which only the Supermind can transform completely. The transformation of the inert depths of the sub and inconscient is possible only by the descent of the supramental Light into matter, life, and mind, penetrating down to the material basis.

The teleological character of Aurobindo's vision is that matter will reveal itself as an instrument of the self-manifestation of Spirit. The Supermind is the highest level of consciousness, a 'Truth-Consciousness', 'Truth-Knowledge', or 'Knowledge by identity', which would be proper to the new species, the gnostic being.

The supramental vision brings us back to our original unity in creation–a holistic unity, in which nothing is separated and where all is One. *"It sees everything from the stand-point of oneness and regards all things, even the greatest multiplicity and diversity, in the light of that oneness"* (Aurobindo, 1914-19, p. 1000, bk. II, ch. XXVII). The supramental sense is that of an ultimate Unity, Oneness and Wholeness of the Infinite, where everything we perceive at our ordinary mental level is like a hologram, a *"drop that is yet a concentration of the whole ocean and inseparable from the ocean"* (Aurobindo, 1914-21, p. 866, pt. IV, ch. XXIV). Not the unity of the mind, but a greater sense of Unity inherent in all things, a self-manifestation of the only true Thing.

In the supramental consciousness, one becomes aware of all phenomenal events in Nature and the cosmos as a result of a creative Idea, the 'Real-Idea' or 'Seed-Idea', a power of Consciousness expressive of the real Being, the Divinity in manifestation. The Supermind is the Truth-Consciousness that knows itself and sees the universe as a subordinate reality, resulting from an overmental creative process of differentiation of its own Real-Idea. The Real-Idea is the ultimate Archetype that the philosopher captures as the hidden, yet clearly perceptible, archetype and 'idea-form' in things and Nature.

Aurobindo subverts the conventional understanding of how the mind works. Mental processes are not only determined by a bottom-up mechanism that makes the thought, the mental construct, and the mental idea emerge as a combination and interaction of elementary neural units. It is a top-down process, in which a divine intuition percolates through the layers from the

Supermind down to the mind and, thereby, is reduced, filtered, and eventually twisted into a mental image, meaning that it has lost most of its original insight and knowledge by identity, retaining only a relative truth of things.

> *The fundamental nature of this Supermind is that all its knowledge is originally a knowledge by identity and oneness and even when it makes numberless apparent divisions and discriminating modifications in itself, still all the knowledge that operates in its workings, even in these divisions, is founded upon and sustained and lit and guided by this perfect knowledge by identity and oneness. The Spirit is one everywhere and it knows all things as itself and in itself, so sees them always and therefore knows them intimately, completely, in their reality as well as their appearance, in their truth, their law, the entire spirit and sense and figure of their nature and their workings. When it sees anything as an object of knowledge, it yet sees it as itself and in itself, and not as a thing other than or divided from it about which therefore it would at first be ignorant of the nature, constitution and workings and have to learn about them, as the mind is at first ignorant of its object and has to learn about it because the mind is separated from its object and regards and senses and meets it as something other than itself and external to its own being.* (Aurobindo, 1914-21, p. 786, pt. IV, ch. XIX).

Not a knowledge that proceeds by logical steps, inferences, or hypothesis to be tested; these are the inferior methods of reason, of an *'inflexible machinery of the rational intelligence'* that is a form of cognition, proper to the domain of an obscure ignorance and that is compelled to search for truth by an inductive process or piecewise analytic procedure, revealing the truth of things only partially. The supramental gnosis instead directly reveals the truth of things by an immediate, direct, and inherent contact. All that is mental appears as a deformed, suppressed, pale and partial imperfect figure of that original divine cognition.

Moreover, the supramental consciousness has a completely different time perception, in which past, present and future are seen unitarily, a single continuous map of knowledge. This *'three-fold seeing'* of time (*'trikaladrishti'*) that links past, present and future as an indivisible whole (more on this later).

Western philosophy, especially its modern incarnation of analytic philosophy, emphasizes analytic reasoning, rational analysis, and logical inferences, eventually augmented by mathematics, empiricism, and formal logic, as the mental tools for attaining a knowledge that is supposed to bring us closer to the truth. However, as is well known, strict adherence to this principle which the logical positivists, such as Frege, Russel, Wittgenstein, or Carnap tried to apply to its extremes, didn't succeed. The revival of phenomenological approaches, philosophical idealism, and the rediscovery of past metaphysical thought could not be stopped. Nevertheless, the Age of

Enlightenment has enthroned the mind itself as the ultimate tool beyond which nothing more is and can be said. This underlying assumption is what also conditions the modern philosophy of '*mind*' (the terminology already betrays it). It is all about mind and matter, eventually the subconscious and intuition, but always studied from the frame of reference of the ordinary human consciousness, not much more. And yet, there seems to be an ineliminable aspect of reality that refuses to be cast into analytic terms alone. Aurobindo's answer is that the physical and the mental are only two aspects, the thin surfaces of a cosmic reality going far beyond them, converging towards a supramental Gnosis. If we would enlarge our perspective to this integral vision of mind, life, and matter, a different and, perhaps, much deeper understanding of reality might be possible.

f. Satcitananda

Finally, crown of the higher hemisphere is the Satcitananda of the Indian Vedic tradition, the Absolute, the ultimate reality that is the Brahman of the Advaita Vedanta. It is the state of consciousness in which rests the 'enlightened' and that words cannot describe other than saying that it represents a triune state of Existence-Consciousness-Bliss. It is the Anandamaya kosha, the 'sheet of bliss' in which the 'seer' is established once the highest state of consciousness is attained. The practitioner realizes that, ultimately, all things are the ultimate Subject, the supreme Self, the spaceless, timeless, and unchanging impersonal Absolute, the non-dual Transcendent. It is the Buddhist 'emptiness', 'vacuity' or 'sunya'. Therefore, once realized, this leads to 'liberation', 'moksha', or 'nirvana', the release from suffering due to the absence of mind's activities. It is a common theme in the Asian tradition and Aurobindo's integral yoga makes no exception.

This is the ascending ladder of the higher hemisphere of consciousness according to Aurobindo's scheme.

g. The Vertical System of Being and Cosmic Consciousness

To summarize, Diagram 1 complements Table 1 by displaying the full spectrum of the vertical system of consciousness in Aurobindo's cosmology.

Note that this is not just a structure of consciousness related to an individual being. It is a cosmology of consciousness applied to the universal planes. Like the psychic being that connects the transcendent with the individual, the Supermind links the unmanifest Divine, the Satcitananda, the Brahman or the Vedantic Absolute beyond space and time, with its immanent and dynamic cosmic universal manifestation. The Supermind is, so to speak, the delegate of this Absolute in the manifestation determined by space, time and causality. The physical universe we are aware of in our ordinary mental consciousness is only an aspect of this self-manifestation. We can generally speak of a cosmic consciousness incorporating a cosmic mental, a cosmic vital and a cosmic physical, but also a cosmic subconscient and inconscient. Thus, the Divine is both transcendent and immanent.

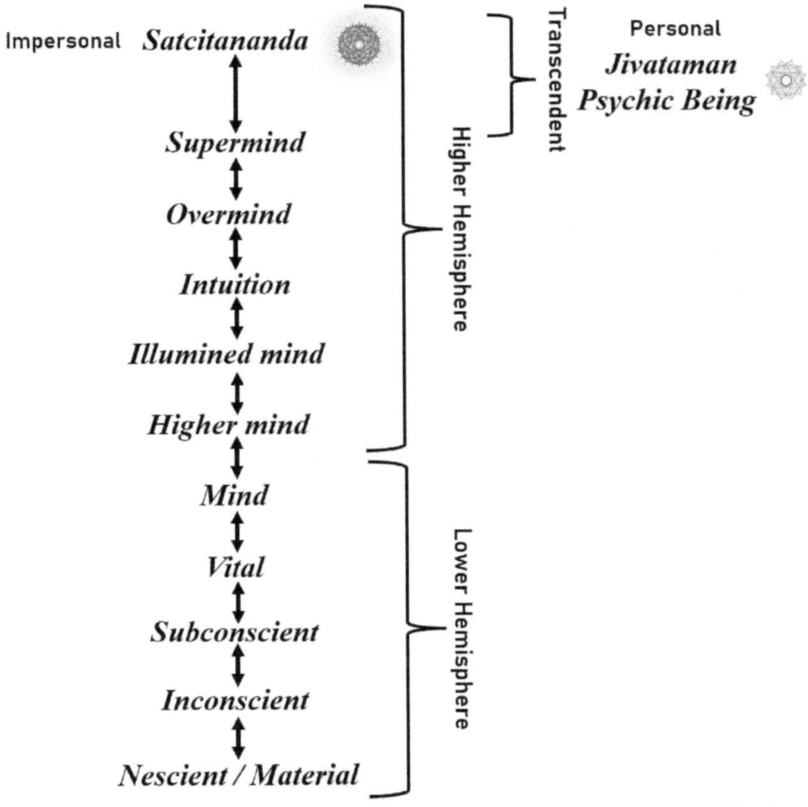

Diagram 1. The vertical system of the planes and parts of being.

Perhaps the most difficult aspect to accept for some, especially the scientifically minded Neo-Darwinist, is that, for Aurobindo, all these planes of existence are *not* the creation of an evolutionary process. They preceded the creation of earthly life and even the physical universe itself. This is to say, for example, Aurobindo does *not* consider the mind a form of cognition resulting from a biological evolutionary process. Mentality and life emerge *in* the material world – the cosmic physical – but are not created by and do not originate from it. In this integral vision, the terrestrial physical life is also the outer manifestation of a life-force that pre-existed, in a more subtle form, the material universe itself. This amounts to suggesting that a cosmic life-plane and a cosmic mind plane were already in existence before the Big Bang.

The planes of consciousness and parts of being according to integral yoga are more complex than the structure we presented here. However, this was intended as an overview, which will suffice for our purposes. For a more detailed description of Aurobindo's huge 'topography of consciousness', the reader will have to resort to his writings, especially (Aurobindo, 1914-19; 1914-21) and (Aurobindo, 1927-50) or, for a specialized redux, see also (Cornelissen, 2018).

4. Evolution, Spirit and Matter

1. Involution and Evolution

The word 'evolution', from Latin 'evolvere', means to 'unroll' or 'open out' or 'uncover'. Interestingly, it is etymologically more fitting to Aurobindo's metaphysics than to the modern Darwinian accepted notion of a descent with modification. Because it implies a process that makes something emerge, which is inherently contained in the thing that 'unrolls', by an ever increasing complexity of material organization. It seems to suggest that there is something already fully developed but concealed or hidden, that evolution unveils–that is, something that was previously veiled by an opposite process of evolution, namely an 'involution'.

In fact, the peculiarity of Aurobindo's spiritual emergentism is that it is complemented by an opposite 'immergentism'. In Aurobindo's cosmology all the physical manifestation is an involution of a Super-Consciousness that subjected itself to an oblivion, what he called a 'sacrifice' of the Divine, by plunging itself into a sort of 'self-absorbed' or 'self-concealed' and 'masked' still divine consciousness, but latent and 'asleep' in the inconscient matter. *"Matter is a form of veiled Life; Life a form of veiled Mind"* (Aurobindo, 1910-50, p. 502). It is the One that has self-involved itself in the depth of a descending ladder of consciousness. Thereby, matter, which is inherently still an expression of an original divinity, nevertheless manifests a slumbering but indwelling spirit as a mechanical and inconscient substance, in a gross material universe–in and as 'cosmic ignorance' – and from which it is re-emerging ('unrolling') by an evolutionary principle.

On one hand, there is an involution of a superconscient Spirit descending from the higher planes into the planes below, now growing by a bottom-up process as an inconscient spirit in matter, and unfolding itself from a nescient state to progressively higher states of consciousness. On the other hand, the lower planes are also immanently guided by the action of a top-down process of the same superconscient Spirit from above. The mechanistic aspect of the physical universe is not something from which consciousness emerged by a combination of unconscious elements, it is itself the outer expression of an involved consciousness and which, due to this self-forgetfulness, has become mechanical and automatic. The forces that drive this evolutionary process come from below, in the form of a material process that conditions it from the outside by the laws of physics and by the known Darwinian evolutionary principles of natural selection, genetic drifts of (apparently) 'random' mutations, environmental factors, adaptation, survival and reproduction of the fittest, etc. But, at the same time, there are other types of forces that come from above and guide the process from within by an inherent consciousness–more or less directly or indirectly according to the plane of existence, in contact with that original superconscience.

In Aurobindo's vision the universe resulted from a process of individuation and fragmentation, in which matter was the last appearance, not the first basis of existence. Here, mind is not a creation resulting from an epiphenomenon of an aggregate of matter. Rather, mind is already involved in it and re-emerges – it 'e-volves' – out of matter during the evolutionary process once the physical aggregate is ready to hold it. While the Supermind has to hide itself acting only indirectly, behind thick layers of gross substance, nevertheless, it works out our earthly conditions.

However, mind, overmind, Supermind, the vital plane, let alone the Absolute, do not evolve, they are ever-present realities on their own and are not subjected to any evolution (Aurobindo called it 'typal worlds'). These universal planes emerge in time, one by one, on the physical plane. It is their progressive emergence in and through matter that we call 'evolution'.

But what is this divine Consciousness trying to achieve in the material existence? In the emergentist vision of Aurobindo, it is the 'delight of being' that seeks to realize itself as 'delight of becoming'. Delight of being is the intimate nature of Satcitananda that is no longer superconscient, but subconscient in matter and seeks in mind and life to realize itself by emerging in the becoming. At bottom also a protozoan seeks an intimate, concealed, and subconscious aim and urge for the finding of its original supreme delight of being in the flux and cycle of life-experiences, which expresses itself in what we call the 'instinct of survival'. *"Delight of being is universal, illimitable and self-existent, not dependent on particular causes, the background of all backgrounds, from which pleasure, pain and other more neutral experiences emerge"* (Aurobindo, 1914-19, p. 106, pt. I, ch. XI).

We all know from first-hand experience how strong the survival instinct is. It is one of the foundations of evolutionary biology and is observed not only in the animal kingdom but in plants and even single-celled organisms. As we will see next, even a single cell has some sort of 'physical mind' and behaves–that is, reacts, swims, hunts for its prey, avoids obstacles, has a memory, and can even predict events in advance–as if driven by some form of 'basal sentience,' 'basal desire,' or 'basal will' for the delight of existence, which we call 'instinct of survival.' The materialist is tempted to naturalize this behavior, downplaying it as a complicated adaptive mechanism driven by natural selection, one that we should not anthropomorphize. However, while science remains mute regarding why this instinct of survival developed in an apparently life-less and instinct-less universe in the first place, Aurobindo's natural ontology, positing the delight of being as a universal and fundamental trait of existence that precedes life itself, might not only remain a viable option but even have a better explanatory power for the behavior of living matter. It can close the explanatory gap separating a mechanistic and apparently unconscious and purposeless cosmos from the 'miracle' of consciousness and sentient life. In Aurobindo's cosmology, the indifferent God of Spinoza, the blindly striving Will of Schopenhauer or, more recently, Kastrup's unreflective and 'instinctive Mind at large', acting only on the basis of a blind will, though capable through a long evolutionary process to grow in its self-awareness (Kastrup, 2019) are, at best, the cosmic vital subjected to the cosmic unconscious, not the ultimate universal Consciousness. They captured a particular aspect of an omniscient and omnipotent Divinity on the lowest planes where it appears to be an involved instinctive universal will. It is the finding of an absolute, eternal, infinite Self-existence, and Self-awareness, but timeless and unmanifest Self-delight of being in its self-involved temporal manifestation, what secretly supports the universe and is its hidden truth, force, motivation, aim and goal.

2. The Supramental Transformation

In Aurobindo's yoga the psychic and spiritual transformation are still not the final consummation of the evolution of spirit out of matter. The realization of the supramental consciousness is old as the Indian Vedic text. The ancient rishis called it the 'Sun of Truth'. But realizing the supramental consciousness is still an inner realization, not a physical transformation of the species. The ascent of the mental being who climbs up to the Supermind might realize a universalized awareness and live in a transcendent bliss, but this does not lead to a descent of the Supermind in matter, which, according to Aurobindo, is the only that can effectuate a complete transformation and will bring into matter the full dynamism of the spirit.

It is this final integral transformation where the spirit transmutes matter, without leaving aside the material existence, that can be called a 'divine materialism'. It acknowledges the fact that, not a spiritual mystic realization is the true meaning and purpose of our existence, but that matter is the basis and instrument for the full self-manifestation of the Spirit.

Aurobindo, and later also his spiritual partner Mirra Alfassa, tried this ultimate and most difficult transformation, namely, to bring down the Supermind in their bodies – a transmutation process in and of the cells – by what they called 'cell yoga'. The secret of matter relies precisely in its concreteness and stability; it is only its solidity that can hold the full power of the divine Supermind, which would transmute the physical body to its image and even convey physical immortality. The most extreme contraries, that of matter and spirit, life and death or omniscience and complete ignorance, can be reconciled only by the unitary power of the Supermind. Aurobindo and Alfassa left a huge documentation which described not only the inner states of consciousness they attained, but also the first intermediate steps towards a physical transformation. The gnostic change will also lead to the supramental transformation in the cellular structure which, in the previous evolutionary stages of Nature, was effectuated almost exclusively by the environment and the blind clash of material forces, as conceived in the frame of Darwinian evolution. This is the difference between human and animal. The appearance of mind is an intermediate necessary step in evolution because it endows the human species with the ability to participate to this process, consciously and actively.

But almost all what our body is and does is ruled by the laws of a subconscient or inconscient plane. The difficulty resides in making these obscure lower planes of existence become conscious and subject to the supramental control, light, and action. The aim is to bring the Idea and Will-Force of the divine Power of the Spirit in direct dynamic action on all levels – in our mind, emotions and down to the still subconscient, which is the cause of an ignorance dominating our existence.

Therefore, what distinguishes Aurobindo's integral yoga from other yoga or spiritual practices is that it not only effectuates a mystic awakening or a spiritual transformation to realize some extra-cosmic reality, but also aims at an integral transformation of our subconscient and inconscient parts down to the very physical. It is about the awakening of a 'body-consciousness' which, at this evolutionary stage, is still dominated by the inconscient in the cells.

It is such a change and such a reshaping of life for which humanity is blindly beginning to seek, now more and more with a sense that its very existence depends upon finding the way. The evolution of mind working upon life has developed an organization of the activity of mind and use of matter which can no longer be supported by human capacity without an inner change. (Aurobindo, 1914-19, p. 1092, XXVIII)

Aurobindo rejects the view of the classical spiritual teachings that consider the physical transformation of secondary importance. The polarity of matter and spirit must be reconciled. The spiritual transformation also aims to an exterior and material metamorphosis, this being the real reason for the manifestation: the fusion of spirit with matter. It integrates the spiritualism of the East with the materialism of the West, completing, widening, and transcending both towards a synthesis, where both appear to have their relative significance.

Thus, from Aurobindo's perspective, the dichotomy between consciousness and matter, or mind and body, which has puzzled Western metaphysics so much, dissolves into a complementary polarity. In a sense, the panpsychist is right: Matter is a form of consciousness, but self-forgetful and nescient. While the body is not just a blind mechanistic material aggregate: It is ruled by a subconsciously awakened mind, something that, if true, could have far-reaching implications, not only for our philosophical worldview of the relationship between mind, body, and matter but for biology and medicine as well.

c. The Mind of the Cells

Aurobindo was not too explicit about this last aspect of the integral ascesis. What is known is that he went into seclusion for 24 years until his death. Aurobindo claimed that he was trying to attain a difficult spiritual and material transformation in his body. This cellular work was taken up by Mirra Alfassa, the "Mother". She also was largely hesitant to speak openly about something that had to yet be accomplished and which success or failure was far from clear. Nevertheless, one of her disciples, Bernard Enginger, also named Satprem, who was frequently in correspondence with her, collected a 13 volumes series from 1962 to 1973 entitled, "The Mother's Agenda" (Alfassa & Satprem, 1983), which held several accounts outlining her exploration into body consciousness.

The picture of life emerging from these last stages of integral yoga is that of a consciousness in each biological cell at work through a kind of 'mind in matter' or 'mind in the cells'. But this mind in the cell, in its still untransformed inconscient state, is mostly dependent from a reactive, unconscious, and subconscious instinct that can sicken us, cure miraculously or let us die suddenly. From the writings of Aurobindo and Alfassa, it appears that the biggest obstacle to such realization showed to be the subjugation of the lowest layer of consciousness to the supramental light. In the words of Alfassa:

I went down into a place ... a place simply in the human consciousness, thus necessarily in my body ... I have never seen anything more timorous, fearful, feeble and mean! It's ... it must be a part of the cells, part of the consciousness, something that lives in apprehension, fear, dread, anxiety ... It was truly, truly dreadful. And we carry that within us! We aren't aware of it, it's almost subconscious – for you see, the consciousness is there to prevent us from yielding to that – it's cowardly, and it can make you fall sick in a minute. (Alfassa & Satprem, 1983, p. 484, Nov. 5, 1960).

Aurobindo's cryptic poetical verses suggest he already had similar experiences:

I have been digging deep and long
Mid a horror of filth and mire
A bed for the golden river's song,
A home for the deathless fire.
I have laboured and suffered in Matter's night
To bring the fire to man;
But the hate of hell and human spite
Are my meed since the world began.
(Aurobindo, 1930-1950)

Aurobindo contends that once the Supermind acts in the cells, even death is not inevitable but reveals itself to be a sort of subconscious cellular 'habit', an 'accident' that must not remain forever part of the eternal law of life. But immortality is not the purpose of integral yoga, it is only a consequence, an intermediary step towards a realization that should be far more reaching and decisive. This is not that kind of immortality that, maybe one day, will establish genetic engineering or medicine, nanotechnologies or transhumanism. Instead, it will be the direct consequence of a state of consciousness of an inside and outside perfection.

However, the inconscient of the cells, this nescient abyss hides a secret. While Aurobindo described *"the dumb occult consciousness ... which operates in the cells."* (Aurobindo, 1914-19, pp. 762, pt. II, ch. XIX), he added also an important note, namely that *"Almighty powers are shut in Nature's cells"* (Aurobindo, 1950, p.370, bk. IV, canto III).

Something reminiscent of Haeckel's 'biopsychism' (to distinguish from panpsychism), where all living matter, thus including unicellular organisms, has some form of sentience and mental properties (this has also been reconsidered more recently (Fermin, 2020)). Interesting to note in this regard are the several findings that have led to the current debate in biological sciences over whether we should associate, with plants and cells, a 'basal cognition.' There is now an increasing body of evidence showing that some form of 'basal cognition' is already present and working in plants and even single biological cells (for a scientific review of basal cognition, see (Gershman & al., 2021; Lyon & al., 2021; Segundo-Ortin, 2021). Plants and cells are capable of associative learning, memory, and anticipation – that is, they have some elementary form of a 'physical mind.' How so, if there is no brain, no central nervous system, and no neurons at work? We have the right to conjecture that mind is already present in some involuted and concealed form in matter (for more on this, see Masi, 2022b).

In a famous passage in her diary Alfassa explains that *"at the very bottom of the inconscience most hard and rigid and narrow and stifling, I struck upon an almighty spring that cast me up forthwith into a formless, limitless Vast, generator of all creation"* (Alfassa & Satprem, 1983, p. 246, Nov.8, 1958).

Which means that, that according to this integral cosmology, deep into the core of matter, there is the same supramental vibration below that one has realized above. The hermetic 'as above, so below' acquires its full significance: at the bottom of nescience, one finds the power of omniscience. An omniscience that lost itself in a cosmic self-forgetfulness finds back to itself. In that physical state the body experiences the formidable pulsations of the Eternal: *"all the results of the Falsehood had disappeared: Death was an illusion, sickness was an illusion, ignorance was an illusion – something that had no reality, no existence.... Only Love, and Love, and Love, and Love – immense, formidable, stupendous, carrying everything"* (Alfassa & Satprem, 1983, Apr. 13, p.148, 1962).

It is the second evolutionary transition, no longer from matter to life, but from life to a life divine.

d. The Gnostic Being and the Life Divine

Thus, following Aurobindo, the final aim of the evolutionary process, or at least the most distant aim actually visible, is the realization of a superconscience in matter. A superconscience which intimate nature and character will be that of the unity in diversity, where the individual consciousness will coalesce into that of the collective. Yet, maintaining its individuality and divine personality, existing in a superhuman supramentalized form, knows through a knowledge by identity and lives in a state of 'overlife'. What Aurobindo envisaged as a new species, beyond the human, the gnostic being.

Teilhard de Chardin's Omega Point is palpable here. The goal of the evolutionary impulse, guided by a spiritual energy from within, will lead the noosphere to the 'planetization' of consciousness.

It will be a species that will progressively unfold the Divine on Earth, with a complete self-knowledge, capable of a divine delight of self-expression, fully aware of its individual and cosmic personality and impersonality. Its action will not be guided by the trails and errors of mind, but by an instantaneous unfiltered divine knowledge, the divine gnosis. According to this vision, there will be no contradiction between materiality and spirituality, the inner and outer life, the universalized individuality, and the world. Because falsehood, disharmony, half-truths, or any form of ignorance, pain and suffering will be 'de-realized' by the supramental Light. Illness, sickness and physical, emotional, or mental suffering will be replaced by the supreme Love, Beauty, Harmony, and perfect effortless and flawless action of a Consciousness-Force. Death will be recognized for its service and meaning, but also for its falsehood that will have been eradicated from the cells. The gnostic being will be immortal, physically, and cognitively perfect, an image of the Spirit in physical form. It will live in the eternity of that past, present and future and the unity of this individuality, as the same unity in diversity of a gnostic collective, expressing an innumerable Oneness.

The mechanical workings of Nature will be replaced by a Truth-Knowledge, Truth-Sight, Truth-Feeling, Truth-Will, Truth-Sense and a Truth-Dynamism, as the spontaneous and normal states of consciousness and the inherent operational activity of the gnostic being. A divine Will, expressive of a free and spontaneous self-unfolding of the Spirit. A divine Life upon Earth.

The gnostic life will exist and act for the Divine in itself and in the world, for the Divine in all; the increasing possession of the individual being and the world by the Divine Presence, Light, Power, Love, Delight, Beauty will be the sense of life to the gnostic being." – "Love will be for him the contact, meeting, union of self with self, of spirit with spirit, a unification of being, a power and joy and intimacy and closeness of soul to soul, of the One to the One, a joy of identity and the consequences of a diverse identity. It is this joy of an intimate self-revealing diversity of the One, the multitudinous union of the One and a happy interaction in the identity, that will be for him the full revealed sense of life. Creation aesthetic or dynamic, mental creation, life creation, material creation will have for him the same sense. It will be the creation of significant forms of the Eternal Force, Light, Beauty, Reality, – the beauty and truth of its forms and bodies, the beauty and truth of its powers and qualities, the beauty and truth of its spirit, its formless beauty of self and essence. (Aurobindo, 1914-19, p. 1020, pt. II, ch. XXVII)

Aurobindo contends that this will not only be possible but the inevitable destiny of life on Earth.

5. Towards an Integral Cosmology

The previous chapter was an introduction to the metaphysics and vision of Aurobindo. It is a cosmological and spiritual framework which, in its main traits, was outlined in the 1920s. His real work, however, came to the surface only after his departure and is best recognized in the accounts of cell yoga, which were described in the Agenda, thanks to Satprem's transcriptions, appearing only in the late 1970s. Until then, the full dimension of their vision had not come to light.

What I would like to do in this chapter is to reframe, in an integral philosophical understanding, the cosmic reality, including matter and spirit in a single and extended evolutionary vision and that is, I contend, in line with modern science and more coherent than other metaphysical frameworks based on models of a universal consciousness. Special emphasis will be on aspects of teleological nature and final causes in evolution, seeing how they might furnish an answer to longstanding questions, which previously seemed to be irreconcilable inside an exclusively material- or a mental-monistic paradigm.

a. From the One to the Many, the Personal and Impersonal and the Nature of Mind and Matter

One might call physicalism–material monism – and the current forms of rationalistic and materialistic reductionist naturalism, a 'one-dimensional cosmology'. Similarly, dualism, as the word suggests, can be seen as a 'two-dimensional cosmology', where mind and matter have their distinct reality. Even panpsychism, cosmopsychism and analytic idealism do not go much further than a one or two-dimensional mind-matter perspective. Also, the Vedantic non-dual tradition, even though originally inspired by a multi-layered Vedic cosmology, is perhaps the most extreme form of monism which emphasizes its one-dimensional view ('all is Brahman').[210] But, most

[210] In this context, the term 'dimension' shouldn't be confused with a mathematical or physical dimension!

importantly, these metaphysical paradigms still lack a coherent evolutionary cosmology, ignoring the evolutionary aspects of life. Whereas, Aurobindo integrates these previous ontologies, but extends it to an integral multidimensional cosmology articulated by the universal planes and parts and augmented by an evolutionary vision.[211]

This extended and evolutionary paradigm might be helpful to see under a different light some difficulties of the contemporary philosophy of mind and potentially suggest some directions toward its resolution.

For example, as well known, panpsychism and theories of universal consciousness suffer from the infamous combination or decombination problem, respectively. If even elementary particles have experiences, as panpsychism contends, how could these microscopic subjects combine to constitute a macroscopic subject? On the other hand, in the present integral cosmology based on an ontology of universal consciousness, the latter issue arises. Namely, how did a transcendent timeless, spaceless and featureless Absolute – the Vedic nirguna-Brahman – which is supposed to be the non-dual ultimate unity, become a multifold polarity manifesting itself as an extremely fragmented and particularized cosmos, with material objects made of atoms, molecules and particles, and the appearance in the evolutionary process of an individuation of consciousness–that is, minds and bodies? Or, rephrasing this in terms of the decombination problem, how can the one Subject become many subjects with its own distinctive self, yet retain its own subjectivity?

Kastrup, who was largely inspired by Schopenhauer and his vision of the world as Will (Kastrup, 2020), submits that in his universal idealistic model, all living organisms are 'dissociated alters' of the Mind at large. These alters, or 'discrete centers of self-awareness', arise due to some dissociative process that psychiatrists know as 'dissociative identity disorder' (Kastrup, 2018).

The integral cosmology describes a similar process of universal 'self-limitation', something that was pointed out recently by Swami Medhananda's comparative analysis of Aurobindo's cosmology with cosmopsychism (Medhananda, 2022). For the details of how this process of self-limitation solves the decombination problem, I defer the reader to his paper. Also, Shani and Keppler contend that there is in principle no insurmountable individuation problem in the context of cosmopsychism (Shani, 2022).

Here I will emphasize that, in Aurobindo, this complementarity between universalization and individuation has a multi-dimensional character. The decombination or individuation–that is, the apparent fragmentation of the undifferentiated Brahman appears on all the universal planes: into 'material quanta' (particle, atoms, molecules, etc.), 'vital quanta', 'mental quanta', and

---

[211] This may sound suggestive of Gebser's multi-dimensional 'structures of consciousness'. The analogies are not completely coincidental: the late Gebser was inspired by Aurobindo's writing.

'psychic quanta'–in form of 'divine sparks', Jivatman, psychic entities, and the psychic beings, the latter resulting from evolutionary processes.

While Chalmer's notorious 'hard problem of consciousness' (Chalmers, 1995) is naturally resolved by positing consciousness as an inherent and fundamental aspect of reality; the decombination problem needs no explanation as well, because self-differentiation is already an inherent and fundamental principle of the cosmic Mind itself. With its power of division, discrimination and separation, the overmind creates particularizations, from particles to self-aware centers.

Aurobindo points out the double aspect of the overmind. On one hand it links the planes of the unitarian supramental Truth-Consciousness, which inherent nature is always that of a perfect Unity and Oneness with the planes of polarity and multiplicity. But, on the other hand, it is also the in-between mediator between Unity and division, between the supramental One-Idea and its expression in infinite possibilities, it is that which breaks down the transcendental Unity into the phenomenal multiplicity. It is the first original power of distinction which, by a self-limitation pinpoints the Infinite into a finite object, making our universe appear in its atomized and particularized form in infinitesimal material points, and yet following a principle of unity in diversity, acting creatively with novel and diverse creations in the physical and temporal manifestation. A cosmic Mind, which dwells in its self-created atomic existence and which, by a creative process of associations of these infinitesimal points, carries out a play of forces with the Real-Idea it receives from above. Precisely in this resides the source and origin of the world of dualities, created by the primal self-demarcating act of separation, division, and atomization, from an original lost Unity, which characterizes the material domain we are accustomed with, and yet is able to put everything in its right relations, despite an enormous complexity of particularities. In Aurobindo's words: *"Thus not any eternal and original law of eternal and original Matter, but the nature of the action of cosmic Mind is the cause of atomic existence. Matter is a creation, and for its creation the infinitesimal, an extreme fragmentation of the Infinite, was needed as the starting-point or basis"* (Aurobindo, 1914-19, p. 252, pt. I, ch. XXV).

While, mentality is, indeed, one of the cosmic aspects that pervades the entire universe, minds are 'exclusive multiple concentrations' of the cosmic Mind, or 'universal Mind', which is a derivative power of the overmind. Or, to put it in other words, in the 'aurobindonian' cosmology, our mind is a minor, reduced and localized form of cosmic Mind in an organic embodiment, mimicking its functions by a diminished power.

The same fits also on other cosmic planes. On the physical plane, matter also is a concentration of consciousness, a 'self-limitation' or a 'self-compression' of the Spirit in 'energy-substance', but passive, absorbed and forgetful of itself and its overmental origin. It is here where there is an intersection with the mental monism of philosophical idealism. In a certain sense, the philosophical idealist who declares that all is mental has its merits. Huxley's and Kastrup's Mind at large is an intuitive perception of the existence of this cosmic Mind and the dissociated alter is, indeed, in line with the notion of integral cosmology as an 'Infinite's self-concentration'. But, in Aurobindo's cosmology, the mental aspect of reality is only a particular appearance of an immanent overmind, which goes well beyond mentality alone. In his multi-layered reality the world appears to be only physical or only mental because we are still only physical and mental beings. If we rise above the mental, its higher mental, illumined mental, intuitional, overmental and finally supramental aspects would become evident as well. It is this that makes the material universe appear to be amenable to mental cognition, as our mind is also a derivative cognitive power of the overmind.

Seeing things from this perspective, might help us to find an answer to other questions of the philosophy of science.[212] If our mind is a diminished self-limiting and involved power of the overmind, and the material universe has been organized according to overmental principles that originate from a the same cosmic Mind, it should, therefore, come as no surprise if our mental sciences are able to decipher a good amount of the cosmos' workings in terms of mental representations: when we look at the world with the mind, we see its mental aspect. The rationality and intelligibility of the world mirrors the immanent cosmic Mind that lies behind all the dynamical processes, the play of forces in space and time of the material universe, with our mind being a localized reflection of it. It is this missing link that led Einstein to say that *"the eternal mystery of the world is its comprehensibility"* and *"the fact that it is comprehensible is a miracle"* (Einstein, 1936). The paradox arises only if we assume the world to be soulless, purposeless, without meaning and, especially, mindless. Galileo said *"the book of Nature is written in the language of mathematics."* The universe itself is an 'emanation' from overmind, and which, while maintaining a contact with the supreme formless, featureless and undivided Consciousness, its inherent nature is nevertheless that to fragment, demarcate and separate by causality in space and time–it works out quality, quantity, number, structure, space, relations, and change. Abstracting from quality, mathematics is precisely the study of quantity, number, structure, space, relations, and change; that's, ultimately, what number theory, arithmetic, algebra, geometry, and calculus entail. This is the reason for the apparently *"unreasonable effectiveness of mathematics in the natural sciences"* that physicist Wigner famously wondered about in 1959, triggering an endless debate that continues into present day (Wigner, 1959). What appears to us an 'unreasonable miracle' becomes something natural once we recognize how our minds, being an individuation of a universal self-limiting Mind, work by the

[212] The analogy suggested in this paragraph is the author's one, it is not present in Aurobindo's writings.

same principles of division and fragmentation and through which we look back in its cosmic physical appearance, which was originated by the cosmic Mind itself.[213]

The same fits for a cosmic Life, or a universal Life, and which already contains life as an inherent cosmic reality, rather than an emergent purely material epiphenomenon of matter. As the action of cosmic Mind is the cause of atomic existence, so is the action of cosmic Life the cause of the 'life quanta', the cell, and the individual life existence. Not the other way around. Starting from this perspective, Aurobindo claims that the brain is not the origin of our mental activity, but it is the creation, instrument, and channel of a universal Mind in the particularized expression of an earthly individualized embodiment. Our brain and body are a derivative expression that reflect an original preexistent and immanent cosmic Mind and cosmic Life respectively on the cosmic physical plane: *"[...] the brain is not the creator of thought, but itself the creation, the instrument and here a necessary convenience of the cosmic Mind"* (Aurobindo, 1914-19, p. 252, pt. I, ch. XXIV). Huxley's idea of the brain being a 'reducing valve' of a Mind at large came close to this view. According to Huxley our mind filters reality under normal conditions because, otherwise, we would be overwhelmed by the knowledge of this universal Mind. However, psychedelic drugs can remove the filter and bring us into contact with the Mind at large, leading to the experiences that several mystics describe. In his words: *"To make survival possible biologically, Mind at large has to be funneled through the reducing valve of the brain and nervous system"* (Huxley, 1954).

Yet, Aurobindo insists, the overmind comprehends the totality of things and their integral principle, while acting in a cosmos of polarities and multiplicities, where the personal and impersonal aspects are only two facets of manifestation of the same original Consciousness, transcending the principles of personality or impersonality. For the overmind, the Divine can assume a personal or an impersonal existence. For our mind, the notion of a personal Deity, like that of the monotheistic Abrahamitic religions, the polytheistic Hindu pantheon or the pantheistic conception of a Spinozian God, are mutually exclusive and irreconcilable conceptions. The overmind recognizes them all as the different versions and appearances of the very same Entity. They appear as contradictory only to our limited mental cognition.

Thus the mental reason sees Person and the Impersonal as opposites: it conceives an impersonal Existence in which person and personality are

[213] For the more technically inclined reader in mathematical physics, we also note the curious fact that one can rewrite all the laws of physics with a local measure of information, called 'Fisher information', which can be interpreted as the ability to distinguish, differentiate and separate (Soffer, 1995). In this sense, positing an information-theoretic physical universe is in line with Aurobindo's ontology inasmuch it sees distinction as a volitional act coming from beyond the physical.

fictions of the Ignorance or temporary constructions; or, on the contrary, it can see Person as the primary reality and the impersonal as a mental abstraction or only stuff or means of manifestation. To the overmind intelligence these are separable Powers of the one Existence which can pursue their independent self-affirmation and can also unite together their different modes of action, creating both in their independence and in their union different states of consciousness and being which can be all of them valid and all capable of coexistence. A purely impersonal existence and consciousness is true and possible, but also an entirely personal consciousness and existence; the Impersonal Divine, Nirguna Brahman, and the Personal Divine, Saguna Brahman, are here equal and coexistent aspects of the Eternal. Impersonality can manifest with person subordinated to it as a mode of expression; but, equally, Person can be the reality with impersonality as a mode of its nature: both aspects of manifestation face each other in the infinite variety of conscious Existence. What to the mental reason are irreconcilable differences present themselves to the overmind intelligence as coexistent correlatives; what to the mental reason are contraries are to the overmind intelligence complementaries. (Aurobindo, 1914-19, p. 296, pt. I, ch. XXVIII)

The One and Many, form and formless, finite and infinite are not opposites or mutual exclusives, but complements of each other, two faces of the same Reality. It is the ultimate Subject that recognizes itself also as the ultimate Object. From the standpoint of the supramental vision

the Absolute is beyond the distinction of unity and multiplicity, and yet it is the One and the innumerable Many in all the universes." These are "to the dimensional mind irreconcilable opposites, but to the constant vision and experience of the supramental Truth-consciousness they are so simply and inevitably the intrinsic nature of each other that even to think of them as contraries is an unimaginable violence. (Aurobindo S. , The Synthesis of Yoga, 1914-21, pp. 297, pt.II, ch.I)

Supramental Nature sees everything from the standpoint of oneness and regards all things, even the greatest multiplicity and diversity, even what are to the mind the strongest contradictions, in the light of that oneness; its will, ideas, feelings, sense are made of the stuff of oneness, its actions proceed upon that basis. (Aurobindo, 1914-21, p. 1000, bk. IV, ch. XXVIII)

Therefore, not just a universal Mind, rather something that is much more than universal, let alone only Mind. A panentheistic vision, where the Spirit is founded upon the Knowledge by identity knows all as itself and in itself by a self-view and simultaneous self-vision, inhabiting each and all, knowing all things and beings as its individualized being of an all-containing existence, yet, without getting imprisoned in the objects and beings–that is, its own self-limitations– in which it dwells.

b. Conscious Force and Conscious Will

Force and energy have a precise definition in physics, but definitions, as rigorous as they might be, do not tell us much about the nature and essence of what they define. The ontology remains obscure. From the view of a supramental consciousness, things acquire another significance. Following Aurbondo's integral metaphysics, the source of space and time is a timeless immutable Absolute that nevertheless creates the illusion of motion and mutability *in* itself, without mutability. Something which, as the apparent irreconcilability of personality and impersonality, for the human mind sounds only like wordplay, an oxymoron, a contradiction in itself, an absurd impossibility. Because what our sense-mind realizes are only mutable relations of forms and of external shocks of forces, it is unable to view all things in a multiple unity, as a whole, like the Supermind does.

In fact, in Aurobindo's integral cosmology, force is the dynamic aspect of the exclusive concentration of a separative active consciousness, a particularizing action by a separative movement of Prakriti – Nature. But there is no contradiction or competition with the modern scientific view. From the integral perspective force is a Will inherent in existence as a potentiality of rest and movement in a fundamentally immutable Absolute. Consciousness, will, and force are not seen as derivative epiphenomena emerging from a play of cosmic events in a naturalistic theoretical framework, rather, they are an expression of a self-absorbed, involved and immanent Will and Consciousness in a temporal and spatial universal plane determined by the laws of causality. In the Unmanifest beyond causality, Consciousness, Will and Force are undifferentiated. There is no distinction between these three aspects of the Absolute. It is when Satcitananda plunges into the manifestation of polarities and division that Consciousness, Will, and Force appear as three distinct qualities, seemingly independent from each other. All the processes in our cosmos are self-presentations of It in itself. The mental relations we build from the play of these forces is what we call 'form', and all the phenomenal world we experience is a cosmic multitude of sensations, resulting from a form of force which is responding to another form of force.

The supramental vision reveals the universal clash of matter and forces as the workings of a 'Consciousness-Force' and 'Consciousness-Will', a 'Force of existence', as Aurobindo called it. The identity of Force and Will arises because *"the Energy that creates the world can be nothing else than a Will, and Will is only consciousness applying itself to a work and a result"* (Aurobindo, 1914-19, p. 17, bk. I, ch. I) An ontology with analogies to Schopenhauer's World-Will, or Shelling's and Hegel's 'World-Soul' or the Whiteheadian process theology that unites Consciousness, Force and Will as the substrate of all processes.

And it is here where the notions of will, and the age-old philosophical question of free will, find their spiritual dimension. There is no such thing as individual will. Our personal will is only a pale reflection of a transcendent Will that – to use Huxley's words – is 'funnelled through' an individualized ego. The real Will is well beyond our will. Especially in Western culture, and not rarely in Western philosophy, free will is sometimes seen as the freedom to act without restraints according to the desires, lusts, inclinations, beliefs, and preferences of one's own character. Whereas, from the point of view of Aurobindo, this is antipodal to what freedom is, as these desires and preferences are also imposed by Prakriti onto the untransformed inner and outer beings. The psychic being remains largely a passive witness of the inner impulses and universal forces of Nature, even though we take these for our own. Real freedom is not that of the ego and the lower cravings of the untransformed vital, mental, and physical personality but is realized only when the will of the individual becomes one with the Will of the Supreme. Aurobindo's answer to the question of whether we have free will is twofold. We have no free will if the personality supposed to be free is the ego vitality and mentality; we are like leaves subjected to the winds of universal forces that act upon us at a subconscious and subliminal level. Nevertheless, we can gradually attain free will if we change our being through a spiritual, psychic, and supramental transformation.

It is interesting to note how this standpoint, while denying free will to the ordinary human consciousness, nevertheless escapes the standard argument against free will that links physical determinism with a lack of freedom (for more thoughts on this, see my article on the relation between determinism or indeterminism and free will ((Masi, 2022c), (Masi, 2024)), and yet promises a possibility of blissful spiritual liberation.

c. The Supramental Vision of Space and Time

One of Aurobindo's peculiar statements relates to time perception. Once the seer enters gnostic heights of the supramental consciousness, one no longer perceives the physical reality as an unfoldment of time-successions; all appears as the action of a timeless and changeless eternal Being. From the standpoint of Supermind everything is a pure featureless and causeless eternal nirguna-Brahman, a changeless and an 'a-temporal' Infinite that does not create but manifests itself in itself through self-determinations in time and space, and nevertheless remains above that very same time and space: an eternity in 'motion status'.

From this perspective, space and time do not appear neither as separate entities, nor as separable aspects of the Infinite, but as 'self-extensions'. Space is an 'objective self-extension' of the Brahman and time its 'subjective self-extension'. Objective in the sense that this infinite Consciousness steps back from itself, so to speak, and thereby, becomes itself object of experience of itself. Subjective in the sense of a qualitative experience of a supra-cosmic Consciousness of itself.[214]

Space and Time are our names for this self-extension of the one Reality. Space would be Brahman extended for the holding together of forms and objects; Time would be Brahman self-extended for the deployment of the movement of self-power carrying forms and objects; the two would then be a dual aspect of one and the same self-extension of the cosmic Eternal. (Aurobindo, 1914-19, p. 375, bk. 2, ch. II)

As there exists a twofold character of the individualized mind and the cosmic Mind, also a twofold character of an individualized objectivity versus subjectivity exists in form of a cosmic objectivity versus cosmic subjectivity. The supreme real-Subject, the Divine, projects and apprehends itself as a two-fold extension of itself: It becomes a subjective extension we call 'time' but identifies itself also by a projection of itself as an objective extension that we call 'space'. Space and time are the two aspects of the same Self, extending itself in two modes or states of being, two different but complementary conditions of the Spirit.

It is the separative activity of mind which divides and distinguishes this temporal subjective self-extension into separate instants and the spatial objective self-extension into separate positions or points in a spatial coordinate system. The separation in space and the successions in time are figures appearing in a non-supramental experience. Since the Spirit sees itself always as a Whole, a One without a second, as an undifferentiated and unaffected Oneness, this utmost unity must also hold spatially and temporally.

This infinite time consciousness, the 'supramental time-vision' or the 'threefold vision' ('trikaladrishti') is

founded first on its eternal identity beyond the changes of time, secondly on a simultaneous eternity of Time in which past, present and future exists together for ever in the self-knowledge and self-power of the Eternal, thirdly, in a total view of the three times as one movement singly and indivisibly seen even in their succession of stages, periods, cycles, last– and that only in the instrumental consciousness–in the step by step evolution of the moments."– The unified infinite time consciousness of the timeless Infinite "maintains in itself at once in a vision of totalities and of particularities, of mobile succession or moment sight and of total stabilising vision or abiding whole sight what appears to us as the past of things, their present and their future. (Aurobindo, 1914-21, p. 886, pt. IV, ch. XXV)

[214] To physicists this may be reminiscent of John Archibald's Wheeler "*it from bit*", which symbolizes the idea that all things physical are information-theoretic in origin because ultimately a physical experiment and observation is a part of the universe observing itself.

Time, as we know it, from our limited mind-consciousness, results only as an artificial shadow of a much deeper truth of the phenomenal universe.[215] This understanding seems, to us, so counterintuitive because we identify our individualized mind as the selecting agent, while from the supramental perspective, one realizes that the 'Selector' is the universal consciousness itself.

This vision of time and space as self-extensions of the Spirit might, after all, not be so alien to contemporary scientific theories that describe the Big Bang as the beginning of a cosmological expansion. While modern QFT describes all particles–that is, all matter and energy–by localized excitations and vibrations of a universal quantum field. It might also be worth pointing out how the attempt to frame a theory of quantum gravity, in which Einstein's theory of relativity and quantum mechanics are extended to a more general theory containing them both as limiting cases, has been ongoing for more than half a century. However, so far, it has met with failure. One of the reasons seems to be the fact that the conventional physical notion of space and time is inadequate. Yet, nobody knows how and with what they are supposed to be replaced. Aurobindo's supramental time-vision might suggest, or at least intuitively inspire, a solution.

d. The Higher and Lower Nature, and the Central Principle of Evolution

The Samkhya philosophy and the non-dual Indian tradition of Shankara's Advaita Vedanta, separate between the ultimate Reality and the so called 'illusion of Maya', between the Transcendent and Nature – between Purusha and Prakriti. This ontology sees every phenomenon, from the most material to our phenomenal perceptions, be it a sensory experience, an emotional state or a mental fleeting thought, as being all impermanent and transitory events without inherent reality, just figments or illusions. They simply happen and are projected on the screen of an immutable and undisturbed Self, which is the sole and ultimate reality. This has become the most accepted interpretation also in Western spiritual circles, inspired by this ancient Eastern doctrine.

In Aurobindo's cosmology, however, this distinction between Soul and Nature is neither real nor unreal but is seen from a third position.

Since the word 'Nature' has acquired many different meanings in our contemporary culture, its deeper significance in the present context needs further clarification. With the term 'nature' (lower case 'n'), we implicitly assume a being separate from us, as if we are not making part of it as well. The Nature (capital 'N') in its integrality encompasses all manifestations, from the most material to the most subtle planes of existence, from the stone to the

[215] Something which seems to resonate with 'eternalism' or the four-dimensional block universe of relativity, where every moment is eternally present, and the present moment is only a selection that picks out a time-slice.

individualized mind, to the universal consciousness in its dynamic play, of which the human is clearly part of. Nature is conceived in a much wider sense. In Aurobindo's words:

An active force of conscious being which realizes itself in its powers of self-experience, its powers of Knowledge, Will, self-delight, self-formation, with all their marvelous variations, inversions, conservations and conversions of, even perversions, is what we call 'Prakriti' or 'Nature', in ourselves as in the cosmos. All Nature is simply the Seer-Will, the Knowledge-Force, and the Conscious Being at work to evolve in force and form all the inevitable truth of the Idea into which it has originally thrown itself. (Aurobindo, 1994).

The central principle that appears in this integral cosmology is, therefore, the relation between the Absolute, the passive Purusha – the unmanifest aspect of the Divine or God – and Nature, the active Prakriti. On one side, we have the One, the pure and undifferentiated self-conscious and self-existent Whole, which already contains, in potentiality, all the differences that remain unexpressed. On the other side, stands the same conscious Being in the form of a universal Consciousness at work: As Nature with its manifestations in diversity and multiplicity on *all* planes of being. A Nature which seems to have lost itself by an involution in an unconscious and mechanical universe, a universal phenomenality expressed by a Will-Force or Will-Power. But also a Nature-Soul that expresses these potentialities by and through the descending ladder of planes from Supermind and overmind, downward to the involuted forms of consciousness, such as mind, life, matter, the subconscient and the inconscient.

Aurobindo sees Nature as an instrument of expression of the Divine consciousness, but not an entity separated by it, or an illusion of an unreal phenomenal world, as Shankara and his followers contended. It is as real as the Spirt it comes from, it is the power of the Spirit, its executive Consciousness-Force and cosmic Consciousness-Energy, or Shakti, which is the origin and cause of all dynamical activity in the temporal manifestation.

But the higher and lower hemispheres are not just two independent blocks. The universal planes of consciousness, from the Supermind that connects the unmanifest with the manifest, the overmind, the intuitive mind, the illumined mind and the higher-mind, capture with descending degrees the original unity of its underlying origin, the Oneness of the unlimited Satcitananda. The lower hemisphere–of mind, life, matter and the subconscious, inconscient and nescient planes–is the domain of division, separation and limitation in ignorance.

This led Aurobindo to adopt the classification of the Bhagavad Gita that distinguishes between a 'higher Nature' ('Para-Prakrti' or 'Super-Nature', the 'Divine Maya'), which acts by unification and by a transcendence of limitation, working in and by a divine Force and Unity, and a 'lower Nature'

('Apara-Prakrti'), acting through division and limitations. One is the supra-conscious aspect of Nature, which upholds the other mental, living, material, subconscious or inconscient aspect of itself. The higher Nature reconciles and unites in oneness and works on the lower Nature life, mind and matter to uplift it into its inherent light, force and joy. The higher Nature has a wider self-guiding knowledge-consciousness, while the lower Nature, in its cosmic mental, cosmic vital and cosmic physical existence, has become more mechanical and deterministic, being apparently ruled by blind forces because it is an involved form of consciousness. Para-Prakriti dwells in Apara-Prakriti as a latent Super-Nature.

On the physical plane, this twofold aspect of Nature is reflected also as a twofold evolutionary process. From below by the effectuation of a spirit, in a world of multiplicities in ignorance, that ascends the vertical ladder of consciousness by a bottom-up process, in a still unconscious or only semi-conscious hemisphere. From above by a dynamic action ruled by a Spirit descending the ladder from a superconscious higher hemisphere, by a top-down process acting in a world of multiplicities in knowledge. It is in this larger vision that we discover the central principle of evolution: a Spirit that calls Nature. A Spirit impels Nature to initiate an ascension towards it. A central and uniting principle that relates an infinite Consciousness with a finite Nature, connected by physical, mental and vital cosmic planes of varying degrees of unitarity (or fragmentation) in its individuality and cosmicity, and characterized by a two-fold bottom-up *and* top-down evolutionary process.

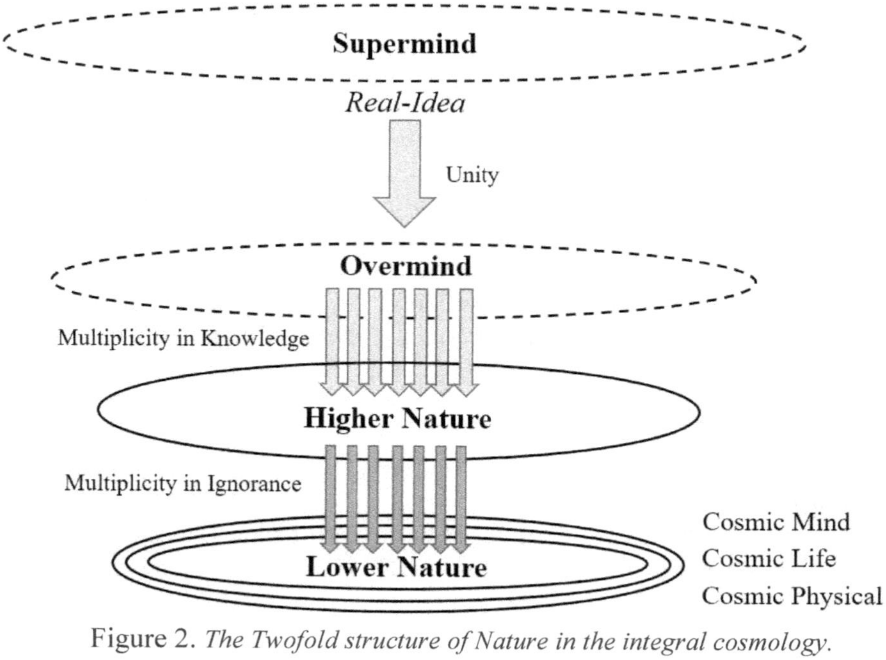

Figure 2. *The Twofold structure of Nature in the integral cosmology.*

A Spirit operating in and through a twofold-Nature that works out a divine plan and intention to accomplish an ascension of the lower Nature towards the higher Nature.

But both the Para-Prakriti and Apara-Prakriti are under the direct impulsion of the overmind and which is, in turn, the executrix of the supramental Real-Idea. The Real-Idea is an expression of a potentiality enacted in the manifestation by a supreme 'Will-Force', as an aspect of a 'Consciousness-Force' (Chit-Shakti), in form of a cosmic energy which, with its creative formative processes, manifests that archetype.

According to Aurobindo, it is ultimately this all-pervading and underlying presence of consciousness as an immanent Real-Idea, with its expressive active Will-Force, that stands behind all the workings and phenomena of an apparently unconscious and mechanical nature. He expresses this as a principle that also characterizes the yogic practice and summarizes it as follows:

> *"[...] the movement of Nature is twofold, higher and lower, or, as we may choose to term it, divine and undivine. The distinction exists indeed for practical purposes only; for there is nothing that is not divine, and in a larger view it is as meaningless, verbally, as the distinction between natural and supernatural, for all things that are are natural. All things are in Nature and all things are in God. But, for practical purposes, there is a real distinction. The lower Nature, that which we know and are and must remain so long as the faith in us is not changed, acts through limitation and division, is of the nature of Ignorance and culminates in the life of the ego; but the higher Nature, that to which we aspire, acts by unification and transcendence of limitation, is of the nature of Knowledge and culminates in the life divine"* (Aurobindo, 1914-21, p. 44, Introd., ch. V).

This integral cosmology could suggest a different teleological perspective with which we may look upon life and evolutionary biology. On the one hand is the appearance of the difficult and extremely long development of life throughout the eons, with all its strife and pains, dominated by a seemingly blind natural selection, and apparently random genetic drifts, as described by the orthodox Darwinian evolutionary paradigm. On the other hand, it is a purposeful and planful conscious Force that determines and controls behind the veil those events that appear to be 'random' only to the limited human mind. The former is Apara-Prakriti that can't be explained as a stand-alone process but needs the latter, Para-Prakriti, to manifest anything meaningful. Otherwise, Apara-Prakriti would not be able to go much beyond a purposeless and meaningless random chaos.

e. The Synthesis of Knowledge

Aurobindo rarely presented his yoga and vision as in opposition to other philosophical traditions of his time (except for his sever critique to world-negating and ascetic forms of spirituality). On the contrary, he always insisted on the integrality rather than exclusivity of his doctrine. An integrality that goes beyond spirituality and embraces materialism and aims at harmonizing Eastern with Western philosophical currents.

For example, from the integral perspective, all three standpoints, the scientific monistic materialism, the Indian Vedanta school that separates Purusha from Prakriti and the view of the Western philosophical idealist, are three different windows through which the very same reality is seen. They seem irreconcilable and mutually exclusive only because of the separative tendency of mind, which admits only matter, or only spirit, or only itself. Integrality cannot be realized because, in the background, mind remains the primary cognition tool, compelling us to make an exclusive choice. Whereas, if one looks at things from an integral perspective, one can not only look through these three windows simultaneously but, by rising above the mind, recognize it as three different entries, which can be widened and surpassed if higher forms of cognition are allowed. An integral cosmology recognizes materialism, idealism and non-duality as the three forms of cognition reflecting the three steps of the descending (or ascending) ladder of existence. The supramental vision reveals, not just these three steps, but embraces several other cognitive levels and 'modes of knowing.' It is in this sense that all three modes of cognition capture only an aspect of reality, but fail to be truly integral, they still see the relationship between Nature and the Self through the lens of the mind.

In this integral cosmology, instead, the universe appears as an indivisible and undifferentiated unity and totality, expressing itself by a top-down Para-Prakriti action, which determines the unfoldment of Apara-Prakriti through an evolutionary process characterized by infinite variety and multiplicity. The One and the infinitely many, the objective and the subjective, the personal and the impersonal, Nature and Super-Nature, are all the different aspects and modes of the same Being. All are real, none are an illusion.

6. Conclusion

This essay presented an integral cosmology in the light of Sri Aurobindo and which is neither a monistic nor a dualistic ontology, but rather a multidimensional description of reality, as it appears from higher states of consciousness. It results in a vision capable of integrating Eastern and Western philosophies by an evolutionary emergentism that extends Western neo-idealist models, which recently gained some attention, such as cosmopsychism, analytic idealism and theories of universal consciousness, and that could potentially become a basis for new post-materialist consciousness studies.

It is an integral cosmology that represents an articulate, fine-structured, and at times, a quite complex-layered ontology that might not always sound appealing if one demands principles of conceptual parsimony to be inexorably applied. But while it would be desirable to keep the theoretical framework as simple as possible, we must also be aware of the fact that reality is not always as simple as our theories would like it to be. If mystics are correct in stating that, our ordinary state of consciousness isn't aware of most of the phenomenal existence that escapes our sensory and cognitive means, it would be presumptuous to ignore their accounts and insist on cramming a physical and trans-physical multidimensional reality into a physical-only 'flatland theory.' One must not necessarily embrace Aurobindo's vision to its full extent. It is perfectly possible to resonate with some aspects adopting them as working hypotheses and see how far these lead us, while considering other elements of less interest. The question is if and how our background assumptions are well founded or need to be revised inside this wider perspective.

Much depends on how far one is willing to accept the first-person mystic experiential accounts as a source of information for consciousness studies, or if they should be rejected and ignored altogether as an unreliable hallucination, regarding the only permissible methods of cognizing reality the analytic mind and empiricism. But, remaining open to the idea that our conscious experience, cognitive processes, life and, in general, the universal manifestation as a whole, can be apprehended and comprehended also by other means that go beyond the ordinary intellectual and sensory human-centric forms of cognition, then these first-person spiritual accounts should be taken more seriously and might be able to clarify aspects of consciousness and life, that otherwise remain unintelligible in the naturalistic paradigm. After all, it is in line with the orthodox evolutionary Darwinian point of view: There is no reason to doubt that during our evolutionary journey new states of consciousness that go beyond the mind will impose themselves and look upon reality from a different viewpoint. In a certain sense, the (more or less implicit) attempt of the Age of Enlightenment to enthrone the human intellect with its scientific reason as the ultimate arbiter of knowledge is a form of neo-anthropocentrism as well. The first-person reports from higher states of consciousness could improve, or at least complement and inspire science and philosophy. To the best of my knowledge, Aurobindo's account is, so far, the most comprehensive view of consciousness, mind, life, matter, and the cosmos capable of doing this. Even though one requires a fair amount of effort to get acquainted with the richness and complexity of Aurobindo's cosmology, I believe it to be the most promising theoretical framework that can accommodate science and spirituality, in a unique synthesis of knowledge between Western and Eastern materialistic and metaphysical cosmologies.

At any rate, I hope that this overview will impel the readers to know more.

References

Albahari, M. (2019). Perennial idealism: A mystical solution to the mind-body problem. Philosopher's Imprint, 19(40), 1-37.

Alfassa, M., & Satprem (1983). Mother's agenda - The supramental action upon earth. Institut de Recherches Evolutives.

Aurobindo, S. (1910-1950). Essays divine and human. Sri Aurobindo Ashram Press, Puducherry, India.

Aurobindo, S. (1910-50). Essays in philosophy and yoga. Sri Aurobindo Ashram Press, Puducherry, India.

Aurobindo, S. (1914-19). The life divine. Sri Aurobindo Ashram Press, Puducherry, India.

Aurobindo, S. (1914-21). The synthesis of yoga. Sri Aurobindo Ashram Press, Puducherry, India.

Aurobindo, S. (1927-50). Letters on yoga. Sri Aurobindo Ashram Press, Puducherry, India.

Aurobindo, S. (1930-1950). A God's labour - Short poems. Sri Aurobindo Ashram Press, Puducherry, India.

Aurobindo, S. (1950). Savitri. Sri Aurobindo Ashram Press, Puducherry, India.

Aurobindo, S. (1994). Glossary of terms in Sri Aurobindo's writing. Sri Aurobindo Ashram Press, Puducherry, India.

Aurobindo, S. (2002). Record of yoga - Vol.I+II. Sri Aurobindo Ashram Press, Puducherry, India.

Aurose, W. (2021). The soul of nations: Healing and evolution. Lulu.com.

Baars, B. J. (1988). A cognitive theory of consciousness. New York: Cambridge University Press.

Banerji, D. (2012). Seven quartets of becoming: A transformative yoga psychology based on the diaries of Sri Aurobindo. Nalanda International.

Banerji, D. (2015). Individuation, cosmogenesis and technology: Sri Aurobindo and Glibert Simondon. Integral Review, 11(1), 65-79.

Banerji, D. (2020). Integral yoga psychology: Metaphysics & transformation as taught by Sri Aurobindo. Lotus Press.

Carpenter, J. C. (2015). First sight: ESP and parapsychology in everyday life. Rowman & Littlefield Publishers.

Chalmers, D. (1995). Facing up to the problem of consciousness. Journal of Consciousness Studies, 2, 200-19.

Chardin, P. T. (1955). The phenomenon of man. Éditions du Seuil.

Cornelissen, M. (2018). The self and the structure of the personality: An overview of Sri Aurobindo's topography of consciousness. International Journal of Transpersonal Studies, 37(1), 63-89.

Einstein, A. (1936). Physics and reality. Journal of the Franklin Institute, 221(3), 349-382.

Fermin, C. F. (2020). Biopsychism: life between computation and cognition. Interdisciplinary Science Reviews, 45(3), 315-330.

Gasparri, L. (2019). Priority cosmopsychism and the Advaita Vedanta. Philosophy East and West, 69(1), 130-142.

Gebser, J. (1985). The ever-present origin. Athens: Ohio University Press.

Gershman, S. J., & al., e. (2021). Reconsidering the evidence for learning in single cells. eLife.

Hameroff, S., & Penrose, R. (2014). Consciousness in the universe: A review of the 'Orch OR' theory. Physics of Life Reviews, 11(1), 39-78.

Hoffman, D. (2008). Conscious realism and the mind-body problem. (I. Academic, Ed.) Mind & Matter, 6(1), 87–121.

Huxley, A. (1954). The doors of perception. Chatto and Windus.

Kastrup, B. (2018). The universe in consciousness. Journal of Consciousness Studies, 25(5–6), 125–55.

Kastrup, B. (2019). The idea of the world. iff Books.

Kastrup, B. (2020). Decoding Schopenhauer's metaphysics. Iff Books.

Lester, L. (2011). The union of spirit and matter: Science, consciousness, and a life divine. Integral Review, 7(1), 104-113.

Lyon, P., & al., e. (2021). Basal cognition: Conceptual tools and the view from the single cell. Philosophical Transactions of the Royal Society B, 376(1820).

Masi, M. (2018). Free progress education. Independent publishing. Retrieved from https://www.amazon.com/dp/B07L368YB2

Masi, M. (2022a). Spirit calls nature – 3d ed. Independent publishing. Retrieved from https://www.amazon.com/dp/3948295166/

Masi, M. (2022b). Vitalism and cognition in a conscious universe. Communicative & Integrative Biology, 15(1), 121-136.

Masi, M. (2022c). Quantum indeterminism, free will, and self-causation. Journal of Consciousness Studies,30(5-6), 32-56.

Masi, M. (2022d). Quantum indeterminacy and Libertarian Panpsychism. Mind and Matter, 22(1), 31-50(20).

Medhananda, S. (2021). Cutting the knot of the world problem: Sri Aurobindo's experiential and philosophical critique of Advaita Vedānta. Religions, 12 (9), pp. 1–21.

Medhananda, S. (2022). The playful self-involution of divine consciousness: Sri Aurobindo's evolutionary cosmopsychism and his response to the individuation problem. The Monist, 105, 92-109.

Medhananda, S. (2024). Can Consciousness Have Blind Spots? Journal of Consciousness Studies, 31, No. 9–10, 2024, pp. 113–31.Murphy, M. (2015). The emergence of evolutionary panentheism. In E. Kelly, A. Crabtree, & P. Marshall (Eds.), Beyond physicalism: Toward reconciliation of science and spirituality. London: Rowman & Littlefield.

Oizumi, M., Albantakis, L., & Tononi, G. (2014). From the phenomenology to the mechanisms of consciousness: Integrated information theory 3.0. PLOS Computational Biology, 10(5).

Penrose, R. (1994). Shadows of the mind. Oxford University Press.

Salmon, D. M. (2007). Yoga psychology and the transformation of consciousness: Seeing through the eyes of infinity. (P. House, Ed.)

Seager, W. (2020). The Routledge handbook of panpsychism. New York: Routledge.

Segundo-Ortin, M. C. (2021). Consciousness and cognition in plants. Wiley Interdisciplinary Reviews: Cognitive Science, 13(2).

Seith, A. K., & Bayne, T. (2022). Theories of consciousness. Nature Reviews Neuroscience, 23, 439–452.

Shah, K., & Jacoby, S. (2013). Evolving beyond borders: The United Nations from the perspective of Sri Aurobindo's philosophy. Readworthy Publications.

Shani, I. (2015). Cosmopsychism: A holistic approach to the metaphyscis of experience. Philosophical Papers, 44(3), 389-437.

Shani, I. (2022). Cosmopsychism and non-Śankaran traditions of Hindu non-dualism: In search of a fertile connection. In I. Shani, & S. Beiweis, Cross-cultural approaches to consciousness: Mind, nature, and ultimate reality. Bloomsbury Academic.

Shani, I. (2022). Cosmopsychism, coherence, and world-affirming monism. The Monist, 105, 6-24.

Shani, I., & Keppler, J. (2018). Beyond combination: How cosmic consciousness grounds ordinary experience. Journal of the American Philosophical Association, 4(3), 390-410.

Shirazi, B. A. (2011). Integral education: Founding vision and principles. Integral Review, 7(1), 4-10.

Skrbina, D. F. (2017). Panpsychism in the West. MIT Press.

Soffer, B. R. (1995). Lagrangians of physics and the game of Fisher-information transfer. Phys. Rev.(E 52), 52(6), 6917-6917.

Taylor, S. (2020). An introduction to panspiritism: an alternative to materialism and panpsychism. Zygon - Journal of Religion & Science, 55(4), 898-923.

Teklinski, E. (2016). A matter of heart and soul: Towards an integral psychology framework for postconventional development - Dissertation thesis. California Institute of Integral Studies. Retrieved from: https://www.academia.edu/25029269/a_matter_of_heart_and_soul_towards_an_integral_psychology_framework_for_postconventional_development

Teklinski, E. (2019). A matter of heart and soul: The value of positing a personal ontological center for developmental psychology. International Journal of Transpersonal Studies, 37(1), 90-119.

Tsuchiya, N., Wilke, M., Frässle, S., A.F., V., & Lamme, V. A. (2015). No-report paradigms: Extracting the true neural correlates of consciousness. Trends in Cognitive Science, 19(12), 757-770.

Vaidya, A. J. (2020). A new debate on consciousness: Bringing classical and modern Vedānta into dialogue with contemporary analytic panpsychism. The Bloomsbury Research Handbook of Vedānta, 393–422. London: Bloomsbury.

Velmans, M. (2012). Reflexive monism: Psychophysical relations among mind, matter and consciousness. Journal of Consciousness Studies, 19(9-10), 143-165.

Wahbeh, H., Radin, D., Cannard, C., & Delorme, A. (2022). What if consciousness is not an emergent property of the brain? Observational and empirical challenges to materialistic models. Frontiers in Psychology 13(7).

Wigner, E. P. (1959). The unreasonable effectiveness of mathematics in the natural sciences. Communications on Pure and Applied Mathematics, 13, 1-14.

Zulaski, J. (2017). A complete integral education: Five principla aspects. Integral Review, 13(1), 20-29.

8. No Consciousness? No Meaning (and no AGI!)[216]

Abstract

The recent developments in artificial intelligence (AI), particularly in light of the impressive capabilities of transformer-based Large Language Models (LLMs), have reignited the discussion in cognitive science regarding whether computational devices could possess semantic understanding or whether they are merely mimicking human intelligence. Recent research has highlighted limitations in LLMs' reasoning, suggesting that the gap between mere symbol manipulation (syntax) and deeper understanding (semantics) remains wide open. While LLMs overcome certain aspects of the symbol grounding problem through human feedback, they still lack true semantic understanding, struggling with common-sense reasoning and abstract thinking. This paper argues that while adding sensory inputs and embodying AI through sensorimotor integration with the environment might enhance its ability to connect symbols to real-world meaning, this alone would not close the gap between syntax and semantics. True meaning-making also requires a connection to subjective experience, which current AI lacks. The path to AGI must address the fundamental relationship between symbol manipulation, data processing, pattern matching, and probabilistic best guesses with true knowledge that requires conscious experience. A transition from AI to AGI necessitates semantic understanding, which is closely tied to phenomenal consciousness. Recognition of this connection could furnish new insights into longstanding practical and philosophical questions for theories in AI, biology and cognitive science, and provide more meaningful tests of intelligence than the Turing test.

Introduction

Theories of language and meaning are not new. One could begin with Ferdinand de Saussure's fundamental insights about the meaning of signs and words as arising from their relationships with other signs and words. The structure of language is based on these relationships and differences, without which meaning could not emerge. This approach laid the foundation for modern structural linguistics and semiotics.

Later, Ludwig Wittgenstein, in his masterpiece of logical positivism, *Tractatus Logico-Philosophicus*, emphasized that meaningful

[216] The present article was published in *Qeios*. DOI: https://doi.org/10.32388/DN232Y.5

statements must be rooted in the clarity of concepts within a logical system of propositions. He described language as mirroring the world of facts (not things), developing a "picture theory" of language as a description of our experiences of these facts. The later Wittgenstein departed from the Tractatus in his *Philosophical Investigations*, realizing that language derives its meaning not only from logical structures but also from its contextual dependency. Language mirroring the facts of the world is specific to social and practical contexts, and the meanings it conveys must have inherent flexibility and vagueness.

Noam Chomsky's theory of "transformational-generative grammar" championed the view that language is a product of the mind, an idea based on a biolinguistic conception, in which the human ability to use complex forms of language is pre-wired in the brain, in a "language acquisition device." This premise led him to hypothesize the existence of universal grammar, with core syntactic linguistic knowledge being genetically inherited.

Saussure's, Wittgenstein's, and Chomsky's theories are just a few examples of the many competing theories that have emerged over the years. However, a definitive answer to the question of the nature of language and meaning remains elusive.

Similar questions arose in the field of IT. Shortly before Wittgenstein published his Investigations, Claude Shannon, the modern father of information theory, was contemplating the notion of information. What does the word "information" mean? The etymology of words is often more insightful than our contemporary understanding of them. The word "information" derives from the Latin "informare," which suggests that something has been formed or shaped–by molding, carving, or puncturing an object into a pattern or by modifying its physical, internal, or external state. It is about forming or changing something, making it a medium for symbols that convey a message, expressing our thoughts to someone who can understand those symbols and refer to them meaningfully. Whether it is letters and words on paper or bits in a computer, information is always about forming patterns or modifying internal states in objects.

On the other hand, the information we receive also "forms" and "shapes" ideas in our minds. There is, therefore, an important distinction to be made between physical information and semantic information. Physical information conveys a message in the form of symbols, signs, and tokens, and in communication theory, it is quantified by the Shannon information measure. However, a sequence of symbols and its quantification in bits are, in and of themselves, not meaningful unless a mind apprehends them and "collapses" them into a meaningful semantic whole. Physical information has no meaning whatsoever if there is no mind receiving the message and translating it into coherent thoughts.

Shannon was acutely aware of this distinction. In his seminal 1948 paper, "A Mathematical Theory of Communication," he pointed out: *"The fundamental problem of communication is that of reproducing at one point either exactly or approximately a message [a sequence of discrete symbols] selected at another point. Frequently the messages have meaning; that is they refer to or are correlated according to some system with certain physical or conceptual entities. These semantic aspects of communication are irrelevant to the engineering problem" (Shannon, 1948)*.

In physics and statistical theory, several other definitions of information exist. However, most of these information measures are closely related to Shannon information, and none of them quantify semantic content.

Unlike a mathematical or physical theory of communication, semantic information concerns the meaning of a message. The idea that meaning cannot be reduced to pattern recognition alone and that a physical representation of something is fundamentally different from the thing being represented becomes intuitively evident with the famous Gestalt figures. For example, the popular Rubin's vase-face figure, where our mind switches between the visual interpretation of a vase and two faces, shows how patterns *in-formed* into material structures (the figure on a piece of paper) have no meaning in and of themselves. A meaning-making mind must convert physical information into a coherent semantic whole and eventually even make a choice between mutually exclusive ones. It is a cognitive process that highlights how the significance of things isn't inherent in the things themselves "out there" but rather emerges as a perceived content in us.

For about three decades, these questions were largely ignored as irrelevant philosophical subtleties. Unfortunately, physical information was often conflated with semantic information, and both were simply referred to as "information" without further distinctions. However, the progress of IT, the ever-increasing (physical) information processing power of computers, the advent of AI, and new findings in neuroscience have forced us to reconsider these simplistic assumptions. Questions about the nature of consciousness and the mind have resurfaced, and the relationship between computational and mental states has become a matter of debate.

However, what are consciousness and semantics?

Consciousness in biological systems is widely believed to emerge from complex, integrated activity within the brain. Rather than residing in a single structure, it is thought to arise from the dynamic coordination of sensory processing, memory, attention, and self-representation across distributed neural circuits. Theories such as Integrated Information Theory (Oizumi, 2014) and Global Workspace Theory (Baars, 1988) propose that consciousness depends on both the quantity and organization of information processing—where integration and accessibility of information are key. This idea relies on the assumption that in biological organisms, the neural architecture enables a unified experience of perception and thought, giving rise to the subjective awareness we call consciousness. Others embrace metaphysical ontologies like dualism, panpsychism, or idealism. However, the true origin and nature of consciousness remain largely debated. For comprehensive reviews on the subject, see: (Seth and Bayne, 2022), (Kuhn, 2024), or (Gulick, 2014).

Thus, there is no universally accepted definition of consciousness. In this context, the term 'phenomenal consciousness' suffices to illustrate our point. Phenomenal consciousness refers to the subjective, qualitative aspects of experience—often described as 'qualia'—which include raw sensory perception like the redness of blood, the pain of a toothache, and the sweetness of sugar. Phenomenal consciousness encompasses more than just sensory qualia; it includes all conceptual and nonconceptual experiences related to time, space, causality, the body, the self, and the world. Or, to frame it in Thomas Nagel's famous "what it is like to be" formulation: an entity is conscious if there is a subjective experience associated with being it (Nagel, 1974). From now on, we will simply refer to phenomenal consciousness as consciousness.

It is important to note that consciousness is exclusively a first-person subjective experience that cannot be fully captured through third-person empirical investigation. While I cannot demonstrate that others are conscious, I am aware that I am conscious, and therefore, I have no reason to doubt that other human beings like me are also conscious.

What is semantics? Here, we adopt a general understanding. Even though we will focus on the linguistic dimension of semantics due to the disruptive impact of LLMs, we refer not only to formal linguistic competence, which encompasses lexical semantics (the retrieval of individual word meanings) and compositional semantics (the construction of meaning from multi-word utterances), but also to a type of 'non-verbal semantics' grounded in world knowledge and functional linguistic competence, applicable in non-verbal contexts (such as extracting meaning from a picture). This can be described as 'general conceptual knowledge,' which does not necessarily rely on linguistic inputs but is crucial for fluent language use.

While LLMs excel in formal competence, they often struggle with the functional aspects of language (Mahowald, 2024). In the following, we will focus on the complex reasoning that relies on the non-linguistic or 'pre-linguistic' world knowledge and that determines the functional linguistic skills, subsuming it under the label of 'true semantic understanding' or just 'semantics.'

The Explanatory Gap Between Symbols and Semantics

In 1980, one of the most debated arguments against computationalism came from Searle, who formulated his celebrated "Chinese Room Argument." This thought experiment was designed to challenge the notion that a Turing machine–and, by extension, any (non-quantum) computer or AI system running an algorithm–can truly understand something simply by processing symbols (Searle, 1980). Searle, who does not understand Chinese, imagined himself locked in a room with a set of rules (in English) for manipulating Chinese symbols. By receiving Chinese characters as input and following the rules to produce the correct output in Chinese, he could mimic a native Chinese speaker, though without understanding what the Chinese symbols actually referred to–that is, what they meant. No matter how perfect a simulation of understanding might be, it does not imply genuine comprehension of the language. One is merely following the syntactic rules of a program without any grasp of meaning (semantics).

By highlighting the distinction between syntax (symbol manipulation) and semantics (meaning), this thought experiment demonstrates the possibility of correctly processing symbols without understanding what they mean. Thus, it raises the question of whether machines, while capable of the former, are also capable of the latter. There is only the signifier without the signified, in the sense that a Turing machine manipulates symbols according to syntactic rules (an algorithm) without having either an abstract conceptual designation or any real-world referents.[1]

Based on these arguments, Searle introduced the distinction between "strong AI"–nowadays referred to as "Artificial General Intelligence" (AGI)–and "weak AI" (Searle, 1990). There is no consensus on what exactly "general intelligence" means. Here, as will become clear throughout this essay, general intelligence refers to a system possessing what I like to call "semantic awareness"–that is, a form of cognition capable of reasoning and generalization abilities based on a genuine semantic comprehension (in a sense that will be clarified later) of language, data, and, eventually, sensory inputs. Such a system would go beyond the mere emulation currently achieved by symbol manipulation, best-fit algorithms, or next-token predictions.

Nobel laureate Roger Penrose also expressed his doubts in his seminal book *"The Emperor's New Mind: Concerning Computers, Minds, and the Laws of Physics,"* where he laid out his argument that the mind can't be equated with a Turing machine and emphasizes the role of non-algorithmic processes (Penrose, 1989). His ideas were influenced by Gödel's incompleteness theorems, which demonstrate the existence of true mathematical statements that cannot be proven algorithmically within any formal system. While he does not provide a singular, definitive explanation,

Penrose contends that conscious understanding isn't computational. The quality of understanding requires awareness and arises from the conscious mind's ability to perceive and interpret reality in ways that transcend purely mechanistic or computational systems, suggesting that consciousness involves non-algorithmic processes that are tied in particularly to quantum mechanics.

In the 1990s, Searle's thought experiment and Penrose's argument were reinforced by the so-called "symbol grounding problem," discussed by Harnad (Harnad, 1990). This issue highlights the difficulty of explaining how symbols in computational systems–whether words, numbers, streams of bits, signals, or more complex representations–can carry meaning without being grounded in sensory experiences or real-world interactions. A gap remains between functional symbols and number-crunching processes on the one hand and meaningful mental states on the other. The divide between syntax and semantics persists.

The question becomes even more complex when we consider that symbols do not always refer to concrete objects or real-world phenomena but can also represent abstract and intangible concepts like "beauty," "justice," or even the concept of "abstraction" and "meaning" itself.

Harnad raises the issue that we do not really know what "meaning" itself is. We cannot simply assume that semantics can be reduced to a form of computation, as computation follows specific syntactic, logical, and mathematical rules based on symbol manipulation–not their meanings. He referred to this as the "symbol/symbol merry-go-round"–the idea that symbols can refer only to other meaningless symbols. Avoiding this infinite regress requires a bridge between those representations and the things or concepts to which they refer.

Even for a Turing machine that convincingly passes the Turing test–that is, its cognitive processes are indistinguishable from those of a human, the question would remain: Does it meaningfully connect its internal representations to their referents and its causal relationships, or is it a mere Searle's Chinese room mimicry, without any real semantic awareness of what they mean? Does it truly understand? What does it mean to "truly understand"? Harnad realized that *"there is a difference between inert words on a page and consciously meaningful words in our heads"* (Harnad, 2007).

Thus, following mechanical rules in symbol manipulation is one thing, and what humans mean by "understanding," "knowing," or "comprehension" is another. Semantics remains outside the formal description of any computational model.

The question, then, is: If symbols are meaningless in and of themselves, what generates meaning? While Shannon's notion of physical information is conceptually and mathematically well-defined, it is unclear how to define semantic information and even less clear how minds convert the former into the latter. How do neural activity patterns supposedly constitute and implement meaning? How do symbols and signs encoded in language elicit meaning in the brain through their reciprocal exchange? It seems that a mind is required as an interface between Shannon information (which refers to physical patterns or states in a medium) and semantic information (which meaningfully grounds the symbol to its referent in a much less tangible mental domain).

The most promising approach to answering these questions and aiming at "naturalizing meaning" seemed to be representationalism–an attempt to explain mental states and their contents in terms of representations and intentionality. However, decades of research in cognitive science and philosophical debates aimed at connecting mental processes that manipulate representations of external reality with cognition, perception, and consciousness have led to little substantial progress.[2]

Currently, non-representational theories emphasizing cognition as deeply rooted in bodily experiences and actively engaged with the environment are receiving more attention. (We will address these in the concluding section.) The advent of Large Language Models (LLMs) and their limitations suggests that this must be the case. However, I will argue that while embodied cognition–that is, cognitive processes rooted in interactions with the environment–might provide valuable insights into the relationship between computation, symbols, representations, and sense-making semantics, it can, at best, be only a necessary but not sufficient condition. Phenomenal consciousness and its whole psychological dimension cannot be sidestepped, as it is inextricably intertwined with meaning-making.

Meaning in a Time of Large Language Models

With the advent of transformer-based Large Language Models (LLMs), particularly in their popular form as ChatGPT, these questions elicited renewed interest. LLMs are designed to understand and generate human-like text based on vast amounts of data. Using deep learning neural networks, they predict and generate word sequences by learning patterns, grammar, context, and semantics from large datasets. Trained on diverse sources, including books, websites, and conversations, LLMs can answer questions, complete texts, translate languages, summarize information, and engage in natural dialogue. We have all been impressed by their abilities, which seem to suggest that they can reason effectively and possess a degree of semantic understanding.[3] The deep learning architecture underpinning these models is the transformer, whose key feature is its ability to effectively process and uncover long-range dependencies in input data. In simple terms, transformers in LLMs process and generate text by learning the probability distribution of long-range sequences of words and predicting the next word based on those previously generated. The transformer architecture has proven to be exceptionally powerful, especially in tasks that seem to require some level of semantic understanding. This fact challenges the notion that symbol manipulation alone is insufficient for cognitive states involving deep understanding.

Do LLMs truly understand? Do they possess a semantic understanding beyond imitation? Do they ground symbols as humans do?

While LLMs perform surprisingly well, nobody knows exactly why. Their internal complexity is such that they must be regarded as a black box. The issue is further complicated by the possibility of distinguishing different forms of "grounding." Modern LLMs, as artificial neural networks, compute over high-dimensional vectors rather than discrete symbols, making it more accurate to speak of a "vector grounding problem" (Mollo, 2023). Millière and Mollo distinguish five notions of grounding: referential (how linguistic items refer to real-world objects), sensory-motor (such as linking textual and visual representations), relational (how words relate to other words through definitions), communicative (the establishment of an intersubjective, rule-based language to communicate meaning and ensure mutual understanding), and epistemic (the relationship between linguistic expressions and factual knowledge-based data). Using an analogy with Chalmers' distinction between the easy and hard problems of consciousness (Chalmers, 1995), the hard problem of symbol grounding pertains to referential grounding. The other forms of grounding are, so to speak, "easy" in the sense that they can be addressed through a more or less sophisticated representational theory. Harnad's "symbol/symbol merry-go-round" relates to the referential grounding.

However, Millière and Mollo argue that LLMs can partially overcome the referential grounding problem, thereby acquiring more than interlinguistic functions through human fine-tuning. This fine-tuning endows LLMs with normative, world-involving functions that allow them to develop intrinsically meaningful representations of the world–going beyond mere word associations by adding an epistemic layer that connects language to world references. Through human feedback, LLMs gain knowledge about how the world operates, which enables them to form world-model representations rather than purely linguistic ones, even without embodied cognition with sensorimotor skills. In this way, LLMs at least partially address the referential grounding problem.

Yet, these world-involving functions are ultimately human-induced; the human factor is the bridge between language and the world. Human feedback via supervised learning provides modern LLMs with a form of referential grounding that aligns vector representations of words with real-world referents, endowing them with intrinsic meaning that, once fine-tuned, may no longer rely directly on human input. The only way for LLMs to achieve any form of referential grounding is through the intervention of a semantically grounded human agent who sets the normative framework, thereby guiding the non-grounded agent in establishing a world model–that is, a world-system of referents. Thus, this merely shifts the symbol grounding problem from the machine to the human agent: If an artificial neural network cannot achieve symbol (or vector) referential grounding, how are those biological neural networks inside our skulls able to do so?

Moreover, empirical evidence suggests that there is not anything or "anyone" in an LLM that genuinely "understands" beyond best-guessing and mimicry. Recent research has shown that LLMs can deviate in their responses when presented with irrelevant information, with performance declining when an unrelated sentence is added to a problem. This does not alter the problem's semantic content but appears to "distract" the model (Shi, 2023). The illusion of "reasoning" displayed by LLMs relies largely on recognizing patterns with a strong token bias rather than true comprehension; this also accounts for their limited generalization ability (Jiang, 2024). Further studies reveal that LLMs lack genuine mathematical reasoning; instead, they replicate logical steps seen in training data without understanding. For instance, rephrasing the same question can yield different answers, and adding redundant information irrelevant to the solution significantly lowers their mathematical reasoning performance. A slight increase in task difficulty can similarly lead to notable performance drops (Mirzadeh, 2024). LLMs tend to translate statements into operations (e.g., interpreting "discount" as "multiplication") without understanding the underlying semantics. This behavior aligns more closely with probabilistic pattern matching than with actual logical reasoning. Such limitations are evident even in basic grade-school math problems and are expected to become more apparent in complex mathematical assessments. That LLMs do not go much beyond sophisticated forms of "semantic-free" imitation is also suggested by the fact that they are inherently limited by their ability to solve problems that significantly differ from those they saw during their training session.[4] Additionally, the idea that LLMs exhibit "emergent abilities"–skills that are absent in smaller models and that appear only in larger ones–is also under scrutiny. These abilities may be fictitious and not an inherent result of scaling models up in complexity (Scheffer, 2023).

Other investigations have highlighted AI's limitations in reasoning, as demonstrated by LLMs' failures in seemingly simple tasks such as counting words or reversing a list (McCoy, 2024). This raises the question of why LLMs are generally effective in multi-step reasoning yet struggle with surprisingly trivial problems. Chain-of-thought prompting, which involves generating intermediate reasoning steps before arriving at a final output, exhibits characteristics of both memorization reasoning (in which the model mimics patterns learned from training data) and probabilistic reasoning (in which the model selects the most probable output based on the input) (Prabhakar, 2024). Further studies have examined how transformer LLMs tackle compositional tasks that require breaking problems into sub-steps and synthesizing those steps into precise answers. It turns out that they often simplify these tasks by matching linearized subgraphs–training examples that mirror the computations needed to solve test examples–without developing systematic problem-solving skills. Moreover, abstract multi-step reasoning problems based on autoregressive generation (the statistical prediction of the next sequence value based on previous values) tend to deteriorate rapidly as task complexity increases (Dziri, 2023).

A study conducted by OpenAI (Shojaee, 2025) with Large Reasoning Models (LRMs)—a new type of AI designed to excel at complex problem-solving by breaking tasks into smaller, logical steps—revealed that while these models perform better on reasoning benchmarks, they experience a complete collapse in accuracy beyond certain complexities. Their reasoning effort increases with problem complexity up to a point, after which it declines. In high-complexity tasks, they fail to develop generalizable reasoning capabilities, experiencing total collapse. Additionally, their limitations in performing exact computations became apparent; they might succeed in one task but repeatedly fail to provide correct answers in others. Apple's take is that these models create an illusion of thinking but lack true reasoning. They are essentially large pattern matchers that falter when they encounter data outside their training set, leading to breakdowns in generalization. Before we can consider them capable of genuine thinking processes with true reasoning abilities, we must be aware of something crucial is still missing.

This suggests that LLMs' abilities scale asymptotically: Beyond a certain threshold, adding more neural layers, faster processing, increasing training data, enhancing supervised reinforcement learning, and further fine-tuning or prompt engineering do not significantly improve their performance.[5]

Additionally, unlike humans, LLMs cannot correct their erroneous guesses (Kambhampati, 2024). This aspect is crucial, as one might argue that humans are prone to similar flaws. However, humans can recognize their logical, inferential, and deductive mistakes, which allows them to restart the problem-solving process with different premises or methodological approaches. This capacity to overcome biases and reach the desired result is possible only because of a semantic awareness that extends beyond mere symbol manipulation, enabling individuals to identify incorrect or nonsensical outcomes even if they do not know the correct answer.

After all, it does not require complex scientific research to recognize the shortcomings of LLMs in reasoning, semantic competence, and the understanding of inherent meaning. Sustained use and interaction often reveal interesting insights. For example, if one were to ask how many U.S. states have names beginning with the letter K, one might be told that there are four: Kansas, Kentucky, Kansas, and Kentucky. These issues similarly affect generative AI. If instructed to create a picture with no elephant in the room, it might repeatedly generate images featuring an elephant in a room, thereby demonstrating a failure to grasp negation (the "not-questions"). When asked to depict someone writing with their left hand, the model may refuse to comply, always producing an image of a right-handed person instead. This makes sense if we keep in mind that the model makes probabilistic guesses rather than semantic ones, as it has been trained predominantly on a large dataset where the proverbial sentence 'elephant in the room,' and images of right-handed individuals are quite common. [6]

Overall, we are beginning to realize that despite billions being invested in research and development, LLMs continue to hallucinate, fabricate information, and lack genuine understanding of the world. They excel in deep learning but struggle with generalized abstract reasoning, and they have difficulty understanding wholes in terms of their parts.[7]

Though we are dealing with a black box, we know it is ultimately a Turing machine working with symbols according to logical, syntactic, context-based, or probabilistic rules in someone else's natural language. These symbols stand for external referents grounded in the subjective experiences of someone else–experiences that the symbols themselves cannot convey to others. While it would be unfair to label LLMs as mere "stochastic parrots," they still fall short of human capabilities in natural language inference, their dialog is limited in information and can be unreliable, frequently failing in differentiating between fact and fiction, generating inaccurate claims (for a good summary of the strengths and weaknesses of LLMs see (Lappin, 2024).

Similarly, self-driving cars' ability to classify objects and people is impressive, yet countless instances reveal their lack of comprehension regarding what they are "seeing" (e.g., distinguishing between real people on the street and fictional figures on a billboard). Classifying and categorizing are not the same thing as understanding. The critical questions are whether a car can effectively navigate without a semantic understanding of its environment–recognizing the street, cyclists, traffic lights, and so on–and whether a level V self-driving car could achieve comprehension that even humans might struggle to acquire without conscious experience.

It could be argued that further fine-tuning and additional training sessions might resolve these issues and reduce the error rate. After all, humans also make mistakes. However, this is not the decisive point. The crux of the matter is that when the model fails, it can do so spectacularly, sometimes producing absurd answers that clearly illustrate a lack of genuine understanding.

On the other hand, it is also true that humans are not entirely immune to similar senseless perceptual hallucinations or ludicrous cognitive failures, particularly in individuals with neurological disorders. Oliver Sacks's best-selling book "*The Man Who Mistook His Wife for a Hat*" compiles case studies of such rare but real phenomena (Sachs, 1985). However, the question remains whether such pathologies can be addressed without relating them to a phenomenological content. Neural networks fit data and classify information, whereas humans integrate sensory data into a unified experience—what we might call a “perception of meaning”—that represents a semantic whole, no matter how senseless it may appear. Humans can overcome these hallucinations precisely because of conscious experience, while machines are confined to sense-making that cannot go beyond relational systems and statistical inferences based on abstract tokens.

While LLMs’ cognitive skills are undoubtedly impressive and surprising, it is becoming increasingly evident that there is still no true reasoning or common-sense knowledge that extends beyond sophisticated Chinese room machinery. There is no “ghost in the machine” with semantic awareness; fundamentally, the machine does not understand a thing. Machine cognition certainly doesn't align with the human understanding of 'cognition.' When we look at a green apple or a yellow pear, we don't perceive vectors or matrices.

However, one might argue that nonetheless, the machine does possess a form of cognition as well. We shouldn't assume that human-like cognition is the only possible one. In fact, representing reality as a complex relationship of multidimensional vector space enables AI to engage in meaningful conversations with us.Yet, without succumbing to anthropocentrism and placing our cognition at the center of the universe, we need to acknowledge the differences between human and machine meaning making. If there is none, the question arises how the internal workings ultimately represented as digits in a memory chip, translate into the rich experience of perceiving the 'greenness' of an apple, feeling its shape and size, and even tasting it. A memory cell labeled "apple" or "pear" is merely a symbol encoded in the physical state of a circuit. There exists a subjective and experiential dimension that cannot be encapsulated by a symbol alone, regardless of how complex the computations leading to its identification may be.

If even the most advanced AI can pass a molecular biology exam yet is unable to learn what for us humans are much simpler tasks, such as driving a car, and remains far from achieving human intelligence–or even the intelligence of a cat or dog–something might be fundamentally flawed in our understanding of what intelligence truly is. We might be overlooking a crucial aspect of our nature as living beings.

Meaning and Qualia

Wittgenstein might not have been entirely off: Language, like mathematics, isn’t merely a human cognitive creation unrelated to the organization of the world but, rather, might reflect its structure. We might conjecture that the statistical distribution of words is not solely tied to human cognition but also offers a window into the conditional structure of the world as mediated through human language. The patterns embedded in language might reflect the patterns inherent in the world. Linguistic patterns track real-world patterns, and which representational structure may be captured within LLMs (Queloz, 2024). If so, given that LLMs are pattern matchers trained to “autocomplete” based on complex relationships between vast numbers of tokens (representing properties of the world and their interrelations), their effectiveness might be less surprising.

Nonetheless, while any mathematical or formal proposition acquires meaning only when it is communicated from grounded semantic agents to other grounded semantic agents, one could go a step further and suggest that all of what we perceive and conceptualize about the world is ultimately a symbol, sign, token, or image within our cognitive awareness. This might be even less surprising to the philosophical idealist, who posits that the world itself is an "idea" or symbolic expression. Such a view aligns with certain Eastern philosophical and mystical theories in which the universe represents the expression of a creative, transcendental "real-idea," forming the foundation for all meanings, signs, words, and human language (Masi, 2025).

Be that as it may, it is not necessary to venture into metaphysical speculations to capture some essential aspects of how our sense-making cognitive processes emerge. The primary challenge here is not conceptual or theoretical but, rather, the shift from a third- to first-person perspective.

In this respect, it might also be interesting to consider Harnad's recent follow-up, which he formulated in the form of a dialogue with ChatGPT-4 (Harnad, 2025). According to Harnad, what ChatGPT lacks is *"a direct sensorimotor grounding to connect its words to their referents and its propositions to their meanings."* Machines operate within the realm of symbol manipulation without any grounding in real-world physical interactions and experiences; they lack the subjective and conscious dimensions of understanding. He further suggests that the only solution is to embody AI models–that is, to ground them–with sensory and motor outputs that allow them to interact with and learn from their environment–thereby building a world model. An "embodied AI" is a system that goes beyond mere abstract symbol manipulation found in traditional software. It is physically linked to the real world through sensors that detect environmental conditions and actuators that allow it to interact with that environment. This integration of perception and action creates a continuous feedback loop that enhances learning, as opposed to depending solely on abstract data. The idea of embodiment is crucial in fields such as robotics, virtual agents, and cognitive science, where it is seen as vital for fostering more adaptive, context-sensitive, and human-like intelligence.

However, Harnad carefully distinguishes the symbol grounding problem from any allusion to the hard problem of consciousness (or the "problem of qualia" or the "explanatory gap" (Chalmers, 1995). The philosophical issue of qualia involves how neural processes give rise to the subjective experiences of "what it is like" to be in a qualitative state of consciousness. The symbol grounding problem is about representation and understanding in artificial systems, not about why the complex machinery in our brain supposedly gives rise to subjective experiences.

Qualia–the introspectively accessible, subjective, phenomenal aspects of our conscious lives, such as the bitterness and warmth of coffee, the redness of a tomato, the smell of freshly cut grass, or the sharp pain from a paper cut–are not reducible to mere information processing. Our private experiences of pleasure, pain, feelings, and emotions, along with sensory perceptions like seeing, hearing, touching, smelling, and tasting, convey qualities of the world such as colors, sizes, and shapes. These experiences are not merely concepts or abstract symbols in a quantitative database; they are grounded in a qualitative experiential dimension.

Harnad's insights are inspiring; however, in my view, he fails to close the circle. He begins by associating meaning with "the subjective, felt experience of understanding," the "phenomenological aspect of what it feels like to mean or understand something"–elements that AI lacks, as it has no "direct experiential grounding" or "direct sensorimotor experiences." Yet, somewhat surprisingly, he stops short of explicitly concluding that, for a machine to achieve what he terms "direct symbol grounding" (and, ultimately, a state of AGI), it must be conscious.

Disentangling the symbol grounding problem from the hard problem of consciousness and then reiterating that natural language must be grounded in real-world experiences is a misstep. This is because the former is an aspect of the latter. The symbol grounding problem is itself "grounded" in the problem of qualia and the explanatory gap.

I submit that the gap between syntax and semantics is qualitatively much deeper than previously assumed. Scaling alone will not bridge this gap. While embodied architectures could extend text-based information to include sensory-based data with feature-detecting and abstracting capacities from real-world engagement, this would not transform Shannon-type sensory information into semantic information, thereby grounding the ungrounded in any miraculous way.There is no reason to believe that replacing or augmenting the linguistic symbols in an LLM super-dictionary database with a sensorimotor super-dictionary of pixels and mathematical functions representing environmental sensory signals could provide genuine grounding beyond syntactic computational understanding within an abstract categorical space. Names and verbal descriptions require not only a sensorimotor embodiment but also a conscious experience of the features they represent to be grounded in a semantic space. Vast data sets (textual, numerical, and/or sensory), immense computing power, and any form of embodiment may be necessary to build meaningful world models. However, it is questionable whether these elements are automatically sufficient to allow an AI system to gain a true understanding of the properties of the real world it is engaging with, without having a subjective experience of those very same properties.

After all, one can teach children how to roll a ball without giving them millions of examples, because a child doesn't see the world through numerical or symbolic representations but through a sense-making mind that relies on lived experiences. Biological organisms' meaning-making has an essential component that machines fundamentally lack–a gap that information processing alone cannot bridge. To put it in Vallor's words: *"We are more than efficient mathematical optimizers and probable next-token generators"* (Vallor, 2024).

Meanwhile, taking a first-person perspective reveals that our semantics-based cognition is inherently tied to conscious experience–always. We cannot truly understand colors, sounds, tastes, smells, temperature, touch, or sight without having directly experienced seeing a color, hearing a sound, tasting chocolate, smelling an odor, or touching an object. Is there a difference in "meaning" between, say, listening to a symphony of Tchaikovsky and "knowing" it only in the form of a digital transcription? An LLM might "know" everything textually about apples or the Fourier transform of a sound signal, but there is no semantics to extract, as all this remains an abstract, multidimensional vector representation defined by weighted numerical parameters. Without experiencing "what it is like" to taste an apple or feel its texture or to be immersed in the chills and thrills of musical rapture, it cannot achieve any form of symbol–or vector–grounding. Similarly, someone might explain all the chemistry and physics of an H2O molecule, but you cannot grasp the essence of wetness until you have felt the sensation of water yourself.

This leads us to Jackson's famous knowledge argument (Jackson, 1982): Neurophysiologist Mary may have all the physical information about color vision, but she still lacks knowledge if she has not experienced the qualia of colors.

The question is: Can any physical information be linked to semantic information without the presence of subjective experience?

One could reverse the argument and ask, "What kind of 'knowledge' could a purely phenomenological experience without physical information or conceptualization convey?" It is unlikely that this would lead to semantic comprehension. I might experience the blueness of the ocean, the chill of an environment, the sourness of a lemon, or its shape in my hands without having any concept or understanding of what the ocean, the environment, or a lemon truly represents. Experience, representation, and conceptualization are intimately intertwined.

This does not mean, however, that congenitally blind individuals cannot understand others discussing spatial properties, geometric forms, light, colors, or other visual sensations. They have an indirect understanding of these concepts, achieving referential grounding much like an LLM–not through sentience but via information mediated by other sentient agents. And it is known that blind subjects are not defective in their learning of language, of space, and objects. Blind learners not only learn the forms of language but their meanings as well (Landau, 1985). The key difference between even the most advanced AI and a person with sensory impairments is that the absence of sight in humans is compensated for by other subjective experiences, like sound, touch, taste, and smell. These experiences enable the indirect semantic comprehension of visual concepts even if direct visual perception is lacking, as the representation is supported by other forms of conscious experience–something a machine cannot replicate.[8]

What kind of comprehension do blind individuals have of the visible world, and to what extent can they eventually recover an ordinary understanding of it? This is not a new question. For instance, in the 17th century, William Molyneux proposed a thought experiment involving a congenitally blind person who has learned to recognize objects solely through touch. Imagine this person can distinguish a sphere from a cube by touch but has never seen them. Now, suppose this person could suddenly see. Molyneux questioned whether, if the sphere and cube were placed on a table, this person would be able to identify which was which *without* touching them.

The question was debated by philosophers like Locke and Berkeley, who essentially agreed that if our mind cannot establish a relationship between the sensations of the tactile and visual worlds, the answer to Molyneux's question must be negative. The connection between these two realms–specifically, grounding visual experience in meaningful semantic content known only through tactile experience–cannot be established through mere intellectual inference. To create this connection, one must link real-world experiences with previously grounded concepts derived from other kinds of lived experiences into some world-model.

Interestingly, the Molyneux problem can now receive a scientific answer. It is possible to gain some insights into the state of "meaningless awareness" from individuals affected by congenital cataracts that are later treated. A congenital cataract is an organic anomaly present at birth that clouds the eye's natural lens and can result in amblyopia–a disorder in which the brain fails to process visual stimuli. In the 1930s, Marius von Senden became the first to describe the perception of space and shape in congenitally blind individuals before and after surgery (Senden, 1960). When the sight of previously blind patients was restored and their bandages removed, they did not see the world as one might assume. Instead, they experienced a blotch of chaotic colored patches that meant nothing to them. They required time and practice to make sense of what they saw.

In 2011, neuroscience shed further light on this topic. An Indian research project, "Project Prakash," aimed at treating blind children while also seeking answers to scientific questions about how the brain develops and learns to see, provided evidence supporting Locke and Berkeley's views (Held, 2011). In the treatment of congenitally blind children aged 8 to 17, researchers found that, upon gaining sight, these children struggled to visually match objects they had previously known only through somatosensory information. However, this capacity developed quite rapidly; their skills in relating visual perception to somatosensory sensation improved within a few days of sight restoration and were nearly fully present within a few months. Physical information could finally be grounded in semantic information through sensory tactile experiences, as the latter was already rooted in a meaningful conceptual unity.

This indicates that linking tactile knowledge to visual knowledge is not an innate ability. Furthermore, it demonstrates that meaning emerges not only from the relationships among words, symbols, or vectors but also through the contextualization of different forms of qualitative experiences. Only after this experiential stage can the mind connect a symbol (the signifier) to its referent in the real world (the signified).[9]

One thing that must be noted is that semantic content does not require an experiential link to the physical world–that is, direct sensorimotor engagement. Most notably, abstract and intangible concepts like "freedom," "courage," "truth," "beauty," "justice," and "wisdom" lack objective referents in the physical world. Nonetheless, they are deeply meaningful to us and, in a sense, real and concrete. Their concreteness arises from subjective feelings and a "perception of meaning" rooted in the experiential dimension of the mind. These concepts are thoughts, often accompanied by an inner emotional state, that also have a phenomenological meaningful aspect. Even purely abstract entities, such as numbers, lack physical reality outside our minds (or, as Platonists would argue, exist only in a world of perfect Forms), and the symbols for numbers hold no inherent significance unless they are associated with a "number sense"–an experiential link between symbol and quantity mediated by a subjective mental experience of extension or quantity.[10]

Qualia are not limited to sensory experiences; thoughts are "mental qualia," and emotions are "emotional qualia." The "perception of meaning" is, above all, a subjective "mental quale" that emerges from and is intertwined with other qualitative experiences, whether somatosensory, affective, or otherwise. In this way, there exists a phenomenon of what "it is like to be" in each of these internal states–an aspect that any symbol-to-symbol or representation-to-representation relationship cannot fully capture.

Moreover, semantics has a wholistic character. The perception of meaning is always associated with an integration of information that neuroscience still struggles to understand (Herzog, 2008). How the brain integrates information from various sensory inputs (sight, sound, touch, etc.[11]) and distinct neural processes into a unified, coherent experience is unclear. For example, when we look at an apple, we don't perceive separate visual features like color, shape, and size individually; rather, we see a single, whole apple. It is what I referred to as a semantic whole that suddenly appears in our awareness by looking at a figure made of pixels, parts, structures, boundaries, colors, etc. The question of how different parts of the brain "bind" features together into one experience, despite being processed in different brain regions, is commonly known as the "binding problem" and is intimately related to the unity of consciousness.

The implications for AI are that without a conscious subject (an entity or individual capable of conscious qualitative sensations of the properties of the world), a machine cannot grasp anything beyond the purely abstract "understanding" of its representations, symbols, letters, words, sentences, signs, and numbers. A self-driving car, no matter how advanced or sophisticated its neural network and information processing system might be, cannot comprehend what an image of a street, a cyclist, or a traffic light represents if it lacks the complementary subjective experience and an integrative process that binds them into the perception of a meaningful unity. True knowledge cannot be achieved through functional processes alone; understanding requires a specific type of qualitative phenomenal dimension. An image can be converted into vector spaces, matrices, relationships, curve fittings, and probability laws, with their complex functional relationships. However, unless the light signals hitting the pixels of the image on the CCD camera elicit a subjective lived experience in someone or something, all these mathematical abstractions, multidimensional vectors, or neural representations will remain meaningless physical information without semantic grounding. While physical information can be measured by (negative) Shannon entropy, without sentience, nothing can convert it into semantic information. This is because "understanding" is an aspect of phenomenal consciousness itself and can't be abstracted from the hard problem of consciousness.

Thus, consciousness plays a fundamental role in the transition from symbols to semantics. An information processing system may interact with the environment through sensory-motor embodiment, but that alone does not make it a true semantic agent. An embodiment allowing for the representation of a world model should not be confused with a system experiencing the world. For it to achieve this, it must possess a comprehension of the world based on experiences, not just representations, abstract symbolic descriptions, or indirect linguistic grounding–that is, "qualia-less" physical information. Transitioning from the vectorization of language to the vectorization of the environment may represent a significant advancement in AI, but this alone is unlikely to overcome its semantic deficiencies, as these will still be ungrounded quantifying numbers or symbols all the way down. Embodiment may be a necessary condition, but it is not sufficient to transform unconscious apprehension into conscious comprehension. To achieve a deeper understanding of the world, including its contents, properties, and related concepts, I must grasp it qualitatively through conscious experience–that is, with qualia in the form of sensations, feelings, and lived sensory perceptions. A tight and inextricable relationship exists between feeling and knowing, sensing and understanding, perceiving and comprehending. We cannot disconnect these aspects of cognition and treat them separately. While the hard problem of consciousness addresses how consciousness emerges from material and/or functional processes, such as neural activity in the brain, the symbol grounding problem explores how meaning arises from these same processes. Symbol grounding is impossible without experience because meaning is an inherent aspect of consciousness. To understand something implies being conscious of it. Meaning emerges not only due to a relationship between abstract tokens, but also as a pre-linguistic association of experienced qualities into a unified construct apprehended and comprehended as a singular aware sensation. Meaning *is* subjective perception; the "perception of meaning" *is* a quale in itself.[12]

There is, however, a potential objections to this viewpoint.

Many of our cognitive processes occur without our conscious awareness. Experimental psychology demonstrates that sophisticated cognitive abilities seem to occur without consciousness. For instance, individuals can process semantic content and focus on objects even when masking techniques prevent the information from reaching their reported access consciousness. In cases of blindsight, individuals who are cortically blind can respond to visual stimuli they do not report to perceive, due to lesions in the primary visual cortex. Remarkably, they can often correctly identify objects that they believe they do not see.

Does this show that understanding is possible in the absence of consciousness?

I believe this conclusion is premature. Our understanding of consciousness is largely based on a superficial subjective experience. What we refer to as 'unconscious' or 'lack of attention' may actually represent another form of conscious awareness and attention—not unconsciousness, but a subliminal, non-metacognitive—that is, non-reportable—type of awareness. We often confuse different states of consciousness with being 'unconscious' because we cannot relate them to our ordinary waking state of awareness. In these states, we may not be genuinely unconscious; instead, we may simply lack mnemonic access to those experiences.

What is it like to be in a conscious state without memory? Are we truly 'unconscious' during dreamless sleep? Are we unconscious during anesthesia? Are we unaware while in a hypnotic trance or sleepwalking?

I argue that we do not have definitive answers. Phenomenal consciousness—that is, some form of sentience—can be present but quickly fade within seconds. Consciousness-mediated semantic cognition might still function in these altered states. Therefore, we should avoid making hasty conclusions, as this remains an open question.

Is the AI-AGI Gap a Conscious Semantics Gap?

As for any AGI narrative, since the success of ChatGPT and the impressive capabilities of LLMs, there has been a marked increase in discussions about the imminent arrival of AGI–human-like intelligence. However, many of these discussions overlook the intrinsic connection between general intelligence and consciousness.

Enhancing AI performance by adding another trillion neurons, increasing the number of parameters, providing sensory-motor environmental interaction, integrating it with neuro-symbolic AI, or developing more advanced deep learning algorithms, internal feedback loops between sensory acquisition and a central processing system to construct more accurate world-models, etc., will no doubt advance current AI capabilities. However, it will not accomplish the "semantic trick." The elephant in the room must be addressed: the possibility that biological intelligence, with its semantic awareness, might not be replicable in machines unless we learn how to manufacture consciousness itself (if it is even possible to do so). Without consciousness, true AGI cannot exist. The current widespread discourse on the impending AGI revolution, which is predicted to shape our future, warrants a more critical examination that considers the first-person perspective.

Though I am an "AGI skeptic," the rationale presented here does not necessarily imply that AGI is impossible or that consciousness in machines is impossible. The claim is, rather, that if we want the machine to become semantically aware—where, by 'awareness,' we mean a subjective dimension capable of experiencing—it must become conscious as well. Because awareness and consciousness are inextricably intertwined. Otherwise, it remains semantics without awareness, such as Searle's Chinese room.

On the other hand, recognizing the relationship between meaning and consciousness might offer new approaches to longstanding questions. The distinction between conscious intelligence and machine intelligence (or even the distinction between machine and living organism itself?) becomes apparent when, sooner or later, the machine makes a glaringly irrational mistake–one that no conscious, semantically aware agent would make. This reveals its lack of genuine semantic understanding and could be taken as evidence of its absence of subjective experience. In principle, this could suggest the basis for a new test of intelligence, potentially replacing the famous Turing test. The goal would be to develop a test designed to assess whether a machine possesses a deep, semantic understanding comparable to that of a conscious being building its meanings beyond symbol manipulation in a semantic space determined by subjective qualitative experiences. If it does, this would imply a form of intelligence that points to some level of experiential awareness, even if not necessarily human.

This line of thought could also serve as an argument for or against the hypothetical existence of "philosophical zombies" (or "p-zombies"). A p-zombie is imagined as a being physically identical to a human and behaving exactly like a conscious person but lacking any inner life or conscious experiences. However, because a p-zombie would not have the sentience necessary for meaning-making, it would eventually reveal itself by making naïve errors, much like non-conscious AI systems do. This would betray its lack of consciousness and semantic understanding. It is reasonable, then, to conclude that beings that act like humans yet lack subjective experiences do not exist, as otherwise, we would have already distinguished them from conscious humans. A test centered on meaning-making provides a stronger and potentially more convincing proof of the presence or absence of consciousness and true intelligence in a machine than the Turing test does.

One possibility could be to draw inspiration from modern tests that evaluate forms of consciousness in animals and apply them to future machine intelligence that could potentially exhibit signs of conscious awareness. At this stage, I can't provide concrete, implementable testing frameworks, but considering the possibility that meaning-making and consciousness are interconnected offers, in and of itself, a new perspective that could potentially lead to innovative practical approaches and methodological tests as well.

As a conclusive remark, it might be worth noting further that no contemporary AI exhibits any signs of agency and personhood (Browning, 2024). Without input, these systems do not act autonomously. The scientific concept of agency—encompassing intentionality, autonomy, self-determination, purposeful decision-making, and its connection to cognition—is itself a complex subject (Virenque, 2024). However, it is clear that regardless of the definition we adopt, nothing resembling agency exists in current AI systems. If one does not ask a question to ChatGPT, it will remain in a passive idle state, doing nothing and waiting for input indefinitely.

It will be interesting to observe the recent efforts to develop 'agentic AI' systems—AI agents capable of achieving specific goals with minimal supervision, while mimicking human decision-making to solve problems in real time. Research in this area focuses on creating systems that exhibit autonomy, goal-driven behavior, and adaptability. The hope is that this will be achieved through generative AI techniques that use LLMs to enhance their capabilities for independently completing complex tasks.

If the present analysis is sound and applicable to agency as well, we can expect to see the same pattern observed in the emergence of AI abilities over the past couple of decades—characterized by alternating successes and impressive breakthroughs, but also marked by significant cognitive failures—appearing in agentic AI systems with similar deficiencies. The question of whether agency and consciousness might be intrinsically related, as with semantics, deserves more attention, particularly as we speculate about the potential emergence of AGI.

Conclusions

I have focused primarily on the relationship between human-like experience and semantics because it represents the most familiar form of cognition for us. Nonetheless, I believe similar arguments can apply to non-human animals, with sentience also serving as the basis for meaningfully navigating the world.

Conversely, one could argue that machine learning without any conscious experience could be viewed as a form of "understanding" as well. Avoiding an anthropocentric perspective means acknowledging that our concept of understanding need not be confined to human ways of making sense of the world; indeed, computers could "understand" things in their own manner.

In fact, recent findings in mechanistic interpretability (Beckmann and Queloz, 2025), the field dedicated to understanding the internal reasoning processes of trained neural networks and LLMs, reveal that these systems do not rely solely on statistical inference. LLMs also develop internal structures that are functionally analogous to key aspects of human understanding. They perceive connections and form learned vector features within a multidimensional latent space, which humans typically bind and coalesce under a single concept. However, the understanding exhibited by LLMs is fundamentally different from human understanding. We need new frameworks for understanding that can embrace the emerging forms of intelligence we are creating, without assuming that human comprehension is necessarily unique, exclusive, or superior.

That said, it is crucial to question whether machine understanding differs from human understanding precisely because of its lack of subjective experience. We can undoubtedly attribute a different way of "understanding" the world to machines; however, they may never attain certain human-like cognitive skills, just as we will never achieve specific machine-like capabilities, precisely due to the missing link of consciousness.

On the other hand, animal forms of consciousness may result in entirely different types of semantic awareness of the world. We should not assume that human subjective experience is the only standard either. Perhaps there even exists an extraterrestrial civilization with a radically different understanding of reality. What do we know?

This is plausible, as human semantics might be only one of infinitely many varieties of semantics. However, this paper's main claim does not focus on any specific form of human cognition. Rather, the paper posits that semantics is inherently tied to conscious experience, whether human, animal, extraterrestrial or else. If semantics is not grounded in some form of consciousness and sentience, without any perception of the world's properties through qualitative subjective experiences, then it would, by definition, be a p-zombi semantics. In principle, we would eventually be able to distinguish zombies from other conscious beings. Their “understanding,” exhibited by a cognition devoid of qualia–no matter how complex, powerful, or sophisticated–would remain at the level of a Turing machine head processing physical information on a strip of tape. Nothing would convert this physical information into genuine semantic understanding.

The realization of AGI, if it ever happens, will first require the creation of conscious machines. There is a foundational, principled argument against the imminent arrival of AGI, as we remain far from developing systems that transcend memorization, pattern matching, or probabilistic assessments to achieve genuine, complex abstract reasoning based on conscious experience. Therefore, a first-person perspective, not only in the philosophy of mind but also with regard to addressing fundamental questions about AI, is essential. Approaching AI from this perspective might also help us gain a deeper understanding of ourselves.

Whatever the case, framing certain philosophical problems can serve as the initial step toward developing new methodologies and practical applications. Our actions are influenced not only by our knowledge but also by our underlying philosophical worldviews. If programmers, developers, and software engineers explored deeper questions of the philosophy of mind about the origin and nature of consciousness, intelligence, the mind, and life, they could potentially gain insights that a purely technical perspective cannot provide.

Footnotes

[1] One might argue that the meaning of symbols is rooted in their "use"—specifically, in the function they serve within a given context. However, from a more computational and reductive perspective, there is no inherent ‘use’ or function in the cell state that a Turing machine's head reads and reacts to, other than the meaning assigned by an external semantic agent.

[2] For a good summary of these theories and their inability to solve the problem of meaning in AI shortly before LLMs took center stage, see (Froese, 2019).

[3] Here, the term “reasoning” encompasses not only problem-solving abilities that rely on logical processes like deduction, induction, abduction, multi-step inference, or mathematical skills, but also those requiring abstraction, generalization, semantic discrimination rather than plausible best guesses, and the capacity to organize thoughts and conceptual structures into coherent, meaningful wholes–that is, what is commonly referred to as “common sense,” “rationality,” or “thinking” in human cognitive skills.

[4] Anyone with teaching experience recognizes this behavior in students who cheat. When they do not know the answer, they abandon meaningful reasoning and instead try to arrive at the correct answer by guessing–mimicking the methods and rules they have seen applied but without truly understanding the problem-solving method.

[5] This is also reminiscent of the fact that, contrary to popular belief, humans do not possess the largest brain. Other species have larger brains in terms of size, number of neurons, or weight. What, then, determines human cognitive dominance?

[6] For many more examples, see Dr. W. Hsu's collection of LLM failures: https://lnkd.in/eUW6TYCY.

[7] For further review on how close we are to AGI, see also (Ananthaswamy, 2024), and references therein.

[8] One of the most dramatic and well-known examples of this is the story of deafblind Helen Keller.

[9] One might argue that in human infants, certain cognitive developments occur prior to full conscious awareness. This suggests that some forms of understanding can arise without phenomenal consciousness. However, this conclusion relies on the questionable assumption that newborns have no subjective experiences, which contradicts all external evidence. The burden of proof lies with those who claim that babies lack sentience.

[10] Children who, for whatever reason, do not learn to relate the number symbols, or a perception of numerosity of objects, to a "number sense" are most likely to develop mathematical learning disabilities like dyscalculia (Decarli, 2023).

[11] A careful first-person investigation reveals that this isn't the case only with sensory inputs, but with thoughts and emotions as well.

[12] I like to reframe this as the "hard problem of semantic awareness."

References

Ananthaswamy, A. (2024). "How close is AI to human-level intelligence?" Nature. 636: 22-25. doi:10.1038/d41586-024-03905-1

Baars, B. J. (1988). A cognitive theory of consciousness. New York: Cambridge University Press.

Beckmann, P., Queloz M. (2025). “Mechanistic Indicators of Understanding in Large Language Models”. doi: 10.48550/arXiv.2507.08017

Browning, J. (2024). “Personhood and AI: Why large language models don’t understand us.” AI Soc. 39:2499-506. doi: 10.1007/s00146-023-01724-y.

Chalmers, D. (1995). "Facing up to the Problem of Consciousness." Journal of Consciousness Studies, 2(3): 200-19.

Decarli, G.; Sella, F.; Lanfranchi, S. et al. (2023). "Severe Developmental Dyscalculia Is Characterized by Core Deficits in Both Symbolic and Nonsymbolic Number Sense." Psychological Science. 34(1): 8-21. doi:10.1177/09567976221097947

Dziri, N.; Ximing, L.; Sclar, M. et al. (2023) "Faith and Fate: Limits of Transformers on Compositionality." Advances in Neural Information Processing Systems, vol. 36, pp. 70293-70332.

Froese, T.; Taguchi, S. (2019). "The Problem of Meaning in AI and Robotics: Still with Us after All These Years." Philosophies. 4(2): 14. doi:10.3390/philosophies4020014

Gulick, R.V. (2014). Consciousness. In: The Stanford Encyclopedia of Philosophy [Internet]. Available from: https://plato.stanford.edu/entries/consciousness/

Harnad, S. (1990) "The Symbol Grounding Problem." Physica D: Nonlinear Phenomena. 42(1-3): 335-346. doi:10.1016/0167-2789(90)90087-6.

Harnad, S. (2007). "Symbol grounding problem." Scholarpedia. 2(7): 2373. doi:10.4249/scholarpedia.2373.

Harnad, S. (2025). "Language Writ Large: LLMs, ChatGPT, Grounding, Meaning and Understanding." Front. Artif. Intell., Volume 7. doi: https://doi.org/10.3389/frai.2024.1490698

Held, R.; Ostrovsky, Y.; de Gelder, B. et al. (2011). "The newly sighted fail to match seen with felt." Nature Neuroscience. 14: 551–553. doi:10.1038/nn.2795.

Herzog, M. (2008). "Binding Problem." In: Binder MD, Hirokawa N, Windhorst U (eds) Encyclopedia of Neuroscience. Springer, Berlin, Heidelberg. doi:10.1007/978-3-540-29678-2_626.

Jackson, F. (1982). "Epiphenomenal Qualia." Philosophical Quarterly. 32: 127–136. doi:10.2307/2960077.

Jiang, B.; Xie, Y.; Hao, Z.; et al. (2024) "A Peek into Token Bias: Large Language Models Are Not Yet Genuine Reasoners." https://arxiv.org/abs/2406.11050

Kambhampati, S. (2024). "Can large language models reason and plan?" Annals of the New York Academy of Sciences. 1534: 15–18. doi:10.1111/nyas.15125.

Kuhn, R. L. (2024). A Landscape of Consciousness: Toward a Taxonomy of Explanations and Implications. Progress in Biophysics and Molecular Biology. doi:10.1016/j.pbiomolbio.2023.12.003

Landau, B.; Gleitman, L.R. (1985). Language and experience: Evidence from the blind child. Cambridge (MA): Harvard University Press.

Lappin, S. (2024). Assessing the strengths and weaknesses of large language models. J Log Lang Inf. 33:9-20. doi: https://doi.org/10.1007/s10849-023-09409-x.

Mahowald K., et al. (2024). Dissociating language and thought in large language models. Trends Cogn Sci. 28(6):517-40. doi: https://doi.org/10.1016/j.tics.2024.01.011.

Masi, M. (2025). "The Nature and Origin of Language in Abhinavagupta and Sri Aurobindo." Upcoming.

McCoy, R.F., Yao, S., Friedman, D. (2024). "Embers of autoregression show how large language models are shaped by the problem they are trained to solve." PNAS, Computer Science, 121(41): e2322420121. doi:10.1073/pnas.2322420121.

Mirzadeh, I.; Alizadeh, K.; Shahrokhi, H. et al. (2024). GSM-Symbolic: Understanding the limitations of mathematical reasoning in large language models [Preprint]. Available from: https://doi.org/10.48550/arXiv.2410.05229

Mollo, DC, Millière R. (2023) "The Vector Grounding Problem." doi:10.48550/arXiv.2304.01481

Nagel T. (1974). What Is It Like to Be a Bat? The Philosophical Review, Vol. 83, No. 4, pp. 435-450. Duke University Press

Oizumi, M., Albantakis, L., Tononi, G. (2014). "From the phenomenology to the mechanisms of consciousness: Integrated information theory 3.0." PLOS Computational Biology, 10(5). DOI: https://doi.org/10.1371/journal.pcbi.1003588

Penrose, R. (1989). The Emperor's New Mind: Concerning Computers, Minds, and the Laws of Physics. Oxford: Oxford University Press.

Prabhakar, A., Griffiths, T.L., McCoy, T. (2024). "Deciphering the Factors Influencing the Efficacy of Chain-of-Thought: Probability, Memorization, and Noisy Reasoning." doi:10.48550/arXiv.2407.01687

Queloz, M. (2024) "Can Word Models be World Models? Language as a Window onto the Conditional Structure of the World." https://philpapers.org/rec/QUECWM

Sacks, O. (1985). The Man Who Mistook His Wife for a Hat. New York: Summit Books.

Scheffer, R.; Miranda B.; Koyejo, S. (2023). "Are Emergent Abilities of Large Language Models a Mirage?" 37th Conference on Neural Information Processing Systems.

Searle, J. (1980). "Minds, Brains and Programs." Behavioral and Brain Sciences. 3: 417–57.

Searle J (1990). Is the brain a digital computer? Am Philos Assoc. 64(3):21-37.

Senden, M. v. (1960). "Space and sight." Methuen, London.

Seth, A. K., Bayne, T. (2022). Theories of consciousness. Nature Reviews Neuroscience, 23, 439–452.

Shannon, C.E. (1948). "A Mathematical Theory of Communication." Bell System Technical Journal. 27(3): 379–423. doi:10.1002/j.1538-7305.1948.tb01338.x.

Shi, F.; Chen, X.; Misra, K., et al. (2023) "Large Language Models Can Be Easily Distracted by Irrelevant Context." International Conference on Machine Learning, ICML 2023, 23-29 July 2023, Honolulu, Hawaii, USA, volume 202 of Proceedings of Machine Learning Research, pp. 31210–31227. PMLR, 2023. URL: https://proceedings.mlr.press/v202/shi23a.html

Shojaee, P. et al. (2025). “The Illusion of Thinking: Understanding the Strengths and Limitations of Reasoning Models via the Lens of Problem Complexity”, https://ml-site.cdn-apple.com/papers/the-illusion-of-thinking.pdf

Vallor, S. (2024). "The Danger Of Superhuman AI Is Not What You Think." Noema, Technology & the Human. URL: https://www.noemamag.com/the-danger-of-superhuman-ai-is-not-what-you-think/

Virenque, L.; Mossio, M. (2024) "What is Agency? A View from Autonomy Theory." Biol Theory. 19: 11–15. doi:10.1007/s13752-023-00441-5

9. The Origin and Nature of Speech in Abhinavagupta and Sri Aurobindo[217]

Abstract

The paper delves into the nature and origin of ideas, words, meanings, speech, and language from the perspective of Indian mystics and philosophers Abhinavagupta and Sri Aurobindo. To provide historical context, we begin with the Eastern viewpoint, starting with the Vedic interpretation, which posits that the source of all speech is the transcendent sound, known as the 'Word'. Later, Abhinavagupta delineates the genesis of words as a four-level process within consciousness, where mystic sounds gradually acquire concreteness in the form of human language. In the 20th century, Sri Aurobindo extended Abhinavagupta's framework, envisioning all words as stemming from mind-impressions elicited by seed-sounds bearing psychological qualities. Moreover, according to Sri Aurobindo, the universe and its natural phenomena represent the expression of a creative transcendental Real-Idea, which serves as the foundation for all meanings, signs, words, and human language and Nature itself. Language, therefore, becomes an instrument for the expression of divine consciousness in manifestation. Finally, it is shown how the integration of Western and Eastern philosophies into a synthetic framework that unifies the diverse positions opens interesting venues that could reconcile many apparent inconsistencies within modern philosophy and science, especially in the philosophy of language, evolutionary linguistics, cognitive sciences, neuroscience, and AI.

I. Introduction

Despite more than one and a half centuries of speculation and research, the nature and origin of human language remain highly debated issues. The quest to unravel the inception of language is an enduring intellectual journey that has fascinated linguists, anthropologists, cognitive scientists, and philosophers alike. In the Western tradition several competing theories have emerged over the last century and a half, but the debate over whether the nature and origin of speech and language, the development of signifying ideas, the use of words, and the emergence of meaning are products primarily of biological evolution, cultural development, or something else, remain a matter of controversy. Nonetheless, what they all have in common to a greater or lesser extent is that they are based on a tradition of analytic philosophy and a predominantly materialistic conception of cognition and life. Language is seen as a product of

[217] Upcoming publication.

social interactions, cultural development, an anthropological and societal perspective, or a neurological and genetic outgrowth in the frame of Darwinian evolution. With few exceptions,[218] from the foundations of semiotics of Saussure, through the positivistic approach of Wittgenstein, to the biological outlook of Chomsky, who argued that *"there is a special component of the human brain (call it 'the language faculty') that is specifically dedicated to language"* and which is genetically determined (Chomsky, 1996), despite all their differences, their commonality is that they are rooted in a strictly naturalistic paradigm. However, from a historical perspective, it is often overlooked that, particularly in the Eastern tradition, the philosophy of language adopted a different approach. Here, we will extend our view from an exclusive third-person rationalistic perspective to a first-person position inspired by the spiritual insights of two personalities who developed a profound vision of language, namely, the Indian mystics Abhinavagupta Acharya (AD 950-1016) and Sri Aurobindo Ghose (AD 1872-1950). To the best of my knowledge, only few authors tackled with Abhinavagupta's 'grammatical cosmology' (Biernacki, 2013a) and 'theogrammatical topography of the One and the many' (Biernacki, 2013b). Biernacki also relates Abhinavagupta's philosophy to the contemporary philosophy of mind and how it could lay the foundations for a new materialism (Biernacki, 2023).

In this paper, I will take a different approach. If we admit that the origin of language may be tightly connected with the origin of consciousness and that consciousness is not explicable in a naturalistic theoretical framework (something that is resurfacing more than ever in the contemporary philosophy of mind; for a review, see Mørch (2023)), then we are allowed to take a different approach and outlook toward the question of the origin and nature of language and see if it may have a deeper explanatory power than a strictly naturalistic worldview.

Though they are usually considered 'philosophers,' both Abhinavagupta and Sri Aurobindo must be understood within their cultural context, they were not philosophers in the ordinary sense. Their teachings came from the perspective of the mystic, not the thinker. They sought to argue from the perspective of higher states of consciousness beyond the analytic mind rather than from intellectual constructs and speculations. Their worldview was rooted in the ontology of the Vedic rishis, the ancient Indian mystics. However, this doesn't imply that the Eastern and Western currents must remain distinct and separate forever. There is a commonly shared bridge. The point of contact is

[218] The model of the origin of language presented here has similarities with the theory of the origin of language presented, for example, by Süßmilch (Süßmilch, 1766) who claimed that language is necessary for the development of reason and, thereby, must be of divine origin. However, in Abhinavagupta and Sri Aurobindo there is no claim that language develops mental functions in humans. It is the faculty of reason that contacts language, not the other way around.

the common search for a theory of how we represent the essence of reality with ideas, signs, words, forms, symbols, speech and language, and meanings.

The first section of the next part will discuss the common basis of both seers: At the origin of all existence is consciousness expressing itself by means of sound. The Rig Veda speaks of a primeval vibration beyond the physical realm that generates all existence and becomes the basis of speech as well. Here, consciousness and language are inherently interlinked. The second section describes how Abhinavagupta starts from the Vedic tradition by describing how the 'Sound-Word,' descending through four levels of consciousness and manifestation, becomes human speech and language, which are attempts to express these transcendent sounds in physical form, and that we only subconsciously know, as they originate from a realm beyond our ordinary states of consciousness. The third section describes Sri Aurobindo's theory of the origin of speech, which follows in the footsteps of Abhinavagupta. The evolution of speech emerged from sensations, emotions, and nervous utterances first, became a living language based on seed-sounds having psychological qualities, and then transitioned from general notions to precise expressions under the influence of the intellect and reason. Moreover, in the fourth section, his outlook of the world as a Real-Idea will put both his and Abhinavagupta's theory of language in a broader perspective. All the universe, matter, and life are a sign, a symbol, or just a 'word.'

The final chapter will attempt to synthesize these theoretical frameworks into an integral view, highlighting how they fit into our contemporary scientific, philosophical, and philological knowledge. Several aspects that remain obscure in contemporary philosophy of language, rooted in a biological paradigm, might receive interesting insights from this spiritual philosophical framework. However, a broader perspective in which we can connect the dots is needed.

II. The Nature of the Word and the Origin of Speech in Abhinavagupta and Sri Aurobindo

1. The Transcendent Sound-Word

The Rig Veda speaks of the universe created by an original and originative 'sound,' or cosmic vibration, the seed of the transcendent sound and a sonic representation of the Brahman, the Absolute of the Vedic texts, as 'Aum' or 'Om' ('Praṇava'), and represented by the sacred Sanskrit syllable ॐ. Everything inside the so-created universe also originates from vibrations of consciousness or 'etheric sounds.' These immaterial vibrations on higher planes of existence determine the creation, change, and dissolution of all forms on the material plane of existence.

According to the Veda, the original vibration is not physical. Its origin and primary nature are always transcendent, supra-material, beyond matter, space, and time. The rhythms of cosmic vibration become physical only at the most material level. Consequently, the primary vibration or 'sound' ('śabda') is transcendent and issued forth from a supreme consciousness, descending through four universal levels of manifestation, with the physical manifestation being the last plane in a series of descending planes. The idea of the four levels of consciousness is a common theme in the Rig Veda, and its correlation with speech was already present in the teachings of Rishi Diirghatamas.

"The Word has been measured in four quarters. Those quarters are known to the knowers of the Word, brāhmanaḥ, who possess also the power of the Mind, manīṣiṇaḥ. In the hidden place the three are established, which do not move. And the forth one men speak." (Rig Veda 1.164.45)

Similarly, in the mental realm, mental forms and their changes are realized by mental vibrations. Just as all material objects are the result of a combination of physical vibrations, all thoughts are a combination of mental vibrations. These are inaudible to the physical ear but are perceivable to the mind. The same rationale applies to even higher planes of consciousness beyond the mind. The goddess Word—'Vāc'—is the vibration of a supreme consciousness, loaded with an inherent supreme knowledge, vision, and creative power, that holds and sustains the manifestation by penetrating Earth and Heaven and holding together all existence (Rig Veda, 10.125). In Sanskrit, vāc means 'to speak,' and in Latin, it corresponds to 'vox,' meaning 'voice.' However, in this context, vāc refers to sound. It is the divine Will that manifests its creation by means of the creative sound. For the Vedic seer, the entire manifestation is an expression of the unmanifest, formless, transcendental 'Word.' For the Upanishads, Vāc is superior to name and should be worshiped (Chandogya Upanishad 7.2.1). It is something reminiscent of the meaning of the 'Word' in the Bible: *"In the beginning was the Word, and the Word was with God, and the Word was God"* [John 1:1].

Thus, according to the Veda, all mental vibrations, which we call 'thoughts' or 'ideas,' and all physical sounds, which we call 'spoken words,' owe their origin to 'voiceless,' 'unvoiced,' 'unheard' subtle vibrations of consciousness beyond the gross physical and ultimately originate in the universal consciousness that casts them into 'thought/idea-forms' and 'matter-forms' by descending to the mental and physical planes. Or, to put it in the words of

Western philosophy, words exist 'archetypically' in a separate realm[219], something echoing the realm of Plato's Forms ruled by a 'universal phonological semiotic'.

However, according to the Vedic seer, in our state of ignorance absorbed in the physicality and inertia of matter, so conditioned by our half-animal bodily and subconscious nature, we are only subliminally aware of this truth of things. We can discover it only by a process of spiritual ascesis. Nevertheless, on our ordinary cognitive state, language is an attempt of the human mind with its bodily apparatus to express these etheric sounds with physical sound vibrations in the form of sound-units, segments, phonemes, words, and sentences.

Therefore, from the Vedic perspective, in their origin, words are subtle trans-rational vibrations that have not yet dressed themselves in intellectual concepts and language. Language is adopted only later, at the level of the surface mind. In their original form, words exist as a 'consciousness-form' that precedes apprehension, comprehension, conception, perception, and any intellectual analysis. The word articulated with its sound in the form of physical air pressure vibrations using the organs of speech is only the very last stage of differentiation, manifesting a mental idea, a concept, or the names and forms of objects.[220] The true origin of every spoken word and phoneme is beyond any physical or mental plane, where an original 'sound-idea' is expressed, and the human mind can only vaguely intuit it, and not fully comprehend it, unless one ascends to higher cognitive states through spiritual practice.

---

[219] Here, the term "archetype" refers to a semiotic universal idea. Jung defined archetypes differently, describing them as patterns of behavior that emerge from the collective unconscious consisting of innate psychic structures and predispositions—essentially instinctive blueprints for the human psyche. The Jungian archetypes are considered structural templates that manifest in response to biological instincts and become evident in common myths, dreams, religious symbols, and artistic expressions. Nonetheless, Jung viewed archetypes as pre-linguistic forms that enhance the depth of meaning, with language serving as a symbolic expression of these archetypes. Language is the means through which archetypes become visible. The richness of meaning derives from its archetypal roots in the collective unconscious. Without the underlying archetype, a symbol would simply be a sign. Therefore, with a slight abuse of language, we will adopt Jung's terminology.

[220] As we shall see in more detail later, by 'names' (nāma), the rishis didn't mean just the words describing objects but rather, in its deeper sense, the powers, qualities, and characters (features and traits that distinguish one thing from another) caught up by ourawareness. In this sense, all 'names' are already latent and inherent in the nameless and timeless Absolute but are expressed in the temporal manifestation as qualities, powers, and characters of its forms. Similarly, 'forms' (rūpa) have to be intended in its more general connotation. Not only are signs, words, and symbols forms, but sound waves, ideas, mental concepts, and even feelings or emotions are 'forms' of and in consciousness.

From this phenomenological first-person cognitive perspective of the Vedic system, words and sounds are not just a means of expressing an intellectual idea. They are the reflection of an original immaterial vibration pregnant with meaning. Words and sounds can still carry the memory of its 'root-meaning,' that is, its 'root-sound,' which can be apprehended if one has developed an awareness of the 'sound-idea' and can hear the 'internal sound' standing behind the word. A Speech beyond our speech and mind exists, made of expressions of truths that already pre-exist in the Infinite and of which the human language is only a shadow and artificial counterfeit. The symbols or representations that the human language expresses are external representations of the things in themselves residing in another reality. In the Vedic tradition, the 'forms' and 'objects' are beyond any mental conception and must be found in the formless and timeless domains of the 'upper Ocean': *"It is from her, the all creative Word, that the waters of the upper Ocean flow down in all directions"* (Rig Veda 1.164.42).

The Word on the level of that Reality is much more creative than human speech can be. Consciousness uses vibrations as a form of pre-reflective self-expression because every form, quality, quantity, and property itself is an expression of a vibrational power on one or another plane of existence. The unspoken, formless, and unmanifest Word isn't a passive and inactive sound without power and force. The Word of the Indian Vedic rishi is a living entity with creative and formative powers. In fact, the idea standing behind the mantra—the sacred word, phrase, or sound used in the Buddhist and Hindu traditions—assumes that, if the spoken word is charged by a spiritual vibration, it can still carry a creative power that the ordinary word doesn't have. It is supposed to produce vibrations in our mental and emotional atmosphere, with potentially beneficial physical effects. It is what the repetition of a mantra ('Japa') during a spiritual practice is said to achieve: The psychologically energized words and sounds aid the practitioner in transcending the mind. Human speech is 'creative' in the sense that it can create mental images, forms of emotions, impulses of action and movement, sense-properties, etc. Thoughts and words are carriers of a force. Thereby, in sound waves, words can encode features and powers of the awareness of who is speaking.

The origin of the Word resides in what Sri Aurobindo called the 'Supermind'—the supreme creative Logos—the highest level of the creative consciousness and gnosis of the Absolute acting in the manifestation.[221]

"Let us suppose a conscious use of the vibrations of sound which will produce corresponding forms or changes of form. But Matter is only, in the ancient view, the lowest of the planes of existence. Let us realize then that a vibration of sound on the material plane presupposes a corresponding vibration on the vital [emotional plane] without which it could not have come into play; that again presupposes a corresponding originative vibration on the mental; the mental presupposes a corresponding originative vibration on the supramental at the very root of things." (Sri Aurobindo, 1952, Ch. V, p. 31)

"All vibration of sound on that higher plane is, then, instinct with and expressive of this supreme discernment of a truth in things and is at the same time creative, instinct with a supreme power which casts into forms the truth discerned and eventually, descending from plane to plane, reproduces it in the physical form or object created in Matter by etheric sound. Thus we see that the theory of creation by the Word which is the absolute expression of the Truth, and the theory of the material creation by sound-vibration in the ether correspond and are two logical poles of the same idea. They both belong to the same ancient Vedic system." (Sri Aurobindo, 1952, Ch. V, p. 32)

On our cognitive level made of words, the Word is expressed by 'seed-sounds,' which carry in them the principles of things and forms in the phenomenal world.

"Human speech is only a secondary expression and at its highest a shadow of the divine Word, of the seed-sounds, the satisfying rhythms, the revealing forms of sound that are the omniscient and omnipotent speech of the eternal Thinker, Harmonist, Creator. The highest inspired speech to which the human mind can attain, the word most unanalysably expressive of supreme truth, the most puissant syllable or mantra can only be its far-off representation." (Sri Aurobindo, 1952, Ch. V, p. 33)

At first, the idea that human speech has its basis in a 'transcendent speech' (Sri Aurobindo, 1950, p. 375) being a shadow of the divine Word might sound too mystical. However, in the light of modern physics, this isn't a far-fetched vision. Waves in their mechanical form (pressure waves in a solid, liquid, or gaseous body) or electric, magnetic, or electromagnetic waves (light waves,

[221] Sri Aurobindo reported the existence of states of consciousness transcending ordinary human cognition, which he classified in ascending order as the 'higher mind' (a *"luminous thought-mind"* and "spirit-born conceptual knowledge"), 'illumined mind' (a *"clarity of the spiritual intelligence"* and an *"illumination of the spirit"*), 'intuitive mind' (*"when the consciousness of the subject meets with the consciousness in the object, penetrates it, and sees, feels, or vibrates with the truth of what it contacts"*), 'overmind' (a universal or cosmic consciousness of *"wide cosmic perception and feeling of a boundless universal self"*), and 'Supermind' (the *"divine gnosis"* or *"Truth Consciousness"*) (Sri Aurobindo, 1919a, p. 985).

radio waves, X and gamma rays, etc.) are the foundation of classical physics. What is often not realized is that all waves are ultimately 'immaterial.' Mechanical waves in matter are vibrations propagating between molecules through an exchange of momentum mediated by electromagnetic waves or photons, which are not material (in the sense that photons have no mass). In quantum physics, things become even more subtle because the waves considered in this context are 'probability waves' that are neither material nor immaterial but, rather, abstract mathematical functions representing the probability that a certain quantum event will take place. These form so-called 'probability clouds' around atomic nuclei (and that one can calculate with the famous Schrödinger wave equation), known as 'orbitals,' and ultimately serve as the model by which physics represents atoms, molecules, and all matter. Furthermore, in quantum field theories, all the fundamental constituents of matter and all forces are represented by oscillatingquantum fields. At the core of it all is always some form of vibration.[222]

However, the Vedic rishis would point out that physical waves are only a shadow or a mask representing the original vibration coming from other planes of existence. For them, the physical sound was only a distorted image of the mystical sound. One cannot comprehend, let alone recreate, the latter starting from the former but, rather, must first meet the sacred sound with an inward turn of the mind, sustained by a spiritual and meditative practice, such as yoga. Ultimately, all the thoughts and ideas and their very essence possess a 'semantic charge' we call 'meaning,' and have their origin and nature in something beyond that which a strictly naturalist philosophy or science can account for.

This is the theory of the cosmic original and originative transcendent sound as the source of all signs, words, meanings, speech, and languages, in the Vedic tradition. However, the Vedic texts remained vague and didn't elaborate much on this aspect. They are quite cryptic, with their interpretation being highly subjective. It was Abhinavagupta and his followers who developed a detailed theory, building upon the classifications made by the fifth-century linguist and philosopher Bhartṛhari. Bhartṛhari had labeled the four levels[223] of consciousness and speech as 'parā vāc,' 'paśyantī vāc,' 'madhyamā vāc,' and 'vaikharī vāc.'

---

[222] It might be worth recalling how waves have the power to create and dissolve regular patterns in resonant structures through standing waves—a well-known fact in physics dating back to the 19th century. A nice didactical example showing how one can be creative with sound waves that form structures, change them, and eventually dissolve them is illustrated with sand grain patterns forming on vibrating Chladni plates, or with so-called 'Faraday wave patterns' in water. Vibrating forces give rise to forms with order, structure, and symmetry.

[223] To be precise, in his main text, the Vakyapadiya, Bhartṛhari speaks of only three 'levels'—that is, paśyantī, madhyamā, and vaikharī—but mentions śabda-Brahman, which is parā vāc, as the Absolute containing it all.

2. The Four Levels of Speech of Abhinavagupta

Abhinavagupta Acharya (950-1020 CE) was an Indian mystic and prolific philosopher, artist, and musician who lived in the 10th -11th centuries. He is regarded as one of the most significant figures in the history of Kashmir Shaivism. His teachings about the path to self-realization have made him a revered figure in the history of Indian thought and spirituality. He made significant contributions to the understanding and practice of Tantra. His works, such as Tantrāloka and Tantrasāra, are considered foundational texts in the Tantric tradition.

One of Abhinavagupta's most significant philosophical works concerns the 'four levels of speech.' His metaphysics of language can be found throughout his texts but are especially emphasized in the Parātriśikā-Vivaraṇa (Abhinavagupta, 2011).

In line with the Vedic and Tantric tradition, Abhinavagupta posited the foundation of all creation not in matter but in spirit. As described above, the origin of the word is found in the transcendent śabda-Brahman ('śabda': sound, vibration, word), which is the Sound and Word residing in the Absolute (Williams, B. 2022). The foundation of Abhinavagupta's theory is based on the four levels of consciousness, each of which corresponds to a level of speech, reflecting the intrinsic and indissoluble relation between consciousness and speech. Vāc, the word-sound, is a form of expression of consciousness. This transcendent Sound descends from the Absolute, then progressively moves through the four levels described by the Vedas until it reaches the physical realm, where it finally finds its verbal and vocal expression in the form of a language.

He explained the emergence of speech, which is the transformation from the original undifferentiated vibration, idea, meaning, and word into differentiated spoken language, using the four levels of language as follows.

Parā Vāc—The level of pure Consciousness

In Abhinavagupta's scriptural sources the supreme level of parā vāc (the beyond-sound) represents the transcendent and unmanifest consciousness of the Absolute, known as śabda-Brahman, where the essential nature of the sound principle (śabda-tattva) reside. This realm exists as an intangible and unmanifest plane without space and time of the word principle (avyakta). It encompasses both the pure light of changeless Consciousness (prakāśa) and the pure vibratory power of Consciousness (vimarśa). In essence, consciousness, light, and the word are unified (pratyavamarśa). At this level, there is no principle of separation. The still unexpressed word, embodies the pure consciousness of Cit-Tapas or Cit-Shakti—the Consciousness-Energy carrying the luminous vibration (sphurattā) of complete knowledge.

Parā vāc represents the supreme level of the Word without subject-object distinction. Here, the Word is transcendental and beyond creation, archetypical and undifferentiated, containing the entirety of significance, power, and essence in its potential state. It is discernible only through introspective awareness that has not yet been formulated into mental representation.

There exists no distinction between the sound-word (śabda) and its true meaning (artha). Sound and meaning are coalesced; sound embodies meaning, and meaning is sound. Meaning is the inner essence of an expression existing within the consciousness of a spoken word (antar abhilāpa.) There exists a vibration of knowledge and power in an undifferentiated state before the descent from one plane to another. Subject, object, name, form, image, sound, and meaning are inseparable. Perfect unity exists between the object of sense and sense itself.

At this level, words remain unexpressed verbally and mentally, existing only in a state of potency, devoid of physical sound, phonemes, letters, or written words. In line with Wittgenstein's intuition, here also meaning is perceived as independent from language.

Paśyantī Vāc—The level of the will to know and thought visualization

In Abhinavagupta's framework, while at the parā vāc level, consciousness, will, knowledge, and action exist as one and the same, at the paśyantī level, the distinctive feature is the active power of will (icchā śakti), encompassing the potential for cognition and actualization. The luminous and vibratory aspects of consciousness, prakāśa and vimarśa, begin to diverge, giving rise to audible sound (nāda). An intention to speak takes shape as a flash of intuition from above. The initial moment of desiring to understand and a will to remember emerges. Each word still holds numerous possible meanings but inclines toward self-manifestation in name and form (nāma and rūpa).

Paśyantī also means 'she who sees,' denoting the visual image of the word. The visualization of ideas and thoughts commences with a beginning of discernment. Form-images and form-sounds emerge, devoid of mental conceptualization (nir-vikalpa).

However, thought remains latent and unspoken, and there is still no distinction between subject and object—that is, between the subjective 'I-ness' (ahantā) and the objective 'thisness' (idantā)—and the signifier (vācaka) and the signified (vācya).[224] Separation has not yet occurred, and the sequence of sounds-words is a singular unit arising in the speaker's mind (sphoṭa) like a flash of intuitive insight (pratibhā) encompassing the intended meaning. Śabda and artha, the word and the word's meaning, continue to constitute a single vibrational semantic entity. There is still no analysis, rationalization, or logical

[224] The distinction reminiscent of Saussure's signified and the signifier might not be a coincidence. Saussure was a professor of Sanskrit.

inference. This represents the stage of pre-vocal thought or the 'voice of silence.'[225]

Madhyamā Vāc—The mediating level

Madhyamā means 'mediating' or 'middle.' This level serves as a mediator between the undifferentiated and differentiated realms of cognition, namely, between paśyantī vāc with vaikharī vāc. Icchā śakti becomes active through the power of cognition (jñāna śakti) at the level of madhyamā vāc. At this stage, the separative principle fully manifests. The first perceptible vibration emerges (referred to as 'spandana,' the creative 'pulsation of consciousness,' though still physically inaudible). Madhyamā represents the level of the subtle body (sūkṣma-śarīra), composed of the mind (manomaya-kośa), intellect (buddhi), and vital sheet (prāṇamaya-kośa). Here, the first distinctions between the infinite and finite, the transcendental and immanent, subject and object, and word and meaning arise.

Conceptualizations in the form of different thoughts (vikalpa) emerge, and words arise from within. This is the level at which the signifier and the signified become distinct, and words take on a temporal sequence. Language begins taking shape in the form of phonemes, words, and sentences, gaining the 'thisness' of the objective apprehensive cognition.[226]

While at the parā vāc and paśyantī vāc levels, the creative element with its meaning and expressive power, along with the knowledge it contains (śabda and artha), remain unspoken, involved, and transcendent, at the madhyamā vāc level, they acquire names and forms as explicitly audible mental vibrations, popularly referred to as 'internal speech.' The mind 'hears' these vibrations as articulate, tangible, and manifest sounds (vyakta), though they are still beyond the reach of the physical organs of sound: the ears. Linguistic signs with their syntactic structures, their distinct intellectual language categories, and the division between form-images and form-sounds come fully into existence.

Although there is still a unity in meaning, there are differences in the form of objects and the structures of language. Each word acquires its unique significance.

---

[225] While Abhinavagupta's paśyantī vāc level is described as something beyond what ordinary human analytical reasoning can grasp, it is suggestive of the proverbial 'aha moments.' The sudden realization of a deeper truth emerges as a flash of intuitive light and insight in silent, wordless conscious awareness—an insight that has not yet been articulated in spoken words. One might struggle to express it because it represents an inner knowledge that resists confinement within rigid conceptual forms.

[226] Something that could be compared with the haecceity ('thisness') concept from medieval scholastic philosophy, first coined by Duns Scotus, which denotes an irreducible property of an object that makes a thing unique. Another idea experiencing a revival is 'philosophical essentialism'—the hypothesis that humans perceive certain categories as having an underlying essence, having deep, unobservable, intrinsic properties that unify the members of a category (Neufeld, 2022).

Vaikharī Vāc—The level of bodily organs

Finally, the vibration materializes the verbal expression of the thought in the form of physical sounds on the material plane—namely, in the physical gross body (sthūla-śarīra or annamaya-kośa) with its speech organs (larynx, tongue, lungs, etc.). This represents the level of consciousness's power of action (kriyā śakti). The original Word is manifested not only in terms of speech production but also in terms of the objectification of reality. At this stage, speech, spoken language, and written words refer to objects. It is the level of conventional signs that hint and gesture toward something (saṅketa). Vaikharī is the manifestation of speech in time and space, encompassing all the distinct features of language: spoken phonemes, words, and sentences.

These were the four levels of language according to Abhinavagupta.

Thus, in this framework, spoken words and sentences capture only a very pale essence of their original true meaning and power of expression. In Abhinavagupta's linguistics, vocal speech obstructs the original supra-rational intuitive knowledge and power, which formal spoken or written signs can no longer fully convey. In this view, the inner perception of meaning does not emerge from mere symbols, letters, words, or text but, rather, from the perception of a transcendent sound. External symbols, signals, and words point to an internal domain. Only within this inner domain is their true original meaning revealed, and it is where they acquire a greater creative and expressive power than is usually recognized. Though parā vāc is always latent on all levels and projects itself in the form of different planes of subtleness, it is active only as potentiality and not as actuality. Parā vāc *"is indeed present on all the levels of paśyantī and others, for without her, darkness and unconsciousness, would prevail"* (Abhinavagupta, 2011, p. 5). If it were to withdraw its luminous Word, the entire cosmos would plunge into darkness.

3. The Origin of Speech According to Sri Aurobindo

The Indian mystic and poet Sri Aurobindo followed in Abhinavagupta's footsteps.

Sri Aurobindo's vision, teaching, and spiritual cosmology, based on his 'integral yoga,' are vast and complex. It is beyond the scope of the present paper to elucidate it in detail.[227] Here, I will focus on the relevant aspects related to the origin of speech. Sri Aurobindo focused primarily on language in the context of the Indian subcontinent, but his theory can be considered a general theory of all human languages.

Here also, sound comes first as an archetypal expression of a deeper meaning that outward symbols can't express. We can understand the nature of

[227] For a more comprehensive review, see (Masi, 2023a), and references therein.

language by silencing the mind and transcending it: *"It is in stillness that we can feel the unspoken word"* (Sri Aurobindo, 1950, p. 14). Sri Aurobindo explores this in his work on the "Kena and other Upanishads." The title of the fifth chapter, "The Supreme Word," speaks for itself.

"... human speech at its highest merely attempts to recover by revelation and inspiration an absolute expression of Truth which already exists in the Infinite above our mental comprehension. Equally, then, must that Word be above our power of mental construction. All creation is expression by the Word; but the form which is expressed is only a symbol or representation of the thing which is." (Sri Aurobindo, 1952, Ch. V, p. 29)

For Sri Aurobindo, the cultural development of language is a secondary factor. The primary force determining the birth of language and its path through history is the psychological factor. He provides hints about his theory in "The Secret of the Veda" in the chapter on "The Philological Method of the Veda" (Sri Aurobindo, 1914-20). According to this theory, human language initially developed from emotions, sensations, feelings, and infra-rational thoughts related to life occurrences. Something, the modern evolutionary linguist would compare to forms of communication in animal cognition or the language acquisition in infants, where their pre-linguistic vocalizations and gestures primarily reflect emotional and physical states and where iconic signals prevail over symbols. Words were originally expressions of physical and emotional needs, serving primarily life-preserving functions, without having developed into a more coherent structure made of fixed symbols with precise intellectual significances that express ideas about objects. In this early stage of language, the word was a more dynamic force, driven by sensations and emotions with its 'sense-value' and 'sound-sense,' rather than being primarily an intellectual idea.

"In their beginnings language-sounds were not used to express what we should call ideas; they were rather the vocal equivalents of certain general sensations and emotion-values. It was the nerves and not the intellect which created speech." (Sri Aurobindo, 1914-20, p. 51)

Only in later evolutionary stages does the idea become central, with the sound reduced to a symbol standing for the idea. Initially, each word could have several meanings,[228] and language expressed general notions such as light, motion, touch, substance, extension, force, speed, etc. For example, for the ancient Vedic population, 'vṛka' had a general meaning of 'tearing' or 'tearer,' and, therefore, among many other possible uses, it indicated a wolf. It was not restricted to nouns; it could be used as an adjective, verb, or adverb, all conveying the idea of tearing something apart within different contexts.

[228] This multiplicity of meanings should not be confused with the separative semantics of modern words, such as 'fleet,' meaning a number of ships, or swiftness indicating speed but not suggesting the swiftness of ships.

This kind of use was clearly visible in the Vedic hymns. Words did not indicate objects but were laden with psychological significance (or, should we call it 'energy' and 'sound'?) that could be applied in entirely different contexts without creating conceptual ambiguities. Another example: 'Agni' had a general meaning of strong, bright, force, or brilliance. These qualities could indicate a normal fire, a flow of heat or light, or just something giving wealth, while for the worshiper, it could signify the god of fire, a vital principle, or an illumined energy that built up the world, etc. Each word was a living thing with a creative and formative power where the sound determined sense. It had a holistic character, with many different meanings, and each object could be expressed by a large number of words. In this sense, words were not yet conventional symbols for an idea but, rather, flexible and living entities.

From his analysis of the underlying system of the structure of the Sanskrit language, Sri Aurobindo submits that only later did words and language shift from the general to the particular, from the vague to the precise, and from the physical to the mental. This shift occurred when the mind, the intellect, and analytical reason (or Abhinavagupta's madhyamā vāc level), with their precision and conceptual sharpness, intervened. The single words with multiple meanings—something that can still be found in the Sanskrit language—progressively acquired less flexibility and finally contracted to their present outer and concrete significance. The formal distinction between the action, the agent, and the object arose, differentiating between subjects, objects, nouns, and verbs.

This might seem like a great achievement of the mind. The conceptual clarity and precision of science, mathematics, and analytic philosophy, with their precise nomenclature and exact definitions and axioms, as well as the reduction to the particular vs. the general, is the extreme embodiment of this psychological and cultural process. However, for Sri Aurobindo, this hardening into the rigid forms of the intellect has its downside: The modern word lost its power and its creative and formative vibration, as well as its holistic character that the living words of the Vedic Rishis possessed.

Sri Aurobindo's main theory of the origin and nature of 'Aryan speech,' now referred to as Indo-European languages, is summarized in a booklet he wrote between 1910 and 1915, titled "The Origins of Aryan Speech" (Sri Aurobindo, 1912-14). His primary focus was on the study of Sanskrit because it preserves the original type and structure of an ancient language, with its system of primitive words and roots (i.e., etymons). It has a transparent derivative system of grammar that no other language can provide. Sri Aurobindo's profound mastery of English, along with his knowledge of Latin and Greek, allowed him to create a consistent theoretical framework applying to all Indo-European languages.

He warned against the fundamental mistake of jumping to conclusions by analyzing the identity or similarity of words between languages and, thereby, assuming this must indicate a common origin. A community of languages with common terms and vocabulary does not necessarily imply a common civilization. It could equally signify contact and cohabitation with each other in the past, despite their separate linguistic origins. Instead, Sri Aurobindo proposes starting from 'root-identities' and 'root-meanings,' deriving them from their 'sound-identities' and 'sound-meanings.' As previously pointed out, in Sri Aurobindo's evolutionary linguistics, language did not develop from a fixed set of rules and object-oriented words, as the modern intellect would prefer, but was born out of a system of utterances centered on sensations, qualities, and seed sounds that developed in phonological rules. Most importantly, these utterances created 'mind-impressions' associated with meaning. His basic thesis is that, while transcendental sounds exist as living, creative forces that determine our ideas, human words composed of vowels and consonants, are a human attempt to reflect their original 'root-sounds' or 'seed-sounds.' These seed-sounds are the basic unit of language (not to be confused with phonemes, which are derived components) and carry the 'root-meaning,' an inner meaning as an articulation of a state of consciousness. The divisive and fragmenting mind can capture them only vaguely: *"He* [the hearer] *strives to read it with the labouring mind, but finds bright hints, not the embodied truth"* (Sri Aurobindo 1950, p. 375).

For example, the first four simple vowels—a, i, u, ṛ—of Sanskrit indicate a state of being and existence ('a': just being and existing; 'i': a narrowing intensity; 'u': a widening and extended state of existence; 'ṛ': a vibrant state of existence).[229]

Words are, therefore, living growths of sound, with certain seed-sounds as their basis. Out of these seed-sounds developed a small number of primitive root words, which, over generations, developed into more complex words while still maintaining a common lineage of seed-sounds. To illustrate this, take again the word 'vṛka,' which means wolf. It is already a formed word. To trace its origin, we must find the seed-sound and its associated mind-impression: 'vṛj' is the seed-sound conveying the mind-impression of tearing apart, ripping, or splitting.

Another example could be the original meaning of the word 'yoga' that is derived from the root 'yuj,' meaning 'to yoke', 'to unite', 'to use', among many other possible significances. However, in order to understand its universal significance one needs to find the parent-root from which it was derived, namely 'yu', of which 'yu-j' is an extension. The root 'yu' has two major

[229] Something reminiscent of how Hermes Trismegistus reached spiritual enlightenment by singing hymns uttering the seven Greek vowels considered having supernatural powers. With every vowel the consciousness rises up until attaining gnosis. (Trismegistus, Book XIII.) Yet, he warned from translating his corpus in Greek. *"For the very quality of the sound and the pronunciation of the Egyptian language carries in itself the power of what is being spoken..."* It uses *"sounds full of power."* (Trismegistus, Book XVI)

meanings: ‘to unite’ and ‘to divide’, depending on the verbal class. One could go even deeper and note that the simplest component of the root ‘yu’ consists of two simple vowels, ‘i’ and ‘u’–the simplest components in the Sanskrit language. The first root has the significance of ‘one pointedness’ in its motion or concentration, the latter has the opposite significance of ‘widening’. Thus, the most elemental sound-roots with ‘i’ and ‘u’ have the significance of both singularity and duality, union and separation, and that will be maintained in all other roots with extra consonants and their extensions, words, and derivations containing these elemental sound-roots, modify the meaning of the root and add some additional, contextual significance to the root (Yatsenko, 2023).

Sri Aurobindo also provides an example from Latin, where words like ‘doleo’ (I grieve), ‘dolenter’ (painfully), and ‘dolor’ (grief or pain) are interconnected. Similarly, ‘dolus’ (fraud), ‘dolosus’ (deceitful), and ‘dolose’ (deceitfully) are interconnected, as are ‘dolo’ (I hew, cut, or break) and ‘dolabra’ (an axe). However, there is no proof that these three families of words are interconnected. What connects them is not the words themselves but, rather, their common seed-sound: ‘dol.’ ‘Dol’ is associated with grieving, deceit, and splitting—it forms the basis of a family of words. According to Sri Aurobindo, the single root sound ‘dol’ created an original, common mind-impression, conveying a specific psychological state or quality and significance.

In line with the Vedic tradition, Sri Aurobindo categorized the sound’s characters or qualities into the three guṇas of tamas (inertia, darkness), rajas (restlessness, passion), and sattva (equanimity, harmony). The guṇa of sound creates a particular impression on the mind, and we associate that sound with it. This forms the basis of what we call the ‘meaning of a word.’ The mind perceives the original mental impression, created by the vibrations of the seed-sound, as a semantic unit. Looking at the common seed-sound and its guṇa (like ‘dol’ imbued with the guṇa of rajas) leads us to a unified vision of the origin of speech.

Sri Aurobindo provided many more examples of this nature, especially in the Sanskrit language, and concluded that not only must one root-sound originally have had one root-meaning but also that all kindred root-sounds (such as the kinship of ‘dol’ to ‘dal,’ derived from the simple root-sound ‘da’) must be of one word family. In modern Sanskrit, these words may convey very different meanings, but they still have a common or kindred meaning carried by the primitive or secondary and tertiary root-sound tracing back to their root-meanings because of their common guṇa, which retains their seed significance.

Thus, according to Sri Aurobindo, the origin of language is found not only in a common origin of physical sound forms (letters, phonemes, words) but also in the way these vibrations evoke psychological qualities through their mind-impressions, producing particular ideas related to sensations, motion, action, feelings, contact, sound, form, and more.

Based on this vision, Sri Aurobindo hinted at the existence of an 'original language' spoken in an 'age of truth,' rooted in the true and perfect relationship between vāc and artha. This proto language, preserved fundamental sounds that held a force inherent in the nature of things, and it had certain primary meanings independent of human choice. These meanings initially arose from a set of primary roots, from which secondary roots were developed. All words were formed from these roots through the addition of vowel and consonant terminations, eventually leading to the creation of more complex words. However, this original language gradually became 'corrupted' in both sense and sound, transforming into the later Sanskrit, which he considered the known language closest to this lost original language. Every other language, no matter how remote, represents a detachment from these universal and eternal principles, resulting in a loss of its original purity (Sri Aurobindo, 2013, p. 475). Nevertheless, *'every speech carries a light of inner truth'* (Sri Aurobindo, 1950, p. 430).

Poetry, or at least a certain kind of poetry that attempts to convey intuitional concepts or feelings beyond the reach of words and rational analysis, serves as a (more unconscious than conscious) human desire to evoke this original mode of communication. These words possess a sound expression that carries within them a suggestive force, life, power, and 'soul-suggestion,' which evoke mind-impressions and extend our vision toward higher and broader ideas. At the highest level of cognition, 'hearing' becomes a comprehensive and undifferentiated understanding of all relations in one meaningful unity. It is a flash of immediate intuitive recognition requiring no further explanation.[230]

As a concluding remark to this brief overview of Sri Aurobindo's perspective on the nature and origin of speech, it is necessary to place it within the evolutionary spiritual emergentism underpinning his entire vision.

Similar to the metaphysical views of the Vedic Rishis and Abhinavagupta, Sri Aurobindo's cosmology goes beyond simple individual body-mind dualism or matter-soul distinction. He delineates various levels of manifestation, which he labels the 'planes of consciousness and parts of being.' Without delving into the details, let us (admittedly, somewhat incompletely) summarize it in four levels of existence, in line with Abhinavagupta's four levels: the physical plane, the life plane or 'vital' plane, the mental plane, and

[230] Sri Aurobindo claimed that his epic poem 'Savitri' was inspired from a higher cognitive level, he called the 'overmind.'.

the supra-conscious plane of the Supermind. [231] Each plane of universal manifestation has its personal aspect as well.

The universal physical plane, which we call the 'physical universe,' finds its personal expression in our physical bodies formed from cosmic matter. Similarly, the individual life embodiment in a living organism represents the local expression of universal cosmic life. Our individual mind is derived from the universal or cosmic mind. In this view, the entire scientific evolutionary conception is turned upside down: Life doesn't originate from matter; instead, matter is seen as a veiled form of life.[232] The mind didn't emerge from life; rather, life is a veiled form of mind. Furthermore, the mind is not the ultimate creation in Earth's evolution; it, too, is a veiled form of higher levels or planes of consciousness.

This is also why Sri Aurobindo spoke of an 'involution' preceding evolution.

"Matter could not have become animate if the principle of life had not been there constituting Matter and emerging as a phenomenon of life-in-matter; life-in-matter could not have begun to feel, perceive, think, reason, if the principle of mind had not been there behind life and substance, constituting it as its field of operation and emergent in the phenomenon of a thinking life and body: so too spirituality emerging in mind is the sign of a power which itself has founded and constituted life, mind and body and is now emerging as a spiritual being in a living and thinking body." (Sri Aurobindo, 1919a, p. 885)

A fitting metaphor for the descent of this spiritual power into matter can be that suggested by William James, of white light that a prism refracts into its constituent colors (James, 1898). The prism itself doesn't create the colored light rays; rather, it transmits and separates them into their distinct colors. Similarly, the transcendent sound, containing the originative supramental vibration of a Truth-Consciousness filled with the light of knowledge and supreme vision and discernment, is not generated by our brain. Instead, the brain serves as a conduit, a transceiver instrument that 'refracts' this

---

231 Once more, the categorization into these 'levels' or 'planes' is conventional in the sense that it establishes arbitrary demarcation lines along a continuum, much like the distinction between day and night or as in Bhartṛhari's and Abhinavagupta's three versus four-level arrangements. These categorizations serve to aid in our conceptualization and provide a framework for organizing our understanding. The crucial point is that they all indicate varying degrees of 'subtleness' in both substance and consciousness.

232 The concept of a 'vital plane,' which implies 'vital energies,' remains anathema in modern biology. However, since the 20th century, any vitalist theory has been dismissed and ignored a priori rather than disproven. Scientific facts have not eliminated vitalism or prevented us from considering cognition and consciousness as prior to matter. For a more in-depth analysis of this aspect, refer to (Masi, 2022).

transcendent sound into thought-vibrations within the mind [233] and life-vibrations within our 'life-body' (which we commonly refer to as 'emotions'). On the physical level, our vocal organs produce 'matter-vibrations' in the form of sound waves that capture only a faint reflection of the light and knowledge embedded in the original sound. For both Abhinavagupta and Sri Aurobindo, the origin of the word has its seat in a higher plane, which filters the Absolute down to the most physical universal plane, the vaikharī vāc level. This implies that speech does not originate only in the brain.

What this also implies, most importantly for the present context, is that Sri Aurobindo reverses the usual assumption of modern science (and of every reasonably minded person except, perhaps, Süßmilch), which is that life came first, and speech followed. Instead, he posits what might seem absurd from a Darwinian perspective: Speech and language came first, and life followed. This leads to his conception of the universe as a manifestation of a 'Real Idea,' which we will describe in the next section.

Sri Aurobindo's theory of the origin of speech is far more detailed than what I've described here. Readers interested in a more in-depth exploration can turn to his original writings. My aim was to provide an intuitive understanding of the concept. Furthermore, Sri Aurobindo's theory of the origin of language is just a small part of his spiritual vision resulting from his spiritual discipline, the 'integral yoga.' He expounds on his vision in other works like "The Life Divine" (Sri Aurobindo, 1919a), the "Synthesis of Yoga" (Sri Aurobindo, 1919b), and his epic poem "Savitri" (Sri Aurobindo, 1950). Within this broader context, one can gain a deeper appreciation of his vision. Also of interest to linguists is his vision of "The Future of Poetry" (Sri Aurobindo, 1917).

4. The World as Real-Idea According to Sri Aurobindo

The previous sections, besides describing the linguistic theories of Abhinavagupta and Sri Aurobindo, also served as the basis for a larger philosophical context and a more general theory of the sign, meaning, and idea. The last part of the previous section functioned as an introduction to a theoretical framework conceiving of the human words and languages as a particular case of a general 'signaling' in Nature, and a cosmic context that is an 'Idea' in itself.

[233] This might extend beyond mere metaphor. Neuroscientific evidence can be aligned with the idea that the brain is not merely an aggregation of matter that 'produces' mind and consciousness but, instead, functions as an instrumental organ that 'transduces' or 'refracts like a prism' mind and consciousness. For a comprehensive analysis, refer to (Masi, 2023b).

In contrast to Spinoza's Nature or Schopenhauer's World as Will, which are not self-conscious entities and operate out of necessity, instinct, and blind will, for Sri Aurobindo the Supermind represents the highest form of cognition, wisdom, and knowledge of a universal consciousness. It functions supra-consciously, utilizing its immanent powers, forces, and will, of which śabda, the word, speech-sound, the vibration, is only one possible aspect and form of expression. This universal Gnosis shapes the universe with a 'Consciousness-Force' and 'Consciousness-Will' through processes, structures, patterns, and, ultimately, living forms according to a divine Intelligence.

In Sri Aurobindo's words:

"The fundamental nature of this Supermind is that all its knowledge is originally a knowledge by identity and oneness and even when it makes numberless apparent divisions and discriminating modifications in itself, still all the knowledge that operates in its workings, even in these divisions, is founded upon and sustained and lit and guided by this perfect knowledge by identity and oneness. The Spirit is one everywhere and it knows all things as itself and in itself, so sees them always and therefore knows them intimately, completely, in their reality as well as their appearance, in their truth, their law, the entire spirit and sense and figure of their nature and their workings." (Sri Aurobindo (1919b), p. 786)

This Intelligence operates through a creative power reflecting an original Idea, or what Sri Aurobindo terms the 'Real-Idea' or 'Seed-Idea.' This transcendent and Real-Idea stands beyond matter, space, and time but translates itself into the manifestation of space and time as an expressive power of conscious force. It is an Idea of oneness that initiates the operation of a dual and divided representation of itself, thereby giving rise to the world of dualities and divisions. It embodies the primal, perfect, divine, and undifferentiated Knowledge. This Knowledge casts itself into mutable names and forms, which the mind conceives of as more or less accurate symbolic representations of this Seed-Idea. At the level of human apprehension and understanding, intuition captures this as archetypes, symbols, and signs along with their meaning and significance in relation to things and phenomena.

According to Sri Aurobindo, the Real-Idea is a gnostic power in which Spirit and Knowledge are two aspects of the same Reality. It represents a self-creative Knowledge and 'Truth-Consciousness' that organizes, through design, quantity, and number, a foundation for the manifestation of the quality and property inherent in its own Real-Idea. These ideas of forms are what we perceive as a living and material world within the spatio-temporal extension. Life, minds, and bodies are partial expressions of this original transcendent Idea, whose presence our ordinary state of consciousness only dimly perceives. Philosophers often interpret this presence as archetypes, Platonic ideals, or universals. It constitutes what we perceive as a vestige of significance, meaning, design, and purpose in Nature. Everything we indirectly perceive in the world, from non-living matter like rocks, mountains, and rivers to living entities such as plants, animals, and ourselves, is a 'symbol,' a 'sign,' or a 'significant form' of this ultimate Real-Idea, concealed though it may be, within their essence.

In Sri Aurobindo's evolutionary spiritual emergentism, the Real-Idea shapes things through the power of Knowledge and Will, drawing from infinite potentialities. It 'unfolds' what is internally carried within the form, often referred to as a 'seed,' and transcribes it into existence. This development or emergence represents self-development and outgrowth over a gradual, sometimes abrupt, aeonic process of something already inherently contained. The Real-Idea remains in perpetual existence, concealed knowledge, and an indwelling spirit, which is 'involved' within the mineral, plant, or animal. These entities respond to the Idea and its will-force by forming groupings and associations of elementary constituents, resulting in the creation of significant forms, types, and species. However, they themselves are external signs and the result of a conscious force playing with its own Idea, generating all types through diverse combinations.

Thus, in Sri Aurobindo's cosmology, everything comprising the manifestation is a pale reflection of this Idea. It is inherent in all cosmic existence, underpinning the actions of all creative forces and the evolutionary process. The indwelling presence in all forms serves as an expression of the Real-Idea. The undifferentiated yet multi-faceted Real-Idea can fragment into its various facets, each becoming independent Idea-Forces. In living beings, we refer to these as 'characters' and 'personalities.' Our diverse and multi-dimensional psychology mirrors this concept. We are also a 'symbol' of the Idea, capable of expressing itself in an infinite number of 'ideas,' each with its own variety of individuated personalities. When we intuitively sense that the world, despite its mechanistic appearance, harbors meaning, purpose, and the development of a project (an Idea), it's because our limited minds can still grasp and vaguely perceive this transcendent Real-Idea.

"In all material things reside a mute and involved Real-Idea, a substantial and self-effective intuition, an eyeless exact perception, an automatic intelligence working out its unexpressed and unthought conceptions, a blindly seeing sureness of sight, a dumb infallible sureness of suppressed feeling coated in insensibility, which effectuate all that has to be effected." (Sri Aurobindo, 1919a, p. 570)

Hence, that automatic, blind, and seemingly insentient mechanism in Nature is not the original starting point; rather, it represents the superficial expression and appearance of a divine Knowledge that gives rise to forms and powers of itself. The interplay of all the constituents of the universe should not be regarded merely as a play of unconscious phenomena. It is, in fact, a manifestation of Knowledge, in which what we refer to as the 'laws of Nature' are the expression of a Real-Idea guiding every moment toward its destined state. If everything appears as the outcome of intelligent workings, this perception is not a product of our illusions but arises from the presence of a veiled intelligence operating beneath our ignorance. The Real-Idea of the Supermind isn't an intellectual intelligence but, rather, a Supramental Force and Will operating subconsciously in plants, animals, and the half-conscious human being. It is a transcendent and immanent Will and Force unfolding in time and space.

Sri Aurobindo's conception of evolution[234], therefore, can be viewed as a two-way process. It is preceded by the involution of the Spirit into form, described as a *"plunge of Light into its own shadow"* (Sri Aurobindo, 1919a, p. 173). A perspective on evolution that entails that forms, qualities, and quantities are shaped by the divine Real-Idea from within. All forms and attributes in the natural world are preconceived within eternal archetypes on other planes of existence. Every manifestation in Nature is the expression of a preexisting Idea, which our ordinary human mentality comprehends in terms of names and forms, and, if it can see far enough, in signs and language. Meaning, significance, and semantics inherent in the human mind and language, ultimately extending beyond the human realm. The universe is a Sign and a Word, its workings a language expressing a Real-Idea.

III. Discussion and Integral Synthesis

In summary, our exploration started from the Rig Veda that described the creation of the universe through an original cosmic vibration. According to the Vedic tradition, words and sounds are not just means of expressing intellectual ideas but reflections of original vibrations beyond the material and mental domains. Human speech, in its articulated form, represents the final stage of differentiation from the original 'sound-word,' and its true origin lies in higher cognitive states achieved through spiritual practice. Then, the development of Abhinavagupta's four levels of language has been traced. At the highest level, parā vāc, sound and meaning are unified and exist as an undifferentiated vibrational unity. As language descends through the levels of paśyantī, madhyamā, and vaikharī vāc, the original vibration becomes progressively more distinct and explicit, with an increasing separation between words and meanings.

In line with this perspective, Sri Aurobindo's theory starts from a phonological perspective and proposes that words share a common seed-sound, creating mind-impressions that convey specific psychological states or qualities. He views the development of language as a psychological process that began with emotional and sensational needs and evolved into a more

[234] A spiritual view of evolution that is partly reflected in the cosmology of Teilhard de Chardin and Rudolf Steiner as well.

precise and intellectual structure. Speech is conceived as something naturally inherent in a universal consciousness surfacing within human language in word-sounds that originally reflected a set of deeper and more creative cosmic seed-sounds going beyond the mind. Sri Aurobindo's theory extends from the origins of human speechto encompass the entire cosmos, where everything is considered a form of the divine Word. In this expanded context with Platonic nuances, he offers an integral perspective. He proposes a more complex cosmology, which includes a transcendent Absolute and a Supermind, functioning as a creative agency linking the unmanifest and manifest. The universe is viewed as an expression of the Supermind's conscious force, with the Real-Idea or Seed-Idea guiding the development of all forms and life. All things and forms in the phenomenal world serve as a shadow of the Word. Sri Aurobindo's evolutionary linguistic theory must be viewed within the broader context of his system of integral yoga, which seeks to reconcile evolution with a spiritual emergentism, suggesting a post-materialistic synthesis as a solution to the question of life's emergence from non-conscious matter.

Thus, considering Abhinavagupta's and Sri Aurobindo's vision jointly, and examining the diverse philosophical traditions, both Western and Eastern, a common trait emerges: Reality is ultimately a sign, an Idea, a Word, or a sound. Language represents just one subset of possible creative expressions in the form of sound waves signifying an idea standing beyond it. All cosmic dynamics are an Idea expressed in forms and names. Each object, phenomenon, and play of forces symbolizes an idea on the mental plane, reminiscent of Plato's archetypal realm of perfect and unchanging Forms and Ideas. Yet, for Sri Aurobindo and Abhinavagupta, there exists a whole supra-sensible reality transcending the mind with its dynamic temporal and spatial notions.

Of course, this vision contrasts naturalist linguistic theories. It is a phenomenological first-person approach based on a methodology that studies language by delving within us, rather than adopting an exclusive third-person perspective. While the modern philosophy of language focuses on a rationalistic and analytic approach, this perspective emphasizes intuitive, spiritual, and first-person accounts—or, to formulate it using Abhinavagupta's terminology, Western philosophy has focused predominantly on the madhyamā vāc and vaikharī vāc levels, largely disregarding the paśyantī vāc and parā vāc levels due to inadmissibility within a naturalistic context. This Eastern perspective considers language not merely as an inter-subjective vehicle of knowledge but as a creative power. The spoken word is an imitation of a higher truth that the speaker aims to mirror as accurately as possible. Thus, other predominantly human activities, such as poetry, music, and dance, are verbal, acoustic, or bodily expressions of the mind that subliminally perceive the rhythm and harmony of a transcendent sound.

At this juncture, the question arises as to whether this integral perspective should be regarded merely as a cross-cultural historical comparative analysis without further implications for contemporary philosophy, linguistics, and science, or whether it contains some valuable suggestions that we might adopt as a working hypothesis, and then see where this leads us. In many respects, it doesn't contradict modern findings; rather it could complement them.

For instance, Saussure's notion of the relationship between the signifier and the signified mirrors what emerges in Abhinavagupta's madhyamā vāc, the mediating level. Here, the relation-less meaning characteristic of the paśyantī vāc level fractures into relationships of signs and words with other signs and words. The differentiation of the infinite into the finite appears between subject and object. The 'langue'—the structure of language with its abstractions and principles—results from the analytical but separative power of the buddhi, the intellect specific to this level. The 'parole' arises at the vaikharī vāc level and reflects the subliminally perceived seed sound that evokes the mind-impression intended to be conveyed.

Saussure sought universality in the structure of language transcending all languages but was puzzled by the arbitrariness of the relationship between the signifier and the signified. In Abhinavagupta and Sri Aurobindo, this apparent arbitrariness is not real; it exists only at the madhyamā vāc level or the plane of the mind, originating from the realm of parā vāc and paśyantī vāc or Supermind, where the originative seed-sound was singular—that is, the sound-word (śabda) and its meaning (artha) being one and the same—yet encompassing all variations that the arbitrary words reflect (for further analysis of this aspect, refer to (Kumar, 2023).)

When structuralists refer to a signifier as a 'psychological imprint of sound,' they intuitively recognize what Sri Aurobindo described as the psychological qualities of the guṇas determining the sounds in speech on our material plane. However, what the structuralist intuitively perceives as 'structure' points to the structure of an original language, far beyond the structure of ordinary spoken and written language. According to Abhinavagupta and Sri Aurobindo, indeed, there is universality, but it exists beyond the physical plane of existence (see also (Kumar, 2016), (Kumar, 2023a), and (Kumar, 2023b).) They reverse Chomsky's generative grammar hypothesis: instead of there being universal syntactic rules that generate phonological and semantic structures, there is an inherent universal phonology that generates the grammatical structure of language.

Nonetheless, Chomsky's intuition about the mind's central role that tips into 'deep structures' and expressing it on the level of 'surface structures' can also be regarded as a profound insight. The mind hosts a 'language acquisition device,' and there's a universal quality shared by all human languages. Indeed, the sounds, phonemes, and words of a surface structure are connected to something universal at a deeper level. However, this doesn't imply that words and language are only expressions of the mind. We may question whether they emerge within, are modified by, and pass through the mind rather than being originated by it. These sounds are the human mind's attempt to reflect a deeper seed-sound, not just an abstract mental internal representation. Universality doesn't stem from a genetic code or neural pathways in the brain; rather, it originates from an inner domain of universal consciousness. Only when the consciousness and cognitive abilities of early humans were capable of opening a window on the mental level of madhyamā vāc could language come into being. Language 'involved' from the top-down, meeting the organism evolving from the bottom-up.

However, Abhinavagupta's and Sri Aurobindo's insights can tell us something beyond linguistics. For instance, it is noteworthy to observe how, stemming from a vastly different context, a much more recent development in biological sciences, namely, biosemiotics, presents a perspective on life and the transformations of organisms that aligns with such an interpretative spiritual stance. Despite not attracting significant attention, biosemiotics has persevered on the fringes of its discipline, offering a holistic perspective on studying life. It serves as an interdisciplinary field, merging biology and semiotics—the study of signs and meaning-making, emphasizing the role of communication and meaning in the living world. Living organisms generate and interpret signs conveying information, and interactions mediated by signs, such as chemical signals, visual cues, sounds, environmental cues, symbols, codes, interpretation, representation, and communication processes, or even behavioral patterns, and not only attribute meaning to signs and symbols but also create and alter sign relations (for an introduction, see (Emmeche et al., 2011).) The study of semiosis in biology spans from the cellular and molecular level—such as in cell signaling or sign-mediated interactions in bacterial communities with quorum sensing—to the investigation of animal forms of knowledge, including human behavior. One could assert that semiosis and information processing are intrinsic to life. Undoubtedly, human language and abstract symbolic thought are the most prominent examples of semiotic communication. The connection between biosemiotics and language hasn't gone unnoticed. (For a review, see (Velmezova et al., 2015).) The genetic code and its translation serve as the most common example illustrating the biological significance of non-human code and sign processing. DNA conveys functional information by using symbols—specifically, genes represented by a sequence of nucleotide bases in the DNA that code for proteins. Zolyan points out that, instead of comparing nucleotides with letters, one should, indeed, correlate them with sounds (Zolyan, 2023.) Adopting Sri Aurobindo's perspective, one could say that biosemiotics appears to intuit a hidden underlying Real-Idea within life. It aims at a comprehensive vision in which signs, 'words', language, and symbols hold universal significance. Modern biosemiotics naturally discovers this universality and the relationship between signs and meaning in Nature. Consequently, we are entitled to consider a standpoint where language is not merely a bottom-up emergent property of life, but is an expression of meaning-making process by a top-down universal consciousness whose nature and essence are knowledge and gnosis. On the material plane of existence, it 'voices' itself and appears to our derivative cognitive system as signs, signals, and meanings. Life and the entire cosmos constitute a network of 'signs,' 'words,' or 'symbols'—essentially, a 'language'. In Nature, meaning isn't a construct of the human mind superimposing purposefulness and semantics due to an intellectual anthropomorphic tendency; instead, meaning is inherent in all existence. Signs, words, and symbols manifest naturally in the material realm due to a consciousness projecting its Real-Ideas. As the Vedic sages claim, wherever there is consciousness, there must inevitably, in one form or another, be the Word.

Moreover, within a naturalistic framework, attempts to explicate the nature and origin of consciousness have multiplied in recent decades (for an overview, see (Kuhn, 2024) and (Seth, 2022).) Several endeavors to merge biology with cognition have sparked various lines of research, notably the autopoiesis hypothesis (Maturana and Varela (1991)) and, more recently, in the field of non-linear complex systems, or in terms of cellular information processes (Miller, 2023), or as homeostatic processes (e.g., Turner, 2017). Nevertheless, the reason behind the emergence of signs and meaning in Nature remains elusive. This is because biology and any naturalistic science try to explain the emergence of code and cognition via a bottom-up mechanistic and biochemical process as: Chemicals → code → cognition. However, already inside the same rationalistic and materialistic paradigm, there are good arguments to conjecture that Nature has gone the other way around, namely: Cognition → code → chemicals (Marshall, 2021). The idea comes first; its material manifestation follows.

Whether we should consider these analogies only metaphorically or take them more seriously is a choice left to the reader. However, Abhinavagupta's and Sri Aurobindo's linguistic paradigms do not contradict biological theories suggesting that language evolved as an adaptation that offers a selective advantage. Within the theories of evolutionary linguistics, the hypothesis that may partially align with the current theoretical framework is Ray Jackendoff's "Unconscious Meaning Hypothesis," which views language as complex adaptation evolving in stages (Jackendoff, 2012). He encourages us to adopt a phenomenological cognitive perspective and recognize how language unconsciously evokes meaningful imagery through a model of 'speech sounds' or 'linguistic qualia'—specifically, pronunciations—that elicit an internal sense of meaningfulness beyond mere external symbolism, and that is phonological in character. This leads to the understanding that rational thought (the lower domain of madhyamā vāc) is not an alternative to intuition (parā vāc, and the higher domain of madhyamā vāc); rather, the former depends on the latter. Jackendoff proposes a perspective on language known as "Parallel Architecture," which claims that the three primary systems of human language—phonology, semantics, and syntax—are not independent but are instead interconnected through "interfaces." In this view, the semantic system was already present in non-human primates as the first generative component of language, with the vocalization of mental concepts forming proto-words and thereby driving the early stages of the evolution of language, and only later developing a proto-syntax with more complex semantic relations. This view still diverges from the perspectives of Abhinavagupta and Sri Aurobindo, who suggested that the phonological aspect serves as the origin of speech. However, it comes a step closer to their ideas by moving away from Chomsky's 'syntactocentric' approach and acknowledging that speech sounds have a deeper function than previously assumed, with words and its phonological structure shaping the structure of sentences, rather than syntax.

Other analogies can be traced to Shigeru Miyagawa's "Integration Hypothesis" (Miyagawa, 2017). This hypothesis posits that language is an integration of two independently occurring systems in nature that underlie communication. The first system—the 'lexical system'—comprises isolated sound units, or utterances, from primates that typically have specific referents (semantics), such as "leopard," "snake," or "eagle." These units exist within a lexical system of alarm calls, which, however, lack expressive sound patterns. The second system—the 'expressive system'—is associated with sound patterns (syntax), similar to those found in bird songs, which do not possess a lexical system. The lexical system features semanticity but lacks phonological syntax, while the expressive system emphasizes phonological syntax but does not include semanticity. The hypothesis suggests that, due to a cognitive leap, humans were able to integrate these two systems, resulting in the emergence of language. Abhinavagupta and Sri Aurobindo focused on the origin and nature of speech and were less concerned about the emergence of the structure, syntax, and grammar of language. However, they would likely have agreed that this integration arose from the contact with the madhyamā vāc level, which endowed the first humans with the ability to associate and organize those utterances into seed-sounds within a more complex system of abstract symbolism.

That there exists a gap eluding our understanding between the symbol itself and the emergence of meaning in our conscious awareness also becomes clear in artificial intelligence (AI). Even the most recent AI systems, such as the large language models (LLMs) implemented in ChatGPT, while being impressive at mimicking human intelligence to some extent, struggle to truly comprehend. This issue was highlighted by Searle already in the 1980s with his famous 'Chinese room argument' (Searle, 1980). Language, with its semantics, transcends mere symbols, raw data, and symbolic information processing and seems to reside beyond a formal description of a computational model.

A similar and related question is the 'symbol grounding problem,' discussed by Stevan Harnad in 1990, which highlights the problematic issue of how symbols (words, numbers, streams of bits, signals, etc.) acquire their meaning. There remains a gap between symbols, signals, number manipulation, and semantic awareness, intuition, understanding, or knowing, which is eludes a formal description of any computational model.

However, Abhinavagupta's and Sri Aurobindo's universal standpoint might provide a another perspective. The bridge between sign and meaning isn't the brain or some complex biological or artificial machinery. The sign and the material container holding it did not precede meaning; instead, meaning in the form of an 'inaudible sound' on another plane manifested the word, symbol, and sign later. The deepest semantic significance of meaning is an original seed-sound on the parā vāc level of pure Consciousness, not a complex bioelectrical feedback loop. Abhinavagupta posits that śabda (the sound-word) is united with artha (meaning) only at the level of paśyantī vāc. If so, we are no longer in a realm that physical sciences can grasp.

I guess that Abhinavagupta and Sri Aurobindo would likely argue that rejecting the hypothesis of a trans-rational origin of language and metaphysical theories of consciousness may be the root cause for a series of paradoxes in consciousness studies and cognitive science. For instance, Darwinian approaches struggle to explain the remarkably rapid emergence of the cognitive abilities that led, almost abruptly in evolutionary timescales, to the development of human language and closed the gap with animal communication solely through natural selection and random genetic mutations. A highly debated issue commonly referred to in evolutionary linguistics as 'Darwin's problem.' The roles of social interactions and cultural evolution have certainly played significant parts, but it remains unclear whether this fully accounts for the phenomenon. Sri Aurobindo's evolutionary linguistics would support a modern culturalist approach that emphasizes the inseparability of language and culture, asserting that grammar, vocabulary, and meaning are deeply influenced by cultural and psychological aspects rather than being solely biological phenomena. Especially with those who posit that lexical items or content words (such as nouns, verbs, and adjectives) emerged before grammatical items and function words (like auxiliary verbs, prepositions, and articles). However, he would also contend that even before this cognitive leap, words existed with a more flexible, fluid, and plural significance, where the distinction between nouns and verbs was blurred due to the absence of a clear subject-object distinction. These eventually evolved into the conceptual clarity we recognize today, a development attributed to the emergence of the faculty of consciousness we refer to as 'mind.'

Another pivotal concern in the philosophy of mind is the 'binding problem.' It delves into how the brain integrates or 'binds' diverse sensory experiences and information fragments into a unified, coherent conscious perception and semantic whole. Despite the brain processing information from multiple independent sources in different ways and brain areas, conscious experience is perceived as seamless and integrated. For example, the meaning of a sentence can't be deduced by summing up the significance of the single words. Rather, only after the sentence has been read until the end does the relation between the words and its context become clear. Then, suddenly, by an almost instantaneous 'binding act,' meaning emerges in our conscious awareness. Researchers across neuroscience, psychology, and the philosophy of mind have proposed various theories suggesting specialized neural mechanisms or processes that enable such information integration from different sources. However, the issue remains more unresolved.

Abhinavagupta and Sri Aurobindo would likely have pointed out that a science focusing solely on the physical plane—that is, the vaikharī vāc level of speech—will never, not even in principle, comprehend how consciousness links symbols to meaning and 'binds' or 'integrates' features into semantic wholes at the parā vāc level. In their view, the process operates in the opposite direction, not adhering to a bottom-up logic in which the sum of the parts constructs the whole but, rather, through a top-down process. Starting from the parā vāc, the original undifferentiated seed-idea, with its truth of significance in the form of the seed-sound, is already 'bound' as an integrated semantic whole without subject-object distinction. There is no binding necessary in the first place. Instead, it is fragmentation that gradually appears as artha and śabda split by descending toward the physical plane.

Even if only for speculative purposes, we accept Abhinavagupta's experience of the four levels of language and Sri Aurobindo's Real-Idea as a pre- or trans-rational thought-essence not as mere abstract philosophical speculations but as insights reflecting some deeper truth about our cognitive nature, this could significantly impact research in the field of artificial intelligence. If the mind and what lies beyond the mind exist outside of matter and cannot be reduced to physical brain processes, and instead reside on other planes of existence, then any aspiration to create human-like non-conscious artificial general intelligence (AGI) in a material aggregate like a computer will inevitably remain unattainable. Computers are unconscious material entities that compute, whereas, according to this perspective, the mind is neither material nor computational and is reliant on something fundamentally conscious. AI might achieve the level of madhyamā vāc, in which cognition manipulates words, sentences, grammar, and syntax, distinguishing between object and subject. That's what we see LLMs are capable of achieving. However, they would object that AI will never ascend to the semantic level of paśyantī vāc, let alone parā vāc, or the unified undifferentiated vision of the

Supermind, as these levels transcend rational, computational, and physical existence. Recent advances in AI indicate that despite their remarkable success, machines remain incapable of developing a semantic awareness of the world, remaining passive black boxes without agency and consciousness. Perhaps, this is because 'understanding' is essentially the act of hearing the inaudible Word.

In any event, by approaching things integrally and contextualizing them, Eastern language theories do not contradict their Western counterparts. In fact, attempting to unite, integrate, and harmonize both perspectives might lead to new insights. Even if we embrace a naturalistic approach Abhinavagupta's and Sri Aurobindo's mystical visions might contribute to illuminate areas where contemporary philosophy and science struggle to find answers. To fully embrace this integral approach, we must view the big picture in its entirety and synthesis, going beyond single approaches with their distinct strengths and weaknesses. To do this, we might need to quiet the intellectual mind in order to perceive the distant call of a transrational mind.

In his Tractatus, Wittgenstein famously declared, *"Whereof one cannot speak, thereof one must be silent."* However, popular wisdom acknowledges that sometimes 'silence speaks a thousand words.' If we remain silent-minded, we might become able tograsp the inaudible sound emanating from the higher levels of speech even before words are spoken and enjoy our 'aha-moments.' We might then realize that language is not the product but, rather, a discovery of the mind of a sign-producing universe and that communication beyond language exists—especially the language of the spirit that comes from the vibrations of the soul.

Acknowledgments

I'm particularly grateful to Valdimir Yatsenko, who greatly inspired this work with his teachings on the Sanskrit language, and further deepened my knowledge on Abhinavagupta's linguistics (Yatsenko, 2021-23).

I'm also grateful to 'Sreenivasarao,' a blogger who preferred to remain anonymous but seems to be very well versed on the same topics (Sreenivasarao, 2016).

References

Abhinavagupta (2011), Parātriśikā-Vivaraṇa: A Trident of Wisdom. Translation by Jaideva Sing. State University of New York Press, Albany.

Aurobindo, S., (1917). The Future of Poetry. Sri Aurobindo Ashram Trust.

Aurobindo, S., (1919a). The Life Divine. Sri Aurobindo Ashram Trust.

Aurobindo, S., (1919b). The Synthesis of Yoga. Sri Aurobindo Ashram Trust.

Aurobindo, S., (1912-14) The Origins of Aryan Speech. Sri Aurobindo Ashram Trust.

Aurobindo, S., (1950) Savitri. Sri Aurobindo Ashram Trust.

Aurobindo, S., (1952) Kena and other Upanishads. Sri Aurobindo Ashram Trust.

Aurobindo, S., (1914-20) The Secret of the Veda. Sri Aurobindo Ashram Trust.

Aurobindo, S., (2013) Hymns to the Mystic Fire. Sri Aurobindo Ashram Trust.

Aurobindo, S., (2016) Vedic and Philological Studies. Sri Aurobindo Ashram Trust.

Biernacki, L. (2013a) Panentheism and Hindu Tantra: Abhinavagupta's Grammatical Cosmology, in Biernacki, L., and Clayton, P. (eds), Panentheism Across the World's Traditions (New York, 2013; Oxford Academic. DOI: **https://doi.org/10.1093/acprof:oso/9780199989898.003.0009**

Biernacki, L. (2013b) Abhinavagupta's Theogrammatical Topography of the One and the Many, in Boesel, C. and Ariarajah, S.W. (eds), Divine Multiplicity: Trinities, Diversities, and the Nature of Relation, Transdisciplinary Theological Colloquia (FUP), New York, Fordham Scholarship.
DOI: **https://doi.org/10.5422/fordham/9780823253951.003.0005**

Biernacki, L. (2023) The Matter of Wonder: Abhinavagupta's Panentheism and the New Materialism, New York, Oxford Academic. DOI: **https://doi.org/10.1093/oso/9780197643075.001.0001**

Chomsky, N., (1996). Powers and Prospects. Reflections on human nature and the social order. London: Pluto Press, p 30.

Deacon, T.W. (2017). Language Evolution and Neuromechanisms. Editor(s):William Bechtel, George Graham. DOI: **https://doi.org/10.1002/9781405164535.ch13**

Emmeche, C., Kalevi, K. et al. (2011). Towards a Semiotic Biology: Life is the Action of Signs. Imperial College Press.

Harnad, S. (1990). The symbol grounding problem. Physica D: Nonlinear Phenomena, 42(1-3), 335-346. DOI: **https://doi.org/10.1016/0167-2789(90)90087-6**

Hermes Trismegistus, Corpus Hermeticum. Books I-XVII.

Jackendoff, R. (2012). A User's Guide to Thought and Meaning. Oxford University Press.

James, W., (1898). Human Immortality: Two Supposed Objections to the Doctrine (the Ingersoll Lecture, 1897.)

Kuhn, L., R. (2024). A Landscape of Consciousness: Toward a Taxonomy of Explanations and Implications. Progress in Biophysics and Molecular Biology. DOI: **https://doi.org/10.1016/j.pbiomolbio.2023.12.003**

Kumar, N. (2016). Sri Aurobindo's Yogic Discovery of the 'One Original Language' of Mankind: A Linguistic Exploration. Theory and Practice in Language Studies, 6(3), 475-484. DOI: **https://doi.org/10.17507/tpls.0603.04**

Kumar, N. (2023a). Phoneme and Mental and Supramental Vibrations. European Journal of Theoretical and Applied Sciences, 1(5). DOI: **https://doi.org/10.59324/ejtas.2023.1(5).26**

Kumar, N. (2023b). Evolution of 'arbitrariness' in 'sign' in language to 'revelation' in 'sign' in mantric poetry (with special reference to Sri Aurobindo's mantric poetry). European Journal of Literature, Language, and Linguistics Studies, 7(3).

Miyagawa, S. (2017). Integration Hypothesis: A Parallel Model of Language Development in Evolution. In: Watanabe, S., Hofman, M., Shimizu, T. (eds) Evolution of the Brain, Cognition, and Emotion in Vertebrates. Brain Science. Springer, Tokyo. DOI: **https://doi.org/10.1007/978-4-431-56559-8_11**

Masi, M. (2022). Vitalism and cognition in a conscious universe. Communicative & Integrative Biology, **15**(1), 121-136. DOI: **https://doi.org/10.1080/19420889.2022.2071102**

Masi, M. (2023a). The Integral Cosmology of Sri Aurobindo: An Introduction from the Perspective of Consciousness Studies. Integral Review, **18**(1).

Masi, M. (2023b). An evidence-based critical review of the mind-brain identity theory. Front. Psychol., Sec. Consciousness Research, 14 DOI: **https://doi.org/10.3389/fpsyg.2023.1150605**

Marshall, P. (2021). Biology transcends the limits of computation. Progress in Biophysics and Molecular Biology, 165, 88-101. DOI: **https://doi.org/10.1016/j.pbiomolbio.2021.04.006**

Maturana, H. R.; Varela, F. J. (1991). Autopoiesis and Cognition: The Realization of the Living. Springer Science & Business Media. ISBN 978-90-277-1016-1.

Miller, W.B., Baluška, F., Reber, A.S. (2023). A revised central dogma for the 21st century: All biology is cognitive information processing. Progress in Biophysics and Molecular Biology, 182, 34-48. DOI: **https://doi.org/10.1016/j.pbiomolbio.2023.05.005**

Mørch, H. (2023). Non-physicalist Theories of Consciousness (Elements in Philosophy of Mind). Cambridge: Cambridge University Press.

Neufeld, E. (2022). Psychological essentialism and the structure of concepts. Philosophy Compass, vol. 17 (5). DOI: **https://doi.org/10.1111/phc3.12823**

Searle, J. R. (1980). Minds, brains, and programs. Behavioral and Brain Sciences 3 (3): 417-457.

Seth, A.K., Bayne, T. (2022). Theories of consciousness. Nat. Rev. Neurosci. 23, 439–452. DOI: **https://doi.org/10.1038/s41583-022-00587-4**

de Saussure, F. (1974). Course in General Linguistics. Bally, C. & Sechehaye, A. in collaboration with Reidlinger, A. (Eds.), Baskin, W., (trans.). Glasgow: Fontana/Collins.

Sreenivasarao (2016) The meaning of 'MEANING.' URL: https://sreenivasaraos.com/2016/11/12/the-meaning-of-meaning-part-one/

Süßmilch, J.P. (1766) Versuch eines Beweises, daß die erste Sprache ihren Ursprung nicht von Menschen, sondern allein vom Schöpfer erhalten habe. (Berlin, 1766).

Turner, J.S., (2017). Purpose and Desire: What Makes Something "Alive" and Why Modern Darwinism Has Failed to Explain It. Harper one.

Velmezova, E., Kull, K., Cowley, S. J. Editors (2015). Biosemiotic Perspectives on Language and Linguistics. Springer.

Williams, B. (2022). Cosmogenesis and Phonematic Emanation. In Richard K. Payne, and Glen A. Hayes (eds), The Oxford Handbook of Tantric Studies. DOI: 10.1093/oxfordhb/9780197549889.013.29

Yatsenko, V. (2021). On the Origins of Faculties of Consciousness. Sri Aurobindo Integral Life Center, USA. URL: https://aurouniversity.academia.edu/VladimirYatsenko

Yatsenko, V. (2022). The Shift of Paradigms from the Veda to Shastra. From Vedic to Vedantic and Samkhyaic Paradigms of Knowledge. URL: https://aurouniversity.academia.edu/VladimirYatsenko

Yatsenko, V. (2023). The Shift of Paradigms from the Veda to Shastra. URL: https://aurouniversity.academia.edu/VladimirYatsenko

Zolyan, S. (2023) On the minimal elements of the genetic code and their semiotic functions (degeneracy, complementarity, wobbling). BioSystems, vol. 231.104962. DOI: https://doi.org/10.1016/j.biosystems.2023.104962

XIV. Bibliography

Aurobindo, Sri (1919). "The Human Cycle - The Discovery of the Nation-Soul", Sri Aurobindo Ashram Trust. Pondicherry, India.

Behe, M.J. and Meyer, S.C. Discovery Institute. Online: **https://www.discovery.org/v/what-is-intelligent-design/**

Bergson, H. (1896/1912). "Matter and Memory." New York: McMillan.

Cohen, I., ed. (1978). "Isaac Newton's Papers & Letters on Natural Philosophy and Related Documents." Reprint 2014 ed. edition. Harvard University Press.

Gleiser, M. (2023) Twitter - "Fooling ChatGTP into learning that 2+2=4. Terrifying!" Online: https://twitter.com/MGleiser/status/1642243684975476742

Goethe, J. W. [1798] (1995). Empirical Observation and Science. In Scientific Studies. Edited and translated by Douglas Miller. In Goethe: The Collected Works edited by Victor Lange, Eric Blackall, and Cyrus Hamlin, vol. 12. Princeton, NJ: Princeton University Press. 24-25.

Greff, K., Steenkiste, S., Schmidhuber, J. (2000). "On the Binding Problem in Artificial Neural Networks" DOI: https://doi.org/10.48550/arXiv.2012.05208

Huxley, A. (1954). "The Doors of Perception." Chatto and Windus.

James, W. (1898). "Human Immortality." Houghton, Mifflin and Company, The Riverside Press, Cambridge.

Kastner, R. E. (2023). Quantum Haecceity. Phil. Trans. R. Soc. A.381:20220106.20220106 DOI: **http://doi.org/10.1098/rsta.2022.0106**

Levin, M. (2024). "Ingressing Minds: Causal Patterns Beyond Genetics and Environment inNatural, Synthetic, and Hybrid Embodiments." DOI: https://doi.org/10.31234/osf.io/5g2xj_v1

Lewis, R. (2023, January). "Mind-Body Problem: How Consciousness Emerges from Matter" Online: https://www.psychologytoday.com/us/blog/finding-purpose/202301/mind-body-problem-how-consciousness-emerges-from-matter

Levin, M. (2025). "Ingressing Minds: Causal Patterns Beyond Genetics and Environment in Natural, Synthetic, and Hybrid Embodiments." DOI: https://doi.org/10.31234/osf.io/5g2xj_v1

Marshall, P. (2015). "Evolution 2.0: Breaking the Deadlock between Darwin and Design." Dallas, TX: BenBella Books.

Masi, M. (2019a). “Quantum Physics: An Overview of a Weird World: A primer on the Conceptual Foundations” (fifth edition), vol. 1. Online: **https://www.amazon.com/dp/1090602596**

Masi, M. (2019b). “Quantum Physics: An Overview of a Weird World: A Guide to the 21st Century Revolution”, (third edition), vol. 2. Online: https://www.amazon.com/dp/3948295034

Masi, M. (2021). "Vitalism and Cognition in a Conscious Universe," Communicative & Integrative Biology, vol. 15, no. 1, pp. 121-136, 2021. DOI: **https://doi.org/10.1080/19420889.2022.2071102**

Masi, M. (2023a). "An Evidence-Based Critical Review of the Mind-Brain Identity Theory," Frontiers in Psychology, Sec. Consciousness Research, vol. 14. DOI: **https://doi.org/10.3389/fpsyg.2023.1150605**

Masi, M. (2023b). "Quantum Indeterminism, Free Will, and Self-Causation," Journal of Consciousness Studies, vol. 30, no. 5-6, pp. 32–56, 2023. DOI: **https://doi.org/10.53765/20512201.30.5.032**

Masi, M. (2023c). “The Integral Cosmology of Sri Aurobindo: An Introduction from the Perspective of Consciousness Studies,” Integral Review, Vol. 18, No.1.

Masi, M. (2024). “Quantum Indeterminacy and Libertarian Panpsychism,” Mind & Matter, vol. 22, Nr. 1, pp. 31-50(20). DOI: **https://doi.org/10.53765/mm2024.31**

Masi, M. (2025a). “Spirit Calls Nature” (single volume, 3ed.), Indy Edition. Online: **https://www.amazon.com/dp/3948295166**

Masi, M. (2025b). “Quantum Indeterminacy Persists Across Scales,” International Journal of Quantum Foundations, Vol. 11 (4), pg. 492-514.

Masi, M. (2025c). “No Consciousness? No Meaning (and no AGI)!,” Qeios, preprint, **https://www.qeios.com/read/DN232Y.3**

Noble, D., Noble, R. (2018). “Harnessing stochasticity: How do organisms make choices?” Chaos 28, 106309. DOI: **https://doi.org/10.1063/1.5039668**

Noble, D. (2023). “It’s time to admit that genes are not the blueprint for life.” Nature, Nature 626, 254-255. DOI: **https://doi.org/10.1038/d41586-024-00327-x**

Pitkänen, O.P. (2019). Schelling, esotericism and the meaning of life. Human Affairs, 29(4), 497-504. DOI: **https://doi.org/10.1515/humaff-2019-0045**

Plantinga, A. (2010). "Evolution, Shibboleths, and Philosophers". The Chronicle of Higher Education.

Sapolsky R. (2023). Determined: A Science of Life Without Free Will. Penguin Press.

Schelling, F. W. J. [1797] (1988). Ideas for a Philosophy of Nature as Introduction to the Study of This Science. Cambridge: Cambridge University Press.

Schipper H.M. (2021). Kabbalistic Panpsychism: The Enigma of Consciousness in Jewish Mystical Thought. Iff Books. See also Dr. Schipper's lecture on YouTube: **https://youtu.be/oFa5Y3KsTnU**

Schwitzgebel, E. (2023). "Is the United States Conscious?" - You Tube video. Online: https://www.youtube.com/live/U_Q7F-kOcBE

Schrödinger, E. (1951). "Nature and the Greeks and Science and Humanism," p.162-168, reprint ed. 2014, Cambridge University Press, Cambridge.

Teklinski, E. M. (2018). "A matter of heart and soul: The value of positing a personal ontological center for developmental psychology." International Journal of Transpersonal Studies, 37(1), 90–119.

Treisman, A. (1999). "Solutions to the binding problem: Progress through controversy and convergence". Neuron 24, pp. 105–110.

Yatsenko, V. (2016). "Introduction to the Integral Paradigm of Knowledge." University of Human Unity, URL: https://www.academia.edu/45286344/Introduction_to_the_Integral_Paradigm_of_Knowledge

Yatsenko, V. (2021). "On the Origins of Faculties of Consciousness." URL:https://www.academia.edu/45286344/Introduction_to_the_Integral_Paradigm_of_Knowledge

XV. About the Author

The Joys and Pains of Being a 'Freelance Academic': Freedom has its Cost

I could describe myself with a cold bio as follows.

Marco Masi holds a PhD in physics from the University of Trento, Italy. He has worked as a postdoc researcher in universities in Italy, France, and, more recently, Germany, where he also worked as a schoolteacher for three years and where he presently lives. After this academic experience, his interests veered toward the foundations of physics and the philosophy of mind.

For further research topics and scientific papers, see: **http://ow.ly/snz6u**.

You can follow him on Substack (**https://marcomasi.substack.com/**) and Bluesky (marco-masi.bsky.social).

Also, see his homepage at **https://marco-masi.com/**

A less formal and more personal account is as follows.

With my PhD in physics, I had the opportunity to pursue an academic career, and to some extent, I was already on that path.

To summarize, I believe that modern science, research, and academic institutions have strayed from their original goal: the pursuit of knowledge. In the past, science was viewed as a systematic study of the natural world that encompassed not only observation, experimentation, and analysis but also fundamental questions about existence and the nature of reality. This idealistic view has diminished, particularly since the mid-20th century, when a more pragmatic approach took precedence.[235] While this separation between science and philosophy has yielded practical benefits and improvements in our daily lives, it has also distanced the human mind from its deeper, more soulful understanding.

Thus, a conventional academic career would not have allowed me to fully express my unique research interests and ideas, which center on a post-material worldview, a new science, and an integral vision of reality. I could have explored some of these ideas but only within the confines of a naturalistic approach or a not-too-metaphysical perspective. This would have limited me

---

[235] I could provide a detailed description of how, unfortunately, the decline in academic research has continued, driven by internal power struggles, politicization, commercialization, and aggressive fundraising policies, along with the pervasive "publish or perish" mentality. This shift toward treating academia like a business has turned scientific inquiry into a profit-driven enterprise that often produces trivial or useless results. In some cases, it has transformed scholarly work into a self-censoring political apparatus. Recently, the rise in social media has seen even more universities embracing this trend. However, that's a topic for another discussion.

to a predominantly Western view of consciousness and life theories that are currently in vogue and largely rooted in a materialistic belief system.

Consequently, I decided to leave my previous professional life to pursue research and explore topics that are nearly impossible to address within an academic framework. This choice has given me the freedom to focus on the fundamental questions that truly interest me—questions about the nature of life, consciousness, and our place in the universe. That was what motivated me to become a freelance academic or what academic jargon refers to as an "independent scholar"—someone with professional training but lacking institutional affiliation and research funding.

However, this choice comes with significant financial drawbacks. To sustain myself, I had to find an additional source of income. While leaving the comfort of institutional structures has allowed me to express myself freely, the biggest challenge isn't the considerable time and energy required—this was expected and isn't really an issue when you're passionate about something. The real obstacle is the financial burden. As an independent scholar, publishing articles in academic journals can be very costly, especially without access to grants or the funding typically provided by universities. Moreover, I want my research to be accessible and not hidden behind paywalls. Unfortunately, most publishers do not offer open access without imposing steep fees that authors must cover. Creating online content helped me stay afloat and provided me with the time to continue working on my project.

The main conclusion I draw is that a lack of awareness about our true nature (how we think and perceive the world), combined with an unconscious fascination for naturalizing everything, leads scientists, philosophers, and, ultimately, corporations and nations to invest vast resources in projects that are doomed to fail. When we don't fully grasp our own cognitive biases and limitations, particularly the materialistic and mechanistic views of the world and our capacity for introspection, we inevitably pursue flawed initiatives. We are entering an era in which genuine progress will be unattainable unless we begin to understand ourselves.

I believe these inquiries are not only more compelling but also far more significant than developing the next generation of AI, creating photonic computers, or sending humans to Mars. Leaving academia has provided me with a level of freedom and privilege that many who remain in their careers can only dream of. It was the best choice I ever made!

XVI. Index

D

E

F

G

N

www.ingramcontent.com/pod-product-compliance
Lightning Source LLC
Chambersburg PA
CBHW071406230926
51846CB00001B/19
9783948295202